Welcome to the 71ˢᵗ International Symposium on Molecular Spectroscopy
June 20-24, 2016
Urbana-Champaign, IL

I0482161

On behalf of the Executive Committee, I extend a heartfelt welcome to all the attendees of the 71ˢᵗ Symposium and welcome you to the University of Illinois at Urbana-Champaign.

The Symposium presents research in fundamental molecular spectroscopy and a wide variety of related fields and applications. The continued vitality and significance of spectroscopy is annually re-affirmed by the number of talks, their variety, and the fact that many are given by students. These presentations are the heart of the meeting and are documented by this Abstract Book. Equally important is the information flowing from informal exchanges and discussions. As organizers, we strive to provide an environment that facilitates both kinds of interactions.

The essence of the meeting lies in the scientific discussions and your personal experiences this week independent of the number of times that you have attended this meeting. It is our sincere hope that you will find this meeting informative and enjoyable both scientifically and personally, whether it is your first or 50th meeting. If we can help to enhance your experience, please do not hesitate to ask the Symposium staff or the Executive Committee.

Ben McCall
Symposium Chair

VENUE AND SPONSOR INFORMATION FOLLOWS AUTHOR INDEX

71st INTERNATIONAL SYMPOSIUM ON MOLECULAR SPECTROSCOPY

International Advisory Committee
Allan Cheung, University of Hong Kong
Emilio Cocinero, Universidad del País Vasco
Stephen Cooke, Purchase College, SUNY
Gary Douberly*, University of Georgia
Brian Drouin, NASA Jet Propulsion Laboratory
Shuiming Hu, University of Science and Technology of China
Laurent Margules, Laboratoire de Physique des Lasers, Atomes et Molécules
Terry Miller, Ohio State University
Hiroyuki Sasada, Keio University
Stephan Schlemmer, University of Cologne
Trevor Sears, Brookhaven National Laboratory, Chair
Tim Steimle, Arizona State University
Jonathan Tennyson, University College London
Jennifer van Wijngaarden*, University of Manitoba
Mathias Weber, University of Colorado and JILA
Susanna Widicus Weaver, Emory University
Yunjie Xu*, University of Alberta
Jun Ye*, JILA, University of Colorado and NIST
* steering committee member

Executive Committee
Ben McCall, Chair
Brian DeMarco
Dana Dlott
Gary Eden
Nick Glumac
Martin Gruebele
So Hirata
Leslie Looney
Josh Vura-Weis
Dave Woon

Please send correspondence to
Ben McCall
International Symposium on Molecular Spectroscopy
Department of Chemistry
600 S. Mathews Avenue
Urbana IL 61801 USA
e-mail: chair@isms.illinois.edu
http://isms.illinois.edu

Mini-Symposia

SPECTROSCOPY IN ATMOSPHERIC CHEMISTRY

Organized by **Steve Brown** (NOAA/University of Colorado) and **Frank Keutsch** (Harvard University). This mini-symposium will focus on advances in different types of spectroscopic methods, in-situ and remote, within the context of their application to atmospheric science. Invited Speakers: **Stan Sander** (Jet Propulsion Laboratory, California Institute of Technology), **Phil Stevens** (Indiana University), **Rainer Volkamer** (University of Colorado, Boulder), and **Gerard Wysocki** (Princeton University)

SPECTROSCOPY IN TRAPS

Organized by **Stephan Schlemmer** (Universität zu Köln). The number of spectroscopy studies in traps rapidly increased over the last decade. This mini-symposium will cover many different scientific and technical aspects of this still unconventional spectroscopy method, including small ions, larger complexes, astrophysically relevant species, and the influence of molecular collisions. Invited Speakers: **Mark Johnson** (Yale University), **Jos Oomens** (Radboud University), **Roland Wester** (Universität Innsbruck)

SPECTROSCOPY OF LARGE AMPLITUDE MOTIONS

Organized by **Isabelle Kleiner** (CNRS et Université Paris Est et Paris Diderot) and **Jon Hougen** (NIST). This mini-symposium will cover all aspects of the spectroscopy of large amplitude motions in molecules, with special emphasis on emerging experimental and theoretical techniques that may trigger rapid advances in the next generation of high-resolution studies. Invited Speakers: **Therese Huet** (Université Lille 1 - CNRS), **Marek Kreglewski** (Adam Mickiewicz University in Poznań), **Stephan Schlemmer** (Universität zu Köln)

Picnic (please note new day of week!)

The Symposium picnic will be held on **Tuesday evening** at Ikenberry Commons.   The cost of the picnic is included in your registration (at below cost to students), so that all may attend the event.   The **Coblentz Society** is the host for refreshments for one hour.   Please see your packet for additional details.

Sponsorship

We are pleased to acknowledge the many organizations that support the 71st Symposium. Principal funding comes from the **Army Research Office** (ARO). We are most grateful to ARO for their long-standing support. We also acknowledge the many efforts and contributions of **The University of Illinois** in hosting the meeting, including financial contributions from the Office of the Vice Chancellor for Research and the Departments of Chemistry, Electrical and Computer Engineering, Astronomy, and Physics.

Our Corporate Sponsors are **BrightSpec, Bristol Instruments, Elsevier/JMS, Ideal Vacuum Products, Journal of Physical Chemistry, and Quantel**. Please see the back of this book for their advertisements.

We are also pleased to acknowledge **JASCO, Lockheed Martin, M Squared Lasers, Market Tech, Menlo Systems, and NRAO** as Contributing Sponsors.

IOS Press has a special insert in your conferee packet. Our sponsors will have exhibits at the Symposium and we encourage you to visit their displays.

Rao Prize

The three Rao Prizes for the most outstanding student talks at the 2015 meeting will be presented. The winners are **Daniel Bakker**, Radboud University; **Scott Dubowsky**, The University of Illinois at Urbana-Champaign; **James McMillan**, The Ohio State University. The Rao Prize was created by a group of spectroscopists who, as graduate students, benefited from the emphasis on graduate student participation, which has been a unique characteristic of the Symposium. This year three more Rao Prize winners will be selected.

The award is administered by a Prize Committee chaired by Gary Douberly, University of Georgia, and comprised of David Anderson (University of Wyoming); Brooks Pate (University of Virginia); Rebecca Peebles (Eastern Illinois University); Jennifer van Wijngaarden (University of Manitoba); and Tim Zwier, (Purdue University). Any questions or suggestions about the Prize should be addressed to the Committee. Anyone (especially post-docs) willing to serve on a panel of judges should contact Gary Douberly (douberly@uga.edu).

Miller Prize

The Miller Prize was created in honor of Professor Terry A. Miller, who served as chair of the International Symposium on Molecular Spectroscopy from 1992 to 2013. The Miller Prize for the best presentation given by a recent PhD at the 2015 meeting will be presented. The winner, **Paul Jansen** (ETH Zürich), will give a lecture on Thursday.

The Miller Prize winner and his or her co-authors will be invited to submit an article to the Journal of Molecular Spectroscopy based on the research in the prize-winning talk. After passing the normal review process, the article will appear in the Journal with a caption identifying the paper with the talk that received the Miller Prize.

The award is administered by a Prize Committee chaired by Jinjun Liu, University of Louisville and comprised of Lan Cheng (Johns Hopkins University); Richard Dawes (Missouri University of Science and Technology); Helen Leung (Amherst College); Trevor Sears (Stony Brook University); Steve Shipman (New College of Florida); John Stanton (University of Texas at Austin); Tim Steimle (Arizona State University); Susanna Widicus Weaver (Emory University); Shanshan Yu (Jet Propulsion Laboratory). Any questions or suggestions about the Prize should be addressed to the Committee. Anyone willing to serve on a panel of judges should contact Jinjun Liu (j.liu@louisville.edu).

Information

ACCOMMODATIONS

The check-in for dormitory accommodations is located in Bousfield Hall, 1214 South First Street, opens at noon on Sunday, June 19th, and remains open 24 hours a day through the Symposium. Hotel information is listed on the ISMS website.

PARKING

Parking permits are for lot E14 (see map at end of book). Please purchase parking as part of your check-in process at the dorm. If you need to purchase meter hang-tags for parking near the meeting rooms, you can do so at the registration desk.

REGISTRATION

The registration desk is located in Room 165 Noyes Lab, and is open on Sunday from 4:00-6:00 PM, and Monday through Friday from 8:00 AM-4:30 PM. Refreshments will be available from 8:00 AM-4:30 PM.

CHEMISTRY LIBRARY

The Chemistry Library will be open and available for your use during Symposium hours. The library has a number of computers, desks and tables to work at, and comfy chairs (and books!).

READY ROOM/STATION

We have set up Noyes Lab 164 as a "Ready Room" with computers that you can use to test your powerpoint presentation. If you have any problems, we will also have a staffed "Ready Station" in Noyes Lab 165 (right next to registration) where you can come for assistance.

COMPUTER LAB (VizLab)

Noyes Lab 151 is a small computer lab with Apple computers that is available for your use during the meeting. Please look in your packet for an access code to enter the room.

INTERNET ACCESS/Wi-Fi

Each attendee will receive a login and password to access campus WiFi (SSID: IllinoisNet) as a guest. This access should work in most locations through campus. Please read the Internet Acceptable Use Policy below.

AUDIO/VIDEO INFORMATION

Each session room is equipped with a computer, onto which presentation files will be pre-loaded by Symposium staff. To submit your presentation file, you must go to the **Manage Presentations** link on our web site and follow the instructions. All files must be submitted by **11:59 PM CDT THE DAY BEFORE** your presentation session. All submitted files will be loaded onto the presentation computer one half-hour prior to the beginning of the session.

ACKNOWLEDGMENTS

The Symposium Chair wishes to acknowledge the hard work of numerous people who made this meeting possible. First and foremost is the Symposium Coordinator Birgit McCall, who has smoothly and single-handedly taken care of almost all of the electronic and logistical aspects of the meeting. Second are our symposium assistants, Scott Dubowsky , Charlie Markus, and Jeff McCollum, who have handled innumerable important details to ensure the sessions and exhibitions go well. The other students in my group also play vital roles in other aspects of the meeting. I wish to acknowledge the hospitality of the Chemistry Department and the School of Chemical Sciences (as well as the School of Molecular and Cell Biology) in tolerating our takeover of their buildings.

LIABILITY

The Symposium fees DO NOT include provisions for the insurance of participants against personal injuries, sickness, theft, or property damage. Participants and companions are advised to obtain whatever insurance they consider necessary. The Symposium organizing committee, its sponsors, and individual committee members DO NOT assume any responsibility for loss, injury, sickness, or damages to persons or belongings, however caused. The statements and opinions stated during oral presentations or in written abstracts are solely the author's responsibilities and do not necessarily reflect the opinions of the organizers.

DISCLAIMER

The views, opinions, and/or findings contained in this report are those of the authors and should not be construed as an official Department of the Army position, policy, or decision, unless so designated by other documentation.

INTERNET ACCEPTABLE USE POLICY

Each attendee will receive a login and password to access campus WiFi (SSID: IllinoisNet) as a guest. Guest accounts are intended to support a broad range of communications. Professional and appropriate etiquette is required. Anonymous access and posting through guest accounts is forbidden. All users must accept that their identity may be associated with any content they provide while using the service. By accessing the campus WiFi network, you expressly acknowledge and agree to the following:

Use of the guest account service is at your sole risk and the entire risk as to satisfactory quality and performance is with you. You agree not to use the guest account intentionally or unintentionally to violate any applicable local, state, national or international law, including, but not limited to, any regulations having the force of law. To the extent not prohibited by law, in no event shall the university be liable for personal injury, or any incidental, special, indirect or consequential damages whatsoever, including, without limitation, damages for loss of profits, loss of data, business interruption or any other commercial damages or losses, arising out of or related to your use or inability to use the guest account, however caused, regardless of the theory of liability (contract, tort or otherwise) and even if the university has been advised of the possibility of such damages. The use of the guest account is subject, but not limited to, all University policies and regulations detailed at the Campus Administrative Manual (http://www.cam.illinois.edu). See the University's Web Privacy Notice (http://www.vpaa.uillinois.edu/policies/web_privacy.cfm) for all applicable laws and policies.

MA. Plenary
Monday, June 20, 2016 – 8:30 AM
Room: Foellinger Auditorium

Chair: Gregory S. Girolami, University of Illinois at Urbana-Champaign, Urbana, IL, USA

Welcome   8:30
Barbara J. Wilson, Chancellor
University of Illinois at Urbana-Champaign

MA01   8:40 – 9:20
ELECTRONIC SPECTROSCOPY OF ORGANIC CATIONS IN GAS-PHASE AT 6 K:
IDENTIFICATION OF C_{60}^+ IN THE INTERSTELLAR MEDIUM , John P. Maier

MA02   9:25 – 10:05
SOME COMPLEX PRESSURE EFFECTS ON SPECTRA FROM SIMPLE CLASSICAL MECHANICS,
Jean-Michel Hartmann

Intermission

MA03   10:35 – 11:15
MOLECULAR SPECTROSCOPY OF LIVING SYSTEMS, Ji-Xin Cheng

MA04   11:20 – 12:00
OPTICAL AND MICROWAVE SPECTROSCOPY OF TRANSIENT METAL-CONTAINING MOLECULES,
Timothy Steimle

MF. Mini-symposium: Spectroscopy of Large Amplitude Motions
Monday, June 20, 2016 – 1:30 PM
Room: 100 Noyes Laboratory

Chair: Isabelle Kleiner, CNRS et Universités Paris Est et Paris Diderot, Créteil, France

MF01   *INVITED TALK*   1:30 – 2:00
ON THE LOWEST RO-VIBRATIONAL STATES OF PROTONATED METHANE:
EXPERIMENT AND ANALYTICAL MODEL , Hanno Schmiedt, Per Jensen, Oskar Asvany, Stephan Schlemmer

MF02   2:05 – 2:20
SYMMETRY IN THE GENERALIZED ROTOR MODEL FOR EXTREMELY FLOPPY MOLECULES, Hanno Schmiedt,
Per Jensen, Stephan Schlemmer

MF03   2:22 – 2:37
AN EFFECTIVE-HAMILTONIAN APPROACH TO CH_5^+, USING IDEAS FROM ATOMIC SPECTROSCOPY,
Jon T. Hougen

MF04   2:39 – 2:54
IMPACT OF ENERGETICALLY ACCESSIBLE PROTON PERMUTATIONS IN THE SPECTROSCOPY AND DYNAM-
ICS OF H_5^+, Zhou Lin, Anne B McCoy

MF05   2:56 – 3:11
VARIATION OF CH STRETCH FREQUENCIES WITH CH_4 ORIENTATION IN THE $CH_4 - F^-$ COMPLEX: MULTI-
PLE RESONANCES AS VIBRATIONAL CONICAL INTERSECTIONS, Bishnu P. Thapaliya, David S. Perry

MF06   3:13 – 3:28
SADDLE POINT LOCALIZATION OF MOLECULAR WAVEFUNCTIONS, Georg Ch. Mellau, Aleksandra Kyuberis,
Oleg Polyansky, Nikolay Fedorovich Zobov, Robert W Field

Intermission

MF07   3:47 – 4:02
SPECTROSCOPIC CHARACTERIZATION OF ISOMERIZATION TRANSITION STATES, Joshua H Baraban, Bryan
Changala, Georg Ch. Mellau, John F. Stanton, Anthony Merer, Robert W Field

MF08   4:04 – 4:19
ELECTRON ANISOTROPY AS A SIGNATURE OF MODE SPECIFIC ISOMERIZATION IN VINYLIDENE[a],
Stephen T Gibson, Benjamin A Laws, Richard Mabbs, Daniel Neumark, Carl Lineberger, Robert W Field

---

[a]Research supported by the Australian Research Council Discovery Project Grant DP160102585.

MF09   4:21 – 4:36
OBSERVING QUANTUM MONODROMY: AN ENERGY-MOMENTUM MAP BUILT FROM EXPERIMENTALLY-
DETERMINED LEVEL ENERGIES OBTAINED FROM THE ν_7 FAR-INFRARED BAND SYSTEM OF NCNCS,
Dennis W. Tokaryk, Stephen Cary Ross, Brenda P. Winnewisser, Manfred Winnewisser, Frank C. De Lucia, Brant E.
Billinghurst

MF10   4:38 – 4:53
SPECTROSCOPY OF H_2^+ AND HD^+ NEAR THE DISSOCIATION THRESHOLD: SHAPE AND FESHBACH RESO-
NANCES, Maximilian Beyer, Frederic Merkt

MF11   4:55 – 5:10
A DIATOMIC MOLECULE WITH EXTREMELY LARGE AMPLITUDE MOTION IN ITS VIBRATIONAL STATES
THAT HAVE LENGTHS OF AT LEAST 12,000 ANGSTROMS. , Nikesh S. Dattani

MF12   5:12 – 5:27
FIT POINT-WISE *AB INITIO* CALCULATION POTENTIAL ENERGIES TO A MULTI-DIMENSION MORSE/LONG-
RANGE MODEL, Yu Zhai, Hui Li, Robert J. Le Roy

MG. Mini-symposium: Spectroscopy in Atmospheric Chemistry
Monday, June 20, 2016 – 1:30 PM
Room: 116 Roger Adams Lab

Chair: Frank Keutsch, Harvard University, Cambridge, MA, USA

MG01   *INVITED TALK*   **1:30 – 2:00**
MAPPING THE WORLD'S LARGEST NATURAL GAS LEAK AND OTHER METHANE SOURCES USING HIGH RESOLUTION SPECTROSCOPY, Stanley P. Sander, Clare Wong, Thomas J Pongetti

MG02   **2:05 – 2:20**
THE ATMOSPHERIC CHEMISTRY EXPERIMENT (ACE): CO, CH_4 AND N_2O ISOTOPOLOGUES, Peter F. Bernath, Eric M. Buzan, Christopher A. Beale, Mahdi Yousefi, Chris Boone

MG03   **2:22 – 2:37**
OPTICAL MEASUREMENTS OF $^{14}CO_2$ USING CAVITY RING-DOWN SPECTROSCOPY, David A. Long, Adam J. Fleisher, Qingnan Liu, Joseph Hodges

MG04   **2:39 – 2:54**
MID-IR CAVITY RINGDOWN SPECTROSCOPY MEASUREMENTS OF ATMOSPHERIC ETHANE TO METHANE RATIO, Linhan Shen, Thinh Quoc Bui, Lance Christensen, Mitchio Okumura

MG05   **2:56 – 3:11**
MEASUREMENTS DOUBLY-SUBSTITUTED METHANE ISOTOPOLOGUE (13CH3D AND 12CH2D2) ABUDANCE USING FREQUENCY STABILIZED MID-IR CAVITY RINGDOWN SPECTROSCOPY, Linhan Shen, Thinh Quoc Bui, Mitchio Okumura

MG06   **3:13 – 3:28**
THE NEAR-IR SPECTRUM OF CH_3D, Shaoyue Yang, Kevin Lehmann, Robert J. Hargreaves, Peter F. Bernath, Michael Rey, Andrei V. Nikitin, Vladimir Tyuterev

MG07   **3:30 – 3:45**
$3\mu m$ - $1.6\mu m$ DOUBLE RESONANCE SPECTROSCOPY OF CH_4, George Schwartz, Erik Belaas, Shaoyue Yang, Kevin Lehmann

Intermission

MG08   **4:04 – 4:19**
GLOBAL ANALYSIS OF THE HIGH TEMPERATURE INFRARED EMISSION SPECTRUM OF $^{12}CH_4$ IN THE DYAD (ν_2/ν_4) REGION, Badr Amyay, Maud Louviot, Olivier Pirali, Robert Georges, Jean Vander Auwera, Vincent Boudon

MG09   **4:21 – 4:36**
GLOBAL ANALYSIS OF SEVERAL BANDS OF THE CF_4 MOLECULE, Mickaël Carlos, Océane Gruson, Vincent Boudon, Robert Georges, Olivier Pirali, Pierre Asselin

MG10   **4:38 – 4:53**
HIGH RESOLUTION FAR INFRARED SPECTROSCOPY OF HFC-134a AT COLD TEMPERATURES, Andy Wong, Chris Medcraft, Christopher Thompson, Evan Gary Robertson, Dominique Appadoo, Don McNaughton

MG11   **4:55 – 5:10**
ON THE USE OF DIFFERENCE BANDS FOR MODELING SF_6 ABSORPTION IN THE $10\mu m$ ATMOSPHERIC WINDOW, Mbaye Faye, Laurent Manceron, P. Roy, Vincent Boudon, Michel Loete

MG12   **5:12 – 5:27**
POSSIBLE MECHANISM FOR SULFUR MASS INDEPENDENT FRACTIONATION IN THE B-X UV TRANSITION OF S_2, Alexander W. Hull, Andrew Richard Whitehill, Shuhei Ono, Robert W Field

MH. Ions
Monday, June 20, 2016 – 1:30 PM
Room: 274 Medical Sciences Building

Chair: Etienne Garand, University of Wisconsin-Madison, Madison, WI, USA

MH01   1:30 – 1:45
COMBINED DUNHAM ANALYSIS OF ROTATIONAL AND ELETRONIC TRANSITIONS OF CH^+, Shanshan Yu, Brian Drouin, John Pearson, Takayoshi Amano

MH02   1:47 – 2:02
PRECISION SATURATED ABSORPTION SPECTROSCOPY OF H_3^+, Yu-chan Guan, Yi-Chieh Liao, Yung-Hsiang Chang, Jin-Long Peng, Jow-Tsong Shy

MH03   2:04 – 2:19
HIGHLY ACCURATE AND PRECISE INFRARED TRANSITION FREQUENCIES OF THE H_3^+ CATION, Adam J. Perry, Charles R. Markus, James N. Hodges, G. Stephen Kocheril, Benjamin J. McCall

MH04   2:21 – 2:36
ROTATIONALLY RESOLVED PHOTOELECTRON SPECTROSCOPIC STUDY OF THE $\tilde{A}^+$ STATE OF H_2O^+, Clément Lauzin, Berenger Gans, Ugo Jacovella, Frederic Merkt

MH05   2:38 – 2:48
STUDIES OF 4-CHLORO-2-FLUOROANISOLE BY TWO-COLOR RESONANT TWO-PHOTON MASS-ANALYZED THRESHOLD IONIZATION SPECTROSCOPY, Pei-Ying Wu, Wen-Bih Tzeng

MH06   2:50 – 3:00
VIBRONIC AND CATION SPECTROSCOPY OF 3,5-DIFLUOROPHENOL, Wei-Chih Peng, Shen-Yuan Tzeng, Wen-Bih Tzeng

Intermission

MH07   3:19 – 3:34
IDENTIFICATION OF STRUCTURAL MOTIFS OF IMIDAZOLIUM BASED IONIC LIQUIDS FROM JET-COOLED INFRARED SPECTROSCOPY., Justin W. Young, Ryan S Booth, Christopher Annesley, Jaime A. Stearns

MH08   3:36 – 3:51
Ce-PROMOTED BOND ACTIVATION OF PROPENE PROBED BY MASS-ANALYZED THRESHOLD IONIZATION SPECTROSCOPY, Yuchen Zhang, Sudesh Kumari, Dong-Sheng Yang

MH09   3:53 – 4:08
SPECTROSCOPIC IDENTIFICATION OF $Y(C_4H_6)$ ISOMERS FORMED BY YTTRIUM-MEDIATED C-H BOND ACTIVATION OF BUTENES, Jong Hyun Kim, Dong-Sheng Yang

MH10   4:10 – 4:25
INFRARED LASER PHOTODISSOCIATION SPECTROSCOPY OF METAL-ACETYLENE CATIONS IN THE GAS PHASE , Antonio David Brathwaite, Timothy B Ward, Michael A Duncan

MH11   4:27 – 4:42
VELOCITY MAP IMAGING STUDIES OF Mg^+-L (L = Ar, N_2, CO_2) COMPLEXES, Jon Maner, Daniel Mauney, Michael A Duncan

MH12   4:44 – 4:59
VIBRATIONAL SIGNATURES OF LARGE AMPLITUDE MOTIONS FOR THE SHACKLED HYDRONIUM ION NESTED IN 18-CROWN-6 ETHER USING D_2 TAGGING, Chinh H. Duong, Fabian Menges, Stephanie Craig, Conrad T. Wolke, Mark Johnson

MH13   5:01 – 5:16
SPECTROSCOPIC OBSERVATION OF WATER-MEDIATED DEFORMATION OF THE CARBOXYLATE-M^{2+} (M= Mg, Ca) CONTACT ION PAIR, Patrick J Kelleher, Joseph W DePalma, Mark Johnson

MH14   5:18 – 5:33
IRMPD ACTION SPECTROSCOPY AND COMPUTATIONAL APPROACHES TO ELUCIDATE GAS-PHASE STRUCTURES AND ENERGETICS OF 2′-DEOXYCYTIDINE AND CYTIDINE SODIUM COMPLEXES, Yanlong Zhu, Lucas Hamlow, Chenchen He, Juehan Gao, Jos Oomens, M T Rodgers

MH15   5:35 – 5:50
INFLUENCE OF 5-HALOGENATION ON THE STRUCTURE OF PROTONATED URIDINE: IRMPD ACTION SPECTROSCOPY AND THEORETICAL STUDIES OF THE PROTONATED 5-HALOURIDINES, Harrison Roy, Lucas Hamlow, Justin Lee, M T Rodgers, Giel Berden, Jos Oomens

MI. Structure determination
Monday, June 20, 2016 – 1:30 PM
Room: B102 Chemical and Life Sciences

Chair: Stewart E. Novick, Wesleyan University, Middletown, CT, USA

MI01   1:30 – 1:45
THE MOLECULAR STRUCTURE OF PHENETOLE STUDIED BY MICROWAVE SPECTROSCOPY AND QUANTUM CHEMICAL CALCULATIONS, Lynn Ferres, Wolfgang Stahl, Ha Vinh Lam Nguyen

MI02   1:47 – 2:02
CP-FTMW SPECTROSCOPY OF A CLAISEN REARRANGEMENT PRECURSOR ALLYL PHENYL ETHER, G. S. Grubbs II, Derek S. Frank, Daniel A. Obenchain, S. A. Cooke, Stewart E. Novick

MI03   2:04 – 2:19
DETERMINATION OF THE PREFERRED STRUCTURE, DYNAMICS, AND PLANARITY OF SUBSTITUTED ANHYDRIDES BY CP-FTMW, Timothy J McMahon, Josiah R Bailey, Ryan G Bird, David Pratt

MI04   2:21 – 2:36
STRUCTURAL ANALYSIS OF 2-FLUOROPHENOL AND 3-FLUOROPHENOL USING FTMW SPECTROSCOPY , Aimee Bell, Omar Mahassneh, James Singer, Jennifer van Wijngaarden

MI05   2:38 – 2:53
STRUCTURAL EXPRESSION OF EXO-ANOMERIC EFFECT, E. R. Alonso, Isabel Peña, Carlos Cabezas, José L. Alonso

MI06   2:55 – 3:10
MICROWAVE SPECTRA, MOLECULAR STRUCTURE AND AROMATIC CHARACTER OF BN-NAPHTHALENE (4A,8A-AZABORANAPHTHALENE) , Aaron M Pejlovas, Stephen G. Kukolich, Arthur J. Ashe III, Adam M Daly

MI07   3:12 – 3:27
THE STRUCTURE AND MOLECULAR PARAMETERS OF CAMPHENE DETERMINED BY FOURIER TRANSFORM MICROWAVE SPECTROSCOPY AND QUANTUM CHEMICAL CALCULATIONS, Elias M. Neeman, Pascal Dréan, T. R. Huet

MI08   3:29 – 3:39
N-METHYL INVERSION IN PSEUDO-PELLETIERINE, Montserrat Vallejo-López, Patricia Ecija, Emilio J. Cocinero, Alberto Lesarri, Francisco J. Basterretxea, José A. Fernández

Intermission

MI09   3:58 – 4:13
MICROWAVE SPECTRA FOR THE THREE $^{13}C_1$ ISOTOPOLOGUES OF PROPENE AND NEW ROTATIONAL CONSTANTS FOR PROPENE AND ITS $^{13}C_1$ ISOTOPOLOGUES , Norman C. Craig, Peter Groner, Andrew R Conrad, Ranil M. Gurusinghe, Michael Tubergen

MI10   4:15 – 4:30
MICROWAVE SPECTRA FOR THE TWO CONFORMERS OF PROPENE-3-d_1 AND NEW ROTATIONAL CONSTANTS FOR THESE SPECIES , Norman C. Craig, Ranil M. Gurusinghe, Michael Tubergen

MI11   4:32 – 4:47
RESOLVING A LONG-STANDING AMBIGUITY: THE NON-PLANARITY OF *gauche*-1,3-BUTADIENE REVEALED BY MICROWAVE SPECTROSCOPY, Marie-Aline Martin-Drumel, Michael C McCarthy, David Patterson, Sandra Eibenberger, Grant Buckingham, Joshua H Baraban, Barney Ellison, John F. Stanton

MI12   4:49 – 5:04
SINGLE-CONFORMATION IR AND UV SPECTROSCOPY OF A PROTOTYPICAL HETEROGENEOUS α/β-PEPTIDE: IS IT A MIXED-HELIX FORMER?, Karl N. Blodgett, Patrick S. Walsh, Timothy S. Zwier

MI13   5:06 – 5:21
STRUCTURES AND NUCLEAR QUADRUPOLE COUPLING TENSORS OF A SERIES OF CHLORINE-CONTAINING HYDROCARBONS, Asela S. Dikkumbura, Erica R Webster, Rachel E. Dorris, Rebecca A. Peebles, Sean A. Peebles, Nathan A Seifert, Brooks Pate

MI14   5:23 – 5:38
INFRA-RED SPECTRA OF SMALL ALKANES INTERACTING WITH ALUMINUM IONS, Muhammad Affawn Ashraf, Christopher Copeland, Ricardo B. Metz

MI15   5:40 – 5:55
QUANTUM CHEMICAL STUDIES ON THE PREDICTION OF STRUCTURES, CHARGE DISTRIBUTIONS AND VIBRATIONAL SPECTRA OF SOME Ni(II), Zn(II), AND Cd(II) IODIDE COMPLEXES, Tayyibe Bardakci, Mustafa Kumru, Ahmet Altun

MJ. Conformers, isomers, chirality, stereochemistry
Monday, June 20, 2016 – 1:30 PM
Room: 217 Noyes Laboratory

Chair: Daniel A. Obenchain, Wesleyan University, Middletown, CT, USA

MJ01   1:30 – 1:45
CHIRPED PULSE MICROWAVE SPECTROSCOPY ON METHYL BUTANOATE, Alicia O. Hernandez-Castillo, Brian M Hays, Chamara Abeysekera, Timothy S. Zwier

MJ02   1:47 – 2:02
LOCAL ANESTHETICS IN THE GAS-PHASE: THE ROTATIONAL SPECTRUM OF BUTAMBEN AND ISOBUTAMBEN, Montserrat Vallejo-López, Patricia Ecija, Walther Caminati, Jens-Uwe Grabow, Alberto Lesarri, Emilio J. Cocinero

MJ03   2:04 – 2:19
WE ARE FAMILY: THE CONFORMATIONS OF 1-FLUOROALKANES, $C_nH_{2n+1}F$ (n = 2,3,4,5,6,7,8), Daniel A. Obenchain, W. Orellana, S. A. Cooke

MJ04   2:21 – 2:36
LARGE MOLECULE STRUCTURES BY BROADBAND FOURIER TRANSFORM MOLECULAR ROTATIONAL SPECTROSCOPY, Luca Evangelisti, Nathan A Seifert, Lorenzo Spada, Brooks Pate

MJ05   2:38 – 2:53
CHIRAL ANALYSIS OF ISOPULEGOL BY FOURIER TRANSFORM MOLECULAR ROTATIONAL SPECTROSCOPY, Luca Evangelisti, Nathan A Seifert, Lorenzo Spada, Brooks Pate

MJ06   2:55 – 3:10
ROTATIONAL SPECTROSCOPY OF TETRAHYDRO-2-FUROIC ACID, ITS CHIRAL AGGREGATES AND ITS COMPLEX WITH WATER, Javix Thomas, Wolfgang Jäger, Yunjie Xu

Intermission

MJ07   3:29 – 3:44
PROBING THE CONFORMATIONAL LANDSCAPE OF POLYETHER BUILDING BLOCKS IN SUPERSONIC JETS, Sebastian Bocklitz, Daniel M. Hewett, Timothy S. Zwier, Martin A. Suhm

MJ08   3:46 – 4:01
MODELING THE CONFORMATION-SPECIFIC INFRARED SPECTRA OF N-ALKYLBENZENES, Daniel P. Tabor, Edwin Sibert, Daniel M. Hewett, Joseph A. Korn, Timothy S. Zwier

MJ09   4:03 – 4:18
WHERE'S THE BEND? LOCATING THE FIRST FOLDED STRUCTURE IN STRAIGHT CHAIN ALKYLBENZENES IN A SUPERSONIC JET EXPANSION, Daniel M. Hewett, Sebastian Bocklitz, Martin A. Suhm, Timothy S. Zwier

MJ10   4:20 – 4:35
INTRINSIC OPTICAL ACTIVITY AND ENVIRONMENTAL PERTURBATIONS: SOLVATION EFFECTS IN CHIRAL BUILDING BLOCKS, Paul M Lemler, Patrick Vaccaro

MJ11   4:37 – 4:52
DOUBLE-RESONANCE FACILITATED DECOMPOSION OF EMISSION SPECTRA, Ryota Kato, Haruki Ishikawa

MJ12   4:54 – 5:09
A CHIRPED PULSE FOURIER TRANSFORM MICROWAVE (CP-FTMW) SPECTROMETER WITH LASER ABLATION SOURCE TO SEARCH FOR ACTINIDE-CONTAINING MOLECULES AND NOBLE METAL CLUSTERS, Frank E Marshall, David Joseph Gillcrist, Thomas D. Persinger, Nicole Moon, G. S. Grubbs II

MJ13   5:11 – 5:26
ISOMERIZATION AND FRAGMENTATION OF CYCLOHEXANONE IN A HEATED MICRO-REACTOR, Jessica P Porterfield, Thanh Lam Nguyen, Joshua H Baraban, Grant Buckingham, Tyler Troy, Oleg Kostko, Musahid Ahmed, John F. Stanton, John W Daily, Barney Ellison

MJ14   5:28 – 5:43
THEORY OF MICROWAVE 3-WAVE MIXING OF CHIRAL MOLECULES, Kevin Lehmann

MJ15   5:45 – 6:00
THEORY OF MICROWAVE 5-WAVE MIXING OF CHIRAL MOLECULES, Kevin Lehmann

MK. Matrix isolation (and droplets)
Monday, June 20, 2016 – 1:30 PM
Room: 140 Burrill Hall

Chair: Paul Raston, James Madison University, Harrisonburg, Virginia, USA

MK01   1:30 – 1:45
IR SPECTROSCOPIC STUDIES ON MICROSOLVATION OF HCl BY WATER, <u>Devendra Mani</u>, Raffael Schwan, Theo Fischer, Arghya Dey, Matin Kaufmann, Britta Redlich, Lex van der Meer, Gerhard Schwaab, Martina Havenith

MK02   1:47 – 2:02
ANETHOLE-WATER: A COMBINED JET, MATRIX, AND COMPUTATIONAL STUDY , <u>Josh Newby</u>, Jackleen Nesheiwat

MK03   *Post-Deadline Abstract*   2:04 – 2:19
VIBRATIONAL SPECTROSCOPY OF CO_2^- RADICAL ANION IN WATER, <u>Ireneusz Janik</u>, G. N. R. Tripathi

MK04   2:21 – 2:31
H-π BEATS n-σ IN PHENYLACETYLENE-HCl HYDROGEN BONDED HETERODIMER: A MATRIX ISOLATION INFRARED AND AB INITIO STUDY, <u>Ginny Karir</u>, K S Viswanathan

MK05   2:33 – 2:43
WHAT IS DIFFERENT BETWEEN BORAZINE-ACETYLENE AND BENZENE-ACETYLENE?
A MATRIX ISOLATION AND *AB-INITIO* STUDY. , <u>Kanupriya Verma</u>, K S Viswanathan

MK06   2:45 – 2:55
INFRARED MATRIX-ISOLATION STUDY OF NEW NOBLE-GAS COMPOUNDS , <u>Cheng Zhu</u>, Markku Räsänen, Leonid Khriachtchev

MK07   2:57 – 3:07
EPR OF CH_3 RADICALS IN SIO_2 CLATHRATE , <u>Yurij Dmitriev</u>, Gianpiero Buscarino, Nikolas Ploutarch Benetis

Intermission

MK08   3:26 – 3:41
OBSERVATION OF TRANS-ETHANOL AND GAUCHE-ETHANOL COMPLEXES WITH BENZENE USING MATRIX ISOLATION INFRARED SPECTROSCOPY, <u>Jay Amicangelo</u>, Matthew J Silbaugh

MK09   3:43 – 3:58
LOW TEMPERATURE THERMODYNAMIC EQUILIBRIUM OF CO_2 DIMER ANION SPECIES IN CRYOGENIC ARGON AND KRYPTON MATRICES, <u>Michael E. Goodrich</u>, David T Moore

MK10   4:00 – 4:15
SIMULTANEOUS DEPOSITION OF MASS SELECTED ANIONS AND CATIONS: IMPROVEMENTS IN ION DELIVERY FOR MATRIX ISOLATION EXPERIMENTS, <u>Michael E. Goodrich</u>, David T Moore

MK11   4:17 – 4:32
PUMP AND PROBE SPECTROSCOPY OF CH_3F-(ortho-H_2)$_n$ CLUSTERS IN SOLID PARAHYDROGEN BY USING TWO CW-IR QUANTUM CASCADE LASERS, <u>Hiroyuki Kawasaki</u>, Asao Mizoguchi, Hideto Kanamori

MK12   4:34 – 4:49
QUANTUM DIFFUSION CONTROLLED CHEMISTRY: THE H + NO REACTION, <u>Morgan E. Balabanoff</u>, David T. Anderson

MK13   4:51 – 5:06
USING INFRARED SPECTROSCOPY TO PROBE THE TEMPERATURE DEPENDENCE OF THE H + N_2O REACTION IN PARAHYDROGEN CRYSTALS, <u>Fredrick M. Mutunga</u>, David T. Anderson

MK14   5:08 – 5:23
INFRARED SPECTROSCOPY OF DEUTERATED ACETYLENE IN SOLID PARAHYDROGEN AND THE HELIUM RECOVERY INITIATIVE, <u>Aaron I. Strom</u>, David T. Anderson

TA. Mini-symposium: Spectroscopy of Large Amplitude Motions
Tuesday, June 21, 2016 – 8:30 AM
Room: 100 Noyes Laboratory

Chair: Hanno Schmiedt, University of Cologne, Cologne, Germany

TA01   *Journal of Molecular Spectroscopy Review Lecture*   8:30 – 9:00
FLOPPY MOLECULES WITH INTERNAL ROTATION AND INVERSION, Marek Kreglewski

TA02   9:05 – 9:20
APPLICATION OF THE HYBRID PROGRAM FOR FITTING MICROWAVE AND FAR-INFRARED SPECTRA OF METHYLAMINE, Isabelle Kleiner, Jon T. Hougen

TA03   9:22 – 9:32
ACCURATE ROVIBRATIONAL ENERGIES FOR THE FIRST EXCITED TORSIONAL STATE OF METHYLAMINE, Iwona Gulaczyk, Marek Kreglewski

TA04   9:34 – 9:49
ANALYSIS OF THE TORSIONAL SPLITTING IN THE ν_8 BAND OF PROPANE NEAR 870.4 cm^{-1} CAUSED BY FERMI RESONANCE WITH THE $2\nu_{14}+2\nu_{27}$ LEVEL, Peter Groner, Agnes Perrin, F. Kwabia Tchana, Laurent Manceron

TA05   9:51 – 10:06
COUPLING OF LARGE AMPLITUDE INVERSION WITH OTHER STATES, John Pearson, Shanshan Yu

TA06   10:08 – 10:23
FINAL RESULTS ON MODELING THE SPECTRUM OF AMMONIA $2\nu_2$ AND ν_4 STATES, Shanshan Yu, John Pearson, Takayoshi Amano, Olivier Pirali

Intermission

TA07   10:42 – 10:52
MILLIMETER WAVE SPECTRUM OF NITROMETHANE, V. Ilyushin

TA08   10:54 – 11:09
IAM(-LIKE) TUNNELING MATRIX FORMALISM FOR ONE- AND TWO-METHYL-TOP MOLECULES BASED ON THE EXTENDED PERMUTATION-INVERSION GROUP IDEA AND ITS APPLICATION TO THE ANALYSES OF THE METHYL-TORSIONAL ROTATIONAL SPECTRA, Nobukimi Ohashi, Kaori Kobayashi, Masaharu Fujitake

TA09   11:11 – 11:26
IS THE COUPLING OF C_{3V} INTERNAL ROTATION AND NORMAL VIBRATIONS A TRACTABLE PROBLEM?, John Pearson, Peter Groner, Adam M Daly

TA10   11:28 – 11:43
A HAMILTONIAN TO OBTAIN A GLOBAL FREQUENCY ANALYSIS OF ALL THE VIBRATIONAL BANDS OF ETHANE, Nasser Moazzen-Ahmadi, Jalal Norooz Oliaee

TA11   11:45 – 12:00
USING SYMMETRY GROUP CORRELATION TABLES TO EXPLAIN WHY ERHAM (AND OTHER PROGRAMS) CANNOT BE USED TO ANALYZE TORSIONAL SPLITTINGS OF SOME MOLECULES, Peter Groner

TB. Mini-symposium: Spectroscopy in Atmospheric Chemistry
Tuesday, June 21, 2016 – 8:30 AM
Room: 116 Roger Adams Lab

Chair: Vincent Boudon, CNRS / Université de Bourgogne, Dijon, France

TB01   *INVITED TALK*   **8:30 – 9:00**
RADICALS AND AEROSOLS IN THE TROPOSPHERE AND LOWER STRATOSPHERE, Rainer Volkamer, Theodore Koenig, Barbara Dix

TB02   **9:05 – 9:20**
APPLICATIONS OF HIGH RESOLUTION MID-INFRARED SPECTROSCOPY FOR ATMOSPHERIC AND ENVIRON-MENTAL MEASUREMENTS, Joseph R Roscioli, J Barry McManus, David Nelson, Mark Zahniser, Scott C Herndon, Joanne Shorter, Tara I Yacovitch, Dylan Jervis, Christoph Dyroff, Charles E Kolb

TB03   **9:22 – 9:37**
A PORTABLE DUAL FREQUENCY COMB SPECTROMETER FOR ATMOSPHERIC APPLICATIONS, Kevin C Cossel, Eleanor Waxman, Gar-Wing Truong, Fabrizio Giorgetta, William C Swann, Sean Coburn, Robert Wright, Greg B Rieker, Ian Coddington, Nathan R. Newbury

TB04   **9:39 – 9:54**
METHANE DETECTION FOR OIL AND GAS PRODUCTION SITES USING PORTABLE DUAL-COMB SPECTROM-ETRY, Sean Coburn, Robert Wright, Kevin C Cossel, Gar-Wing Truong, Esther Baumann, Ian Coddington, Nathan R. Newbury, Caroline Alden, Subhomoy Ghosh, Kuldeep Prasad, Greg B Rieker

TB05   **9:56 – 10:11**
TIME-RESOLVED FREQUENCY COMB SPECTROSCOPY FOR STUDYING THE KINETICS AND BRANCHING RA-TIO OF OD+CO, Thinh Quoc Bui, Bryce J Bjork, Oliver H Heckl, Bryan Changala, Ben Spaun, Mitchio Okumura, Jun Ye

Intermission

TB06   **10:30 – 10:45**
DEMONSTRATION OF A RAPIDLY-SWEPT EXTERNAL CAVITY QUANTUM CASCADE LASER FOR ATMO-SPHERIC SENSING APPLICATIONS, Brian E Brumfield, Matthew S Taubman, Mark C Phillips, Jonathan D Suter

TB07   **10:47 – 11:02**
DEVELOPMENT OF A QUANTUM CASCADE LASER-BASED SPECTROMETER FOR MEASUREMENTS OF BIO-GENIC VOLATILE ORGANIC COMPOUNDS, Jacob Stewart

TB08   **11:04 – 11:14**
RAMAN LIDAR PROFILING OF TROPOSPHERIC WATER VAPOR , Watheq Al-Basheer

TB09   **11:16 – 11:31**
PHOTOCHEMICAL FORMATION OF AEROSOL IN PLANETARY ATMOSPHERES: PHOTON AND WATER MEDI-ATED CHEMISTRY OF SO_2, Jay A Kroll, D. J. Donaldson, Veronica Vaida

TB10   **11:33 – 11:48**
GAS PHASE HYDRATION OF METHYL GLYOXAL TO FORM THE GEMDIOL, Jay A Kroll, Jessica L Axson, Veronica Vaida

TC. Instrument/Technique Demonstration
Tuesday, June 21, 2016 – 8:30 AM
Room: 274 Medical Sciences Building

Chair: David A. Long, National Institute of Standards and Technology, Gaithersburg, MD, USA

TC01   **8:30 – 8:40**
HIGH HARMONIC GENERATION XUV SPECTROSCOPY FOR STUDYING ULTRAFAST PHOTOPHYSICS OF CO-ORDINATION COMPLEXES, Elizabeth S Ryland, Ming-Fu Lin, Max A Verkamp, Josh Vura-Weis

TC02   **8:42 – 8:57**
PHOTOELECTRON VELOCITY MAP IMAGING OF VIBRATIONALLY EXCITED, GAS-PHASE BIOMOLECULES AND THEIR ANIONS , Daniël Bakker, Sjors Bakels, Rutger van der Made, Atze Peters, Anouk Rijs

TC03   **8:59 – 9:14**
AN OPTICALLY ACCESSIBLE PYROLYSIS MICROREACTOR, Joshua H Baraban, Donald E David, Barney Ellison, John W Daily

TC04   **9:16 – 9:31**
BOHENDI@FELIX: PROBING THE FAR-INFRARED FINGERPRINT OF SMALL CLUSTERS IN HELIUM NANODROPLETS WITH A FREE ELECTRON LASER, Gerhard Schwaab, Raffael Schwan, Devendra Mani, Arghya Dey, Theo Fischer, Matin Kaufmann, Britta Redlich, Lex van der Meer, Martina Havenith

TC05   **9:33 – 9:43**
SUPERCONTINUUM CAVITY ENHANCED ABSORPTION SPECTROSCOPY FOR H_2O/D_2O SOLUTIONS, Mingyun Li, Kevin Lehmann

Intermission

TC06   **10:02 – 10:17**
MID-INFRARED FREQUENCY-AGILE DUAL-COMB SPECTROSCOPY, Pei-Ling Luo, Ming Yan, Kana Iwakuni, Guy Millot, Theodor W. Hänsch, Nathalie Picqué

TC07   **10:19 – 10:34**
IMPROVED SPECTROSCOPY OF MOLECULAR IONS IN THE MID-INFRARED WITH UP-CONVERSION DETECTION, Charles R. Markus, Adam J. Perry, James N. Hodges, Benjamin J. McCall

TC08   **10:36 – 10:51**
HIGH RESOLUTION ROVIBRATIONAL SPECTROSCOPY OF LARGE MOLECULES USING INFRARED FREQUENCY COMBS AND BUFFER GAS COOLING, Bryan Changala, Ben Spaun, David Patterson, Bryce J Bjork, Oliver H Heckl, John M. Doyle, Jun Ye

TC09   **10:53 – 11:08**
PROGRESS OF A NEW INSTRUMENT TO STUDY MOLECULAR DYNAMICS OF INTERSTELLAR ION-NEUTRAL REACTIONS , Kevin Roenitz, Ben Lamm, Lydia Rudd, Andy Justl, Steven Landeweer, Danny Roadman, Justyna Koscielniak, Andrew Sonnenberger, Manori Perera

TC10   **11:10 – 11:25**
CAVITY-ENHANCED ULTRAFAST SPECTROSCOPY: ULTRAFAST MEETS ULTRASENSITIVE, Thomas K Allison, Melanie Roberts Reber, Yuning Chen

TD. Mini-symposium: Spectroscopy in Traps
Tuesday, June 21, 2016 – 8:30 AM
Room: B102 Chemical and Life Sciences

Chair: Stephan Schlemmer, I. Physikalisches Institut, Köln, Germany

TD01   *INVITED TALK*   **8:30 – 9:00**
BOUND-FREE AND BOUND-BOUND SPECTROSCOPY OF COLD TRAPPED MOLECULAR IONS, Roland Wester

TD02   **9:05 – 9:20**
EQUATION-OF-MOTION COUPLED-CLUSTER CALCULATIONS OF PHOTODETACHMENT CROSS SECTIONS FOR ATOMIC NEGATIVE IONS ACROSS THE PERIODIC TABLE, Takatoshi Ichino, Lan Cheng, John F. Stanton

TD03   **9:22 – 9:37**
THE ELECTRONIC SPECTRUM OF CRYOGENIC RUTHENIUM-TRIS-BIPYRIDINE DICATIONS, Shuang Xu, James E. T. Smith, J. Mathias Weber

TD04   **9:39 – 9:54**
ELECTRONIC SPECTRA OF TRIS(2,2'-BIPYRIDINE)-METAL COMPLEX IONS IN GAS PHASE, Shuang Xu, James E. T. Smith, J. Mathias Weber

Intermission

TD05   **10:13 – 10:28**
ELECTRONIC SPECTROSCOPY OF TRAPPED PAH PHOTOFRAGMENTS, Christine Joblin, Anthony Bonnamy

TD06   **10:30 – 10:45**
ELECTRONIC SPECTRA OF BARE AND SOLVATED RUTHENIUM POLYPYRIDINE COMPLEXES, Shuang Xu, James E. T. Smith, J. Mathias Weber

TD07   **10:47 – 11:02**
CONFORMATIONAL SPECIFIC INFRARED AND ULTRAVIOLET SPECTROSCOPY OF COLD YA(D-Pro)AA·H$^+$ IONS: A STEROCHEMICAL "TWIST" ON THE PROLINE EFFECT, Christopher P Harrilal, Andrew F DeBlase, Nicole L Burke, Scott A McLuckey, Timothy S. Zwier

TD08   **11:04 – 11:19**
ALKALI CATION CHELATION IN COLD β-O-4 TETRALIGNOL COMPLEXES, Andrew F DeBlase, Eric T Dziekonski, John R. Hopkins, Nicole L Burke, Hilkka I Kenttamaa, Scott A McLuckey, Timothy S. Zwier

TD09   **11:21 – 11:31**
SINGLE MOLECULAR ION SPECTROSCOPY: TOWARDS PRECISION MEASUREMENTS ON CaH$^+$, Kenneth R Brown, Ncamiso B Khanyile, Rene Rugango, Gang Shu, Aaron Calvin

TE. Fundamental interest
Tuesday, June 21, 2016 – 8:30 AM
Room: 217 Noyes Laboratory

Chair: S. A. Cooke, Purchase College SUNY, Purchase, NY, USA

TE01　　　　　　　　　　　　　　　　　　　　　　　　　　　　　　　8:30–8:45
HIGH RESOLUTION SPECTROSCOPY OF $A^1B_{1u} \leftarrow X^1A_g 8_0^1 4_0^1$ BAND OF NAPHTHALENE REFERENCED TO AN OPTICAL FREQUENCY COMB , Kazuki Nakashima, Akiko Nishiyama, <u>Masatoshi Misono</u>

TE02　　　　　　　　　　　　　　　　　　　　　　　　　　　　　　　8:47–9:02
BONDING OF ALKALI-ALKALINE EARTH MOLECULES IN THE LOWEST Σ^+ STATES OF DOUBLET AND QUARTET MULTIPLICITY, <u>Johann V. Pototschnig</u>, Andreas W. Hauser, Wolfgang E. Ernst

TE03　　　　　　　　　　　　　　　　　　　　　　　　　　　　　　　9:04–9:19
A CANONICAL APPROACH TO MULTI-DIMENSIONAL VAN DER WAALS, HYDROGEN-BONDED, AND HALOGEN-BONDED POTENTIALS, Jay R. Walton, <u>Luis A. Rivera-Rivera</u>, Robert R. Lucchese, John W. Bevan

TE04　　　　　　　　　　　　　　　　　　　　　　　　　　　　　　　9:21–9:31
OBSERVATION OF BROADBAND ULTRAVIOLET EMISSION FROM Hg_3^*, <u>Wenting Wendy Chen</u>, Thomas C. Galvin, J. Gary Eden

TE05　　　　　　　　　　　　　　　　　　　　　　　　　　　　　　　9:33–9:48
QUANTITATIVE DETERMINATION OF LINESHAPE PARAMETERS FROM VELOCITY MODULATION SPECTROSCOPY, <u>James N. Hodges</u>, Benjamin J. McCall

TE06　　　　　　　　　　　　　　　　　　　　　　　　　　　　　　　9:50–10:05
USING NICE-OHVMS LINESHAPES TO STUDY RELAXATION RATES AND TRANSITION DIPOLE MOMENTS, <u>James N. Hodges</u>, Benjamin J. McCall

TE07　　　　　　　　　　　　　　　　　　　　　　　　　　　　　　　10:07–10:12
CO-ASSIGNMENT OF THE MOLECULAR VIBRATIONAL FREQUENCIES IN DIFFERENT ELECTRONIC STATES, <u>Yurii Panchenko</u>, Alexander Abramenkov

Intermission

TE08　　　　　　　　　　　　　　　　　　　　　　　　　　　　　　　10:31–10:46
DOPPLER BROADENING THERMOMETRY BASED ON CAVITY RING-DOWN SPECTROSCOPY, Jin Wang, Yu Robert Sun, Cunfeng Cheng, Lei-Gang Tao, Yan Tan, Peng Kang, An-Wen Liu, <u>Shui-Ming Hu</u>

TE09　　　　　　　　　　　　　　　　　　　　　　　　　　　　　　　10:48–11:03
CHEMICAL SYNTHESIS AND HIGH RESOLUTION SPECTROSCOPIC CHARACTERIZATION OF 1-AZA-ADAMANTANE-4-ONE $C_9H_{13}NO$ FROM THE MICROWAVE TO THE INFRARED, <u>Olivier Pirali</u>, Manuel Goubet, Vincent Boudon, Lucia D'accolti, Cosimo Annese, Caterina Fusco

TE10　　　　　　　　　　　　　　　　　　　　　　　　　　　　　　　11:05–11:20
ASSIGNMENT OF THE PERFLUOROPROPIONIC ACID-FORMIC ACID COMPLEX AND THE DIFFICULTIES OF INCLUDING HIGH K_a TRANSITIONS., <u>Daniel A. Obenchain</u>, Wei Lin, Stewart E. Novick, S. A. Cooke

TE11　　　　　　　　　　　　　　　　　　　　　　　　　　　　　　　11:22–11:37
INFRARED SPECTROSCOPIC INVESTIGATION ON CH BOND ACIDITY IN CATIONIC ALKANES, <u>Yoshiyuki Matsuda</u>, Min Xie, Asuka Fujii

TE12　　　　　　　　　　　　　　　　　　　　　　　　　　　　　　　11:39–11:54
DIELECTRIC STUDY OF ALCOHOLS USING BROADBAND TERAHERTZ TIME DOMAIN SPECTROSCOPY (THz-TDS)., <u>Sohini Sarkar</u>, Debasis Saha, Sneha Banerjee, Arnab Mukherjee, Pankaj Mandal

TF. Mini-symposium: Spectroscopy of Large Amplitude Motions
Tuesday, June 21, 2016 – 1:30 PM
Room: 100 Noyes Laboratory

Chair: David S. Perry, The University of Akron, Akron, OH, USA

TF01   1:30 – 1:45
HIGH-RESOLUTION INFRARED SPECTRSCOPY OF THE HYDROXYMETHYL RADICAL IN SOLID PARAHY-DROGEN, Morgan E. Balabanoff, David T. Anderson

TF02   1:47 – 2:02
INFRARED SPECTROSCOPIC STUDIES OF OCS TRAPPED IN SOLID PARAHYDROGEN: INDIRECT EVIDENCE OF LARGE AMPLITUDE MOTIONS, David T. Anderson

TF03   2:04 – 2:19
HYPERCONJUGATION IN THE S_1 STATE OF SUBSTITUTED TOLUENE PROBED BY INFRARED SPECTROSCOPY, Takashi Chiba, Katsuhiko Okuyama, Asuka Fujii

TF04   2:21 – 2:36
HIGH RESOLUTION DIRECT FREQUENCY COMB SPECTROSCOPY OF VINYL BROMIDE (C_2H_3Br) AND NI-TROMETHANE (CH_3NO_2) IN THE CH STRETCH REGION, Bryan Changala, Ben Spaun, David Patterson, Jun Ye

TF05   2:38 – 2:53
FTIR SYNCHROTRON SPECTROSCOPY OF THE ASYMMETRIC C-H STRETCHING BANDS OF METHYL MER-CAPTAN (CH_3SH) – A PERPLEXITY OF PERTURBATIONS, Ronald M. Lees, Li-Hong Xu, Elias M. Reid, Bishnu P. Thapaliya, Mahesh B. Dawadi, David S. Perry, Sylvestre Twagirayezu, Brant E. Billinghurst

TF06   2:55 – 3:10
THE TORSIONAL FUNDAMENTAL BAND AND ROTATIONAL SPECTRA UP TO 940 GHz OF THE GROUND, FIRST AND SECOND EXCITED TORSIONAL STATES OF ACETONE, V. Ilyushin, Iuliia Armieieva, Olga Dorovskaya, E. A. Alekseev, Marcela Tudorie, R. A. Motiyenko, L. Margulès, Olivier Pirali, Brian Drouin

TF07   3:12 – 3:22
ASSIGNING THE VIBRATION-ROTATION SPECTRA USING THE LWW PROGRAM PACKAGE, Wiesław Łodyga, Marek Kreglewski

Intermission

TF08   3:41 – 3:56
FAR-IR SPECTROSCOPY OF NEUTRAL GAS PHASE PEPTIDES: SIGNATURES FROM COMBINED EXPERIMENTS AND SIMULATIONS , Jérôme Mahé, Marie-Pierre Gaigeot, Daniël Bakker, Sander Jaeqx, Anouk Rijs

TF09   3:58 – 4:13
INVERSION VIBRATIONAL ENERGY LEVELS OF $AsH_3{}^+$ STUDIED BY ZERO-KINETIC-ENERGY PHOTOELEC-TRON SPECTROSCOPY, Yuxiang Mo

TF10   4:15 – 4:30
MICROWAVE MEASUREMENTS OF CYCLOPROPANECARBOXYLIC ACID AND ITS DOUBLY HYDROGEN BONDED DIMER WITH FORMIC ACID*, Aaron M Pejlovas, Wei Lin, Stephen G. Kukolich

TF11   4:32 – 4:47
GAS PHASE MEASUREMENTS OF MONO-FLUORO-BENZOIC ACIDS AND THE DIMER OF 3-FLUORO-BENZOIC ACID, Adam M Daly, Spencer J Carey, Aaron M Pejlovas, Kexin Li, Lu Kang, Stephen G. Kukolich

TF12   4:49 – 5:04
MICROWAVE MEASUREMENTS OF THE TROPOLONE-FORMIC ACID DOUBLY HYDROGEN BONDED DIMER*, Aaron M Pejlovas, Agapito Serrato III, Wei Lin, Stephen G. Kukolich

TF13   5:06 – 5:21
VIBRATION-ROTATION-TUNNELING SPECTRUM OF FORMIC ACID DIMER IN THE 7.3μm REGION, Chuanxi Duan

TF14   5:23 – 5:38
MICROWAVE MEASUREMENTS OF MALEIMIDE AND ITS DOUBLY HYDROGEN BONDED DIMER WITH FORMIC ACID*, Aaron M Pejlovas, Lu Kang, Stephen G. Kukolich

14

TG. Mini-symposium: Spectroscopy in Atmospheric Chemistry
Tuesday, June 21, 2016 – 1:30 PM
Room: 116 Roger Adams Lab

Chair: Rainer Volkamer, University of Colorado Boulder, Boulder, CO, USA

TG01   *INVITED TALK*   1:30 – 2:00
NOVEL IMPLEMENTATIONS OF FARADAY ROTATION SPECTROSCOPY - FROM IN-SITU RADICAL DETECTION TO STUDIES OF ENVIRONMENTAL NITROGEN CYCLING, Eric Zhang, Jonas Westberg, Gerard Wysocki

TG02   2:05 – 2:20
WITHDRAWN TALK - Check mobile app and last-minute changes for a replacement

TG03   2:22 – 2:37
USING MULTI RESONANCE EFFECTS TOWARDS SINGLE CONFORMER MICROWAVE SPECTROSCOPY , Chamara Abeysekera, Alicia O. Hernandez-Castillo, Brian M Hays, Timothy S. Zwier

TG04   2:39 – 2:54
ISOMER SPECIFIC MICROWAVE SPECTRUM OF (E)- AND (Z)- PHENYLVINYLNITRILE. IMPLEMENTING A NEW MULTI-RESONANT SPECTRAL ANALYSIS TOOL. , Alicia O. Hernandez-Castillo, Brian M Hays, Chamara Abeysekera, Timothy S. Zwier

TG05   2:56 – 3:11
PHOTOELECTRON IMAGING OF OXIDE.VOC CLUSTERS, Kellyn M. Patros, Jennifer Mann, Caroline Chick Jarrold

Intermission

TG06   3:30 – 3:45
FT-IR MEASUREMENTS OF MID-IR PROPENE (C_3H_6) CROSS SECTIONS FOR TITAN STRATOSPHERE , Keeyoon Sung, Geoffrey C. Toon, Brian Drouin, Timothy J. Crawford, Arlan Mantz, Mary Ann H. Smith

TG07   3:47 – 4:02
THE UV SPECTROSCOPY OF JET-COOLED 3-PHENYL-2-PROPYNENITRILE, Khadija M. Jawad, Timothy S. Zwier

TG08   4:04 – 4:19
THEORETICAL STUDIES OF THE RELAXATION MATRIX FOR MOLECULAR SYSTEMS, Qiancheng Ma, C. Boulet

TG09   4:21 – 4:36
CALCULATED VIBRATIONAL STATES OF OZONE UP TO DISSOCIATION, Richard Dawes, Steve Alexandre Ndengue, Xiao-Gang Wang, Tucker Carrington, Hua Guo

TG10   4:38 – 4:53
ROTATIONAL QUENCHING STUDY IN ISOVALENT H^+ + CO AND H^+ + CS SYSTEMS, Rajwant Kaur, T. J. Dhilip Kumar

TG11   4:55 – 5:10
MODELING PHOTODETACHMENT FROM HO_2^- USING THE pd CASE OF THE GENERALIZED MIXED CHARACTER MOLECULAR ORBITAL MODEL, Christopher C Blackstone, Andrei Sanov

TG12   5:12 – 5:27
HITRAN APPLICATION PROGRAMMING INTERFACE (HAPI): EXTENDING HITRAN CAPABILITIES, Roman V Kochanov, Iouli E Gordon, Laurence S. Rothman, Piotr Wcislo, Christian Hill, Jonas Wilzewski

TH. Astronomy
Tuesday, June 21, 2016 – 1:30 PM
Room: 274 Medical Sciences Building

Chair: Michael C McCarthy, Harvard-Smithsonian CfA, Cambridge, MA, USA

TH01   1:30 – 1:45
LABORATORY MEASUREMENTS AND ASTRONOMICAL SEARCH OF THE HSO RADICAL, Gabriele Cazzoli, Cristina Puzzarini, Valerio Lattanzi, Belén Tercero, Jose Cernicharo

TH02   1:47 – 2:02
THE BENDING VIBRATIONS OF THE C_3-ISOTOPOLOGUES IN THE 1.9 TERAHERTZ REGION, A. Breier, Thomas Büchling, Volker Lutter, Rico Schnierer, Guido W Fuchs, Thomas Giesen

TH03   2:04 – 2:19
LABORATORY MEASUREMENTS OF SMALL SILICON BEARING MOLECULES OF ASTROPHYSICAL INTEREST, Carl A Gottlieb, Michael C McCarthy

TH04   2:21 – 2:36
MILLIMETER-WAVE SPECTROSCOPY OF METHOXYMETHANOL, R. A. Motiyenko, L. Margulès, J.-C. Guillemin, Didier Despois

TH05   2:38 – 2:53
SUBMILLIMETER SPECTRUM OF THE METHOXY RADICAL TO GUIDE INTERSTELLAR SEARCHES, Jacob Laas, Susanna L. Widicus Weaver

TH06   2:55 – 3:10
THE COMPLETE, TEMPERATURE RESOLVED SPECTRUM OF METHYL CYANIDE BETWEEN 200 AND 277 GHZ, James P. McMillan, Christopher F. Neese, Frank C. De Lucia

TH07   3:12 – 3:27
COMPREHENSIVE ANALYSIS OF INTERSTELLAR Iso-PROPYL CYANIDE UP TO 480 GHZ, Lucie Kolesniková, E. R. Alonso, Carlos Cabezas, Santiago Mata, José L. Alonso

TH08   3:29 – 3:44
SEARCHING FOR AMINOMETHANOL AMONGST THE REACTION PRODUCTS OF O(^{1}D) INSERTION INTO METHYLAMINE, Morgan N McCabe, Carson Reed Powers, Brian M Hays, Samuel Zinga, Susanna L. Widicus Weaver

Intermission

TH09   4:03 – 4:18
SEGMENTED CHIRPED-PULSE MILLIMETER-WAVE SPECTROSCOPY FOR ASTROCHEMISTRY , Benjamin E Arenas, Amanda Steber, Sébastien Gruet, Melanie Schnell

TH10   4:20 – 4:35
THE MILLIMETER-WAVE SPECTROSCOPY OF HYDANTOIN, A POTENTIAL PRECURSOR OF GLYCINE, Hiroyuki Ozeki, Rio Miyahara, Hiroto Ihara, Satoshi Todaka, Kaori Kobayashi, Masatoshi Ohishi

TH11   4:37 – 4:52
THE MICROWAVE SPECTROSCOPY OF AMINOACETONITRILE IN THE VIBRATIONAL EXCITED STATES 2, Chiho Fujita, Haruka Higurashi, Hiroyuki Ozeki, Kaori Kobayashi

TH12   4:54 – 5:09
THE STUDY OF ACENAPHTHENE AND ITS COMPLEXATION WITH WATER, Amanda Steber, Cristobal Perez, Anouk Rijs, Melanie Schnell

TH13   5:11 – 5:26
THE RADIO SPECTRA AND -VE INERTIAL DEFECTS BEHAVIOR OF PLANAR AROMATIC HETEROCYCLES, Don McNaughton, Michaela K Jahn, Jens-Uwe Grabow, Peter Godfrey, Michael Travers, Dennis Wachsmuth

TH14   5:28 – 5:43
A NEW LABORATORY FOR TERAHERTZ CHARACTERIZATION OF COSMIC ANALOG DUSTS, Thushara Perera

TH15   5:45 – 5:55
MM/SUBMM STUDY OF GAS-PHASE PHOTOPRODUCTS FROM METHANOL INTERSTELLAR ICE ANALOGUES, AJ Mesko, Houston Hartwell Smith, Stefanie N Milam, Susanna L. Widicus Weaver

TI. Clusters/Complexes
Tuesday, June 21, 2016 – 1:30 PM
Room: B102 Chemical and Life Sciences

Chair: Zbigniew Kisiel, Institute of Physics, Polish Academy of Sciences, Warszawa, Poland

TI01   1:30 – 1:45
INFRARED SPECTROSCOPIC STUDY FOR THE HYDRATED CLUSTERS OF PENTANE CATION, Tomoya Endo, Yoshiyuki Matsuda, Asuka Fujii

TI02   1:47 – 2:02
INFLUENCE OF AROMATIC MOLECULES ON THE STRUCTURE AND SPECTROSCOPY OF WATER CLUSTERS, Daniel P. Tabor, Edwin Sibert, Patrick S. Walsh, Timothy S. Zwier

TI03   2:04 – 2:19
VIBRATIONAL COUPLING IN SOLVATED FORM OF EIGEN PROTON: TUNING THE COUPLING VIA ISOTOPO-LOGUES, Jheng-Wei Li, Jer-Lai Kuo

TI04   2:21 – 2:36
GAS-PHASE MOLECULAR STRUCTURE OF NOPINONE AND ITS WATER COMPLEXES STUDIED BY MICROWAVE FOURIER TRANSFORM SPECTROSCOPY AND QUANTUM CHEMICAL CALCULATIONS, Elias M. Neeman, Juan-Ramon Aviles Moreno, T. R. Huet

TI05   2:38 – 2:53
CHARACTERIZATION OF MICROSOLVATED CROWN ETHERS FROM BROADBAND ROTATIONAL SPEC-TROSCOPY, Cristobal Perez, Melanie Schnell, Susana Blanco, Juan Carlos Lopez

TI06   2:55 – 3:10
CONCERTED BREAKING OF TWO HYDROGEN BONDS IN WATER HEXAMER PRISM REVEALED FROM BROADBAND ROTATIONAL SPECTROSCOPY , Jeremy O Richardson, Cristobal Perez, Simon Lobsiger, Adam A. Reid, Berhane Temelso, George C. Shields, Zbigniew Kisiel, David J. Wales, Brooks Pate, Stuart C. Althorpe

TI07   3:12 – 3:27
STRUCTURES OF MICROSOLVATED CAMPHOR FROM BROADBAND ROTATIONAL SPECTROSCOPY, Cristobal Perez, Anna Krin, Amanda Steber, Juan Carlos Lopez, Zbigniew Kisiel, Melanie Schnell

Intermission

TI08   3:46 – 4:01
ROTATIONAL INVESTIGATION OF THE ADDUCTS OF FORMIC ACID WITH ALCOHOLS, ETHERS AND ESTERS, Luca Evangelisti, Lorenzo Spada, Weixing Li, Walther Caminati

TI09   4:03 – 4:18
MICROWAVE SPECTRUM OF THE ETHANOL-METHANOL DIMER, Ian A Finneran, Brandon Carroll, Griffin Mead, Geoffrey Blake

TI10   4:20 – 4:35
HYDROPEROXIDES AS HYDROGEN BOND DONORS, Kristian H. Møller, Camilla M. Tram, Anne S. Hansen, Henrik G. Kjaergaard

TI11   4:37 – 4:52
GAZ PHASE IR AND UV SPECTROSCOPY OF NEUTRAL CONTACT ION PAIRS, Sana Habka, Valerie Brenner, Michel Mons, Eric Gloaguen

TI12   *Post-Deadline Abstract*   4:54 – 5:09
DETERMINATION OF STRUCTURAL AND ELECTRONIC PARAMETERS OF ANTIMONY COMPLEX, FROM THE-ORETICAL CALCULATIONS, Berna Catikkas, Ismail Kosar

TI13   5:11 – 5:26
EXPERIMENTAL DETERMINATION OF GAS PHASE THERMODYNAMIC PROPERTIES OF BIMOLECULAR COM-PLEXES, Anne S. Hansen, Zeina Maroun, Kasper Mackeprang, Henrik G. Kjaergaard

TI14   *Post-Deadline Abstract*   5:28 – 5:43
PHOTOELECTRON IMAGING OF TaBO$^-$: OBSERVATION OF A BORONYL TRANSITION METAL COMPLEX, Joseph Czekner, Lai-Sheng Wang

TJ. Vibrational structure/frequencies
Tuesday, June 21, 2016 – 1:30 PM
Room: 217 Noyes Laboratory

Chair: Gary E. Douberly, The University of Georgia, Athens, GA, USA

TJ01   1:30 – 1:45
EXPLORING THE RELATIONSHIPS BETWEEN ANHARMONICITY AND OH BOND LENGTHS IN HYDROGEN BONDED COMPLEXES, Anne B McCoy, Sotiris Xantheas

TJ02   1:47 – 2:02
SPECTROSCOPIC MANIFESTATION OF VIBRATIONALLY-MEDIATED STRUCTURE CHANGE IN THE ISOLATED FORMATE MONOHYDRATE, Joanna K. Denton, Conrad T. Wolke, Olga Gorlova, Helen Gerardi, Anne B McCoy, Mark Johnson

TJ03   2:04 – 2:19
THEORETICAL INVESTIGATION OF ANHARMONIC EFFECTS OBSERVED IN THE INFRARED SPECTRA OF THE FORMALDEHYDE CATION AND ITS HYDROXYMETHYLENE ISOMER, Lindsey R Madison, Jonathan Mosley, Daniel Mauney, Michael A Duncan, Anne B McCoy

TJ04   2:21 – 2:36
SPECTROSCOPIC SIGNATURES AND STRUCTURAL MOTIFS OF DOPAMINE: A COMPUTATIONAL STUDY , Santosh Kumar Srivastava, Vipin Bahadur Singh

TJ05   2:38 – 2:53
SPECTROSCOPIC SIGNATURES AND STRUCTURAL MOTIFS IN ISOLATED AND HYDRATED XANTHINE: A COMPUTATIONAL STUDY, Vipin Bahadur Singh

TJ06   2:55 – 3:10
VIBRATIONAL SPECTROSCOPY AND THEORY OF $Fe_x^+(CH_4)_n$ (x =2,3) (n = 1–3), Christopher Copeland, Muhammad Affawn Ashraf, Ricardo B. Metz

TJ07   3:12 – 3:27
VIBRATIONAL ANALYSIS OF THE SiCN $\tilde{X}$ $^2\Pi$ SYSTEM, Masaru Fukushima, Takashi Ishiwata

TJ08   3:29 – 3:44
ELECTRONIC STRUCTURE OF SMALL LANTHANIDE CONTAINING MOLECULES, Jared O. Kafader, Manisha Ray, Josey E Topolski, Caroline Chick Jarrold

Intermission

TJ09   4:03 – 4:18
EXTENSIVE MEASUREMENTS OF VIBRATION-INDUCED PERMANENT ELECTRIC DIPOLE MOMENTS OF METHANE, Shoko Okuda, Hiroyuki Sasada

TJ10   4:20 – 4:35
GLOBAL FREQUENCY AND INTENSITY ANALYSIS OF THE $\nu_{10}/\nu_7/\nu_4/\nu_{12}$ BANDS SYSTEM OF $^{12}C_2H_4$ at 10 μm USING THE D_{2h} TOP DATA SYSTEM, Abdulsamee Alkadrou, Maud Rotger, Vincent Boudon, Jean Vander Auwera

TJ11   4:37 – 4:52
HIGH-RESOLUTION STIMULATED RAMAN SPECTROSCOPY AND ANALYSIS OF ν_2 AND ν_3 BANDS OF of $^{13}C_2H_4$ USING THE D_{2h} TOP DATA SYSTEM, Abdulsamee Alkadrou, Maud Rotger, Dionisio Bermejo, Jose Luis Domenech, Vincent Boudon

TJ12   4:54 – 5:09
OBSERVATION OF THE LOW-LYING $a^3\Delta$ AND $A^1\Delta$ STATES IN JET-COOLED TANTALUM MONONITRIDE, Sheo Mukund, Soumen Bhattacharyya, Sanjay G. Nakhate

TJ13   5:11 – 5:26
APPLICATION OF MULTIVALUED HIGH ORDER PADE-HERMITE APPROXIMANTS TO RESUMMATION OF PERTUBATION SERIES. VIBRATIONAL AND ROVIBRATIONAL ENERGY SPECTRUM OF H_2CO MOLECULE., Andrey Duchko, Sergei N. Yurchenko, Alexandr Bykov

TJ14   5:28 – 5:43
CALCULATIONS OF THE ELECTRONIC STRUCTURE AND VIBRATIONAL STATES OF THE EXCITED a^3A" TRIPLET AND A^1A" SINGLET STATES OF FORMALDEHYDE, Bradley Welch, Richard Dawes, Vladimir Tyuterev, Ludovic Daumont

TK. Theory and Computation
Tuesday, June 21, 2016 – 1:30 PM
Room: 140 Burrill Hall

Chair: Jonathan Tennyson, University College London, London, United Kingdom

TK01   1:30 – 1:45
IS WATSON'S "CHARGE-MODIFIED" REDUCED MASS ALWAYS BEST FOR DIATOMIC IONS ?, Robert J. Le Roy, Nikesh S. Dattani

TK02   1:47 – 2:02
INELASTIC SCATTERING OF H+CO: INFLUENCE OF RENNER-TELLER COUPLING, Steve Alexandre Ndengue, Richard Dawes

TK03   2:04 – 2:14
ROOM TEMPERATURE LINE LISTS FOR CO_2 ISOTOPOLOGUES WITH *AB INITIO* COMPUTED INTENSITIES, Emil Zak, Jonathan Tennyson, Oleg Polyansky, Lorenzo Lodi, Nikolay Fedorovich Zobov, Sergey Tashkun, Valery Perevalov

TK04   2:16 – 2:31
ANALYSIS OF THE VIBRATIONAL SPECTRA TO CALCULATE THE THERMODYNAMIC QUANTITIES CLOSE TO PHASE TRANSITIONS IN NH_4F, Hamit Yurtseven, Ozlem Tari

TK05   2:33 – 2:48
THE EFFECT OF INTERMOLECULAR MODES ON THE XH-STRETCHING VIBRATIONS IN HYDROGEN BONDED COMPLEXES, Kasper Mackeprang, Henrik G. Kjaergaard

TK06   2:50 – 3:00
THEORETICAL ANALYSIS OF VCD SPECTRA OF α AND β L-FUCOPYRANOSIDE IN THE CH STRETCHING REGION, Sofiane Moussi, Ourida Ouamerali

Intermission

TK07   3:19 – 3:34
FULL CI BENCHMARK POTENTIALS FOR THE $6e^-$ SYSTEM Li_2 WITH A CBS EXTRAPOLATION FROM aug-cc-pCV5Z AND aug-cc-pCV6Z BASIS SETS USING FCIQMC AND DMRG, Nikesh S. Dattani, Sandeep Sharma, Ali Alavi

TK08   3:36 – 3:51
AN IMPROVED EMPIRICAL POTENTIAL FOR THE HIGHLY MULTI-REFERENCE SEXTUPLY BONDED TRANSITION METAL BENCHAMRK MOLECULE Cr_2, Nikesh S. Dattani, Michał Tomza, Giovanni Li Manni

TK09   3:53 – 4:03
VIBRONIC TRANSITIONS IN THE X-Sr SERIES (X=Li, Na, K, Rb): ON THE ACCURACY OF NUCLEAR WAVE-FUNCTIONS DERIVED FROM QUANTUM CHEMISTRY, Ralf Meyer, Johann V. Pototschnig, Andreas W. Hauser, Wolfgang E. Ernst

TK10   4:05 – 4:20
AB INITIO INVESTIGATIONS OF THE EXCITED ELECTRONIC STATES OF CaOCa , Wafaa M Fawzy, Michael Heaven

TK11   4:22 – 4:37
AB INITIO CALCULATION OF NH_3 SPECTRUM, Oleg Polyansky, Roman I. Ovsyannikov, Aleksandra Kyuberis, Lorenzo Lodi, Jonathan Tennyson, Sergei N. Yurchenko, Andrey Yachmenev, Nikolay Fedorovich Zobov

TK12   4:39 – 4:54
AB INITIO EXPLORATION OF THE POTENTIAL ENERGY SURFACE OF THE O_2-SO_2 OPEN-SHELL COMPLEX., Wafaa M Fawzy, Jon T. Hougen

TK13   4:56 – 5:11
APPLYING QUANTUM MONTE CARLO TO THE ELECTRONIC STRUCTURE PROBLEM, Andrew D Powell, Richard Dawes

TK14   5:13 – 5:28
CAN WE PREDICT QUANTUM YIELDS USING EXCITED STATE DENSITY FUNCTIONAL THEORY FOR NEW FAMILIES OF FLUORESCENT DYES?, Alexander W. Kohn, Zhou Lin, James J. Shepherd, Troy Van Voorhis

TK15   5:30 – 5:45
INCORPORATION OF A ROVIBRATIONAL ANALYSIS OF OC-H_2O INTO 6-D MORPHED POTENTIALS OF THE COMPLEX, Luis A. Rivera-Rivera, Sean D. Springer, Blake A. McElmurry, Igor I Leonov, Robert R. Lucchese, John W. Bevan, L. H. Coudert

WA. Mini-symposium: Spectroscopy of Large Amplitude Motions
Wednesday, June 22, 2016 – 8:30 AM
Room: 100 Noyes Laboratory

Chair: Mahesh B. Dawadi, The University of Akron, Akron, OH, USA

WA01 *INVITED TALK* **8:30 – 9:00**
TORSION - ROTATION - VIBRATION EFFECTS IN THE GROUND AND FIRST EXCITED STATES OF METHACROLEIN AND METHYL VINYL KETONE, Olena Zakharenko, R. A. Motiyenko, Juan-Ramon Aviles Moreno, T. R. Huet

WA02 **9:05 – 9:20**
SYNCHROTRON SPECTROSCOPY AND TORSIONAL STRUCTURE OF THE CSH-BENDING AND CH_3-ROCKING BANDS OF METHYL MERCAPTAN, Ronald M. Lees, Li-Hong Xu, Brant E. Billinghurst

WA03 **9:22 – 9:37**
VIBRATIONAL CONICAL INTERSECTIONS IN CH_3SH: IMPLICATIONS FOR SPECTROSCOPY AND DYNAMICS IN THE CH STRETCH REGION, David S. Perry, Bishnu P. Thapaliya, Mahesh B. Dawadi, Ram Bhatta

WA04 **9:39 – 9:49**
THE EQUIVALENCE OF THE METHYL GROUPS IN PUCKERED 3,3-DIMETHYL OXETANE, Alberto Macario, Susana Blanco, Juan Carlos Lopez

Intermission

WA05 **10:08 – 10:23**
SPECTROSCOPIC STUDY OF METHYLGLYOXAL AND ITS HYDRATES : A GASEOUS PRECURSOR OF SECONDARY ORGANIC AEROSOLS., Sabath Bteich, Manuel Goubet, L. Margulès, R. A. Motiyenko, T. R. Huet

WA06 **10:25 – 10:40**
TORSION - VIBRATION COUPLINGS IN THE $CH_3OO\cdot$ RADICAL, Meng Huang, Terry A. Miller, Anne B McCoy, Kuo-Hsiang Hsu, Yu-Hsuan Huang, Yuan-Pern Lee

WA07 **10:42 – 10:57**
WEAK INTRAMOLECULAR INTERACTIONS EFFCTS ON THE STRUCTURE AND THE TORSIONAL SPECTRA OF ETHYLENE GLYCOL, AN ASTROPHYSICAL SPECIES, Maria Luisa Senent, Rahma Boussessi

WA08 **10:59 – 11:14**
HIGH-ACCURATE INTERMOLECULAR POTENTIAL ENERGY SURFACE OF $HCN - H_2$ COMPLEX WITH INTRAMOLECULAR VIBRATIONAL MODE OF HCN INCLUDED, Yu Zhai, Hui Li

WA09 **11:16 – 11:31**
THE ROTATIONAL SPECTRA OF CYANOACETYLENE DIMER, H-C-C-C-N ••• H-C-C-C-N, Lu Kang, Philip Davis, Ian Dorell, Kexin Li, Adam M. Daly, Stephen G. Kukolich, Stewart E. Novick

WA10 **11:33 – 11:48**
INFRARED SPECTRUM OF $CO-O_2$, A 'NEW' WEAKLY-BOUND COMPLEX, Bob McKellar, A. J. Barclay, K. H. Michaelian, Nasser Moazzen-Ahmadi

WB. Mini-symposium: Spectroscopy in Atmospheric Chemistry
Wednesday, June 22, 2016 – 8:30 AM
Room: 116 Roger Adams Lab

Chair: Gerard Wysocki, Princeton University, Princeton, NJ, USA

WB01   *INVITED TALK*   **8:30 – 9:00**
OH WHERE OH WHERE IS OH? MEASURING THE ELUSIVE HYDROXYL RADICAL IN THE ATMOSPHERE USING LASER-INDUCED FLUORESCENCE, Philip S. Stevens

WB02   **9:05 – 9:20**
EXPERIMENTAL AND THEORETICAL He-BROADENED LINE PARAMETERS OF CARBON MONOXIDE IN THE FUNDAMENTAL BAND, Adriana Predoi-Cross, Hoimonti Rosario, Koorosh Esteki, Shamria Latif, Hossein Naseri, Franck Thibault, V. Malathy Devi, Mary Ann H. Smith, Arlan Mantz

WB03   **9:22 – 9:37**
SELF- AND H_2-BROADENED LINE PARAMETERS OF CARBON MONOXIDE IN THE FIRST OVERTONE BAND, Adriana Predoi-Cross, Koorosh Esteki, Hossein Naseri, V. Malathy Devi, Mary Ann H. Smith, Arlan Mantz, Sergei V Ivanov

WB04   **9:39 – 9:54**
MULTISPECTRUM ANALYSIS OF THE OXYGEN A-BAND, Brian Drouin, Linda R. Brown, Matthew J. Cich, Timothy J. Crawford, Alexander Guillaume, Fabiano Oyafuso, Vivienne H Payne, Keeyoon Sung, Shanshan Yu, D. Chris Benner, V. Malathy Devi, Joseph Hodges, Eli J Mlawer, David Robichaud, Edward H Wishnow

WB05   **9:56 – 10:11**
HIGH RESOLUTION PHOTOACOUSTIC SPECTROSCOPY OF THE OXYGEN A-BAND, Matthew J. Cich, Elizabeth M Lunny, Gautam Stroscio, Thinh Quoc Bui, Caitlin Bray, Daniel Hogan, Priyanka Rupasinghe, Timothy J. Crawford, Brian Drouin, Charles Miller, David A. Long, Joseph Hodges, Mitchio Okumura

Intermission

WB06   **10:30 – 10:45**
MEASUREMENTS @ MM-/SUB-MM-WAVE SPECTROSCOPY LABORATORY OF BOLOGNA: ROTATIONAL SPECTROSCOPY APPLIED TO ATMOSPHERIC STUDIES, Cristina Puzzarini

WB07   **10:47 – 11:02**
THz AND FT-IR STUDY OF 18-O ISOTOPOLOGUES OF SULFUR DIOXIDE: $^{32}S^{16}O^{18}O$ AND $^{32}S^{18}O_2$, L. Margulès, R. A. Motiyenko, J. Demaison, Agnes Perrin, F. Kwabia Tchana, Laurent Manceron

WB08   **11:04 – 11:19**
IMPROVE THE ABSOLUTE ACCURACY OF OZONE INTENSITIES IN THE 9-11 μm REGION VIA MW/IR MULTIWAVELENGTH SPECTROSCOPY, Shanshan Yu, Brian Drouin

WB09   **11:21 – 11:36**
FIRST HIGH-RESOLUTION ANALYSIS OF PHOSGENE $^{35}CL_2CO$ AND $^{35}CL^{37}CLCO$ FUNDAMENTALS IN THE 250 - 480 cm^{-1} SPECTRAL REGION , F. Kwabia Tchana, M. Ndao, Laurent Manceron, Agnes Perrin, Jean-Marie Flaud, Walter Lafferty

WC. Chirped pulse
Wednesday, June 22, 2016 – 8:30 AM
Room: 274 Medical Sciences Building

Chair: Robert W Field, MIT, Cambridge, MA, USA

WC01   8:30 – 8:45
CHIRPED PULSE ROTATIONAL SPECTROSCOPY OF A SINGLE THUJONE+WATER SAMPLE, Zbigniew Kisiel, Cristobal Perez, Melanie Schnell

WC02   8:47 – 9:02
MICROWAVE SPECTRUM AND MOLECULAR STRUCTURE OF THE ARGON-*CIS*-1,2-DICHLOROETHYLENE COMPLEX, Mark D. Marshall, Helen O. Leung, Craig J. Nelson, Leonard H. Yoon

WC03   9:04 – 9:19
INFLUENCE OF HALOGEN VARIATION ON STRUCTURE AND INTERACTIONS IN VINYL HALIDE $(H_2C=CHX)\cdots CO_2$ (X = F, Cl, Br) COMPLEXES, Ashley M. Anderton, Cori L. Christenholz, Rachel E. Dorris, Rebecca A. Peebles, Sean A. Peebles

WC04   9:21 – 9:36
H-BONDING NETWORKS IN SUGAR ALCOHOLS: IDENTIFYING GLUCOPHORES?, E. R. Alonso, Santiago Mata, Carlos Cabezas, Isabel Peña, José L. Alonso

WC05   9:38 – 9:48
SEVEN CONFORMERS OF PIPECOLIC ACID IDENTIFIED IN THE GAS PHASE, Carlos Cabezas, Alcides Simao, José L. Alonso

WC06   9:50 – 10:05
PREFERRED CONFORMERS OF NON-PROTEINOGENIC AMINO ACIDS HOMOSERINE AND HOMOCYSTEINE, Verónica Díez, Miguel A. Rodríguez, Santiago Mata, E. R. Alonso, Carlos Cabezas, José L. Alonso

WC07   10:07 – 10:22
THE ROTATIONAL SPECTRUM OF THE UREA$\cdots$ISOCYANIC ACID COMPLEX, John C Mullaney, Chris Medcraft, Nick Walker, Anthony Legon, Luke Lewis-Borrell, Bernard T Golding

Intermission

WC08   10:41 – 10:56
GEOMETRY OF AN ISOLATED DIMER OF IMIDAZOLE CHARACTERISED BY ROTATIONAL SPECTROSCOPY AND AB INITIO CALCULATIONS, John C Mullaney, Daniel P. Zaleski, David Peter Tew, Nick Walker, Anthony Legon

WC09   10:58 – 11:13
MICROWAVE SPECTRUM OF THE ACETALDEHYDE-WATER DIMER, Griffin Mead, Ian A Finneran, Brandon Carroll, Geoffrey Blake

WC10   11:15 – 11:30
THE CONFORMATIONAL BEHAVIOUR OF THE ODORANT DIHYDROCARVEOL, Donatella Loru, Natasha Jarman, M. Eugenia Sanz

WC11   11:32 – 11:47
STRUCTURAL CHARACTERISATION OF FENCHONE AND ITS COMPLEXES WITH ETHANOL BY BROADBAND ROTATIONAL SPECTROSCOPY , Donatella Loru, M. Eugenia Sanz

WC12   11:49 – 12:04
BROADBAND MICROWAVE SPECTROSCOPY AS A TOOL TO STUDY DISPERSION INTERACTIONS IN CAMPHOR-ALCOHOL SYSTEMS, Mariyam Fatima, Cristobal Perez, Melanie Schnell

WD. Mini-symposium: Spectroscopy in Traps
Wednesday, June 22, 2016 – 8:30 AM
Room: B102 Chemical and Life Sciences

Chair: Roland Wester, Universität Innsbruck, Innsbruck, Austria

WD01   *INVITED TALK*   **8:30 – 9:00**
INFRARED ION SPECTROSCOPY AT FELIX: APPLICATIONS IN PEPTIDE DISSOCIATION AND ANALYTICAL CHEMISTRY, Jos Oomens

WD02   **9:05 – 9:20**
INFRARED PREDISSOCIATION SPECTROSCOPY OF He-TAGGED SMALL MOLECULAR CATIONS OF ASTRO-CHEMICAL INTEREST, Alexander Stoffels, Britta Redlich, Jos Oomens, Oskar Asvany, Sandra Brünken, Pavol Jusko, Sven Thorwirth, Stephan Schlemmer

WD03   **9:22 – 9:37**
PROBING THE VIBRATIONAL SPECTROSCOPY OF THE DEPROTONATED THYMINE RADICAL BY PHOTODE-TACHMENT AND STATE-SELECTIVE AUTODETACHMENT PHOTOELECTRON SPECTROSCOPY VIA DIPOLE-BOUND STATES, Dao-Ling Huang, Guo-Zhu Zhu, Lai-Sheng Wang

WD04   **9:39 – 9:54**
QUANTIFICATION OF STRUCTURAL ISOMERS VIA MODE-SELECTIVE IRMPD, Nicolas C Polfer

WD05   **9:56 – 10:11**
URIDINE NUCLEOSIDE THIATION: GAS-PHASE STRUCTURES AND ENERGETICS, Lucas Hamlow, Justin Lee, M T Rodgers, Giel Berden, Jos Oomens

Intermission

WD06   **10:30 – 10:45**
STRUCTURE DETERMINATION OF ORNITHINE-LINKED CISPLATIN BY INFRARED MULTIPLE PHOTON DIS-SOCIATION ACTION SPECTROSCOPY, Chenchen He, Bett Kimutai, Lucas Hamlow, Harrison Roy, Y-W Nei, Xun Bao, Juehan Gao, Jonathan K Martens, Giel Berden, Jos Oomens, Philippe Maitre, Vincent Steinmetz, Christopher P McNary, Peter B Armentrout, C S Chow, M T Rodgers

WD07   **10:47 – 11:02**
CONTROLLED FORMATION AND VIBRATIONAL CHARACTERIZATION OF LARGE SOLVATED IONIC CLUS-TERS IN CRYOGENIC ION TRAPS, Etienne Garand, Brett Marsh, Jonathan Voss, Erin M. Duffy

WD08   **11:04 – 11:19**
VIBRATIONAL CHARACTERIZATION OF CATALYTIC INTERMEDIATES IN A DUAL CRYOGENIC ION TRAP SPECTROMETER, Erin M. Duffy, Jonathan Voss, Brett Marsh, Etienne Garand

WD09   **11:21 – 11:36**
MODELING AND OPTIMIZING RF MULTIPOLE ION TRAPS, Sven Fanghaenel, Oskar Asvany, Stephan Schlemmer

WE. (Hyper)fine structure, tunneling
Wednesday, June 22, 2016 – 8:30 AM
Room: 217 Noyes Laboratory

Chair: Trevor Sears, Brookhaven National Laboratory, Upton, NY, USA

WE01   8:30 – 8:45
IODINE: MANY ELECTRONS AND MUCH TO DISCUSS. . . THE NUCLEAR QUADRUPOLE COUPLING, NUCLEAR SPIN-ROTATION, CONFORMATIONAL ANALYSIS, AND STRUCTURAL DETERMINATION OF 2-IODOBUTANE, Eric A. Arsenault, Yoon Jeong Choi, Daniel A. Obenchain, S. A. Cooke, Thomas A. Blake, Stewart E. Novick

WE02   8:47 – 9:02
A STUDY OF THE CONFORMATIONAL ISOMERISM OF 1-IODOBUTANE BY MICROWAVE SPECTROSCOPY, Eric A. Arsenault, Daniel A. Obenchain, S. A. Cooke, Thomas A. Blake, Stewart E. Novick

WE03   9:04 – 9:19
^{14}N QUADRUPOLE COUPLING IN THE MICROWAVE SPECTRA OF N-VINYLFORMAMIDE, Raphaela Kannengießer, Wolfgang Stahl, Ha Vinh Lam Nguyen, William C. Bailey

WE04   9:21 – 9:36
SOLVING THE TAUTOMERIC EQUILIBRIUM OF PURINE THROUGH THE ANALYSIS OF THE COMPLEX HYPERFINE STRUCTURE OF THE FOUR ^{14}N NUCLEI, Emilio J. Cocinero, Iciar Uriarte, Patricia Ecija, Laura B. Favero, Lorenzo Spada, Camilla Calabrese, Walther Caminati

WE05   9:38 – 9:53
CP-FTMW SPECTRUM OF BROMOPERFLUOROACETONE, Frank E Marshall, Nicole Moon, Thomas D. Persinger, David Joseph Gillcrist, G. S. Grubbs II

Intermission

WE06   10:12 – 10:27
ROTATIONAL SPECTROSCOPY OF CF_2ClCCl_3 AND ANALYSIS OF HYPERFINE STRUCTURE FROM FOUR QUADRUPOLAR NUCLEI, Zbigniew Kisiel, Ewa Białkowska-Jaworska, Iciar Uriarte, Francisco J. Basterretxea, Emilio J. Cocinero

WE07   10:29 – 10:44
AB INITIO CALCULATIONS OF SPIN-ORBIT COUPLING FOR HEAVY-METAL CONTAINING RADICALS , Lan Cheng

WE08   10:46 – 11:01
MOLECULAR BEAM OPTICAL ZEEMAN SPECTROSCOPY OF VANADIUM MONOXIDE, VO, Trung Nguyen, Ruohan Zhang, Timothy Steimle

WE09   11:03 – 11:18
MOLECULAR BEAM OPTICAL STUDY OF GOLD SULFIDE AND GOLD OXIDE, Ruohan Zhang, Yuanqin Yu, Timothy Steimle

WE10   11:20 – 11:35
HYPERFINE SPLITTINGS IN THE NEAR-INFRARED SPECTRUM OF $^{14}NH_3$, Sylvestre Twagirayezu, Trevor Sears, Gregory Hall

WE11   11:37 – 11:52
HYPERFINE STRUCTURE IN ROTATIONAL SPECTRA OF DEUTERATED MOLECULES: THE HDS AND ND_3 CASE STUDIES, Gabriele Cazzoli, Cristina Puzzarini

WF. Mini-symposium: Spectroscopy of Large Amplitude Motions
Wednesday, June 22, 2016 – 1:30 PM
Room: 100 Noyes Laboratory

Chair: Jon T. Hougen, National Institute of Standards and Technology, Gaithersburg, MD, USA

WF01   1:30 – 1:45
THE ROLE OF SYMMETRIC-STRETCH VIBRATION IN ASYMMETRIC-STRETCH VIBRATIONAL FREQUENCY SHIFT: THE CASE OF 2CH EXCITATION INFRARED SPECTRA OF ACETYLENE-HYDROGEN VAN DER WAALS COMPLEX, Dan Hou, Yong-Tao Ma, Xiao-Long Zhang, Yu Zhai, Hui Li

WF02   1:47 – 2:02
HALOGEN BONDING VS HYDROGEN BONDING IN CHF_2I COMPLEXES WITH NH_3 AND $N(CH_3)_3$, Chris Medcraft, Yannick Geboes, Anthony Legon, Nick Walker

WF03   2:04 – 2:19
THE CURIOUS CASE OF PYRIDINE - WATER, Becca Mackenzie, Chris Dewberry, CJ Smith, Ryan D. Cornelius, Ken Leopold

WF04   2:21 – 2:36
LABORATORY ROTATIONAL SPECTRUM AND ASTRONOMICAL SEARCH OF S-METHYL THIOFORMATE, Atef Jabri, Isabelle Kleiner, R. A. Motiyenko, L. Margulès, J.-C. Guillemin, E. A. Alekseev, Belén Tercero, Jose Cernicharo

WF05   2:38 – 2:53
LOW BARRIER METHYL ROTATION IN 3-PENTYN-1-OL AS OBSERVED BY MICROWAVE SPECTROSCOPY, Konrad Eibl, Raphaela Kannengießer, Wolfgang Stahl, Ha Vinh Lam Nguyen, Isabelle Kleiner

WF06   2:55 – 3:10
MILLIMETER WAVE SPECTRA OF METHYL CYANATE, METHOXYAMINE AND N-METHYLHYDROXYLAMINE: LABORATORY STUDIES AND ASTRONOMICAL SEARCH IN SPACE, Lucie Kolesniková, José L. Alonso, Celina Bermúdez, E. R. Alonso, Belén Tercero, Jose Cernicharo, J.-C. Guillemin

WF07   3:12 – 3:27
FURTHER ANALYSIS OF THE LABORATORY ROTATIONAL SPECTRUM OF CH_3NCO, Zbigniew Kisiel, Lucie Kolesniková, E. R. Alonso, José L. Alonso, Manfred Winnewisser, Frank C. De Lucia, Ivan Medvedev, Belén Tercero, Jose Cernicharo, J.-C. Guillemin

Intermission

WF08   3:46 – 3:56
MOLECULAR ELECTRONIC ENVIRONMENT FROM METHYL TORSION AND ^{14}N QUADRUPOLE COUPLING, Ranil M. Gurusinghe, Michael Tubergen

WF09   3:58 – 4:13
THz SPECTROSCOPY OF EXCITED TORSIONAL STATES OF MONODEUTERATED METHYL FORMATE ($DCOOCH_3$), Miguel Carvajal, Chuanxi Duan, Shanshan Yu, John Pearson, Brian Drouin, Isabelle Kleiner

WF10   4:15 – 4:30
TWO EQUIVALENT METHYL INTERNAL ROTATIONS IN 2,5-DIMETHYLTHIOPHENE INVESTIGATED BY MICROWAVE SPECTROSCOPY, Vinh Van, Wolfgang Stahl, Ha Vinh Lam Nguyen

WF11   4:32 – 4:47
PROBING THE METHYL TORSIONAL BARRIERS OF THE E AND Z ISOMERS OF BUTADIENYL ACETATE BY MICROWAVE SPECTROSCOPY, Atef Jabri, Ha Vinh Lam Nguyen, Isabelle Kleiner, Vinh Van, Wolfgang Stahl

WF12   4:49 – 5:04
PROGRESS IN THE ROTATIONAL ANALYSIS OF THE GROUND AND LOW-LYING VIBRATIONALLY EXCITED STATES OF MALONALDEHYDE, E. S. Goudreau, Dennis W. Tokaryk, Stephen Cary Ross, Brant E. Billinghurst

WF13   5:06 – 5:21
THE ORIGINS OF INTRA- AND INTER-MOLECULAR VIBRATIONAL COUPLINGS: A CASE STUDY OF H_2O – Ar ON FULL AND REDUCED-DIMENSIONAL POTENTIAL ENERGY SURFACE , Dan Hou, Yong-Tao Ma, Xiao-Long Zhang, Hui Li

WF14   5:23 – 5:38
ROTATIONAL SPECTRA OF T-SHAPED CYANOACETYLENE – CARBON DIOXIDE COMPLEX, $HCCCN — CO_2$, Lu Kang, Ian Dorell, Philip Davis, Onur Oncer, Stephen G. Kukolich, Stewart E. Novick

WG. Clusters/Complexes
Wednesday, June 22, 2016 – 1:30 PM
Room: 116 Roger Adams Lab

Chair: J. Mathias Weber, University of Colorado, Boulder, CO, USA

WG01   1:30 – 1:45
SOLVENT-INDUCED REDUCTIVE ACTIVATION IN GAS PHASE $[Bi(CO_2)_n]^-$ CLUSTERS, Michael C Thompson, Jacob Sondergaard Ramsay, J. Mathias Weber

WG02   1:47 – 2:02
STRUCTURES AND SOLVATION EFFECTS OF $[Fe(CO_2)_n]^-$ CLUSTER ANIONS, Michael C Thompson, J. Mathias Weber

WG03   2:04 – 2:19
NITROGEN MOLECULE-ETHYLENE SULFIDE COMPLEX INVESTIGATED BY FOURIER TRANSFORM MICROWAVE SPECTROSCOPY AND AB INITIO CALCULATION, Sakae Iwano, Yoshiyuki Kawashima, Eizi Hirota

WG04   2:21 – 2:36
NITROGEN MOLECULE-DIMETHYL SULFIDE COMPLEX INVESTIGATED BY FOURIER TRANSFORM MICROWAVE SPECTROSCOPY AND AB INITIO CALCULATION, Yoshiyuki Kawashima, Sakae Iwano, Eizi Hirota

WG05   2:38 – 2:53
PLANAR CoB_{18}^- CLUSTER: A NEW MOTIF FOR HETERO- AND METALLO-BOROPHENES, Teng-Teng Chen, Tian Jian, Gary Lopez, Wan-Lu Li, Xin Chen, Jun Li, Lai-Sheng Wang

WG06   2:55 – 3:10
CO_2 DIMER: FOUR INTERMOLECULAR MODES OBSERVED VIA INFRARED COMBINATION BANDS, Jalal Norooz Oliaee, Mehdi Dehghany, Mojtaba Rezaei, Bob McKellar, Nasser Moazzen-Ahmadi

WG07   3:12 – 3:27
JET-COOLED HIGH RESOLUTION INFRARED SPECTROSCOPY OF SMALL VAN DER WAALS SF_6 CLUSTERS, Pierre Asselin, Vincent Boudon, Alexey Potapov, Laurent Bruel, Marc-André Gaveau, Michel Mons

Intermission

WG08   3:46 – 4:01
DOES A SECOND HALOGEN ATOM AFFECT THE NATURE OF INTERMOLECULAR INTERACTIONS IN PROTIC ACID-HALOETHYLENE COMPLEXES? IN (E)-1-CHLORO-2-FLUOROETHYLENE-HYDROGEN CHLORIDE IT DEPENDS ON HOW YOU LOOK AT IT, Helen O. Leung, Mark D. Marshall

WG09   4:03 – 4:18
DOES A SECOND HALOGEN ATOM AFFECT THE NATURE OF INTERMOLECULAR INTERACTIONS IN PROTIC ACID-HALOETHYLENE COMPLEXES? IN (Z)-1-CHLORO-2-FLUOROETHYLENE-HYDROGEN CHLORIDE IT MOST CERTAINLY DOES!, Hannah K. Tandon, Helen O. Leung, Mark D. Marshall

WG10   4:20 – 4:35
FIRST OBSERVATION OF THE N_2O-OC VAN DER WAALS COMPLEX AND NEW SET OF EXPERIMENTAL MEASUREMENTS ON THE N_2O-CO COMPLEX., Clément Lauzin, A. J. Barclay, S. Sheybani-Deloui, Nasser Moazzen-Ahmadi

WG11   4:37 – 4:52
LASER-INDUCED FLUORESCENCE SPECTRA OF C_3Ar NEAR 25400-25600 cm^{-1}, Yi-Jen Wang, Yen-Chu Hsu

WG12   4:54 – 5:09
A MICROWAVE STUDY OF 3,5 DIFLUOROPYRIDINE$\cdots CO_2$: THE EFFECT OF META-FLUORINATION ON INTERMOLECULAR INTERACTIONS OF PYRIDINE, Chris Dewberry, Ryan D. Cornelius, Becca Mackenzie, CJ Smith, Michael A. Dvorak, Ken Leopold

WG13   5:11 – 5:26
HIGH RESOLUTION INFRARED SPECTROSCOPY OF THE CO_2-CO DIMERS AND $(CO_2)_2$-CO TRIMER, A. J. Barclay, S. Sheybani-Deloui, K. H. Michaelian, Bob McKellar, Nasser Moazzen-Ahmadi

WG14   5:28 – 5:43
MICROWAVE OBSERVATION OF THE VAN DER WAALS COMPLEX O_2-CO , Frank E Marshall, Thomas D. Persinger, David Joseph Gillcrist, Nicole Moon, Steve Alexandre Ndengue, Richard Dawes, G. S. Grubbs II

WH. Astronomy
Wednesday, June 22, 2016 – 1:30 PM
Room: 274 Medical Sciences Building

Chair: Brett A. McGuire, National Radio Astronomy Observatory, Charlottesville, VA, USA

WH01   1:30 – 1:45
SPECTROSCOPIC STUDY AND ASTRONOMICAL DETECTION OF VIBRATIONALLY EXCITED n-PROPYL CYANIDE, Holger S. P. Müller, Nadine Wehres, Olivia H. Wilkins, Frank Lewen, Stephan Schlemmer, Adam Walters, Rémi Vicente, Delong Liu, Robin T. Garrod, Arnaud Belloche, Karl M. Menten

WH02   1:47 – 2:02
THE INTERSTELLAR DETECTION OF CH_3NCO IN Sgr B2(N), DeWayne T Halfen, V. Ilyushin, Lucy M. Ziurys

WH03   2:04 – 2:19
A SURVEY OF HNCO AND CH_3NCO IN MOLECULAR CLOUDS, DeWayne T Halfen, Lucy M. Ziurys

WH04   2:21 – 2:36
MODELLING STUDY OF INTERSTELLAR ETHANIMINE ISOMERS, Donghui Quan, Eric Herbst, Joanna F. Corby, Allison Durr, George Hassel

WH05   2:38 – 2:53
COMPLETE RESULTS FROM A SPECTRAL-LINE SURVEY OF Sgr B2(N), DeWayne T Halfen, Lucy M. Ziurys

WH06   2:55 – 3:10
DISCOVERY OF THE FIRST INTERSTELLAR CHIRAL MOLECULE: PROPYLENE OXIDE, Brandon Carroll, Brett A. McGuire, Ryan Loomis, Ian A Finneran, Philip Jewell, Anthony Remijan, Geoffrey Blake

WH07   3:12 – 3:27
THE CO TRANSITION FROM DIFFUSE MOLECULAR GAS TO DENSE CLOUDS: PRELIMINARY RESULTS, Johnathan S Rice, Steven Federman

WH08   3:29 – 3:44
CENTRAL 300 PC OF THE GALAXY PROBED BY THE INFRARED SPECTRA OF H_3^+ AND CO: I. PREDOMINANCE OF WARM AND DIFFUSE GAS AND HIGH H_2 IONIZATION RATE , Takeshi Oka, Thomas R. Geballe, Miwa Goto, Tomonori Usuda, Nick Indriolo

Intermission

WH09   4:03 – 4:18
THE PRECISE RADIO OBSERVATION OF THE ^{13}C ISOTOPIC FRACTIONATION FOR CARBON CHAIN MOLECULE HC_3N IN THE LOW-MASS STAR FORMING REGION L1527, Mitsunori Araki, Shuro Takano, Nami Sakai, Satoshi Yamamoto, Takahiro Oyama, Nobuhiko Kuze, Koichi Tsukiyama

WH10   4:20 – 4:35
CARMA OBSERVATIONS OF L1157: CHEMICAL COMPLEXITY IN THE SHOCKED OUTFLOW, Andrew M Burkhardt, Niklaus M Dollhopf, Joanna F. Corby, Brandon Carroll, Christopher N Shingledecker, Ryan Loomis, S. Tom Booth, Geoffrey Blake, Anthony Remijan, Brett A. McGuire

WH11   4:37 – 4:52
MODELING THE AFTER-EFFECTS OF SHOCKS TOWARD L1157, Andrew M Burkhardt, Brett A. McGuire, Niklaus M Dollhopf, Eric Herbst

WH12   4:54 – 5:09
FILAMENTARY STRUCTURE OF SERPENS MAIN AND SERPENS SOUTH SEEN IN N_2H^+, HCO^+, AND HCN, Erin Guilfoil Cox, Manuel Fernandez-Lopez, Leslie Looney, Héctor Arce, Lee Mundy, Shaye Storm, Robert J Harris, Peter J. Teuben

WH13   5:11 – 5:26
SULFUR CHEMISTRY IN VY CANIS MAJORIS REVEALED BY ALMA, Andrew M Burkhardt, Brett A. McGuire, Gilles Adande, Lucy M. Ziurys, Anthony Remijan

WH14   5:28 – 5:43
TRACING THE ORIGINS OF NITROGEN BEARING ORGANICS TOWARD ORION KL WITH ALMA, Brandon Carroll, Nathan Crockett, Edwin Bergin, Geoffrey Blake

WH15   5:45 – 6:00
SPECTROSCOPIC FITS TO THE ALMA SCIENCE VERIFICATION BAND 6 SURVEY OF THE ORION HOT CORE AND COMPACT RIDGE, Satyakumar Nagarajan, James P. McMillan, Andrew M Burkhardt, Christopher F. Neese, Frank C. De Lucia, Anthony Remijan

WI. Mini-symposium: Spectroscopy in Traps
Wednesday, June 22, 2016 – 1:30 PM
Room: B102 Chemical and Life Sciences

Chair: Jos Oomens, Radboud University, Nijmegen, The Netherlands

WI01   *INVITED TALK*   1:30 – 2:00
ISOLATING SITE-SPECIFIC SPECTRAL SIGNATURES OF INDIVIDUAL WATER MOLECULES IN H-BONDED NETWORKS WITH ISOTOPOMER-SELECTIVE, IR-IR DOUBLE RESONANCE VIBRATIONAL PREDISSOCIATION SPECTROSCOPY, Conrad T. Wolke, Mark Johnson

WI02   2:05 – 2:20
IR-UV DOUBLE RESONANCE SPECTROSCOPY OF A COLD PROTONATED FIBRIL-FORMING PEPTIDE: NNQQNY·H$^+$, Andrew F DeBlase, Christopher P Harrilal, Patrick S. Walsh, Scott A McLuckey, Timothy S. Zwier

WI03   2:22 – 2:37
VIBRATIONAL AND ROTATIONAL SPECTROSCOPY OF CD$_2$H$^+$, Oskar Asvany, Pavol Jusko, Sandra Brünken, Stephan Schlemmer

WI04   2:39 – 2:54
INFRARED PREDISSOCIATION SPECTROSCOPY OF THE HYDROCARBON CATIONS C$_3$H$^+$, C$_2$H$^+$, and C$_3$H$_2^+$, Sandra Brünken, Filippo Lipparini, Jürgen Gauss, Alexander Stoffels, Britta Redlich, Lex van der Meer, Giel Berden, Jos Oomens, Stephan Schlemmer

WI05   2:56 – 3:11
FREQUENCY COMB ASSISTED IR MEASUREMENTS OF H$_3^+$, H$_2$D$^+$ AND D$_2$H$^+$ TRANSITIONS , Pavol Jusko, Oskar Asvany, Stephan Schlemmer

Intermission

WI06   3:30 – 3:45
INFRARED SPECTROSCOPY OF IONS IN SELECTED ROTATIONAL AND SPIN-ORBIT STATES, Ugo Jacovella, Josef A. Agner, Hansjürg Schmutz, Frederic Merkt

WI07   3:47 – 4:02
PRECISION SPECTROSCOPY ON SINGLE COLD TRAPPED MOLECULAR NITROGEN IONS, Gregor Hegi, Kaveh Najafian, Matthias Germann, Ilia Sergachev, Stefan Willitsch

WI08   4:04 – 4:19
UV PHOTODISSOCIATION SPECTROSCOPY OF TEMPERATURE CONTROLLED HYDRATED PHENOL CLUSTER CATION, Itaru Kurusu, Reona Yagi, Yasutoshi Kasahara, Haruki Ishikawa

WI09   4:21 – 4:36
STRUCTURAL INFORMATION INFERENCE FROM LANTHANOID COMPLEXING SYSTEMS: PHOTOLUMINES-CENCE STUDIES ON ISOLATED IONS, Jean Francois Greisch, Michael E. Harding, Jiri Chmela, Willem M. Klopper, Detlef Schooss, Manfred M Kappes

WI10   4:38 – 4:53
PHOTO-IONIZATION AND PHOTO-DISSOCIATION OF TRAPPED PAH CATIONS, Christine Joblin, Junfeng Zhen, Sarah Rodriguez Castillo, Giacomo Mulas, Hassan Sabbah, Aude Simon, Alexandre Giuliani, Laurent Nahon, Serge Martin, Jean-Philippe Champeaux, Paul M. Mayer

WI11   4:55 – 5:10
PROGRESS OF THE JILA ELECTRON EDM EXPERIMENT, Yan Zhou, William Cairncross, Matt Grau, Dan Gresh, Kia Boon Ng, Yiqi Ni, Jun Ye, Eric Cornell

WJ. Small molecules
Wednesday, June 22, 2016 – 1:30 PM
Room: 217 Noyes Laboratory

Chair: Colan Linton, University of New Brunswick, Fredericton, NB, Canada

WJ01   1:30 – 1:45
SPECTRAL LINE SHAPE PARAMETERS FOR THE ν_1, ν_2, and ν_3 BANDS OF HDO: SELF AND CO_2 BROADENED, V. Malathy Devi, D. Chris Benner, Keeyoon Sung, Timothy J. Crawford, Robert R. Gamache, Candice L. Renaud, Arlan Mantz, Mary Ann H. Smith, Geronimo L. Villanueva

WJ02   1:47 – 2:02
SPECTRAL LINE SHAPES IN THE $2\nu_3$ Q BRANCH OF $^{12}CH_4$, V. Malathy Devi, D. Chris Benner, Keeyoon Sung, Linda R. Brown, Timothy J. Crawford, Shanshan Yu, Mary Ann H. Smith, Syed Ismail, Arlan Mantz, Vincent Boudon

WJ03   2:04 – 2:19
ABSORPTION CROSS SECTIONS OF HOT HYDROCARBONS IN THE 3 μm REGION, Christopher A. Beale, Robert J. Hargreaves, Eric M. Buzan, Peter F. Bernath

WJ04   2:21 – 2:36
LINE LISTS AND ASSIGNMENTS OF HOT AMMONIA IN THE INFRARED, Christopher A. Beale, Robert J. Hargreaves, Andy Wong, Peter F. Bernath

WJ05   2:38 – 2:53
THE EFFECT OF TERMINAL SUBSTITUTION ON THE HELICAL CARBON STRUCTURE OF FLUORO-ALKANE CHAINS: A PURE ROTATIONAL STUDY OF $CH_2OH-C_{n-1}F_{2n-1}$ (n = 4, 5,& 6), Aaron Z. A. Schwartz, Mark P. Maturo, Daniel A. Obenchain, S. A. Cooke

WJ06   2:55 – 3:05
DEVELOPMENT OF A NEW DETECTION SCHEME TO PROBE PREDISSOCIATED LEVELS OF THE S_1 STATE OF ACETYLENE, Jun Jiang, Trevor J. Erickson, Anthony Merer, Robert W Field

WJ07   3:07 – 3:22
FOURIER TRANSFORM MICROWAVE SPECTROSCOPY OF $Sc^{13}C_2$ AND $Sc^{12}C^{13}C$: ESTABLISHING AN ACCURATE STRUCTURE OF ScC_2 ($\tilde{X}^2A_1$), Mark Burton, DeWayne T Halfen, Jie Min, Lucy M. Ziurys

WJ08   3:24 – 3:39
ANALYSIS OF QUARTET AND DOUBLET STATES OF NO MOLECULE EXCITED BY GLOW DISCHARGE, Mohammed A Gondal

Intermission

WJ09   3:58 – 4:13
SPECTROSCOPY OF THE $X^1\Sigma^+$, $A^1\Pi$ and $B^1\Sigma^+$ ELECTRONIC STATES OF MgS, Nicholas Caron, Dennis W. Tokaryk, Allan G. Adam, Colan Linton

WJ10   4:15 – 4:30
IDENTIFICATION OF TWO NEW ELECTRONIC STATES OF NiCl USING INTRACAVITY LASER SPECTROSCOPY AND THE CORRELATION BETWEEN THEORETICAL PREDICTIONS AND EXPERIMENTAL OBSERVATIONS, Jack C Harms, Ethan M Grames, Shu Han, Leah C O'Brien, James J O'Brien

WJ11   4:32 – 4:47
NEW EMPIRICAL POTENTIAL ENERGY FUNCTIONS FOR THE HEAVIER HOMONUCLEAR RARE GAS PAIRS: Ne_2, Ar_2, Kr_2, and Xe_2, Philip Thomas Myatt, Matthew T. Baker, Ju-Hee Kang, Andres Escobar Moya, Frederick R. W. McCourt, Robert J. Le Roy

WJ12   4:49 – 5:04
SPECTROSCOPIC LINE PARAMETERS OF HELIUM- AND HYDROGEN-BROADENED $^{12}C^{16}O$ TRANSITIONS IN THE 3–0 BAND FROM 6270 cm^{-1} TO 6402 cm^{-1}., Zachary Reed, Joseph Hodges

WJ13   5:06 – 5:21
AN *AB INITIO* STUDY OF SbH_2 AND BiH_2: THE RENNER EFFECT, SPIN-ORBIT COUPLING, LOCAL MODE VIBRATIONS AND ROVIBRONIC ENERGY LEVEL CLUSTERING IN SbH_2, Bojana Ostojic, Peter Schwerdtfeger, Phil Bunker, Per Jensen

WJ14   5:23 – 5:38
THE BICHROMATIC FORCE ON SMALL MOLECULES, Leland M. Aldridge, Scott E. Galica, Donal Sheets, Edward E. Eyler

WK. Spectroscopy as an analytical tool
Wednesday, June 22, 2016 – 1:30 PM
Room: 140 Burrill Hall

Chair: Kyle N. Crabtree, University of California, Davis, CA, USA

WK01   1:30 – 1:45
IDENTIFICATION AND CHARACTERIZATION OF 1,2-BN CYCLOHEXENE USING MICROWAVE SPECTROSCOPY, Stephen G. Kukolich, Ming Sun, Adam M. Daly, Jacob S. A. Ishibashi, Shih-Yuan Liu

WK02   1:47 – 2:02
AUTOMATED MICROWAVE DOUBLE RESONANCE SPECTROSCOPY: A TOOL TO IDENTIFY AND CHARACTERIZE CHEMICAL COMPOUNDS, Marie-Aline Martin-Drumel, Michael C McCarthy, David Patterson, Brett A. McGuire, Kyle N. Crabtree

WK03   2:04 – 2:19
UTILIZATION OF MICROWAVE SPECTROSCOPY TO IDENTIFY AND PROBE REACTION DYNAMICS OF HSNO, A CRUCIAL BIOLOGICAL SIGNALING MOLECULE, Matthew Nava, Marie-Aline Martin-Drumel, John F. Stanton, Christopher Cummins, Michael C McCarthy

WK04   2:21 – 2:36
ROTATIONAL SPECTRUM AND CARBON ATOM STRUCTURE OF DIHYDROARTEMISINIC ACID, Luca Evangelisti, Nathan A Seifert, Lorenzo Spada, Brooks Pate

WK05   2:38 – 2:53
PROBING THE $CH_3SH + N_2O_3$ REACTION BY AUTOMATED MICROWAVE DOUBLE RESONANCE SPECTROSCOPY, Michael C McCarthy, Marie-Aline Martin-Drumel, Matthew Nava, Sven Thorwirth

WK06   2:55 – 3:10
MILLIMETER-WAVE SPECTROSCOPY FOR ANALYTICAL CHEMISTRY: THERMAL EVOLUTION OF LOW VOLATILITY IMPURITIES AND DETECTION WITH A FOURIER TRANSFORM MOLECULAR ROTATIONAL RESONANCE SPECTROMETER (TEV FT-MRR), Brent Harris, Shelby S. Fields, Justin L. Neill, Robin Pulliam, Matt Muckle, Brooks Pate

Intermission

WK07   3:29 – 3:44
DETECTION OF *in vitro* S-NITROSYLATED COMPOUNDS WITH CAVITY RING-DOWN SPECTROSCOPY , Mary Lynn Rad, Monique Michele Mezher, Benjamin M Gaston, Kevin Lehmann

WK08   3:46 – 4:01
CAVITY RING DOWN ABSORPTION OF OXYGEN IN AIR AS A TEMPERATURE SENSOR, Carlos Manzanares, Parashu R Nyaupane

WK09   4:03 – 4:18
ATMOSPHERIC REMOTE SENSING VIA INFRARED-SUBMILLIMETER DOUBLE RESONANCE , Sree Srikantaiah, Jennifer Holt, Christopher F. Neese, Dane Phillips, Henry O. Everitt, Frank C. De Lucia

WK10   4:20 – 4:35
MOLECULAR STRUCTURE AND REACTIVITY IN THE PYROLYSIS OF ALDEHYDES, Eric Sias, Sarah Cole, John Sowards, Brian Warner, Emily Wright, Laura R. McCunn

WK11   4:37 – 4:52
VUV FLUORESCENCE OF JET-COOLED WATER AS A VEHICLE FOR SATELLITE THRUSTER PLUME CHARACTERIZATION., Justin W. Young, Jaime A. Stearns

WK12   4:54 – 5:04
INFLUENCE OF BIODEGRADATION ON THE ORGANIC COMPOUNDS COMPOSITION OF PEAT., Olga Serebrennikova, Lidiya Svarovskaya, Maria Duchko, Evgeniya Strelnikova, Irina Russkikh

WK13   5:06 – 5:21
NON-LINEAR THERMAL LENS SIGNAL OF THE ($\Delta v = 6$) C-H VIBRATIONAL OVERTONE OF BENZENE IN LIQUID SOLUTIONS OF HEXANE, Parashu R Nyaupane, Carlos Manzanares

WK14   5:23 – 5:38
SCREENING OF POLY CYSTIC OVARIAN SYNDROME BY MID INFRARED SPECTROMETRY, Mohammadreza Khanmohammadi, Fatemeh Golpour, Amir Bagheri Garmarudi, Fahimeh Ramezani Tehrani

WK15   5:40 – 5:55
OBSERVATION OF ORTHO-PARA DEPENDENCE OF PRESSURE BROADENING COEFFICIENT IN ACETYLENE $\nu_1+\nu_3$ VIBRATION BAND USING DUAL-COMB SPECTROSCOPY , Kana Iwakuni, Sho Okubo, Hajime Inaba, Atsushi Onae, Feng-Lei Hong, Hiroyuki Sasada, Koichi MT Yamada

RA. Plenary
Thursday, June 23, 2016 – 8:30 AM
Room: Foellinger Auditorium

Chair: Brian D. Fields, The University of Illinois, Urbana, IL, USA

RAO AWARDS 8:30
Presentation of Awards by Gary Douberly, University of Georgia

2015 Rao Award Winners
Daniel Bakker, Radboud University
Scott Dubowsky, University of Illinois at Urbana-Champaign
James McMillan, The Ohio State University

MILLER PRIZE 8:40
Introduction by Jinjun Liu, University of Louisville

RA01 *Miller Prize Lecture* 8:55 – 9:10
HIGH-RESOLUTION SPECTROSCOPY OF He_2^+ USING RYDBERG-SERIES EXTRAPOLATION AND ZEEMAN-DECELERATED SUPERSONIC BEAMS OF METASTABLE He_2, Paul Jansen, Luca Semeria, Frederic Merkt

COBLENTZ AWARD 9:15
Presentation of Award by Linda Kidder, Coblentz Society

RA02 *Coblentz Society Award Lecture* 9:20 – 10:00
EXAMINING THE NANOWORLD USING A MOLECULAR SPECTROSCOPIST'S TOOLBOX, Kenneth L. Knappenberger, Jr.

Intermission

RA03 10:30 – 11:10
TWO DECADES OF ADVANCES IN HIGH-RESOLUTION SPECTROSCOPY OF LARGE-AMPLITUDE MOTIONS IN N-FOLD POTENTIAL WELLS, AS ILLUSTRATED BY METHANOL, Li-Hong Xu

RA04 11:15 – 11:55
MOLECULAR SPECTROSCOPY IN SPACE: DISCOVERING NEW MOLECULES FROM LINE SURVEYS AND LABORATORY SPECTROSCOPY, Jose Cernicharo

RF. Radicals
Thursday, June 23, 2016 – 1:30 PM
Room: 100 Noyes Laboratory

Chair: Terry A. Miller, The Ohio State University, Columbus, OH, USA

RF01   *Post-Deadline Abstract*   1:30 – 1:45
2C-R4WM SPECTROSCOPY OF JET COOLED NO_3, Masaru Fukushima, Takashi Ishiwata, Eizi Hirota

RF02   1:47 – 2:02
HIGH-RESOLUTION LASER SPECTROSCOPY OF THE $\tilde{B} \leftarrow \tilde{X}$ TRANSITION OF $^{14}NO_3$ RADICAL: VIBRATIONALLY EXCITED STATES OF THE $\tilde{B}$ STATE, Shunji Kasahara, Kohei Tada, Michihiro Hirata, Takashi Ishiwata, Eizi Hirota

RF03   2:04 – 2:19
QUANTIFYING THE EFFECTS OF HIGHER ORDER JAHN-TELLER COUPLING TERMS ON A QUADRATIC JAHN-TELLER HAMILTONIAN IN THE CASE OF NO_3 AND Li_3., Henry Tran, John F. Stanton, Terry A. Miller

RF04   2:21 – 2:36
ANALYSIS OF THE ROTATIONALLY RESOLVED, NON-DEGENERATE (a_1'') AND DEGENERATE (e') VIBRONIC BANDS IN THE $\tilde{A}^2 E'' \leftarrow \tilde{X}^2 A_2'$ TRANSITION OF NO_3., Henry Tran, Terry A. Miller

RF05   2:38 – 2:53
NEAR-INFRARED SPECTROSCOPY OF ETHYNYL RADICAL, C_2H, Anh T. Le, Gregory Hall, Trevor Sears

RF06   2:55 – 3:10
STUDY OF INFRARED EMISSION SPECTROSCOPY FOR THE $B^1\Delta_g$-$A^1\Pi_u$ AND $B'^1\Sigma_g^+$-$A^1\Pi_u$ SYSTEMS OF C_2, Jian Tang, Wang Chen, Kentarou Kawaguchi, Peter F. Bernath

RF07   3:12 – 3:27
A ZERO-ORDER PICTURE OF THE INFRARED SPECTRUM FOR THE METHOXY RADICAL: ASSIGNMENT OF STATES, Britta Johnson, Edwin Sibert

Intermission

RF08   3:46 – 4:01
INFRARED IDENTIFICATION OF THE CRIEGEE INTERMEDIATE $(CH_3)_2COO$, Yi-Ying Wang, Yuan-Pern Lee

RF09   4:03 – 4:18
ANALYSES OF THE $\tilde{A}$-$\tilde{X}$ ELECTRONIC TRANSITIONS OF THE $CH_2XOO\cdot$(X = I, Br, Cl) RADICALS, Neal Kline, Meng Huang, Terry A. Miller

RF10   4:20 – 4:35
NOO PEROXY ISOMER EXPOSED WITH VELOCITY-MAP IMAGING, Benjamin A Laws, Steven J Cavanagh, Brenton R Lewis, Stephen T Gibson

RF11   4:37 – 4:52
INFRARED LASER SPECTROSCOPY OF THE n-PROPYL AND i-PROPYL RADICALS IN HELIUM DROPLETS: SIGNIFICANT BEND-STRETCH COUPLING REVEALED IN THE CH STRETCH REGION , Christopher P. Moradi, Gary E. Douberly, Daniel P. Tabor, Edwin Sibert

RF12   4:54 – 5:09
INFRARED SPECTRUM OF FULVENALLENE AND FULVENALLENYL, Alaina R. Brown, Joseph T. Brice, Gary E. Douberly

RF13   5:11 – 5:26
TWO-CENTER THREE-ELECTRON BONDING IN $ClNH_3$ REVEALED VIA HELIUM DROPLET INFRARED SPECTROSCOPY: ENTRANCE CHANNEL COMPLEX ALONG THE $Cl + NH_3 \rightarrow ClNH_2 + H$ REACTION, Peter R. Franke, Christopher P. Moradi, Matin Kaufmann, Changjian Xie, Hua Guo, Gary E. Douberly

RF14   5:28 – 5:43
JET-COOLED CHLOROFLUOROBENZYL RADICALS: SPECTROSCOPY AND MECHANISM, Young Yoon, Sang Lee

RF15   5:45 – 6:00
MECHANISM OF THE THERMAL DECOMPOSITION OF ETHANETHIOL AND DIMETHYLSULFIDE, William Francis Melhado, Jared Connor Whitman , Jessica Kong, Daniel Easton Anderson, AnGayle (AJ) Vasiliou

RG. Clusters/Complexes
Thursday, June 23, 2016 – 1:30 PM
Room: 116 Roger Adams Lab

Chair: Brooks Pate, The University of Virginia, Charlottesville, VA, USA

RG01   1:30 – 1:45
CHARACTERIZATION OF AMMONIA-WATER CLUSTERS BY BROADBAND ROTATIONAL SPECTROSCOPY, Luca Evangelisti, Cristobal Perez, Berhane Temelso, George C. Shields, Brooks Pate

RG02   1:47 – 2:02
2OH OVERTONE SPECTROSCOPY OF WATER-CONTAINING VAN DER WAALS SPECIES, Thomas Vanfleteren, Tomas Földes, Michel Herman, Jacques Liévin, Jérome Loreau, L. H. Coudert

RG03   2:04 – 2:19
AN INFRARED SPECTROSCOPIC STUDY ON THE FORMATION OF THE HYDROGEN BONDED INCLUSION-STRUCTURES IN THE PROTONATED METNANOL WATER CLUSTERS, Marusu Katada, Po-Jen Hsu, Asuka Fujii, Jer-Lai Kuo

RG04   2:21 – 2:36
STEPWISE INTERNAL ENERGY CONTROL FOR PROTONATED METHANOL CLUSTERS BY USING THE INERT GAS TAGGING, Takuto Shimamori, Jer-Lai Kuo, Asuka Fujii

RG05   2:38 – 2:53
INFRARED ABSORPTION OF METHANOL-WATER CLUSTERS $M_n(H_2O)$, n = 1-4, RECORDED WITH THE VUV-IONIZATION/IR-DEPLETION TECHNIQUES, Yu-Fang Lee, Yuan-Pern Lee

RG06   2:55 – 3:10
PREDICTING CERIUM + H_2O CLUSTER FORMATION WITH SIMULATED AND EXPERIMENTAL SPEC-TROSCOPY, Josey E Topolski, Jared O. Kafader, Manisha Ray, Caroline Chick Jarrold

Intermission

RG07   3:29 – 3:44
STRONG QUANTUM COUPLING BETWEEN O-H$^+$-O STRETCH AND FLANKING GROUP MOTIONS IN $(CH_3OH)_2H^+$ PART I: UNMASKING THE 800-1200 cm^{-1}PEAKS, Jake Acedera Tan, Jer-Lai Kuo

RG08   3:46 – 4:01
STRONG QUANTUM COUPLING BETWEEN O-H$^+$-O STRETCH AND FLANKING GROUP MOTIONS IN $(CH_3OH)_2H^+$ Part II: TUNING THE COUPLING VIA ISOTOPOLOGUES , Jake Acedera Tan, Jer-Lai Kuo

RG09   4:03 – 4:18
INFRARED SPECTROSCOPY AND ANHARMONIC VIBRATIONAL ANALYSIS OF THE AR·H$^+$·AR PROTON-BOUND DIMER CATION, David C McDonald II, Jake Acedera Tan, Joshua H Marks, Jer-Lai Kuo, Michael A Duncan

RG10   4:20 – 4:35
THE FORMAMIDE$_2$-H$_2$O COMPLEX: STRUCTURE AND HYDROGEN BOND COOPERATIVE EFFECTS, Susana Blanco, Pablo Pinacho, Juan Carlos Lopez

RG11   4:37 – 4:52
A STUDY OF THE FORMAMIDE-(H$_2$O)$_3$ COMPLEX BY MICROWAVE SPECTROSCOPY, Pablo Pinacho, Juan Carlos Lopez, Susana Blanco

RG12   *Post-Deadline Abstract*   4:54 – 5:09
PURE ROTATIONAL SPECTRUM AND MOLECULAR GEOMETRY OF AN ISOLATED COMPLEX OF IMIDAZOLE AND UREA, Susana Blanco, John C Mullaney, Chris Medcraft, Nick Walker, Anthony Legon

RG13   5:11 – 5:26
AN AB INITIO APPROACH TO ANALYZE FERMI RESONANCE IN AMMONIA CLUSTERS, Kun-Lin Ho, Marusu Katada, Jer-Lai Kuo, Asuka Fujii

RG14   5:28 – 5:43
MILLIMETER/SUBMILLIMETER SPECTRA OF WEAKLY-BOUND CLUSTERS, Luyao Zou, Susanna L. Widicus Weaver

RH. Astronomy
Thursday, June 23, 2016 – 1:30 PM
Room: 274 Medical Sciences Building

Chair: Susanna L. Widicus Weaver, Emory University, Atlanta, GA, USA

RH01   1:30–1:40
DUO: A GENERAL MULTI-STATE PROGRAM FOR SOLVING THE NUCLEAR MOTION SCHRÖDINGER EQUATION FOR OPEN SHELL DIATOMIC MOLECULES, Sergei N. Yurchenko, Lorenzo Lodi, Jonathan Tennyson, Andrey Stolyarov

RH02   1:42–1:57
ExoMol: MOLECULAR LINE LIST FOR EXOPLANETS AND OTHER ATMOSPHERES, Jonathan Tennyson, Sergei N. Yurchenko

RH03   1:59–2:14
INFRARED SPECTROSCOPY OF HOT METHANE: EMPIRICAL LINE LISTS WITHIN THE 1 - 2 μm REGION, Andy Wong, Robert J. Hargreaves, Peter F. Bernath

RH04   2:16–2:31
QUANTUM CHEMISTRY MEETS ROTATIONAL SPECTROSCOPY FOR ASTROCHEMISTRY: INCREASING MOLECULAR COMPLEXITY , Cristina Puzzarini

RH05   2:33–2:48
MOLECULES IN LABORATORY AND IN INTERSTELLAR SPACE? , Venkatesan S. Thimmakondu

RH06   2:50–3:05
THE CENTER FOR ASTROCHEMICAL STUDIES AT THE MAX PLANCK INSTITUTE FOR EXTRATERRESTRIAL PHYSICS., Valerio Lattanzi, Luca Bizzocchi, Jacob Laas, Barbara Michela Giuliano, Silvia Spezzano, Christian Endres, Paola Caselli

RH07   3:07–3:22
MILLIMETER/SUBMILLIMETER SPECTROSCOPY TO MEASURE THE BRANCHING RATIOS FOR METHANOL PHOTOLYSIS, Morgan N McCabe, Carson Reed Powers, Samuel Zinga, Susanna L. Widicus Weaver

Intermission

RH08   3:41–3:56
HIGH-RESOLUTION SPECTROSCOPY OF THE $A^1\Pi(v'=0\text{-}10)-X^1\Sigma^+(v''=0)$ BANDS IN $^{13}C^{18}O$, Jean Louis Lemaire, Michele Eidelsberg, Alan Heays, Lisseth Gavilan, Steven Federman, Glenn Stark, James R Lyons, Nelson de Oliveira, Denis Joyeux

RH09   3:58–4:13
HIGH-RESOLUTION INFRARED SPECTROSCOPY OF CARBON-SULFUR CHAINS: I. C_3S AND SC_7S, John B Dudek, Thomas Salomon, Sven Thorwirth

RH10   4:15–4:30
HIGH-RESOLUTION INFRARED SPECTROSCOPY OF CARBON-SULFUR CHAINS: II. C_5S AND SC_5S, Sven Thorwirth, Thomas Salomon, John B Dudek

RH11   4:32–4:47
ASTROCHEMISTRY LECTURE AND LABORATORY COURSES AT THE UNIVERSITY OF ILLINOIS: APPLIED SPECTROSCOPY, David E. Woon, Benjamin J. McCall

RH12   4:49–5:04
ALMA DATA MINING TOOLKIT, Douglas Friedel, Leslie Looney, Peter J. Teuben, Marc W. Pound, Kevin P. Rauch, Lee Mundy, Robert J Harris, Lisa Xu

RH13   5:06–5:21
THE INFRARED DETECTION OF DEUTERATED PAHS IN HII REGIONS, Kirstin D Doney, Alessandra Candian, Tamami Mori, Takashi Onaka, Xander Tielens

RH14   5:23–5:38
LIGHT ON THE 3 μm EMISSION BAND FROM SPACE WITH MOLECULAR BEAM SPECTROSCOPY, Elena Maltseva, Cameron J. Mackie, Alessandra Candian, Annemieke Petrignani, Xander Tielens, Jos Oomens, Xinchuan Huang, Timothy Lee, Wybren Jan Buma

RH15   5:40–5:55
TOP DOWN CHEMISTRY VERSUS BOTTOM UP CHEMISTRY, Takeshi Oka, Adolf N. Witt

RI. Metal containing
Thursday, June 23, 2016 – 1:30 PM
Room: B102 Chemical and Life Sciences

Chair: Anthony Merer, University of British Columbia, Vancouver, BC, Canada

RI01   1:30 – 1:45
HIGHLY UNSATURATED PLATINUM AND PALLADIUM CARBENES PtC_3 AND PdC_3 ISOLATED AND CHARAC-TERIZED IN THE GAS PHASE, <u>Dror M. Bittner</u>, Daniel P. Zaleski, David Peter Tew, Nick Walker, Anthony Legon

RI02   1:47 – 2:02
THE PURE ROTATIONAL SPECTRA OF FCPtF AND FPtI, <u>Dror M. Bittner</u>, Nick Walker, Anthony Legon

RI03   2:04 – 2:19
SPECTROSCOPIC STUDY OF LOCAL INTERACTIONS OF PLATINUM IN SMALL $[Ce_xO_y]Pt_{x'}{}^-$ CLUSTERS , <u>Manisha Ray</u>, Jared O. Kafader, Caroline Chick Jarrold

RI04   2:21 – 2:36
ROTATIONAL SPECTROSCOPY OF $ClZnCH_3$ ($\tilde{X}^1A_1$): CHARACTERIZATION OF A MONOMERIC GRIGNARD-TYPE REAGENT, <u>K. M. Kilchenstein</u>, Jie Min, Matthew Bucchino, Lucy M. Ziurys

RI05   2:38 – 2:53
THRESHOLD IONIZATION OF $La(C_5H_8)$ FORMED BY La-MEDIATED DEHYDROGENATION OF 1-PENTENE, <u>Wenjin Cao</u>, Yuchen Zhang, Dong-Sheng Yang

RI06   2:55 – 3:10
YTTRIUM-ASSISTED C-H AND C-C BOND ACTIVATION OF ETHYLENE PROBED BY MASS-ANALYZED THRESHOLD IONIZATION SPECTROSCOPY, <u>Jong Hyun Kim</u>, Dong-Sheng Yang

RI07   3:12 – 3:27
THE EFFECT OF ARGON TAGGING ON THE $Ti^+(H_2O)$ SYSTEM OBSERVED THROUGH INFRARED SPEC-TROSCOPY, <u>Timothy B Ward</u>, Prosser Carnegie, Michael A Duncan

Intermission

RI08   3:46 – 4:01
THE MYSTERY OF THE ELECTRONIC SPECTRUM OF RUTHENIUM MONOPHOSPHIDE, <u>Allan G. Adam</u>, Ryan M Christensen, Jacob M Dore, Ricarda M. Konder, Dennis W. Tokaryk

RI09   4:03 – 4:18
LASER SPECTROSCOPY OF IRIDIUM MONOCHLORIDE, <u>Colan Linton</u>, Allan G. Adam, Samantha Foran, Tongmei Ma, Timothy Steimle

RI10   4:20 – 4:35
LASER INDUCED FLUORESCENCE SPECTROSCOPY OF JET-COOLED CaOCa, <u>Michael N. Sullivan</u>, Daniel J. Frohman, Michael Heaven, Wafaa M Fawzy

RI11   4:37 – 4:52
ELECTRONIC BANDS OF ScC IN THE REGION 620 - 720 NM, Chiao-Wei Chen, <u>Anthony Merer</u>, Yen-Chu Hsu

RI12   4:54 – 5:09
LASER SPECTROSCOPY AND AB INITIO CALCULATIONS ON THE TaF MOLECULE , Kiu Fung Ng, Wenli Zou, Wenjian Liu, <u>Allan S.C. Cheung</u>

RI13   5:11 – 5:26
SPECTROSCOPIC STUDY OF $ThCl^+$ BY TWO-PHOTON IONIZATION, <u>Joshua Bartlett</u>, Robert A. VanGundy, Michael Heaven, Kirk Peterson

RJ. Comparing theory and experiment
Thursday, June 23, 2016 – 1:30 PM
Room: 217 Noyes Laboratory

Chair: Marie-Aline Martin-Drumel, Harvard-Smithsonian CfA, Cambridge, MA, USA

RJ01   **1:30 – 1:45**
TO KINK OR NOT: THE SEARCH FOR LONG CHAIN CUMULENONES USING MICROWAVE SPECTRAL TAXON-OMY, Michael C McCarthy, Marie-Aline Martin-Drumel

RJ02   **1:47 – 2:02**
CONFORMATION-SPECIFIC SPECTROSCOPY OF ALKYL BENZYL RADICALS: EFFECTS OF A RADICAL CEN-TER ON THE CH STRETCH INFRARED SPECTRA OF ALKYL CHAINS, Joseph A. Korn, Timothy S. Zwier, Daniel P. Tabor, Edwin Sibert

RJ03   **2:04 – 2:19**
ANALYSIS OF THE SPECTRUM OF CH3OOH USING SECOND-ORDER PERTURBATION THEORY, Laura C. Dzugan, Anne B McCoy, Amitabha Sinha, Jamie Matthews

RJ04   **2:21 – 2:36**
DOING THE LIMBO WITH A LOW BARRIER: HYDROGEN BONDING AND PROTON TRANSFER IN HYDROXYFORMYLFULVENE , Zachary Vealey, Deacon Nemchick, Patrick Vaccaro

RJ05   **2:38 – 2:53**
ROTATIONAL AND FINE STRUCTURE OF PSEUDO-JAHN-TELLER MOLECULES WITH C_1 SYMMETRY, Jinjun Liu

RJ06   **2:55 – 3:10**
DISPERSED-FLUORESCENCE SPECTROSCOPY OF JET-COOLED CALCIUM ETHOXIDE RADICAL ($CaOC_2H_5$), Anam C. Paul, Md Asmaul Reza, Jinjun Liu

Intermission

RJ07   **3:29 – 3:44**
PHOTODETACHMENT OF O^- YIELDING $O(^1D_2, ^3P)$ ATOMS, VIEWED WITH VELOCITY MAP IMAGING, Stephen T Gibson, Benjamin A Laws, Brenton R Lewis, Ly Duong

RJ08   **3:46 – 4:01**
HIGH RESOLUTION VELOCITY MAP IMAGING PHOTOELECTRON SPECTROSCOPY OF THE BERYLLIUM OX-IDE ANION, BeO^-, Amanda Reed Dermer, Kyle Mascaritolo, Michael Heaven

RJ09   **4:03 – 4:18**
CHARACTERIZING MOLECULAR STRUCTURE BY COMBINING EXPERIMENTAL MEASUREMENTS WITH DENSITY FUNCTIONAL THEORY COMPUTATIONS, Juan M Lopez-Encarnacion

RJ10   **4:20 – 4:35**
SPECTROSCOPIC STUDY OF TORSIONAL POTENTIALS, MOLECULAR STRUCTURE, NBO ANALYSIS AND OTHER MOLECULAR PARAMETERS OF SOME BIPYRIDINE-DICARBOXYLIC ACIDS USING FTIR AND FT-RAMAN SPECTRA AND THEORETICAL METHODS (DFT and IVP), Byru Venkatram Reddy, Jyothi Prashanth, G. Ra-mana Rao

RJ11   *Post-Deadline Abstract*   **4:37 – 4:52**
ELECTRONIC STRUCTURE AND SPECTROSCOPY OF HBr and HBr^+, Gabriel J. Vazquez, H. P. Liebermann, H. Lefebvre-Brion

RJ12   **4:54 – 5:09**
THEORETICAL STUDY ON SERS OF WAGGING VIBRATIONS OF BENZYL RADICAL ADSORBED ON SILVER ELECTRODES, De-Yin Wu, Yan-Li Chen, Zhong-Qun Tian

RJ13   **5:11 – 5:26**
THEORETICAL STUDY ON SURFACE-ENHANCED RAMAN SPECTRA OF WATER ADSORBED ON NOBLE METAL CATHODES OF NANOSTRUCTURES, De-Yin Wu, Ran Pang, Zhong-Qun Tian

FA. Fundamental physics
Friday, June 24, 2016 – 8:30 AM
Room: 100 Noyes Laboratory

Chair: Shui-Ming Hu, University of Science and Technology of China, Hefei, China

FA01   **8:30 – 8:45**
TOWARD PRECISION MID-INFRARED SPECTROSCOPY ON THE OH RADICAL, Arthur Fast, John Furneaux, Samuel Meek

FA02   **8:47 – 9:02**
STUDY OF LASER PRODUCED PLASMA OF LIMITER OF THE ADITYA TOKOMAK FOR DETECTION OF MOLECULAR BANDS, Awadhesh Kumar Rai

FA03   **9:04 – 9:19**
CAVITY RING-DOWN SPECTROSCOPY OF HYDROGEN IN THE 784-852 NM REGION AND CORRESPONDING LINE SHAPE IMPLEMENTATION INTO HITRAN, Yan Tan, Jin Wang, Cunfeng Cheng, An-Wen Liu, Shui-Ming Hu, Piotr Wcislo, Roman V Kochanov, Iouli E Gordon, Laurence S. Rothman

FA04   **9:21 – 9:36**
CARRIER DYNAMICS IN $CsPbBr_3$ NANOCRYSTALS IN PRESENCE OF ELECTRON AND HOLE ACCEPTORS: A TIME RESOLVED TERAHERTZ SPECTROSCOPY STUDY., Sohini Sarkar, Sneha Banerjee, Yettapu Gurivi Reddy, Vikash Kumar Ravi, Angshuman Nag, Pankaj Mandal

FA05   **9:38 – 9:53**
TOWARD ROTATIONAL STATE-SELECTIVE PHOTOIONIZATION OF ThF^+ IONS, Yan Zhou, Kia Boon Ng, Dan Gresh, William Cairncross, Matt Grau, Yiqi Ni, Eric Cornell, Jun Ye

Intermission

FA06   **10:12 – 10:27**
HIGH RESOLUTION GHZ AND THZ (FTIR) SPECTROSCOPY AND THEORY OF PARITY VIOLATION AND TUNNELING FOR 1,2-DITHIINE ($C_4H_4S_2$) AS A CANDIDATE FOR MEASURING THE PARITY VIOLATING ENERGY DIFFERENCE BETWEEN ENANTIOMERS OF CHIRAL MOLECULES, Sieghard Albert, Irina Bolotova, Ziqiu Chen, Csaba Fábri, Lubos Horny, Martin Quack, Georg Seyfang, Daniel Zindel

FA07   **10:29 – 10:44**
A GLOBAL MODEL FOR LONG-RANGE INTERACTION 'DAMPING FUNCTIONS', Philip Thomas Myatt, Frederick R. W. McCourt, Robert J. Le Roy

FA08   **10:46 – 11:01**
DIRECT-POTENTIAL-FIT (DPF) ANALYSIS FOR THE $A\,^3\Pi_1 - X\,^1\Sigma^+$ SYSTEM OF $I^{35/37}Cl.$, Shinji Kobayashi, Nobuo Nishimiya, Tokio Yukiya, Masao Suzuki, Robert J. Le Roy

FB. Instrument/Technique Demonstration
Friday, June 24, 2016 – 8:30 AM
Room: 116 Roger Adams Lab

Chair: Christopher F. Neese, The Ohio State University, Columbus, OH, USA

FB01   8:30 – 8:45
DUAL EXCITATION-EMISSION PROPAGATION (DEEP) IMPACT- FTMW SPECTROMETER , Dennis Wachsmuth, Domenico Prudenzano, Jens-Uwe Grabow

FB02   8:47 – 9:02
A CMOS MILLIMETER-WAVE TRANSCEIVER EMBEDDED IN A SEMI-CONFOCAL FABRY-PEROT CAVITY, Brian Drouin, Adrian Tang, Erich T Schlecht, Emily Brageot, Adam M Daly, Qun Jane Gu, Yu Ye, Ran Shu, M.-C. Frank Chang, Rod M. Kim

FB03   9:04 – 9:19
FINITE-DIFFERENCE TIME-DOMAIN MODELING OF FREE INDUCTION DECAY SIGNAL IN CHIRPED PULSE MILLIMETER WAVE SPECTROSCOPY , Alexander Heifetz, Sasan Bakhtiari, Hual-Teh Chien, Kirill Prozument, Stephen K Gray, Richard M Williams

FB04   9:21 – 9:36
HETERODYNE RECEIVER FOR LABORATORY SPECTROSOCPY OF MOLECULES OF ASTROPHYSICAL IMPORTANCE, Nadine Wehres, Frank Lewen, Christian Endres, Marius Hermanns, Stephan Schlemmer

FB05   9:38 – 9:53
^{13}C-TRIPLY LABELED ETHYL CYANIDE SUBMILLIMETERWAVE STUDY WITH LILLE'S FAST SCAN DDS-BASED SPECTROMETER, A. Pienkina, R. A. Motiyenko, L. Margulès, Holger S. P. Müller, J.-C. Guillemin

Intermission

FB06   10:12 – 10:22
THE KASSEL LABORATORY ASTROPHYSICS THZ SPECTROMETRS, Johanna Chantzos, Doris Herberth, Pia Kutzer, Christoph Muster, Guido W Fuchs, Thomas Giesen

FB07   10:24 – 10:39
DETERMINING CONCENTRATIONS AND TEMPERATURES IN SEMICONDUCTOR MANUFACTURING PLASMAS VIA SUBMILLIMETER ABSORPTION SPECTROSCOPY, Yaser H. Helal, Christopher F. Neese, Frank C. De Lucia, Paul R. Ewing, Ankur Agarwal, Barry Craver, Phillip J. Stout, Michael D. Armacost

FB08   10:41 – 10:56
AUTOFIT AND THE SPECTRUM OF EUGENOL, Erika Riffe, Sawyer Welden, Emma Cockram, Katherine Ervin, Steven Shipman, Cameron M Funderburk, Gordon G Brown, Susanna L. Widicus Weaver

FB09   10:58 – 11:13
COHERENT NONLINEAR TERAHERTZ SPECTROSCOPY OF HALOMETHANE LIQUIDS, Ian A Finneran, Ralph Welsch, Marco A. Allodi, Thomas F. Miller III, Geoffrey Blake

FB10   11:15 – 11:30
SUB-THZ VIBRATIONAL SPECTROSCOPY FOR ANALYSIS OF OVARIAN CANCER CELLS, Jerome P. Ferrance, Igor Sizov, Amir Jazaeri, Aaron Moyer, Boris Gelmont, Tatiana Globus

FB11   *Post-Deadline Abstract*   11:32 – 11:47
THE DATABASE FOR ASTRONOMICAL SPECTROSCOPY - UPDATES, ADDITIONS AND PLANS FOR SPLATALOGUE FOR ALMA FULL SCIENCE OPERATIONS, Anthony Remijan, Nathan A Seifert, Brett A. McGuire

FC. Chirped pulse
Friday, June 24, 2016 – 8:30 AM
Room: 274 Medical Sciences Building

Chair: Justin L. Neill, BrightSpec, Inc., Charlottesville, VA, USA

FC01   8:30 – 8:45
IT IS ALL ABOUT PHASE AND IT IS NOT STAR TREK, Robert W Field, David Grimes, Timothy J Barnum, Stephen Coy, Yan Zhou

FC02   8:47 – 9:02
OBSERVATION OF SUPERRADIANCE IN MMWAVE SPECTROSCOPY OF RYDBERG STATES: BAD IS THE NEW GOOD, David Grimes, Timothy J Barnum, Yan Zhou, Stephen Coy, Robert W Field

FC03   9:04 – 9:19
INTENSITIES OF WEAKLY-ALLOWED RYDBERG-RYDBERG TRANSITIONS MEASURE CORE MULTIPOLES: WHY AND HOW, Stephen Coy, Timothy J Barnum, David Grimes, Robert W Field

FC04   9:21 – 9:36
CPMMW SPECTROSCOPY OF RYDBERG STATES OF NITRIC OXIDE, Timothy J Barnum, Catherine A. Saladrigas, David Grimes, Stephen Coy, Edward E. Eyler, Robert W Field

FC05   9:38 – 9:48
A 75–110 GHz CP-FTmmW SPECTROMETER FOR REACTION DYNAMICS AND KINETICS STUDIES, Daniel P. Zaleski, Kirill Prozument

FC06   9:50 – 10:05
MICROWAVE SPECTRAL TAXONOMY AND ASTRONOMICAL SEARCHES FOR VIBRATIONALLY-EXCITED C_2S AND C_3S, Brett A. McGuire, Marie-Aline Martin-Drumel, John F. Stanton, Michael C McCarthy

Intermission

FC07   10:24 – 10:34
PHOTODISSOCIATION OF ISOXAZOLE AND PYRIDINE STUDIED USING CHIRPED PULSE MICROWAVE SPECTROSCOPY IN PULSED UNIFORM SUPERSONIC FLOWS , Nuwandi M Ariyasingha, Baptiste Joalland, Alexander M Mebel, Arthur Suits

FC08   10:36 – 10:46
CHIRPED PULSE MICROWAVE SPECTROSCOPY IN PULSED UNIFORM SUPERSONIC FLOWS: OBSERVATION OF K-DEPENDENT RATES IN THE CL + PROPYNE REACTION, Nuwandi M Ariyasingha, Bernadette M. Broderick, James O. F. Thompson, Arthur Suits

FC09   10:48 – 11:03
SOME SIGNAL PROCESSING TECHNIQUES FOR USE IN BROADBAND TIME DOMAIN MICROWAVE SPECTROSCOPY, S. A. Cooke

FC10   11:05 – 11:20
FREQUENCY BAND PERFORMANCE COMPARISONS FOR ROOM-TEMPERATURE CHIRPED PULSE MILLIMETER WAVE SPECTROSCOPY, Justin L. Neill, Brent Harris, Robin Pulliam, Matt Muckle, Brooks Pate

FC11   11:22 – 11:32
CHIRPED-PULSE FOURIER TRANSFORM MICROWAVE SPECTROSCOPY OF DIFLUOROBENZALDEHYDES, Gordon G Brown, Sydney A Gaster, Deondre L Parks, Brandon J Yarbrough

FC12   11:34 – 11:49
MICROWAVE SPECTROSCOPY AND STRUCTURE DETERMINATION OF H_2S – MI (M=Cu,Ag,Au), Chris Medcraft, Anthony Legon, Nick Walker

FD. Dynamics and kinetics
Friday, June 24, 2016 – 8:30 AM
Room: B102 Chemical and Life Sciences

Chair: Anh T. Le, Brookhaven National Laboratory, Upton, Ny, USA

FD01　　　　　　　　　　　　　　　　　　　　　　　　　　　　　　　8:30 – 8:45
COOLING OF ELECTRONICALLY-EXCITED He_2 MOLECULES IN A MICROCAVITY PLASMA JET, Rui Su, Thomas J. Houlahan, Jr., J. Gary Eden

FD02　　　　　　　　　　　　　　　　　　　　　　　　　　　　　　　8:47 – 9:02
NON-ADIABATIC DYNAMICS OF $ICN^-(Ar)_n$ and $BrCN^-(Ar)_n$, Bernice Opoku-Agyeman, Anne B McCoy

FD03　　　　　　　　　　　　　　　　　　　　　　　　　　　　　　　9:04 – 9:19
ROVIBRATIONAL LEVELS AND INELASTIC SCATTERING OF THE H_2O-Ar CLUSTER IN FULL AND REDUCED DIMENSIONALITY, Steve Alexandre Ndengue, Moumita Majumder, Richard Dawes, Fabien Gatti, Hans-Dieter Meyer

FD04　　　　　　　　　　　　　　　　　　　　　　　　　　　　　　　9:21 – 9:36
TWO-TONE FREQUENCY MODULATION ABSORPTION SPECTROSCOPY STUDYING HO_2 FORMATION FROM THE OXIDATION OF TETRAHYDROFURAN, Ming-Wei Chen, Ivan Antonov, Brandon Rotavera, Leonid Sheps, Craig A. Taatjes

FD05　　　　　　　　　　　　　　　　　　　　　　　　　　　　　　　9:38 – 9:53
PYROLYSIS OF TROPYL RADICAL (C_7H_7) AND BENZYL RADICAL ($C_6H_5CH_2$) IN A HEATED MICRO-REACTOR, Grant Buckingham, Barney Ellison, Jessica P Porterfield, John W Daily, Musahid Ahmed, David Robichaud, Mark R Nimlos

Intermission

FD06　　　　　　　　　　　　　　　　　　　　　　　　　　　　　　　10:12 – 10:27
SINGLE PHOTON INITIATED DECOMPOSITION REARRANGEMENT REACTIONS (SPIDRR) OF ORGANIC MOLECULES MEDIATED BY THE Ni^+ CATION, Darrin Bellert, Adam Mansell, Zachary Theis, Michael Gutierrez

FD07　　　　　　　　　　　　　　　　　　　　　　　　　　　　　　　10:29 – 10:44
FLUORESCENCE MICROSPECTROSCOPY FOR TESTING THE DIMERIZATION HYPOTHESIS OF BACE1 PROTEIN IN CULTURED HEK293 CELLS, Spencer Gardeen, Joseph L. Johnson, Ahmed A Heikal

FD08　　　　　　　　　　　　　　　　　　　　　　　　　　　　　　　10:46 – 11:01
ULTRAFAST TRANSIENT ABSORPTION SPECTROSCOPY INVESTIGATION OF PHOTOINDUCED DYNAMICS IN NOVEL DONOR-ACCEPTOR CORE-SHELL NANOSTRUCTURES FOR ORGANIC PHOTOVOLTAICS, Jacob Strain, Abdelqader Jamhawi, Thulitha M Abeywickrama, Wendy Loomis, Hemali Rathnayake, Jinjun Liu

FD09　　　　　　　　　　　　　　　　　　　　　　　　　　　　　　　11:03 – 11:08
DIFFUSION ASSISTED ELECTRON INJECTION TO CDS QUANTUM DOTS: A CONCLUSIVE FITTING WITH STATIC QUENCHING COLLINS KIMBALL MODEL, Subhadip Ghosh

FD10　　　　　　　　　　　　　　　　　　　　　　　　　　　　　　　11:10 – 11:20
ULTRAFAST EXTREME ULTRAVIOLET ABSORPTION SPECTROSCOPY OF METHYLAMMONIUM LEAD IODIDE PEROVSKITE, Max A Verkamp, Ming-Fu Lin, Elizabeth S Ryland, Josh Vura-Weis

40

FE. Synchrotron
Friday, June 24, 2016 – 8:30 AM
Room: 217 Noyes Laboratory

Chair: Sven Thorwirth, University of Cologne, Cologne, Germany

FE01
                                                                    8:30 – 8:45
THE SOLEIL VIEW ON PROTOTYPICAL ORGANIC NITRILES: SELECTED VIBRATIONAL MODES OF ETHYL CYANIDE, C_2H_5CN, AND SPECTROSCOPIC ANALYSIS USING AN AUTOMATED SPECTRAL ASSIGNMENT PROCEDURE (ASAP), Christian Endres, Paola Caselli, Marie-Aline Martin-Drumel, Michael C McCarthy, Olivier Pirali, Nadine Wehres, Stephan Schlemmer, Sven Thorwirth

FE02
                                                                    8:47 – 9:02
FAR-INFRARED SPECTROSCOPY OF *SYN*-VINYL ALCOHOL, Paul Raston, Hayley Bunn

FE03
                                                                    9:04 – 9:19
FAR-INFRARED SPECTROSCOPY OF *ANTI*-VINYL ALCOHOL, Paul Raston, Hayley Bunn

FE04
                                                                    9:21 – 9:36
A COMBINED GIGAHERTZ AND TERAHERTZ SYNCHROTRON-BASED FOURIER TRANSFORM INFRARED (TERAHERTZ) SPECTROSCOPIC INVESTIGATION OF META- AND ORTHO-D-PHENOL: OBSERVATION OF TUNNELING SWITCHING, Ziqiu Chen, Sieghard Albert, Csaba Fábri, Robert Prentner, Martin Quack

Intermission

FE05
                                                                    9:55 – 10:10
FAR INFRARED SYNCHROTRON SPECTRUM OF TRIMETHLYENE OXIDE, Omar Mahassneh, Jennifer van Wijngaarden

FE06
                                                                    10:12 – 10:27
WAKEFIELDS IN COHERENT SYNCHROTRON RADIATION , Brant E. Billinghurst, J. C. Bergstrom, C. Baribeau, T. Batten, L. Dallin, Tim E May, J. M. Vogt, Ward A. Wurtz, Robert L. Warnock, D. A. Bizzozero, S. Kramer, K. H. Michaelian

FE07
                                                                    10:29 – 10:44
ULTRAFAST MOLECULAR THREE-ELECTRON COLLECTIVE AUGER DECAY, Raimund Feifel

FE08
                                                                    10:46 – 11:01
$^{14}NH_3$ LINE POSITIONS AND INTENSITIES IN THE FAR-INFRARED: COMPARISON OF FT-IR MEASUREMENTS TO EMPIRICAL HAMILTONIAN MODEL PREDICTIONS, Keeyoon Sung, Shanshan Yu, John Pearson, Olivier Pirali, F. Kwabia Tchana, Laurent Manceron

FE09
                                                                    11:03 – 11:18
INFRARED SPECTROSCOPY OF THE H_2/HD/D_2-O_2 VAN DER WAALS COMPLEXES , Paul Raston, Hayley Bunn

FE10   *Post-Deadline Abstract*   **11:20 – 11:35**
VACUUM ULTRAVIOLET SPECTROSCOPY OF THE LOWEST-LYING ELECTRONIC STATE IN SUB-CRITICAL AND SUPERCRITICAL WATER, Timothy W Marin, Ireneusz Janik, David M Bartels, Dan Chipman

MA. Plenary
Monday, June 20, 2016 – 8:30 AM
Room: Foellinger Auditorium

Chair: Gregory S. Girolami, University of Illinois at Urbana-Champaign, Urbana, IL, USA

Welcome   8:30
Barbara J. Wilson, Chancellor
University of Illinois at Urbana-Champaign

MA01   8:40 – 9:20

ELECTRONIC SPECTROSCOPY OF ORGANIC CATIONS IN GAS-PHASE AT 6 K: IDENTIFICATION OF C_{60}^+ IN THE INTERSTELLAR MEDIUM

JOHN P. MAIER, *Department of Chemistry, University of Basel, Basel, Switzerland.*

After the discovery of C_{60},[a] the question of its relevance to the diffuse interstellar bands was raised. In 1987 H. W. Kroto wrote: "The present observations indicate that C_{60} might survive in the general interstellar medium (probably as the ion C_{60}^+)".[b] In 1994 two diffuse interstellar bands (DIBs) at 9632 and 9577 Å were detected and proposed to be the absorption features of C_{60}^+.[c] This was based on the proximity of these wavelengths to the two prominent absorption bands of C_{60}^+ measured by us in a neon matrix in 1993.[d] Confirmation of the assignment required the gas phase spectrum of C_{60}^+ and has taken 20 years. The approach which succeeded confines C_{60}^+ ions in a radiofrequency trap, cools them by collisions with high density helium allowing formation of the weakly bound C_{60}^+–He complexes below 10 K. The photofragmentation spectrum of this mass-selected complex is then recorded using a cw laser. In order to infer the position of the absorption features of the bare C_{60}^+ ion, measurements on C_{60}^+–He$_2$ were also made. The spectra show that the presence of a helium atom shifts the absorptions by less than 0.2 Å, much less than the accuracy of the astronomical measurements. The two absorption features in the laboratory have band maxima at 9632.7(1) and 9577.5(1) Å, exactly the DIB wavelengths, and the widths and relative intensities agree. This leads to the first definite identification of now five bands among the five hundred or so DIBs known and proves the presence of gaseous C_{60}^+ in the interstellar medium.[e][f] The absorption of cold C_{70}^+ has also been obtained by this approach. In addition the electronic spectra of a number of cations of astrophysical interest ranging from those of carbon chains including oxygen to larger polycyclic aromatic hydrocarbon could be measured in the gas phase at around 10 K in the ion trap but using an excitation-dissociation approach. The implications of these laboratory spectra in relation to the diffuse interstellar band absorptions can be discussed.

[a]H. W. Kroto, J. R. Heath, S. C. O'Brian, R. E. Curl & R. E. Smalley, Nature, 318, 162, 1985

[b]H. W. Kroto in "Polycyclic aromatic hydrocarbons and astrophysics", eds. A. Leger, L. B. d'Hendecourt & N. Boccara, Reidel, Dordrecht, 1987, p.197

[c]B. H. Foing & P. Ehrenfreund, Nature, 369, L296, 1994

[d]J. Fulara, M. Jakobi & J. P. Maier, Chem. Phys. Lett., 211, 227, 1993

[e]E. K. Campbell, M. Holz, D. Gerlich & J. P. Maier, Nature, 523, 323, 2015

[f]G. A. H. Walker, D. A. Bohlender, J. P. Maier & E. Campbell, Astrophys. J. Lett., 812, L8, 2015

MA02   9:25 – 10:05

SOME COMPLEX PRESSURE EFFECTS ON SPECTRA FROM SIMPLE CLASSICAL MECHANICS

JEAN-MICHEL HARTMANN, *LISA, UPEC, Université Paris Est, Creteil, France.*

I will first recall how [the two Newton's equations, 1rst year of university] one can very easily compute the rotational and translational classical dynamics of an ensemble of linear molecules interacting through an (input) pair-wise intermolecular potential. These Classical Molecular Dynamics Simulations (CMDS), which provide the time dependence of the positions and axis-orientations of gas phase molecules, are then used to calculate a number of pressure effects manifesting in absorption and scattering spectra. The cases of CO2, O2 and N2 will be considered, systems for which fully quantum approaches are intractable, and comparisons with measured data will be made, free of any adjusted parameter. I will show that, with a few input ingredients from literature (molecule geometry, electric multipoles, polarizabilities, ...) an no adjusted parameter, excellent agreements with various measurements are obtained. Examples will be given for: (1) Collision induced absorption (due to the interaction induced dipole) ; (2) The far wings of absorption (due to the dipole) and light scattering (due to polarizability) bands ; (3) The broadening and shapes (with their deviations from the Voigt profile) of individual absorption lines for both "free" and spatially tightly confined gases. If times allows, additional demonstrations of the interest of CMDS will be given by considering line-mixing effects and the relaxation of laser-kicked molecules.

Intermission

MA03 10:35 – 11:15

MOLECULAR SPECTROSCOPY OF LIVING SYSTEMS

JI-XIN CHENG, *Department of Chemistry, Purdue University, West Lafayette, IN, USA.*

Molecular spectroscopy has been a powerful tool in the study of molecules in gas phase, condensed phase, and at interfaces. The transition from in vitro spectroscopy to spectroscopic imaging of living systems is opening new opportunities to reveal cellular machinery and to enable molecule-based diagnosis (Science 2015, 350: 1054). Such a transition involves more than a simple combination of spectrometry and microscopy. In this presentation, I will discuss the most recent efforts that have pushed the physical limits of spectroscopic imaging in terms of spectral acquisition speed, detection sensitivity, spatial resolution and imaging depth. I will further highlight significant applications in functional analysis of single cells and in label-free detection of diseases.

MA04 11:20 – 12:00

OPTICAL AND MICROWAVE SPECTROSCOPY OF TRANSIENT METAL-CONTAINING MOLECULES

TIMOTHY STEIMLE, *School of Molecular Sciences, Arizona State University, Tempe, AZ, USA.*

Small metal containing molecules are ideal venues for testing Fundamental Physics, investigating relativistic effects, and modelling spin-orbit induced unimolecular dynamics. Electronic spectroscopy is an effective method for probing these phenomena because such spectra are readily recorded at the natural linewidth limited resolution and accuracy of 0.0001 cm^{-1}. The information garnered includes fine and hyperfine interactions, magnetic and electric dipoles, and dynamics. With this in mind, three examples from our recent (unpublished) studies will be highlighted.

SiHD: Long ago Duxbury et al.[a] developed a semi-quantitative model invoking Renner-Teller and spin-orbit coupling of the $\tilde{a}^3B_1$, $\tilde{X}^1A_1$, and $\tilde{A}^1B_1$, states to explain the observed local perturbations and anomalous radiative lifetimes in the visible spectrum. More recently, the $\tilde{a}^3B_1$ to $\tilde{A}^1B_1$ intersystem crossing has been modeled using both semi-classical transition state theory and quantum trajectory surface hopping dynamics [b]. Here we investigate the effects of the reduced symmetry of SiHD on the spectroscopy and dynamics using 2D spectroscopy [c]. Rotationally resolved lines in the origin $\tilde{X}^1A' \rightarrow \tilde{A}^1A''$ band are assigned to both c-type transitions and additional axis-switching [d] induced transitions.

AuO and AuS: The observed markedly different bonding of thiols and alcohols to gold clusters should be traceable to the difference in Au-O and Au-S bonding. To investigate this difference we have used optical Stark and Zeeman spectroscopy to determine the permanent electric dipole moments and magnetic g-factors. The results are rationalized using simple m.o. correlation diagrams and compared to ab initio predictions.

TaN: TaN is the best candidate to search for a T,P- violating nuclear magnetic quadrupole moment[e]. Here we report on the optical 2D, Stark, and Zeeman spectra, and our efforts to record the pure rotational spectrum using the separated field pump/probe microwave-optical double resonance.Implications for T,P- violating experiments will be presented.

---

[a]G. Duxbury, A. Alijah and R. R. Trieling, J. Chem. Phys. 98, 811 (1993)

[b]R. R. Zaari and S. A. Varganov, JPCA 119 , 1332 (2015)

[c]N. J. Reilly, T. W. Schmidt and S. H. Kable, JPCA 110, 12355(2006)

[d]J. T. Hougen and J. K. G. Watson, Can. J. Phys. 43 , 298 (1965)

[e]L. V. Skripnikov, et.al. Phys. Rev. A: 92, 1 (2015)

MF. Mini-symposium: Spectroscopy of Large Amplitude Motions
Monday, June 20, 2016 – 1:30 PM
Room: 100 Noyes Laboratory

Chair: Isabelle Kleiner, CNRS et Universités Paris Est et Paris Diderot, Créteil, France

---

MF01   *INVITED TALK*   1:30 – 2:00

ON THE LOWEST RO-VIBRATIONAL STATES OF PROTONATED METHANE:
EXPERIMENT AND ANALYTICAL MODEL

HANNO SCHMIEDT, *I. Physikalisches Institut, University of Cologne, Cologne, Germany*; PER JENSEN, *Faculty of Mathematics and Natural Sciences, University of Wuppertal, Wuppertal, Germany*; OSKAR ASVANY, STEPHAN SCHLEMMER, *I. Physikalisches Institut, Universität zu Köln, Köln, Germany.*

Protonated methane, CH_5^+, is the prototype of an extremely floppy molecule. To the best of our knowledge all barriers are surmountable in the rovibrational ground state; the large amount of zero-point vibrational energy leads to large amplitude motions for many degrees of freedom. Low resolution but broad band vibrational spectroscopy [1] revealed an extremely wide range of C-H stretching vibrations. Comparison with theoretical IR spectra supported the structural motif of a CH_3 tripod and an H_2 moiety, bound to the central carbon atom by a 3c2e bond. In a more dynamic picture the five protons surround the central carbon atom without significant restrictions on the H-C-H bending or H_n-C torsional motions. The large-amplitude internal motions preclude a simple theoretical description of the type possible for more conventional molecules, such as the related spherical-top methane molecule. Recent high-resolution ro-vibrational spectra obtained in cold ion trap experiments [2] show that the observed CH_5^+ transitions belong to a very well-defined energy level scheme describing the lowest rotational and vibrational states of this enigmatic molecule. Here we analyse the experimental ground state combination differences and associate them with the motional states of CH_5^+ allowed by Fermi-Dirac statistics. A model Hamiltonian for unrestricted internal rotations in CH_5^+ yields a simple analytical expression for the energy eigenvalues, expressed in terms of new quantum numbers describing the free internal rotation. These results are compared to the experimental combination differences and the validity of the model will be discussed together with the underlying assumptions.

[1] O. Asvany, P. Kumar, I. Hegemann, B. Redlich, S. Schlemmer and D. Marx, Science **309**, (2005) 1219-1222
[2] O. Asvany, K.M.T. Yamada, S. Brünken, A. Potapov, S. Schlemmer, Science **347** (2015) 1346-1349

MF02   2:05 – 2:20

SYMMETRY IN THE GENERALIZED ROTOR MODEL FOR EXTREMELY FLOPPY MOLECULES

HANNO SCHMIEDT, *I. Physikalisches Institut, University of Cologne, Cologne, Germany*; PER JENSEN, *Faculty of Mathematics and Natural Sciences, University of Wuppertal, Wuppertal, Germany*; STEPHAN SCHLEMMER, *I. Physikalisches Institut, University of Cologne, Cologne, Germany.*

Protonated methane CH_5^+ is unique: It is an extremely fluxional molecule. All attempts to assign quantum numbers to the high-resolution transitions obtained over the last 20 years have failed because molecular rotation and vibration cannot be separated in the conventional way[a][b]. The first step towards a theoretical description is to include internal rotational degrees of freedom into the overall ones, which can be used to formulate a fundamentally new zero order approximation for the (now) generalized rotational states and energies. Predictions from this simple five-dimensional rotor model compare very favorably with the combination differences of protonated methane found in recent low temperature experiments[c]. This talk will focus on symmetry aspects and implications of permutation symmetry for the generalized rotational states. Furthermore, refinements of the theory will be discussed, ranging from the generalization to even higher-dimensional rotors to explicit symmetry breaking and corresponding energy splittings. The latter includes the link to well-known theories of internal rotation dynamics and will show the general validity of the presented theory.

---

[a]Schmiedt, H., et al.; *J. Chem. Phys.* **143** (15), 154302 (2015)
[b]Wodraszka, R. et al.; *J. Phys. Chem. Lett.* **6**, 4229-4232 (2015)
[c]Asvany, O. et al.; *Science*, **347**, (6228), 1346-1349 (2015)

MF03

AN EFFECTIVE-HAMILTONIAN APPROACH TO CH_5^+, USING IDEAS FROM ATOMIC SPECTROSCOPY

JON T. HOUGEN, *Sensor Science Division, National Institute of Standards and Technology, Gaithersburg, MD, USA.*

In this talk we present the first steps in the design of an effective Hamiltonian for the vibration-rotation energy levels of CH_5^+. Such a Hamiltonian would allow calculation of energy level patterns anywhere along the path travelled by a hypothetical CH_5^+ (or CD_5^+) molecule as it passes through various coupling cases, and might thus provide some hints for assigning the observed high-resolution spectra. The steps discussed here, which have not yet addressed computational problems, focus on mapping the vibration-rotation problem in CH_5^+ onto the five-electron problem in the boron atom, using ideas and mathematical machinery from Condon and Shortley's book on atomic spectroscopy. The mapping ideas are divided into: (i) a mapping of particles, (ii) a mapping of coordinates (i.e., mathematical degrees of freedom), and (iii) a mapping of quantum mechanical interaction terms. The various coupling cases along the path correspond conceptually to: (i) the analog of a free-rotor limit, where the H atoms see the central C atom but do not see each other, (ii) the low-barrier and high-barrier tunneling regimes, and (iii) the rigid-molecule limit, where the H atoms remain locked in some fixed molecular geometry. Since the mappings considered here often involve significant changes in mathematics, a number of interesting qualitative changes occur in the basic ideas when passing from B to CH_5^+, particularly in discussions of: (i) antisymmetrization and symmetrization ideas, (ii) n, l, m_l, m_s or n, l, j, m_j quantum numbers, and (iii) Russell-Saunders computations and energy level patterns. Some of the mappings from B to CH_5^+ to be discussed are as follows. Particles: the atomic nucleus is replaced by the C atom, the electrons are replaced by protons, and the empty space between particles is replaced by an "electron soup." Coordinates: the radial coordinates of the electrons map onto the five local C-H stretching modes, the angular coordinates of the electrons map onto three rotational degrees of freedom and seven bending vibrational degrees of freedom. The half-integral electron spins map onto half-integral proton spins or onto integral deuterium spins (for CD_5^+). Interactions: the Coulomb attraction between nucleus and electrons maps onto a Morse-oscillator C-H stretching potential, spin-orbit interaction maps onto proton-spin-overall-rotation interaction, and Coulomb repulsion between electrons maps onto some kind of proton repulsion that leads to the equilibrium geometry.

MF04

IMPACT OF ENERGETICALLY ACCESSIBLE PROTON PERMUTATIONS IN THE SPECTROSCOPY AND DYNAMICS OF H_5^+

ZHOU LIN[a], *Department of Chemistry, Massachusetts Institute of Technology, Cambridge, MA, USA*; ANNE B McCOY, *Department of Chemistry, University of Washington, Seattle, WA, USA.*

H_5^+ has been proposed to be the intermediate of the astrochemically interesting proton transfer reaction $H_3^+ + H_2 \rightarrow H_2 + H_3^+$. The scrambling of five protons in this floppy, "structureless" ion introduces complications to its high-resolution rovibrational spectroscopy and the proton transfer dynamics between H_3^+ and H_2. Quantum chemical studies are performed to predict and interpret the spectroscopic and dynamical properties of H_5^+, with special consideration paid to the group theoretical aspects. If the full permutation of protons were allowed in H_5^+, just like in CH_5^+, the system should have been characterized by the G_{240} complete permutation-inversion group.[b] However, our diffusion Monte Carlo calculations indicate that such a full permutation is not allowed for most of the molecular configurations sampled by the reaction path of the proton transfer process in question, and the energetically accessible permutations are functions of the distance between the H_3^+ and H_2 fragments.[c] In the present study, we investigate two extreme geometries of H_5^+, the $[H_2\text{-}H\text{-}H_2]^+$ shared-proton intermediate and the $H_3^+ \cdots H_2$ long-range complex, using two subgroups of G_{240}, G_{16} and G_{24}, respectively. In these two limiting circumstances, we derive the symmetry-adapted basis functions for the energy levels that describe the nuclear spins and the rovibrational motions of H_5^+. Based on the results of these derivations, we discuss the spectroscopic properties of H_5^+, including the coupling between different rovibrational degrees of freedom in the effective nuclear motion Hamiltonian, the electric-dipole selection rules for rovibrational spectroscopy, and correlations of energy levels between $[H_2\text{-}H\text{-}H_2]^+$ and $H_3^+ \cdots H_2$. Our study can be considered as the first step towards the implementation of future quantitative theoretical investigations for comparison with spectroscopic and dynamical experiments. [d]

[a] The study was performed when Z. Lin was a graduate student with A. B. McCoy at The Ohio State University.

[b] X.-G. Wang and T. Carrington Jr., *J. Chem. Phys.*, **129**, 234102 (2008).

[c] Z. Lin and A. B. McCoy, *J. Phys. Chem. A*, **119**, 12109 (2015).

[d] Z. Lin, *submitted to J. Mol. Spec.*

MF05 2:56 – 3:11

VARIATION OF CH STRETCH FREQUENCIES WITH CH_4 ORIENTATION IN THE $CH_4 - F^-$ COMPLEX: MULTIPLE RESONANCES AS VIBRATIONAL CONICAL INTERSECTIONS

BISHNU P. THAPALIYA, DAVID S. PERRY, *Department of Chemistry, The University of Akron, Akron, OH, USA.*

In the $CH_4 - F^-$ complex, an adiabatic separation of the CH stretch frequencies from the CH_4 orientational coordinates allows the calculation of the four adiabatic CH stretch surfaces. These ab initio calculations reveal (i) a large variation of CH stretch frequencies (>100 cm^{-1}) in the orientational space and (ii) the existence of four symmetrically equivalent sets of vibrational conical intersections (CIs). Two sets of symmetry-allowed CIs are identified in addition to the symmetry-required CIs at the front- and back-side C_{3v} geometries. These results have implications for the evolution of excited CH vibrations in methane during its approach to a potentially reactive surface.

MF06 3:13 – 3:28

SADDLE POINT LOCALIZATION OF MOLECULAR WAVEFUNCTIONS

GEORG CH. MELLAU, *Physikalisch Chemisches Institut, Justus Liebig Universitat Giessen, Giessen, Germany*; ALEKSANDRA KYUBERIS, *Microwave Spectroscopy, Institute of Applied Physics, Nizhny Novgorod, Russia*; OLEG POLYANSKY, *Department of Physics and Astronomy, University College London, London, IX, United Kingdom*; NIKOLAY FEDOROVICH ZOBOV, *Microwave Spectroscopy, Institute of Applied Physics, Nizhny Novgorod, Russia*; ROBERT W FIELD, *Department of Chemistry, MIT, Cambridge, MA, USA.*

The quantum mechanical description of isomerization is based on bound eigenstates of the molecular potential energy surface. For the near-minimum regions there is a textbook-based relationship between the potential and eigenenergies. Here we show how the saddle point region that connects the two minima is encoded in the energy levels and wave functions of the potential energy surface.

Intermission

MF07 3:47 – 4:02

SPECTROSCOPIC CHARACTERIZATION OF ISOMERIZATION TRANSITION STATES

JOSHUA H BARABAN, *Department of Chemistry, University of Colorado, Boulder, CO, USA*; BRYAN CHANGALA, *Department of Physics, JILA - University of Colorado, Boulder, CO, USA*; GEORG CH. MELLAU, *Physikalisch Chemisches Institut, Justus Liebig Universitat Giessen, Giessen, Germany*; JOHN F. STANTON, *Department of Chemistry, The University of Texas, Austin, TX, USA*; ANTHONY MERER, *Department of Chemistry, University of British Columbia, Vancouver, BC, Canada*; ROBERT W FIELD, *Department of Chemistry, MIT, Cambridge, MA, USA.*

Transition state theory is central to our understanding of chemical reaction dynamics. We demonstrate here a method for extracting transition state energies and properties from a characteristic pattern found in frequency domain spectra of isomerizing systems. This pattern, a dip in the spacings of certain barrier-proximal vibrational levels, can be understood using the concept of effective frequency, ω^{eff}. The method is applied to the *cis-trans* conformational change in the S_1 state of C_2H_2 and the bond-breaking HCN-HNC isomerization. In both cases, the barrier heights derived from spectroscopic data agree extremely well with previous *ab initio* calculations. We also show that it is possible to distinguish between vibrational modes that are actively involved in the isomerization process and those that are passive bystanders. (This work has been published in J. H. Baraban, P. B. Changala, G. Ch. Mellau, J. F. Stanton, A. J. Merer, and R. W. Field. Spectroscopic characterization of isomerization transition states. *Science*, 350(6266):1338–1342, 2015.)

MF08   4:04 – 4:19

ELECTRON ANISOTROPY AS A SIGNATURE OF MODE SPECIFIC ISOMERIZATION IN VINYLIDENE[a]

STEPHEN T GIBSON, BENJAMIN A LAWS, *Research School of Physics and Engineering, Australian National University, Canberra, ACT, Australia*; RICHARD MABBS, *Department of Chemistry, Washington University, St. Louis, MO, USA*; DANIEL NEUMARK, *Department of Chemistry, The University of California, Berkeley, CA, USA*; CARL LINEBERGER, *Department of Chemistry and Biochemistry, JILA - University of Colorado, Boulder, CO, USA*; ROBERT W FIELD, *Department of Chemistry, MIT, Cambridge, MA, USA*.

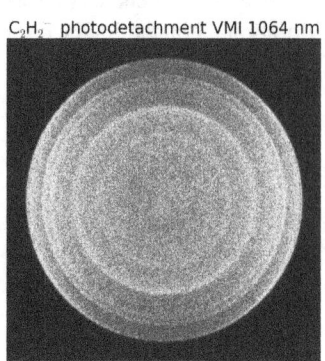

C₂H₂ photodetachment VMI 1064 nm

The nature of the isomerization process that turns vinylidene into acetylene has been awaiting advances in experimental methods, to better define fractionation widths beyond those available in the seminal 1989 photoelectron spectrum measurement.[b] This has proven a challenge. The technique of velocity-map imaging (VMI) is one avenue of approach. Images of electrons photodetached from vinylidene negative-ions, at various wavelengths, 1064 nm shown, provide more detail, including unassigned structure, but only an incremental improvement in the instrument line width. Intriguingly, the VMIs demonstrate a mode dependent variation in the electron anisotropy. Most notable in the figure, the inner-ring transition clusters are discontinuously, more isotropic. Electron anisotropy may provide an alternative key to examine the character of vinylidene transitions, mediating the necessity for an extreme resolution measurement. Vibrational dependent anisotropy has previously been observed in diatomic photoelectron spectra, associated with the coupling of electronic and nuclear motions.[c]

[a]Research supported by the Australian Research Council Discovery Project Grant DP160102585.
[b]K. M. Ervin, J. Ho, and W. C. Lineberger, *J. Chem. Phys.* **91**, 5974 (1989). doi:10.1063/1.457415
[c]M. van Duzor *et al. J. Chem. Phys.* **133**, 174311 (2010). doi:10.1063/1.3493349

MF09   4:21 – 4:36

OBSERVING QUANTUM MONODROMY: AN ENERGY-MOMENTUM MAP BUILT FROM EXPERIMENTALLY-DETERMINED LEVEL ENERGIES OBTAINED FROM THE ν_7 FAR-INFRARED BAND SYSTEM OF NCNCS

DENNIS W. TOKARYK, STEPHEN CARY ROSS, *Department of Physics, University of New Brunswick, Fredericton, NB, Canada*; BRENDA P. WINNEWISSER, MANFRED WINNEWISSER, FRANK C. DE LUCIA, *Department of Physics, The Ohio State University, Columbus, OH, USA*; BRANT E. BILLINGHURST, *EFD, Canadian Light Source Inc., Saskatoon, Saskatchewan, Canada*.

The concept of Quantum Monodromy (QM) provides a fresh insight into the structure of rovibrational levels in those flexible molecules for which a bending mode can carry the molecule through the linear configuration. To confirm the existence of QM in a molecule required the fruits of several strands of development: the formulation of the abstract mathematical concept of monodromy, including the exploration of its relevance to systems described by classical mechanics and its manifestation in quantum molecular applications; the development of the required spectroscopic technology and computer-aided assignment; and the development of a theoretical model to apply in fitting to the observed data. We present a timeline for each of these strands, converging in our initial confirmation of QM in NCNCS from pure rotational data alone.[a] In that work a Generalised SemiRigid Bender (GSRB) Hamiltonian was fitted to the experimental rotational structure. Rovibrational energies calculated from the fitted GSRB parameters allowed us to construct an "Energy-Momentum" map and confirm the presence of QM in NCNCS. In further experimental work at the Canadian Light Source Synchrotron we have identified a network of transitions directly connecting the relevant energy levels and thereby have produced a refined Energy Momentum map for NCNCS from experimental measurements alone. This map extends from the ground vibrational level to well above the potential energy barrier, beautifully illustrating the characteristic signature of QM in a system uncomplicated by interaction with other vibrational modes.

[a]B. P. Winnewisser *et al.*, Phys. Rev. Lett. **95**, 243002 (2005).

MF10 4:38 – 4:53

SPECTROSCOPY OF H_2^+ AND HD^+ NEAR THE DISSOCIATION THRESHOLD: SHAPE AND FESHBACH RESONANCES

MAXIMILIAN BEYER, <u>FREDERIC MERKT</u>, *Laboratorium für Physikalische Chemie, ETH Zurich, Zurich, Switzerland.*

We use high Rydberg states to measure the properties of H_2^+ and HD^+ in the vicinity of their dissociation limits H^++H, H^++D and $H+D^+$, with particular emphasis on quasibound rovibrational levels above the dissociation threshold of the X^+ $^2\Sigma_g^+$ ground state.

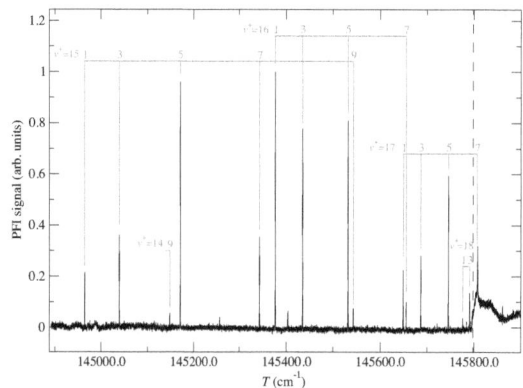

PFI-ZEKE photoelectron spectrum of H₂ near the dissociation threshold (dashed line).

Although the existence of these quasibound levels has been predicted a long time ago, they have never been observed. Positions and widths of the lowest resonances have not been calculated either. Given the role that such states play in the three-body and radiative recombination of H(1s) and H^+ to form H_2^+, this lack of data may be regarded as one of the largest unknown aspects of this otherwise accurately known fundamental molecular cation.

We present measurements of the positions and widths of the lowest-lying quasibound rotational levels (shape resonances) of H_2^+, located close to the top of the centrifugal barriers and which decay by quantum-mechanical tunneling. For HD^+ we present measurements of rovibrational levels of the A^+ $^2\Sigma_u^+$ state, located between the two dissociation limits. Because of the g-u-symmetry breaking in HD^+, these levels are coupled to the H^++D continuum by nonadiabatic interactions (Feshbach resonances).

The experimental results will be compared with the positions and widths we calculate for these levels using a potential model for the X^+ and the A^+ state of H_2^+ and HD^+ which includes adiabatic, nonadiabatic, relativistic and radiative corrections to the Born-Oppenheimer potential energies.

MF11 4:55 – 5:10

A DIATOMIC MOLECULE WITH EXTREMELY LARGE AMPLITUDE MOTION IN ITS VIBRATIONAL STATES THAT HAVE LENGTHS OF AT LEAST 12,000 ANGSTROMS.

<u>NIKESH S. DATTANI</u>, *Department of Chemistry, Kyoto University, Kyoto, Japan.*

The state-of-the-art empirical potential, *and* the state-of-the-art *ab initio* potential for the $b(1^3\Pi_{2_u})$ state of $^{7,7}Li_2$ agree with each other that the $(v = 100, J = 0)$ ro-vibrational state has an outer classical turning point larger than the diameter of most bacteria and many animal cells. The 2015 empirical potential[a] based on a significant amount of spectroscopic data, predicts the $(v = 100, J = 0)$ level to be bound by only $0.000\,000\,000\,004$ cm^{-1} (< 0.2 Hz). The outer turning point of the vibrational wavefunction is about $671\,000$ Å or 0.07 mm. Here, the two Li atoms are bound to each other, despite being nearly as far apart as the lines on a macroscopic ruler. The 2014 *ab initio* calculation based on a powerful Fock space MRCC method[b] and with the long-range tail anchored by C_3^{7Li}/r^3 with the ultra-high precision 2015 value of C_3^{7Li}, has this same level bound by $0.000\,000\,000\,1$ cm^{-1} (< 3 Hz), with an outer turning point of > 0.01 mm. While this discovery occurred during a study of Li_2, the $b(1^3\Pi_{2_u})$ states of heavier alkali diatomics are expected to have even larger amplitude vibrational states. While it might be tempting to call these very large molecules "Rydberg molecules", it is important to remember that this term is already used to describe highly excited *electronic* states whose energy levels follow a formula similar to that for the famous Rydberg series. The highly delocalized *vibrational* states are a truly unfamiliar phenomenon.

[a]Dattani (2015) http://arxiv.org/abs/1508.07184v1
[b]Musial & Kucharski (2014) Journal of Chemical Theory and Computation, **10**, 1200.

FIT POINT-WISE *AB INITIO* CALCULATION POTENTIAL ENERGIES TO A MULTI-DIMENSION MORSE/LONG-RANGE MODEL

YU ZHAI, HUI LI, *Institute of Theoretical Chemistry, Jilin University, Changchun, China*; ROBERT J. LE ROY, *Department of Chemistry, University of Waterloo, Waterloo, ON, Canada.*

A potential energy surface (PES) is a fundamental tool and source of understanding for theoretical spectroscopy and for dynamical simulations. Making correct assignments for high-resolution rovibrational spectra of floppy polyatomic and van der Waals molecules often relies heavily on predictions generated from a high quality *ab initio* potential energy surface. Moreover, having an effective analytic model to represent such surfaces can be as important as the *ab initio* results themselves. For the one-dimensional potentials of diatomic molecules, the most successful such

$$V = V_{\text{MLR}}(r) = \mathfrak{D}_e \left\{ 1 - \frac{u_{\text{LR}}(r)}{u_{\text{LR}}(r_e)} e^{-\beta(r) \cdot y_p^{\text{eq}}(r)} \right\}^2 + V_{\text{min}}$$

parameters set as angular functions

$$V = V_{\text{MD-MLR}}(r; \theta_i, \phi_j \ldots) = V_{\text{MLR}}^{\theta_i, \phi_j \cdots}(r)$$

model to date is arguably the "Morse/Long-Range" (MLR) function developed by R. J. Le Roy and coworkers.[a] It is very flexible, is everywhere differentiable to all orders. It incorporates correct predicted long-range behaviour, extrapolates sensibly at both large and small distances, and two of its defining parameters are always the physically meaningful well depth D_e and equilibrium distance r_e. Extensions of this model, called the Multi-Dimension Morse/Long-Range (MD-MLR) function, have been applied successfully to atom-plus-linear molecule, linear molecule–linear molecule and atom–non-linear molecule systems.[b] However, there are several technical challenges faced in modelling the interactions of general molecule-molecule systems, such as the absence of radial minima for some relative alignments, difficulties in fitting short-range potential energies, and challenges in determining relative-orientation dependent long-range coefficients. This talk will illustrate some of these challenges and describe our ongoing work in addressing them.

---

[a]*Mol. Phys.* **105**, 663 (2007); *J. Chem. Phys.* **131**, 204309 (2009); *Mol. Phys.* **109**, 435 (2011).

[b]*Phys. Chem. Chem. Phys.* **10**, 4128 (2008); *J. Chem. Phys.* **130**, 144305 (2009); *J. Chem. Phys.* **132**, 214309 (2010); *J. Chem. Phys.* **140**, 214309 (2014); *J. Chem. Phys.* **144**, 014301 (2016).

MG. Mini-symposium: Spectroscopy in Atmospheric Chemistry

Monday, June 20, 2016 – 1:30 PM

Room: 116 Roger Adams Lab

Chair: Frank Keutsch, Harvard University, Cambridge, MA, USA

MG01   *INVITED TALK*   1:30 – 2:00

MAPPING THE WORLD'S LARGEST NATURAL GAS LEAK AND OTHER METHANE SOURCES USING HIGH RESOLUTION SPECTROSCOPY

STANLEY P. SANDER, CLARE WONG, THOMAS J PONGETTI, *Jet Propulsion Laboratory, California Institute of Technology, Pasadena, CA, USA.*

CH_4 is a potent greenhouse gas with a 100-year Global Warming Potential more than thirty times larger than CO_2 if carbon-climate feedbacks are considered. In urban areas such as Los Angeles, anthropogenic methane emissions are poorly characterized because of the large diversity of sources: landfills, sewage treatment plants, agriculture, leaks in the natural gas distribution system, cattle and dairy farms, thermogenic emissions from oil fields and seeps. The California Laboratory for Atmospheric Remote Sensing (CLARS), operated by the Jet Propulsion Laboratory, is a mountaintop facility overlooking most of the Los Angeles basin, equipped with JPL-built Fourier transform spectrometers for measurements of the slant column abundances of several greenhouse gases including methane with high spatial and temporal resolution. This presentation will cover several topics including the design features of the two FTS instruments, spectroscopic issues associated with the retrieval of slant column abundances, and uncertainty analysis. One FTS has been in continuous operation since 2011, providing sufficient data to identify several CH_4 emission hot spots in the LA basin. On October 23, 2015, a well pipe suffered a failure in a natural gas storage facility in Aliso Canyon, northwest of downtown Los Angeles resulting in a massive CH_4 plume transported by winds throughout the LA basin. The CLARS FTS captured the plume propagation throughout the 4-month duration of the leak. We will show how the emission ratio method may be employed to derive a lower bound to the CH_4 emission rate from the leaking well without the use of complex atmospheric transport models. The CLARS measurement system provides a small-scale example of the data that would be acquired by an imaging FTS on a geostationary space platform.[a]

---

[a]copyright 2016, California Institute of Technology. Government sponsorship acknowledged.

MG02   2:05 – 2:20

THE ATMOSPHERIC CHEMISTRY EXPERIMENT (ACE): CO, CH_4 AND N_2O ISOTOPOLOGUES

PETER F. BERNATH, ERIC M. BUZAN, *Department of Chemistry and Biochemistry, Old Dominion University, Norfolk, VA, USA*; CHRISTOPHER A. BEALE, *Department of Ocean, Earth and Atmospheric Sciences, Old Dominion University, Norfolk, VA, USA*; MAHDI YOUSEFI, *Department of Physics, Old Dominion University, Norfolk, Virginia, USA*; CHRIS BOONE, *Department of Chemistry, University of Waterloo, Waterloo, ON, Canada.*

ACE (also known as SCISAT) is making a comprehensive set of simultaneous measurements of numerous trace gases, thin clouds, aerosols and temperature by solar occultation from a satellite in low earth orbit. A high inclination orbit gives ACE coverage of tropical, mid-latitudes and polar regions. The primary instrument is a high-resolution (0.02 cm^{-1}) infrared Fourier Transform Spectrometer (FTS) operating in the 750–4400 cm^{-1} region, which provides the vertical distribution of trace gases, and the meteorological variables of temperature and pressure. Aerosols and clouds are being monitored through the extinction of solar radiation using two filtered imagers as well as by their infrared spectra. Although now in its thirteenth year, the ACE-FTS is still operating nominally. A short introduction and overview of the ACE mission will be presented (see http://www.ace.uwaterloo.ca for more information). This talk will focus on ACE observations of the CO, CH_4 and N_2O isotopologues, and comparisons with chemical transport models.

MG03   2:22 – 2:37

OPTICAL MEASUREMENTS OF $^{14}CO_2$ USING CAVITY RING-DOWN SPECTROSCOPY

DAVID A. LONG, ADAM J. FLEISHER, QINGNAN LIU, JOSEPH HODGES, *Material Measurement Laboratory, National Institute of Standards and Technology, Gaithersburg, MD, USA.*

Measurements of radiocarbon (^{14}C) provide a unique platform in order to determine the age of a material or alternatively for source attribution between biogenic and petrochemical sources. We describe a cavity ring-down spectrometer which uses an infrared quantum cascade laser to probe the fundamental of $^{14}CO_2$. This instrument offers one of the highest sensitivities which has been reported in the mid-infrared and has fully automated spectral scanning for continuous data acquisition. Despite the ultra-low abundance of $^{14}CO_2$ in the atmosphere (1.2 pmol/mol relative to $^{12}CO_2$) we have been able to rapidly record the $^{14}CO_2$ transition and determine the origin of carbon dioxide samples. Our experimental approach and future improvements to the instrument will be discussed as well as selected measurement targets.

MG04   2:39 – 2:54

MID-IR CAVITY RINGDOWN SPECTROSCOPY MEASUREMENTS OF ATMOSPHERIC ETHANE TO METHANE RATIO

LINHAN SHEN, THINH QUOC BUI, *Division of Chemistry and Chemical Engineering, California Institute of Technology, Pasadena, CA, USA*; LANCE CHRISTENSEN, *Science Division, Jet Propulsion Laboratory/Caltech, Pasadena, CA, USA*; MITCHIO OKUMURA, *Division of Chemistry and Chemical Engineering, California Institute of Technology, Pasadena, CA, USA.*

In this work, we demonstrated a mid-IR (3.3 μm) cw cavity ringdown spectrometer capable of measuring atmospheric ethane abundance and ethane to methane ratio. This technique can measure atmospheric ethane concentration as low as 70 ppb. Since ethane is a tracer for thermogenic methane emissions, this technique could be used to identify sources of atmospheric methane. We have demonstrated the capability of this instrument by measuring the atmospheric ethane composition and ethane to methane ratio in ambient air in Pasadena, California.

MG05   2:56 – 3:11

MEASUREMENTS DOUBLY-SUBSTITUTED METHANE ISOTOPOLOGUE (13CH3D AND 12CH2D2) ABUDANCE USING FREQUENCY STABILIZED MID-IR CAVITY RINGDOWN SPECTROSCOPY

LINHAN SHEN, THINH QUOC BUI, MITCHIO OKUMURA, *Division of Chemistry and Chemical Engineering, California Institute of Technology, Pasadena, CA, USA.*

In this work, we demonstrated a spectroscopic method of measuring abundances of doubly-substituted methane isotopologues ($^{13}CH_3D$, $^{12}CH_2D2$). In this method, we use a frequency stabilized cavity ringdown spectroscopy (FS-CRDS) technique to measure $\Delta^{12}CH_2D2$ in naturally abundant methane to sub 0.1% level within one hour of average. Compare to traditional isotope-ratio mass spectrometer, which requires more than 24 hours of average to achieve comparable precision, this method provides a fast way of measuring clumped isotopologue abundance optically without destroying samples.

MG06   3:13 – 3:28

THE NEAR-IR SPECTRUM OF CH_3D

SHAOYUE YANG, *Department of Physics, The University of Virginia, Charlottesville, VA, USA*; KEVIN LEHMANN, *Departments of Chemistry and Physics, University of Virginia, Charlottesville, VA, USA*; ROBERT J. HARGREAVES, PETER F. BERNATH, *Department of Chemistry and Biochemistry, Old Dominion University, Norfolk, VA, USA*; MICHAEL REY, *Groupe de Spectrométrie Moléculaire et Atmosphérique, UMR CNRS 7331, Université de Reims, Reims Cedex 2, France*; ANDREI V. NIKITIN, *Atmospheric Spectroscopy Div., Institute of Atmospheric Optics, Tomsk, Russia*; VLADIMIR TYUTEREV, *Laboratoire GSMA, CNRS / Université de Reims Champagne-Ardenne, REIMS, France*.

The near-IR spectrum, from 5000-8960 cm^{-1}, of isotopically pure CH_3D was taken at temperatures of 294, 400, 500, 600, 700, 800, and 900K with a high resolution Fourier Transform machine at Old Dominion University. The spectra where analyzed to give the wavenumbers, integrated line intensities, and lower state term values (using lines observed in at least 3 different spectra). For the 294 K spectrum 12080 lines with S between 3.6×10^{-22} and 1×10^{-27} cm (not corrected for CH_3D natural abundance) were determined for this spectral interval.

A theoretical spectrum of CH_3D has also been calculated at the Univ. of Reins, with ¿400,000 transitions predicted between 5000-6300 cm^{-1} with S values between 2.1×10^{-22} and 1×10^{-27} cm at 294K. Comparison of the predictions with 175 J" = 0 and 1 transitions previously assigned by the ETH group[1] shows that for 130 of these the absolute difference between the observed and predicted line wavenumbers is less than 0.1 cm^{-1} and for all but one transition the absolute difference is less than 1 cm^{-1}.

In this project, we are combining the temperature dependence of the line intensities, combination differences, and comparisons of line positions and strengths with the theoretical spectrum to extend the assignments of CH_3D lines in this spectral region. Selected assignments will be confirmed by IR-IR double resonance measurements at the University of Virginia. Ultimately, we hope to give a global analysis of CH_3D spectrum using a global effective Hamiltonian model.

1. Ulenikov, O.N. *et al.*, Molecular Physics **108**, 1209-1240 (2010)

MG07   3:30 – 3:45

3μm - 1.6μm DOUBLE RESONANCE SPECTROSCOPY OF CH_4

GEORGE SCHWARTZ, *Department of Physics, The University of Virginia, Charlottesville, VA, USA*; ERIK BELAAS, *Department of Chemistry, The University of Virginia, Charlottesville, VA, USA*; SHAOYUE YANG, *Department of Physics, The University of Virginia, Charlottesville, VA, USA*; KEVIN LEHMANN, *Department of Chemistry and Physics, The University of Virginia, Charlottesville, VA, USA*.

The Near-IR Spectrum of CH_4 is dense with many overlapping bands that perturb each other by vibrational and ro-vibrational interactions. Assignments of the individual lines are needed in order to simulate the spectrum as a function of pressure and temperature, as needed in the search for CH_4 in extrasolar planets. Both the group at the University College, London[1] and that at the University of Reins[2] have produced theoretical spectra that allows simulation up to the high temperatures expected on "Hot Jupiters". The accuracy of these theoretical spectra need to be further tested.

Because CH_4 is a light spherical top, assignment of its perturbed spectra is a formable challenge as none of the lines allowed in the rigid rotor approximation have ground vibrational state combination differences. We are using IR-IR double resonance to observe modulation in the strength of near-IR absorption caused by a modulation of a 3 μm OPO beam that is tuned to a particular transition in the C-H stretching fundamental of CH_4. This produces V-type double resonance transitions (which share the lower state with the pump transition), which provides firm assignments for lines normally observed in absorption in the near-IR. We also observe sequential double resonance which reveals transitions that have a known rotational level of the ν_3 fundamental as the lower state and reaches final states in the 9000 cm^{-1} spectral region. These are states of A, E, F_1 vibrational symmetries which are forbidden in transitions from the ground vibrational state. These 3 level double resonance transitions are Doppler Free and have a linewidth of $\sim$10 MHz due to a combination of near-IR laser jitter and power broadening of the mid-IR transition. We also observed many 4-level double resonance transitions that we have tentatively assigned as arising from the ν_4 fundamental level. These are distinguished from the 3-level double resonance transitions by they being Doppler broadened and having a large phase shift relative to the intensity modulation.

1. S.N. Yurchenko, PNAS **111** 9379-83 (2014); 2. M. Rey, JQSRT **18**, 207-220 (2015), PCCP **18**, 176-189 (2016)

Intermission

MG08

GLOBAL ANALYSIS OF THE HIGH TEMPERATURE INFRARED EMISSION SPECTRUM OF $^{12}CH_4$ IN THE DYAD (ν_2/ν_4) REGION

BADR AMYAY, MAUD LOUVIOT, *Laboratoire ICB, CNRS/Université de Bourgogne, DIJON, France*; OLIVIER PIRALI, *AILES beamline, Synchrotron SOLEIL, Saint Aubin, France*; ROBERT GEORGES, *IPR UMR6251, CNRS - Université Rennes 1, Rennes, France*; JEAN VANDER AUWERA, *Service de Chimie Quantique et Photophysique, Université Libre de Bruxelles, Brussels, Belgium*; <u>VINCENT BOUDON</u>, *Laboratoire ICB, CNRS/Université de Bourgogne, DIJON, France*.

We report new assignments of vibration-rotation line positions of methane ($^{12}CH_4$) in the so-called Dyad (ν_2/ν_4) region (1000 – 1500 cm^{-1}), and the resulting update of the vibration-rotation effective model of methane, previously reported by Nikitin *et al.* [A.V. Nikitin *et al.* PCCP, **15**, (2013), 10071], up to and including the Tetradecad. High resolution (0.01 cm^{-1}) emission spectra of methane have been recorded up to about 1400 K using the high-enthalpy source developed at IPR associated with the Fourier transform spectrometer of the SOLEIL synchrotron facility (AILES beamline). Analysis of these spectra allowed extending rotational assignments in the well-known cold band (Dyad−GS) and related hot bands in the Pentad−Dyad system (3000 cm^{-1}) up to $J_{max} = 30$ and 29, respectively. In addition, 8512 new transitions belonging to the Octad−Pentad (up to $J = 28$) and Tetradecad−Octad (up to $J = 21$) hot band systems were successfully identified. As a result, the MeCaSDa database of methane was significantly improved. The line positions assigned in this work, together with the information available in the literature, were fitted using 1096 effective parameters with a dimensionless standard deviation $\sigma = 2.09$. The root mean square deviations d_{RMS} are 3.60 $\times 10^{-3}$ cm^{-1} for Dyad−GS cold band, 4.47 $\times 10^{-3}$ cm^{-1} for the Pentad−Dyad, 5.43 $\times 10^{-3}$ cm^{-1} for the Octad−Pentad and 4.70 $\times 10^{-3}$ cm^{-1} for the Tetradecad−Octad hot bands. The resulting new line list will contribute to improve opacity and radiative transfer models for hot atmospheres, such as those of hot-Jupiter type exoplanets.

MG09

GLOBAL ANALYSIS OF SEVERAL BANDS OF THE CF$_4$ MOLECULE

MICKAËL CARLOS, OCÉANE GRUSON, <u>VINCENT BOUDON</u>, *Laboratoire ICB, CNRS/Université de Bourgogne, DIJON, France*; ROBERT GEORGES, *IPR UMR6251, CNRS - Université Rennes 1, Rennes, France*; OLIVIER PIRALI, *AILES beamline, Synchrotron SOLEIL, Saint Aubin, France*; PIERRE ASSELIN, *MONARIS UMR8233, CNRS - UNiversité Paris 6 UPMC, Paris, France*.

Carbon tetrafluoride is a powerful greenhouse gas, mainly of anthropogenic origin. Its absorption spectrum is, however, still badly modeled, especially for hot bands in the strongly absorbing ν_3 region. To overcome this problem, we have undertaken a systematic study of all the lower rovibrational transitions of this molecule. In particular, new far-infrared spectra recorded at the SOLEIL Synchrotron facility give access to bands implying the "forbidden" modes ν_1 and ν_2 which have only been investigated previously thanks to stimulated Raman spectroscopy[a], that is with a lower accuracy and much less data. Combined with the previous analyses performed in our group[b], we thus report here a new global fit of line positions of CF$_4$ by considering several transitions altogether: $\nu_2, 2\nu_2 - \nu_2, \nu_4, 2\nu_4, \nu_3$ and $\nu_3 - 2\nu_2$. This gives a consistent set of molecular parameters that will be of great help for the analysis of hot bands like $\nu_3 + \nu_2 - \nu_2$. A second separate global fit including the $\nu_1, \nu_1 - \nu_4$ and $2\nu_1 - \nu_1$ bands will also be presented.

[a]V. Boudon, D. Bermejo, R. Z. Martínez, J. Raman Spectrosc. **44**, 731?738 (2013).

[b]V. Boudon, J. Mitchell, A. Domanskaya, C. Maul, R; Georges, A. Benidar, W. G. Harter, Mol. Phys. **109**, 17–18 (2011).

MG10 4:38 – 4:53

HIGH RESOLUTION FAR INFRARED SPECTROSCOPY OF HFC-134a AT COLD TEMPERATURES

ANDY WONG[a], CHRIS MEDCRAFT[b], CHRISTOPHER THOMPSON, *School of Chemistry, Monash University, Melbourne, Victoria, Australia*; EVAN GARY ROBERTSON, *Department of Chemistry and Physics, La Trobe Institute for Molecular Sciences, Latrobe University, Melbourne, Australia*; DOMINIQUE APPADOO, *800 Blackburn Road, Australian Synchrotron, Melbourne, Victoria, Australia*; DON McNAUGHTON, *School of Chemistry, Monash University, Melbourne, Victoria, Australia*.

Since the signing of the Montreal protocol, long-lived chlorofluorocarbons have been banned due to their high ozone depleting potential. In order to minimise the effect of such molecules, hydrofluorocarbons (HFCs) were synthesized as replacement molecules to be used as refrigerants and foam blowing agents. HFC-134a, or 1,1,1,2-tetrafluoroethane, is one of these molecules. Although HFCs do not cause ozone depletion, they are typically strong absorbers within the 10 micron atmospheric window, which lead to high global warming potentials. A high resolution FT-IR analysis of the ν_8 band (near $665\ cm^{-1}$) of HFC-134a has been performed to help understand the intermode coupling between the ν_8 vibrational state and unobserved dark states.

[a]Current address: Department of Chemistry and Biology, Old Dominion University, Norfolk, VA, 23519, USA
[b]Current address: Max-Planck-Institut für Struktur und Dynamik der Materie, 22761 Hamburg, Germany

MG11 4:55 – 5:10

ON THE USE OF DIFFERENCE BANDS FOR MODELING SF_6 ABSORPTION IN THE $10\mu m$ ATMOSPHERIC WINDOW

MBAYE FAYE, *AILES beamline, Synchrotron SOLEIL, Saint Aubin, France*; LAURENT MANCERON, *Beamline AILES, Synchrotron SOLEIL, Saint-Aubin, France*; P. ROY, *AILES beamline, Synchrotron SOLEIL, Saint Aubin, France*; VINCENT BOUDON, MICHEL LOETE, *Laboratoire ICB, CNRS/Université de Bourgogne, DIJON, France*.

To model correctly the SF_6 atmospheric absorption requires the knowledge of the spectroscopic parameters of all states involved in the numerous hot bands in the $10,5\mu m$ atmospheric transparency window. However, due to their overlapping, a direct analysis of the hot bands near the $10,5\mu m$ absorption of SF_6 in the atmospheric window is not possible. It is necessary to use another strategy, gathering information in the far and mid infrared regions on initial and final states to compute the relevant total absorption.

In this talk, we present new results from the analysis of spectra recorded at the AILES beamline at the SOLEIL Synchrotron facility. For these measurements, we used a IFS125HR interferometer combined with the synchrotron radiation in the 100-3200 cm^{-1}range, coupled to a cryogenic multiple pass cell [a]. The optical path length was varied from 45 to 141m with measuring temperatures between 223 and 153+/-5 K. The new information obtained on $\nu_2+\nu_4-\nu_5$, $2\nu_5-\nu_6$ and $\nu_3+\nu_6-\nu_4$ allowed to derive improved parameters for ν_5, $2\nu_5$ and $\nu_3+\nu_6$. In turn, they are used to model the more important $\nu_3+\nu_5-\nu_5$ and $\nu_3+\nu_6-\nu_6$ hot band contributions. By including these new parameters in the XTDS model [b], we substantially improved the SF_6 parameters used to model the atmosphere.

[a]F. Kwabia Tchana, F. Willaert, X. Landsheere, J. M. Flaud, L. Lago, M. Chapuis, P. Roy, L. Manceron. A new, low temperature long-pass cell for mid-IR to THz Spectroscopy and Synchrotron Radiation Use. Rev. Sci. Inst. 84, 093101, (2013)
[b]C. Wenger, V. Boudon, M. Rotger, M. Sanzharov, and J.-P. Champion,"XTDS and SPVIEW: Graphical tools for Analysis and Simulation of High Resolution Molecular Spectra", J. Mol. Spectrosc. 251, 102 (2008)

MG12   **5:12 – 5:27**

POSSIBLE MECHANISM FOR SULFUR MASS INDEPENDENT FRACTIONATION IN THE B-X UV TRANSITION OF S_2

ALEXANDER W. HULL, *Department of Chemistry, MIT, Cambridge, MA, USA*; ANDREW RICHARD WHITEHILL, SHUHEI ONO, *Earth, Atmospheric, and Planetary Sciences, MIT, Cambridge, MA, USA*; ROBERT W FIELD, *Department of Chemistry, MIT, Cambridge, MA, USA.*

Anomalous sulfur isotope ratios, called mass independent fractionation (MIF), are commonly observed in sedimentary rocks formed more than 2.5 billion years ago. These anomalies likely originated from photochemistry of small, sulfur-containing molecules in the early Earth's atmosphere. The disappearance of MIF in rocks younger than 2.5 billion years is thought to be evidence of rising O_2 concentrations during the Great Oxygenation Event (GOE), an important milestone in the development of life on Earth. However, the photochemical origin of the pre-GOE anomaly is not well understood. Here, we use a model of the X, B, and B" states of S_2, originally developed by Western, to determine a possible mechanism for isotopologue-selective photodissociation. A model of the rotation-vibration structure of the B-X UV transition shows small perturbations between the bright B state and the dark B" state that vary depending on isotopologue. These perturbations suggest a sequential two-photon mechanism for selective photodissociation. Symmetry (e.g., 32-32 vs. 32-34) may also contribute to MIF. This presentation will primarily focus on the UV spectra of the 32-32 and 32-34 isotopologues. We also examine the possibility that a similar mechanism may be at work in the B-X transition of SO.

MH. Ions

Monday, June 20, 2016 – 1:30 PM

Room: 274 Medical Sciences Building

Chair: Etienne Garand, University of Wisconsin-Madison, Madison, WI, USA

MH01   1:30 – 1:45

COMBINED DUNHAM ANALYSIS Of ROTATIONAL AND ELETRONIC TRANSITIONS OF CH$^+$

SHANSHAN YU, BRIAN DROUIN, JOHN PEARSON, <u>TAKAYOSHI AMANO</u>, *Jet Propulsion Laboratory, California Institute of Technology, Pasadena, CA, USA.*

A Dunham analysis of the $A^1\Pi - X^1\Sigma^+$ band was carried out by Müller, and predictions of the pure rotational transition frequencies were made.[a] More recently, in submillimeter to THz region, several rotational lines were observed for ^{12}CH$^+$, ^{13}CH$^+$, and CD$^+$.[bcd] In this investigation, those newly obtained rotational lines are incorporated in the Dunham analysis. The Λ-doubling splittings in $^1\Pi$ electronic states have been expressed as $(1/2)qJ(J+1)$ in most investigations. However, it should be noted that the e-levels of $^1\Pi$ state interact with $^1\Sigma^+$ states, while the f-levels with $^1\Sigma^-$ states. For CH$^+$, the e-levels of $A^1\Pi$ are pushed upward from the interaction with the ground $X^1\Sigma^+$ state. The $^1\Sigma^-$ states are not known experimentally and they, if any, should lie high over the $A^1\Pi$ state. In this analysis, only the f-levels are included in the least-squares analysis by neglecting the Λ-doubling. The mass independent parameters have been obtained, and the conventional spectroscopic parameters are derived for each isotopologue. These results should be useful for determining the potential energies of this fundamental ion.[e]

---

[a]H. S. P. Müller,*A&A,514*, L6(2010)

[b]T. Amano, *Ap.J.Lett.*, **716**, L1 (2010)

[c]T. Amano, *J. Chem. Phys.*, **133**, 244305 (2010)

[d]S. Yu et al,*The 70th International Symposium on Moecular Spectroscopy*, **RD06**(2015)

[e]Y. S. Cho and R. J. LeRoy,*J. Chem. Phys.*,**144**, 024311(2016)

MH02   1:47 – 2:02

PRECISION SATURATED ABSORPTION SPECTROSCOPY OF H$_3^+$

<u>YU-CHAN GUAN</u>, YI-CHIEH LIAO, YUNG-HSIANG CHANG, *Institute of Photonics Technologies, National Tsing Hua University, Hsinchu, Taiwan*; JIN-LONG PENG, *Center for Measurement Standards, Industrial Technology Research Institute, Hsinchu, Taiwan*; JOW-TSONG SHY, *Department of Physics, National Tsing Hua University, Hsinchu, Taiwan.*

In our previous work on the Lamb dips of the ν_2 fundamental band of H$_3^+$, the saturated absorption spectrum was obtained by the third-derivative spectroscopy using frequency modulation [1]. However, the frequency modulation also causes error in absolute frequency determination. To solve this problem, we have built an offset-locking system to lock the OPO pump frequency to an iodine-stabilized Nd:YAG laser. With this modification, we are able to scan the OPO idler frequency precisely and obtain the profile of the Lamb dips. Double modulation (amplitude modulation of the idler power and concentration modulation of the ion) is employed to subtract the interference fringes of the signal and increase the signal-to-noise ratio effectively. To Determine the absolute frequency of the idler wave, the pump wave is offset locked on the R(56) 32-0 a$_{10}$ hyperfine component of ^{127}I$_2$, and the signal wave is locked on a GPS disciplined fiber optical frequency comb (OFC). All references and lock systems have absolute frequency accuracy better than 10 kHz. Here, we demonstrate its performance by measuring one transition of methane and sixteen transitions of H$_3^+$. This instrument could pave the way for the high-resolution spectroscopy of a variety of molecular ions.

[1] H.-C. Chen, C.-Y. Hsiao, J.-L. Peng, T. Amano, and J.-T. Shy, Phys. Rev. Lett. 109, 263002 (2012).

MH03   2:04 – 2:19

HIGHLY ACCURATE AND PRECISE INFRARED TRANSITION FREQUENCIES OF THE H_3^+ CATION

ADAM J. PERRY, CHARLES R. MARKUS, JAMES N. HODGES, G. STEPHEN KOCHERIL, *Department of Chemistry, University of Illinois at Urbana-Champaign, Urbana, IL, USA*; BENJAMIN J. McCALL, *Departments of Chemistry and Astronomy, University of Illinois at Urbana-Champaign, Urbana, IL, USA.*

Calculation of *ab initio* potential energy surfaces for molecules to high accuracy is only manageable for a handful of molecular systems. Among them is the simplest polyatomic molecule, the H_3^+ cation. In order to achieve a high degree of accuracy (<1 cm^{-1}) corrections must be made to the to the traditional Born-Oppenheimer approximation that take into account not only adiabatic and non-adiabatic couplings, but quantum electrodynamic corrections as well. For the lowest rovibrational levels the agreement between theory and experiment is approaching 0.001 cm^{-1}, whereas the agreement is on the order of $0.01 - 0.1$ cm^{-1} for higher levels which are closely rivaling the uncertainties on the experimental data[a][b]. As method development for calculating these various corrections progresses it becomes necessary for the uncertainties on the experimental data to be improved in order to properly benchmark the calculations.

Previously we have measured 20 rovibrational transitions of H_3^+ with MHz-level precision, all of which have arisen from low lying rotational levels[c][d]. Here we present new measurements of rovibrational transitions arising from higher rotational and vibrational levels. These transitions not only allow for probing higher energies on the potential energy surface, but through the use of combination differences, will ultimately lead to prediction of the "forbidden" rotational transitions with MHz-level accuracy.

[a]L.G. Diniz, J.R. Mohallem, A. Alijah, M. Pavanello, L. Adamowicz, O.L. Polyansky, J. Tennyson *Phys. Rev. A* (2013), **88**, 032506.

[b]O.L. Polyansky, A. Alijah, N.F. Zobov, I.I. Mizus, R.I. Ovsyannikov, J. Tennyson, L. Lodi, T. Szidarovszky, A.G. Császár *Phil. Trans. R. Soc. A* (2012), **370**, 5014.

[c]J.N. Hodges, A.J. Perry, P.A. Jenkins II, B.M. Siller, B.J. McCall *J. Chem. Phys.* (2013), **139**, 164201.

[d]A.J. Perry, J.N. Hodges, C.R. Markus, G.S. Kocheril, B.J. McCall *J. Molec. Spectrosc.* (2015), **317**, 71-73.

MH04   2:21 – 2:36

ROTATIONALLY RESOLVED PHOTOELECTRON SPECTROSCOPIC STUDY OF THE $\tilde{A}^+$ STATE OF H_2O^+

CLÉMENT LAUZIN[a], BERENGER GANS[b], UGO JACOVELLA, FREDERIC MERKT, *Laboratorium für Physikalische Chemie, ETH Zurich, Zurich, Switzerland.*

This talk will present the analysis of the rotationally resolved pulsed-field-ionization zero-kinetic-energy (PFI-ZEKE) photoelectron spectrum of H_2O and will be focussed on the $\tilde{A}^+ \leftarrow \tilde{X}$ transitions. H_2O^+ in the $\tilde{A}^+$ state is predicted to be linear [c]. The sensitivity and the high resolution of PFI- ZEKE photoelectron spectroscopy allowed us to observe the rotational structure of low bending vibrational levels of the $\tilde{A}^+$ state of H_2O^+ from the $\tilde{X}$ ground electronic state of H_2O. The assignment of the rotational structure of ionic levels previously observed by optical spectroscopy of the $\tilde{A}^+$ - $\tilde{X}^+$ band system of H_2O^+ [d][e] will be presented and the intensity distribution of the photoelectron spectrum will be discussed in terms of the even or odd nature of the orbital angular momentum quantum number l of the photoelectron. Tentative assignments will be presented for several low-lying vibrational levels of the $\tilde{A}^+$ state and compared with theoretical predictions [c]. They will also be discussed in terms of the rotational structure of higher $\tilde{A}^+$ vibrational levels of the same symmetry.

[a]Institue of Condensed Matter and Nanosciences (IMCN), Université catholique de Louvain, Louvain-la-Neuve, Belgium.

[b]Institut des Sciences Moléculaires d'Orsay (ISMO), CNRS, Univ. Paris-Sud, Université Paris-Saclay, F-91405 Orsay (France)

[c]M. Brommer, B. Weis, B. Follmeg, P. Rosmus, S. Carter, N. C. Handy, H. J. Werner, and P. J. Knowles, J. Chem. Phys. 98, 5222 (1993)

[d]T .Huet, I. H. Bachir, J. L. Destombes, and M. Vervloet, J. Chem. Phys. 107,5645 (1997).

[e]H. Lew, Can. J. Phys. 54, 2028 (1976).

MH05   2:38 – 2:48

STUDIES OF 4-CHLORO-2-FLUOROANISOLE BY TWO-COLOR RESONANT TWO-PHOTON MASS-ANALYZED THRESHOLD IONIZATION SPECTROSCOPY

PEI-YING WU, WEN-BIH TZENG, *Institute of Atomic and Molecular Sciences, Academia Sinica, Taipei, Taiwan.*

We applied the two-color resonant two-photon mass analyzed threshold ionization technique to record the cation spectra of 4-chloro-2-fluoroanisole by ionizing via five intermediate vibronic levels. The excitation and adiabatic ionization energies were determined to be 35 227, and 67 218 cm^{-1}, respectively. Spectral analysis and theoretical calculation suggest that the geometry of the aromatic ring of the neutral species in the S_1 state is non-planar, but that of the cation in the D_0 state is planar.

MH06   2:50 – 3:00

VIBRONIC AND CATION SPECTROSCOPY OF 3,5-DIFLUOROPHENOL

WEI-CHIH PENG, SHEN-YUAN TZENG, WEN-BIH TZENG, *Institute of Atomic and Molecular Sciences, Academia Sinica, Taipei, Taiwan.*

The vibronic, photoionization efficiency, and cation spectra of 3,5-difluorophenol have been recorded by using the resonant two-photon ionization and mass-analyzed threshold ionization techniques . The distinct spectral features mainly result from the in-plane substituent-sensitive bending and ring deformation vibrations of these species in the electronically excited S_1 and cationic ground D_0 states. Comparing these results with those of previously reported 2,4-difluorophenol, 2,5-difluorophenol, and 3,5-difluorophenol, we gain knowledge about the substituent effect on the transition energy and molecular vibration of these positional isomers.

Intermission

MH07   3:19 – 3:34

IDENTIFICATION OF STRUCTURAL MOTIFS OF IMIDAZOLIUM BASED IONIC LIQUIDS FROM JET-COOLED INFRARED SPECTROSCOPY.

JUSTIN W. YOUNG, RYAN S BOOTH, CHRISTOPHER ANNESLEY, JAIME A. STEARNS, *Space Vehicles Directorate, Air Force Research Lab, Kirtland AFB, NM, USA.*

Highly variable and potentially revolutionary, ionic liquids (IL) are a class of molecules with potential for numerous Air Force applications such as satellite propulsion, but the complex nature of IL structure and intermolecular interactions makes it difficult to adequately predict structure-property relationships in order to make new IL-based technology a reality. For example, methylation of imidazolium ionic liquids leads to a substantial increase in viscosity but the underlying physical mechanism is not understood. In addition the role of hydrogen bonding in ILs, especially its relationship to macroscopic properties, is a matter of ongoing research. Here, structural motifs are identified from jet-cooled infrared spectra of different imidazolium based ionic liquids, such as 1-ethyl-3-methylimidazolium bis(trifluoromethyl-sulfonyl)imide. Measurements of the C-H stretches indicate three structural families present in the gas phase.

MH08   3:36 – 3:51

Ce-PROMOTED BOND ACTIVATION OF PROPENE PROBED BY MASS-ANALYZED THRESHOLD IONIZATION SPECTROSCOPY

YUCHEN ZHANG, SUDESH KUMARI, DONG-SHENG YANG, *Department of Chemistry, University of Kentucky, Lexington, KY, USA.*

The reaction of Ce + propene (CH_2=CH-CH_3) was carried out in a laser-ablation supersonic molecular beam source. CeC_2H_2, CeC_3H_4, CeC_3H_6, CeC_4H_6, CeC_6H_{10}, and CeC_6H_{12} were identified by photoionization time-of-flight mass spectrometric measurements, and their structures and electronic states were investigated with mass-analyzed threshold ionization (MATI) spectroscopy and theoretical calculations. The metal complexes containing two or three carbon atoms were formed by the C-C bond breakage (CeC_2H_2), dehydrogenation (CeC_3H_4), or metal insertion into a C-H bond (CeC_3H_6) of a propene molecule. The larger complexes with four to six carbons are formed through secondary reactions involving C-C bond coupling and dehydrogenation. The ground electronic states of the neutral CeC_2H_2, CeC_3H_4, CeC_3H_6, and CeC_4H_6 complexes are triplets with a $4f^1 6s^1$ electron configuration on the Ce center, and those of the corresponding ions are doublet with a $4f^1$ configuration. Their MATI spectra are much more complex than those of the corresponding La species formed in the La + propene reaction previously observed by our group. The spectral complexity arises from possibly multiple electronic transitions due to the existence of a 4f electron of the Ce atom which could be located in any one of the seven f-atomic orbitals or involved in considerable spin-orbit interactions.

MH09   **3:53 – 4:08**

SPECTROSCOPIC IDENTIFICATION OF Y(C$_4$H$_6$) ISOMERS FORMED BY YTTRIUM-MEDIATED C-H BOND AC-
TIVATION OF BUTENES

JONG HYUN KIM, DONG-SHENG YANG, *Department of Chemistry, University of Kentucky, Lexington, KY, USA.*

Y(C$_4$H$_6$) was observed from the reactions of laser-vaporized Y atom with 1-butene (CH$_2$=CHCH$_2$CH$_3$) and iso-butene (CH$_2$=C(CH$_3$)$_2$) in a pulsed molecular beam source, and its structural isomers were investigated with mass-analyzed threshold ionization spectroscopy combined with electronic structure calculations and spectral simulations. Y(C$_4$H$_6$) was identified as a five-membered metallacycle [Y(CH$_2$-CH=CH-CH$_2$)] from the Y + 1-butene reaction and a tetrahedral structure [YC(CH$_2$)$_3$] from the Y + iso-butene reaction. The metallacycle has a C$_s$ structure with Y binding to the two terminal carbon atoms, whereas the tetrahedron has C$_{3v}$ symmetry with Y binding to the tertiary carbon atom of trimethylenemethane. Both isomers have a doublet ground state with the highest molecular orbital being largely a Y 5s character. Ionization removes the metal based electron, and the resultant singlet ion has a similar structure to the neutral complex. However, the adiabatic ionization energy [46309(5) cm^{-1}] of the tetrahedron is considerably higher than that [43473(5) cm^{-1}] of the cyclic structure.

MH10   **4:10 – 4:25**

INFRARED LASER PHOTODISSOCIATION SPECTROSCOPY OF METAL-ACETYLENE CATIONS IN THE GAS
PHASE

ANTONIO DAVID BRATHWAITE, *College of Science and Mathematics, University of the Virgin Islands, St. Thomas, USVI*; TIMOTHY B WARD, MICHAEL A DUNCAN, *Department of Chemistry, University of Georgia, Athens, GA, USA.*

Mass-selected metal-acetylene cation complexes of the form M(C$_2$H$_2$)$_n$$^+$ (M = Cu, Au and V) are produced by laser ablation and studied via infrared laser photodissociation spectroscopy in the C-H stretching region. Spectra for larger species are measured via ligand elimination, whereas argon tagging is employed to enhance dissociation yields in smaller complexes. The number of infrared active bands, their frequency positions and their relative intensities provide insight into the structure, bonding and reactivity of these ions. Density functional theory calculations are carried out in support of this work. For Cu(C$_2$H$_2$)$_n$$^+$, the combined data shows that cation-π bonds are formed for the n=1-3 species, resulting in red-shifted C-H stretches on the acetylene ligands. Three acetylene ligands complete the coordination of the copper cation. Additional ligands (n=4-6) solvate the n=3 core by forming CH-π bonds. Comparison of experimental and theoretical spectra for V(C$_2$H$_2$)$_n$$^+$ complexes in the C-H stretching region suggests the formation of exotic metallacycles.

MH11   **4:27 – 4:42**

VELOCITY MAP IMAGING STUDIES OF Mg$^+$-L (L = Ar, N$_2$, CO$_2$) COMPLEXES

JON MANER, DANIEL MAUNEY, MICHAEL A DUNCAN, *Department of Chemistry, University of Georgia, Athens, GA, USA.*

Mg$^+$-L (L = Ar, N$_2$, CO$_2$) complexes are formed by laser vaporization of a magnesium rod in a supersonic expansion of argon, nitrogen, or carbon dioxide. Complexes are pulse-extracted into a time-of-flight mass spectrometer, mass selected, and dissociated with 266 nm light. The ionic fragments are then accelerated toward a two-dimensional detector using an assembly of electrostatic lenses configured for velocity map imaging. While dissociation of Mg$^+$Ar and Mg$^+$N$_2$ results in simple ligand elimination, dissociation of Mg$^+$CO$_2$ produces both Mg$^+$ and MgO$^+$ fragments. Images of these fragments are collected using the DC-slice technique and compared with images for dissociation of Mg$^+$Ar and Mg$^+$N$_2$ to explain the photochemistry of the Mg$^+$CO$_2$ system. Upper limits for the binding energies of these systems are estimated by measuring the maximum kinetic energy released in their photofragments and applying a thermochemical cycle.

MH12

VIBRATIONAL SIGNATURES OF LARGE AMPLITUDE MOTIONS FOR THE SHACKLED HYDRONIUM ION NESTED IN 18-CROWN-6 ETHER USING D_2 TAGGING

CHINH H. DUONG, FABIAN MENGES, STEPHANIE CRAIG, CONRAD T. WOLKE, MARK JOHNSON, *Department of Chemistry, Yale University, New Haven, CT, USA.*

The diffuse spectra arising from the excess proton in dilute acids suggests that its behavior is highly dependent on the local environment surrounding it. In this work, we report how the spectra of the H_3O^+, NH_4^+, and $CH_3NH_3^+$ ions respond when docked to the rigid, tri-coordinated binding pocket of the 18-crown-6 ether using cryogenic ion vibrational predissociation (CIVP) spectroscopy with D_2 tagging at 10 K. The $H_3O^+ \cdot 18$-crown-6 ether complex displays a broad (350 cm^{-1} FWHM) unstructured band arising from the OH stretching fundamentals, which is significantly broader than the corresponding band (125 cm^{-1} FWHM) in the Eigen cation ($H_9O_4^+$) spectrum. Perdeuterated isotopologue studies for both systems yield sharper bands with clear multiplet structures, indicating that the broadening arises from nuclear quantum effects. The key displacements underlying this coupling were explored using the vibrationally adiabatic scheme introduced by McCoy in the context of similar broadening in the $Ca_2^+OH^-(H_2O)_n$ system.[a]

---

[a]Christopher J. Johnson, Laura C. Dzugan, Arron B. Wolk, Christopher M. Leavitt, Joseph A. Fournier, Anne B. McCoy, Mark A. Johnson, *J. Phys. Chem. A* **118**, 2014.

MH13

SPECTROSCOPIC OBSERVATION OF WATER-MEDIATED DEFORMATION OF THE CARBOXYLATE-M^{2+} (M= Mg, Ca) CONTACT ION PAIR

PATRICK J KELLEHER, JOSEPH W DePALMA, MARK JOHNSON, *Department of Chemistry, Yale University, New Haven, CT, USA.*

The binding of alkaline earth dications to the biologically relevant carboxylate ligand has previously been studied using vibrational sum frequency generation (VSFG) spectroscopy of the air-water interface, infrared multiple photon dissociation (IRMPD) spectroscopy of clusters, and DFT methods. These results suggest the presence of both monodentate and bidentate binding motifs of the M^{2+} ions to the cayboxyl head groups depending on the extent of solvation. We revisit these systems using vibrational predissociation spectroscopy to measure the gas-phase vibrational spectra of the D_2-tagged microhydrated $[MgOAc(H_2O)_{n=1-5}]^+$ and $[CaOAc(H_2O)_{n=1-6}]^+$ clusters. The spectra show that $[MgOAc(H_2O)_n]^+$ switches from bidentate to monodentate binding promptly at n = 5, while $[CaOAc(H_2O)_n]^+$ retains its bidentate attachment such that the sixth water molecule initiates the second solvation shell. The difference in binding behavior between these two divalent metal ions is analyzed in the context of the local acidity of the solvent water molecules and the strength of the metal-carboxylate and metal-water interactions. This cluster study provides insight into the chemical physics underlying the unique and surprising impacts of Mg^{2+} and Ca^{2+} on the chemistry mediated by sea spray aerosols. Funding for this work was provided by the NSF's Center for Aerosol Impacts on Climate and the Environment.

MH14
5:18 – 5:33

IRMPD ACTION SPECTROSCOPY AND COMPUTATIONAL APPROACHES TO ELUCIDATE GAS-PHASE STRUCTURES AND ENERGETICS OF 2′-DEOXYCYTIDINE AND CYTIDINE SODIUM COMPLEXES

YANLONG ZHU, LUCAS HAMLOW, CHENCHEN HE, *Department of Chemistry, Wayne State University, Detroit, MI, USA*; JUEHAN GAO, *Institute for Molecules and Materials (IMM), Radboud University Nijmegen, Nijmegen, Netherlands*; JOS OOMENS, *FELIX Laboratory, Radboud University, Nijmegen, The Netherlands*; M T RODGERS, *Department of Chemistry, Wayne State University, Detroit, MI, USA*.

The local structures of DNA and RNA are influenced by protonation, deprotonation and noncovalent interactions with cations. In order to determine the effects of Na^+ cationization on the gas-phase structures of 2′-deoxycytidine, $[dCyd+Na]^+$, and cytidine, $[Cyd+Na]^+$, infrared multiple photon dissociation (IRMPD) action spectra of these sodium cationized nucleosides are measured over the range extending from 500 to 1850 cm^{-1} using the FELIX free electron laser. Complementary electronic structure calculations are performed to determine the stable low-energy conformations of these complexes. Geometry optimizations, frequency analyses, and IR spectra of these species are determined at the B3LYP/6-311+G(d,p) level of theory. Single-point energies are calculated at the B3LYP/6-311+G(2d,2p) level of theory to determine the relative stabilities of these conformations. Comparison of the measure IRMPD action spectra and computed linear IR spectra enable the conformations accessed in the experiments to be elucidated. For both cytosine nucleosides, tridentate binding of the Na^+ cation to the O2, O4′ and O5′ atoms of the nucleobase and sugar is observed. Present results for the sodium cationized nucleosides are compared to results for the analogous protonated forms of these nucleosides to elucidate the effects of multiple chelating interactions with the sodium cation vs. hydrogen bonding interactions in the protonated systems on the structures and stabilities of these nucleosides.

MH15
5:35 – 5:50

INFLUENCE OF 5-HALOGENATION ON THE STRUCTURE OF PROTONATED URIDINE: IRMPD ACTION SPECTROSCOPY AND THEORETICAL STUDIES OF THE PROTONATED 5-HALOURIDINES

HARRISON ROY, LUCAS HAMLOW, JUSTIN LEE, M T RODGERS, *Department of Chemistry, Wayne State University, Detroit, MI, USA*; GIEL BERDEN, JOS OOMENS, *Institute for Molecules and Materials (IMM), Radboud University Nijmegen, Nijmegen, Netherlands*.

The chemical and structural diversity and the extent of post-transcriptional modification of RNA is remarkable! Presently, there are 142 different naturally-occurring and many more synthetically modified nucleosides known. Uridine (Urd) is the most commonly modified nucleoside among those that occur naturally, but has also been an important target for synthesis and development of modified nucleosides for pharmaceutical applications. Indeed, modified nucleosides are of pharmaceutical interest due to their bioactivities. In particular, 5-bromouridine (br^5Urd) has been shown to exhibit antiviral activity to human immunodeficiency virus and has been used in RNA labeling studies. Halogenation is a common modification employed in pharmaceutical studies that enables systematic variation is the electronic properties of the molecule of interest due to the availability of halogen substituents that vary in size, dipole moment, polarizability, and electron withdrawing properties. In order to elucidate the influence of 5-halogenation on the intrinsic gas-phase structure and stability on the protonated form of Urd, synergistic spectroscopic and theoretical studies of the protonated forms of the 5-halouridines are performed here, where x^5Urd = 5-fluorouridine (f^5Urd), 5-chlorouridine (cl^5Urd), br^5Urd, and 5-iodouridine (i^5Urd). Infrared multiple photon dissociation (IRMPD) action spectra of the protonated forms of the 5-halouridines, $[x^5Urd+H]^+$, are measured over the IR fingerprint region using the FELIX free electron laser and the hydrogen stretching region using an OPO/OPA laser from 3300-3800 cm^{-1}. Complementary electronic structure calculations are performed to determine the stable low-energy conformations available to these species and to predict their IR spectra. Comparative analyses of the measured IRMPD spectra and predicted IR spectra are performed to elucidate the preferred sites of protonation, and the low-energy tautomeric conformations that are populated by electrospray ionization to be determined. Comparisons among these systems and to results previously reported for the protonated form of uridine, $[Urd+H]^+$, provides insight into the impact of the 5-halogen substituent on the structures and IR signatures.

MI. Structure determination
Monday, June 20, 2016 – 1:30 PM
Room: B102 Chemical and Life Sciences

Chair: Stewart E. Novick, Wesleyan University, Middletown, CT, USA

MI01   1:30 – 1:45

THE MOLECULAR STRUCTURE OF PHENETOLE STUDIED BY MICROWAVE SPECTROSCOPY AND QUANTUM CHEMICAL CALCULATIONS

LYNN FERRES, WOLFGANG STAHL, *Institute for Physical Chemistry, RWTH Aachen University, Aachen, Germany*; HA VINH LAM NGUYEN, *Laboratoire Interuniversitaire des Systèmes Atmosphériques (LISA), CNRS et Universités Paris Est et Paris Diderot, Créteil, France.*

A pulsed molecular beam Fourier transform microwave spectrometer operating in the frequency range 2 - 26.5 GHz was used to measure the spectrum of phenetole (ethyl phenyl ether or ethoxybenzene, $C_6H_5OC_2H_5$). The conformational landscape is completely determined by the orientations of the phenyl ring and the ethyl group. A two-dimensional potential energy surface was calculated at the MP2/6-311++G(d,p) level of theory. Two conformers were found: The trans conformer has a C_s symmetry, and the gauche conformer has the ethyl group tilted out of the phenyl plane by about $70°$.

Totally 186 rotational transitions were assigned to the more stable planar trans conformer, and fitted using a semi-rigid rotor model to measurement accuracy of 2 kHz. Highly accurate rotational and centrifugal distortion constants were determined. Several method and basis set combinations were applied to check for convergence and to compare with the experimentally deduced molecular parameters. The inertial defect of the observed conformer $\Delta_c = (I_c - I_a - I_b) = -6.718$ u$Å^2$ confirms that the heavy atom skeleton is planar with two pairs of hydrogen atoms out of plane. All lines in the spectrum could be assigned to the trans conformer, which confirms that the gauche conformer cannot be observed under our measurement conditions. In agreement with the rather high torsional barrier of the methyl group ($V_3 = 1168$ cm^{-1}) calculated by quantum chemical methods, all assigned lines appeared sharp and no signs of splittings were observed for the methyl internal rotation.

MI02   1:47 – 2:02

CP-FTMW SPECTROSCOPY OF A CLAISEN REARRANGEMENT PRECURSOR ALLYL PHENYL ETHER

G. S. GRUBBS II, *Department of Chemistry, Missouri University of Science and Technology, Rolla, MO, USA*; DEREK S. FRANK, DANIEL A. OBENCHAIN, *Department of Chemistry, Wesleyan University, Middletown, CT, USA*; S. A. COOKE, *Natural and Social Science, Purchase College SUNY, Purchase, NY, USA*; STEWART E. NOVICK, *Department of Chemistry, Wesleyan University, Middletown, CT, USA.*

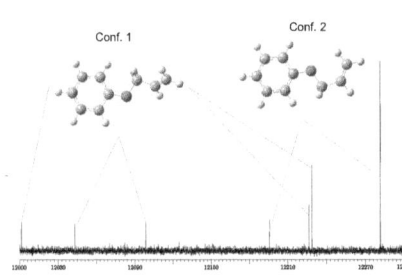

The pure rotational spectrum of a Claisen rearrangement precursor, allyl phenyl ether (APE), has been measured on a chirped pulse Fourier transform microwave (CP-FTMW) spectrometer in the 8-14 GHz region. Rotational and centrifugal distortion constants for multiple conformations have been determined for the first time and will be discussed. This is the first study of a phenyl-containing ether where multiple conformers were experimentally observed all within their ground vibrational states. Quantum chemical calculations have been performed to isolate low energy geometries of APE and are implemented to aid in spectral assignment. Other structural parameters such as planar moments and inertial defects for the APE conformers are presented and compared to similar molecules for discussion.

MI03   2:04 – 2:19

DETERMINATION OF THE PREFERRED STRUCTURE, DYNAMICS, AND PLANARITY OF SUBSTITUTED ANHYDRIDES BY CP-FTMW

TIMOTHY J McMAHON, JOSIAH R BAILEY, RYAN G BIRD, *Chemistry, University of Pittsburgh Johnstown, Johnstown, PA, USA*; DAVID PRATT, *Chemistry, University of Vermont, Burlington, VT, USA.*

The planarity of five-membered rings is derived from a competition between ring-angle strain and stability of the torsional angles. The planar form maximizes the already stressed, smaller-than-normal, C-C bond angles, while puckering reduces the unfavorable eclipsed interactions. The structure, dynamics, and planarity of three anhydrides, succinic, methylsuccinic, and methylene (itaconic) anhydride, were studied and compared using chirped-pulse Fourier transform microwave spectroscopy.

MI04 2:21 – 2:36

STRUCTURAL ANALYSIS OF 2-FLUOROPHENOL AND 3-FLUOROPHENOL USING FTMW SPECTROSCOPY

AIMEE BELL, OMAR MAHASSNEH, JAMES SINGER, JENNIFER VAN WIJNGAARDEN, *Department of Chemistry, University of Manitoba, Winnipeg, MB, Canada.*

The ground states of 2-fluorophenol (2-FPh) and 3-fluorophenol (3-FPh) were studied using Fourier transform microwave (FTMW) techniques to record their rotational spectra in the range of 6 to 26 GHz. Two planar conformers of similar energy were observed for 3-FPh (cis, trans) while only the lowest energy conformer (cis) of 2-FPh was observed due to the stabilizing effect of an intramolecular interaction between fluorine and OH. Rotational constants derived from spectra of the ^{13}C analogs, observed in natural abundance, were used to calculate the substitution (r_s) and effective ground state (r_0) parameters for cis-2-FPh and trans-3-FPh in order to study the effect of fluorination at sites along the benzene backbone. Geometry optimization at the MP2/6-311++G(2d,2p) level was used obtain the equilibrium (r_e) structures for comparison. Furthermore, natural bond order (NBO) calculations provided supporting information of a OH•••F interaction in cis-2-FPh.

MI05 2:38 – 2:53

STRUCTURAL EXPRESSION OF EXO-ANOMERIC EFFECT

E. R. ALONSO, ISABEL PEÑA, CARLOS CABEZAS, JOSÉ L. ALONSO, *Grupo de Espectroscopia Molecular, Lab. de Espectroscopia y Bioespectroscopia, Unidad Asociada CSIC, Universidad de Valladolid, Valladolid, Spain.*

Structural signatures for exo–anomeric effect have been extracted from the archetypal methyl–β–D–xyloside using broadband Fourier transform microwave spectroscopy combined with laser ablation. Spectrum analysis allows the determination of a set of rotational constants, which has been unequivocally attributed to conformer cc–β-^{4}C$_1$ g-, corresponding to the global minimum of the potential energy surface, where the aglycon residue (CH$_3$) orientation contributes towards maximization of the exo-anomeric effect. Further analysis allowed the determination of the r_s structure, based on the detection of eleven isotopologues - derived from the presence of six ^{13}C and five ^{18}O atoms - observed in their natural abundances. The observed glycosidic C_1-O_1 bond length decrease (1.38 Å) can be interpreted in terms of the exo–anomeric effect. As such, the exo–anomeric effect presents itself as one of the main driving forces controlling the shape of many biologically important oligosaccharides.

MI06 2:55 – 3:10

MICROWAVE SPECTRA, MOLECULAR STRUCTURE AND AROMATIC CHARACTER OF BN-NAPHTHALENE (4A,8A-AZABORANAPHTHALENE)[a]

AARON M PEJLOVAS, STEPHEN G. KUKOLICH, *Department of Chemistry and Biochemistry, University of Arizona, Tucson, AZ, USA*; ARTHUR J. ASHE III, *Department of Chemistry, University of Michigan, Ann Arbor, MI, USA*; ADAM M DALY, *Department of Chemistry and Biochemistry, University of Arizona, Tucson, AZ, USA.*

The microwave spectra for seven unique isotopologues of BN-naphthalene (4a,8a-azaboranaphthalene) were measured using a pulsed-beam Fourier transform microwave spectrometer. Spectra were obtained for the normal isotopologues with ^{10}B, ^{11}B, all unique single ^{13}C and the ^{15}N isotopologue, in natural abundance. The rotational, centrifugal distortion and quadrupole coupling constants determined for the ^{11}B^{14}N isotopologue are A = 3042.7128(4) MHz, B = 1202.7066(4) MHz, C = 862.2201(4) MHz, D$_J$ = 0.06(1) kHz, 1.5 eQq$_{aa}$ (^{14}N) = 2.578(6) MHz, 0.25(eQq$_{bb}$- eQq$_{cc}$) (^{14}N) = -0.119(2) MHz, 1.5 eQq$_{aa}$ (^{11}B) = -3.922(8) MHz, and 0.25(eQq$_{bb}$- eQq$_{cc}$) (^{11}B) = -0.907(2) MHz. The experimental inertial defect is Δ = -0.159 amu Å^2, which is consistent with a planar structure. The B-N bond length is 1.47 Å, indicating π-bonding character. The results are compared with similar results for B-N bonding in 1,2-dihydro-1,2-azaborine and BN-cyclohexene.

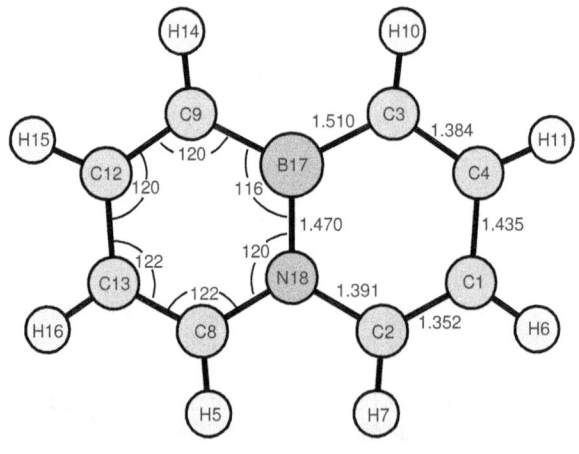

[a] Supported by the NSF CHE-1057796

MI07

THE STRUCTURE AND MOLECULAR PARAMETERS OF CAMPHENE DETERMINED BY FOURIER TRANSFORM
MICROWAVE SPECTROSCOPY AND QUANTUM CHEMICAL CALCULATIONS

ELIAS M. NEEMAN, PASCAL DRÉAN, T. R. HUET, *UMR 8523 CNRS - Universités des Sciences et Technologies de Lille, Laboratoire PhLAM, 59655 Villeneuve d'Ascq, France.*

The emission of volatile organic compounds, from plants has strong revelance for plant physiology, plant ecology and atmospheric chemistry.[a] Camphene ($C_{10}H_{16}$) is a bicyclic monoterpene which is emitted in the atmosphere by biogenic sources.[b,c] The structure of the unique stable conformer was optimized using density functional theory and *ab initio* calculations. The rotational spectrum of camphene was recorded in a supersonic jet expansion with a Fourier transform microwave spectrometer over the range 2-20 GHz. Signals from the parent species and from the ten ^{13}C isotopomers were observed in natural abundance. The rotational and centrifugal distortion parameters were fitted to a Watson's Hamiltonian in the A-reduction. A magnetic hyperfine structure associated with the pairs of hydrogen nuclei in the methylene groups was observed and modeled.

The rotational constants coupled to the equilibrium structure calculations were used to determine the r_0 and the $r_m^{(1)}$ gas-phase geometries of the carbon skeleton. The present work provides the first spectroscopic characterization of camphene in the gas phase and these results are also relevant for ozonolysis kinetics study through Criegee intermediates.[d]

[a]R. Baraldi, F. Rapparini, O. Facini, D. Spano and P. Duce, *Journal of Mediterranean Ecology*, **Vol.6**, No.1, (2005).
[b]A. Bracho-Nunez, N. M. Knothe, S. Welter, M. Staudt, W. R. Costa, M. A. R. Liberato, M. T. F. Piedade, and J. Kesselmeier *Biogeosciences*, **10**, 5855-5873, (2013).
[c]Minna Kivimäenpää, Narantsetseg Magsarjav, Rajendra Ghimire, Juha-Matti Markkanen, Juha Heijari, Martti Vuorinen and Jarmo K. Holopainen, *Atmospheric Environment*, **60**, 477-485, (2012).
[d]R.C. de M. Oliveira and G. F. Bauerfeldt, *J. Phys. Chem. A*, **119** 2802-2812 (2015).

MI08

N-METHYL INVERSION IN PSEUDO-PELLETIERINE

MONTSERRAT VALLEJO-LÓPEZ, *Physical Chemistry, University of the Basque Country, Leioa Bilbao, Spain*; PATRICIA ECIJA, EMILIO J. COCINERO, *Physical Chemistry Department, Universidad del País Vasco, Bilbao, Spain*; ALBERTO LESARRI, *Departamento de Química Física y Química Inorgánica, Universidad de Valladolid, Valladolid, Spain*; FRANCISCO J. BASTERRETXEA, JOSÉ A. FERNÁNDEZ, *Physical Chemistry Department, Universidad del País Vasco, Bilbao, Spain.*

We have previously conducted rotational studies of several tropanes, [a,b,c] since this bicyclic structural motif forms the core of different alkaloids of pharmaceutical interest. Now we report on the conformational properties and molecular structure of pseudo-pelletierine (9-methyl-9-azabicyclo[3.3.1]nonan-3-one), probed in a jet expansion with Fourier-transform microwave spectroscopy. Pseudo-pelletierine is an azabicycle with two fused six-membered rings, where the *N*-methyl group can produce inverting axial o equatorial conformations. The two conformations were detected in the rotational spectrum, recorded in the region 6-18 GHz. Unlike tropinone and *N*-methylpiperidone, where the most stable conformer is equatorial, the axial species was found dominant for pseudo-pelletierine. All monosubstituted isotopic species (^{13}C, ^{15}N and ^{18}O) were identified for the axial conformer, leading to an accurate determination of the effective and substitution structures. An estimation of conformational populations was derived from relative intensities. The experimental results will be compared with *ab initio* (MP2) and DFT (M06-2X, B3LYP) calculations.

[a]E. J. Cocinero, A. Lesarri, P. Écija, J.-U. Grabow, J. A. Fernández, F. Castaño, *Phys. Chem. Chem. Phys.* **2010**, *49*, 4503
[b]P. Écija, E. J. Cocinero, A. Lesarri, F. J. Basterretxea, J. A. Fernández, F. Castaño, *Chem. Phys. Chem.* **2013**, *14*, 1830
[c]P. Écija, M. Vallejo-Lopez, I. Uriarte, F. J. Basterretxea, A. Lesarri, J. A. Fernández, E. J. Cocinero, *submitted* **2016**

Intermission

MI09   3:58 – 4:13

MICROWAVE SPECTRA FOR THE THREE $^{13}C_1$ ISOTOPOLOGUES OF PROPENE AND NEW ROTATIONAL CONSTANTS FOR PROPENE AND ITS $^{13}C_1$ ISOTOPOLOGUES

NORMAN C. CRAIG, *Department of Chemistry and Biochemistry, Oberlin College, Oberlin, OH, USA*; PETER GRONER, *Department of Chemistry, University of Missouri - Kansas City, Kansas City, MO, USA*; ANDREW R CONRAD, RANIL M. GURUSINGHE, MICHAEL TUBERGEN, *Department of Chemistry and Biochemistry, Kent State University, Kent, OH, USA.*

New measurements of microwave lines (A and E) of propene and its three $^{13}C_1$ isotopologues have been made in the 10-22 GHz region with FT accuracy. The revised lines for propene along with many hundreds from the literature were fitted with the ERHAM program for internal rotors to give improved rotational constants. The new constants for propene are A_0 = 46280.2904(16), B_0 = 9305.24260(30), and C_0 = 8134.22685(28) MHz. Lines for the 3-$^{13}C_1$ species were observed in a pure sample; lines for the 1-$^{13}C_1$ and 2-$^{13}C_1$ species were observed in natural abundance. In fitting the limited sets of lines for the $^{13}C_1$ species, many of the centrifugal distortion constants and most of the tunneling parameters were transferred from the fit of propene itself with 27 parameters. Improved rotational constants for the $^{13}C_1$ species are reported.

MI10   4:15 – 4:30

MICROWAVE SPECTRA FOR THE TWO CONFORMERS OF PROPENE-3-d_1 AND NEW ROTATIONAL CONSTANTS FOR THESE SPECIES

NORMAN C. CRAIG, *Department of Chemistry and Biochemistry, Oberlin College, Oberlin, OH, USA*; RANIL M. GURUSINGHE, MICHAEL TUBERGEN, *Department of Chemistry and Biochemistry, Kent State University, Kent, OH, USA.*

Propene-3-d_1 has two conformers: a single sy conformer with the CD bond eclipsing the C=C bond; two equivalent asy conformers with the CD bond out of the symmetry plane of the sy conformer. In the MW spectrum the sy conformer has single transitions of a semi-rigid rotor. The asy conformer has (+) and (-) transitions as a consequence of tunneling. A MW investigation with FT accuracy and sensitivity was done in the 10-22 GHz region, where the light propene molecule has relatively few transitions strong enough to be observed. The lines were split by deuterium quadrupolar coupling. Rotational constants for the sy species are: A_0 = 40582.1(2), B_0 = 9067.04(1), and C_0 = 7766.02(1) MHz. We have observed the (+) and (-) transitions for the asy conformer in the same spectral region. Rotational constants fit to the averages of these lines are: A_0 = 43403.8(1), B_0 = 8659.00(2), and C_0 = 7718.24(2) MHz. In fitting for both conformers, the cw lines observed by Lide and Christensen for J = 1 to 2; K_a = 0, 1 transitions were used to strengthen the data sets.[a] These new rotational constants will be used in determining a semiexperimental equilibrium structure for propene.

---

[a]D. R. Lide, Jr and D. Christensen, J. Chem. Phys. 35, 1374-1378 (1961).

MI11   4:32 – 4:47

RESOLVING A LONG-STANDING AMBIGUITY: THE NON-PLANARITY OF *gauche*-1,3-BUTADIENE REVEALED BY MICROWAVE SPECTROSCOPY

MARIE-ALINE MARTIN-DRUMEL, MICHAEL C McCARTHY, *Atomic and Molecular Physics, Harvard-Smithsonian Center for Astrophysics, Cambridge, MA, USA*; DAVID PATTERSON, SANDRA EIBENBERGER, *Department of Physics, Harvard University, Cambridge, MA, USA*; GRANT BUCKINGHAM, JOSHUA H BARABAN, BARNEY ELLISON, *Department of Chemistry and Biochemistry, University of Colorado, Boulder, CO, USA*; JOHN F. STANTON, *Department of Chemistry, The University of Texas, Austin, TX, USA.*

The preferred conformation of *cis*-1,3-butadiene ($CH_2=CH-CH=CH_2$) has been of long-standing importance in organic chemistry because of its role in Diels-Alder transition states. The molecule could adopt a planar *s-cis* conformation, in favor of conjugations in the carbon chain, or a non-planar *gauche* conformation, as a result of steric interactions between the terminal H atoms.

To resolve this ambiguity, we have now measured the pure rotational spectrum of this isomer in the microwave region, unambiguously establishing a significant inertial defect, and therefore a *gauche* conformation. Experimental measurements of *gauche*-1,3-butadiene and several of its isotopologues using cavity Fourier-transform microwave (FTMW) spectroscopy in a supersonic expansion and chirped-pulse FTMW spectroscopy in a 4 K buffer gas cell will be summarized, as will new quantum chemical calculations.

MI12 4:49 – 5:04

SINGLE-CONFORMATION IR AND UV SPECTROSCOPY OF A PROTOTYPICAL HETEROGENEOUS α/β-PEPTIDE: IS IT A MIXED-HELIX FORMER?

KARL N. BLODGETT, PATRICK S. WALSH, TIMOTHY S. ZWIER, *Department of Chemistry, Purdue University, West Lafayette, IN, USA.*

Synthetic foldamers are non-natural polymers designed to fold into unique secondary structures that either mimic nature's preferred secondary structures, or expand their possibilities. Among the most studied synthetic foldamers are β-peptides, which lengthen the distance between amide groups from the single substituted carbon spacer in α-peptides by one additional carbon. We present data on a mixed α/β tri-peptide in which a single β-residue with a conformationally constrained cis-2-aminocyclohexanecarboxylic acid (cis-ACHC) substitution is inserted in an α-peptide backbone to form Ac-Ala-β-ACHC-Ala-NHBn. This $\alpha\beta\alpha$ structure is known in longer sequences to prefer formation of a 9/11 mixed helix. Under isolated, jet cooled conditions, four unique conformers were observed in the expansion. The dominant conformer is configured in a tetramer cycle with every amide carbonyl and amine group involved in hydrogen bonding, giving rise to a tightly folded C12/C7/C8/C7 structure reminiscent of a β-turn. This talk will describe the conformation specific IR and UV spectroscopy methods used to study this mixed peptide, as well as its experimentally observed conformational preferences.

MI13 5:06 – 5:21

STRUCTURES AND NUCLEAR QUADRUPOLE COUPLING TENSORS OF A SERIES OF CHLORINE-CONTAINING HYDROCARBONS

ASELA S. DIKKUMBURA, ERICA R WEBSTER, RACHEL E. DORRIS, REBECCA A. PEEBLES, SEAN A. PEEBLES, *Department of Chemistry, Eastern Illinois University, Charleston, IL, USA;* NATHAN A SEIFERT, BROOKS PATE, *Department of Chemistry, The University of Virginia, Charlottesville, VA, USA.*

Rotational spectra for *gauche*-1,2-dichloroethane (12DCE), *gauche*-1-chloro-2-fluoroethane (1C2FE) and both *anti*- and *gauche*-2,3-dichloropropene (23DCP) have been observed using chirped-pulse Fourier-transform microwave (FTMW) spectroscopy in the 6-18 GHz region. Although the *anti* conformers for all three species are predicted to be more stable than the *gauche* forms, they are nonpolar (12DCE) or nearly nonpolar (predicted dipole components for *anti*-1C2FE: $\mu_a = 0.11$ D, $\mu_b = 0.02$ D and for *anti*-23DCP: $\mu_a = 0.25$ D, $\mu_b = 0.02$ D); nevertheless, it was also possible to observe and assign the spectrum of *anti*-23DCP. Assignments of parent spectra and ^{37}Cl and ^{13}C substituted isotopologues utilized predictions at the MP2/6-311++G(2d,2p) level and Pickett's SPCAT/SPFIT programs. For the weak *anti*-23DCP spectra, additional measurements also utilized a resonant-cavity FTMW spectrometer. Full chlorine nuclear quadrupole coupling tensors for *gauche*-12DCE and both *anti*- and *gauche*-23DCP have been diagonalized to allow comparison of coupling constants. Kraitchman's equations were used to determine r_s coordinates of isotopically substituted atoms and r_0 structures were also deduced for *gauche* conformers of 12DCE and 1C2FE. Structural details and chlorine nuclear quadrupole coupling constants of all three molecules will be compared, and effects of differing halogen substitution and carbon chain length on molecular properties will be evaluated.

MI14 5:23 – 5:38

INFRA-RED SPECTRA OF SMALL ALKANES INTERACTING WITH ALUMINUM IONS

MUHAMMAD AFFAWN ASHRAF, CHRISTOPHER COPELAND, RICARDO B. METZ, *Department of Chemistry, University of Massachusetts, Amherst, MA, USA.*

Herein we present experimentally determined infra-red spectra of entrance channel complexes of various small alkanes interacting with alkanes. The entrance channel complexes of ethane, propane and n-butane with positively charged aluminum ions were studied in the gas phase. The spectra are compared with theoretical calculations, and the structures of various products are elucidated from the spectra.

MI15   **5:40 – 5:55**

QUANTUM CHEMICAL STUDIES ON THE PREDICTION OF STRUCTURES, CHARGE DISTRIBUTIONS AND VI-
BRATIONAL SPECTRA OF SOME Ni(II), Zn(II), AND Cd(II) IODIDE COMPLEXES

TAYYIBE BARDAKCI[a], *Faculty of Medicine, Istanbul University, Istanbul, Turkey*; MUSTAFA KUMRU, *Department of Physics, Fatih University, Istanbul, Turkey*; AHMET ALTUN, *Department of Genetics and Bioengineering, Fatih University, Istanbul, Turkey.*

Transition metal complexes play an important role in coordination chemistry as well as in the formation of metal-based drugs. In order to obtain accurate results for studying these type of complexes quantum chemical studies are performed and especially density functional theory (DFT) has become a promising choice. This talk represents molecular structures, charge distributions and vibrational analysis of Ni(II), Zn(II), and Cd(II) iodide complexes of p-toluidine and m-toluidine by means of DFT. Stable structures of the ligands and the related complexes have been obtained in the gas phase at B3LYP/def2-TZVP level and calculations predict Ni(II) complexes as distorted polymeric octahedral whereas Zn(II) and Cd(II) complexes as distorted tetrahedral geometries. Charge distribution analysis have been performed by means of Mulliken, NBO and APT methods and physically most meaningful method for our compounds is explained. Vibrational spectra of the title compounds are computed from the optimized geometries and theoretical frequencies are compared with the previously obtained experimental data. Since coordination occurs via nitrogen atoms of the free ligands, N-H stretching bands of the ligands are shifted towards lower wavenumbers in the complexes whereas NH_2 wagging and twisting vibrations are shifted towards higher wavenumbers.

---

[a]tayyibe.b@gmail.com

MJ. Conformers, isomers, chirality, stereochemistry

Monday, June 20, 2016 – 1:30 PM

Room: 217 Noyes Laboratory

Chair: Daniel A. Obenchain, Wesleyan University, Middletown, CT, USA

MJ01   1:30 – 1:45

CHIRPED PULSE MICROWAVE SPECTROSCOPY ON METHYL BUTANOATE

ALICIA O. HERNANDEZ-CASTILLO, <u>BRIAN M HAYS</u>, CHAMARA ABEYSEKERA, TIMOTHY S. ZWIER, *Department of Chemistry, Purdue University, West Lafayette, IN, USA.*

The microwave spectrum of methyl butanoate has been taken from 8-18 GHz using a chirped pulse spectrometer. This molecule is a model biofuel, and its thermal decomposition products are of interest due to its many dissociation channels. As a preliminary step before such pyrolysis studies, we have examined the jet cooled spectrum of methyl butanoate in a chirped pulse spectrometer, which shows a very rich spectrum. Several conformers have been identified, each with tunneling splittings in the methyl ester group due to internal rotation. These spectra have been fit to obtain rotational constants, relative populations, and methyl rotor barriers for each conformational isomer. The results of these studies are compared to high level calculations.

MJ02   1:47 – 2:02

LOCAL ANESTHETICS IN THE GAS-PHASE: THE ROTATIONAL SPECTRUM OF BUTAMBEN AND ISOBUTAMBEN

<u>MONTSERRAT VALLEJO-LÓPEZ</u>, *Physical Chemistry, University of the Basque Country, Leioa Bilbao, Spain*; PATRICIA ECIJA, *Physical Chemistry Department, Universidad del País Vasco, Bilbao, Spain*; WALTHER CAMINATI, *Dep. Chemistry 'Giacomo Ciamician', University of Bologna, Bologna, Italy*; JENS-UWE GRABOW, *Institut für Physikalische Chemie und Elektrochemie, Gottfried-Wilhelm-Leibniz-Universität, Hannover, Germany*; ALBERTO LESARRI, *Departamento de Química Física y Química Inorgánica, Universidad de Valladolid, Valladolid, Spain*; EMILIO J. COCINERO, *Physical Chemistry Department, Universidad del País Vasco, Bilbao, Spain.*

Benzocaine (BZ), butamben (BTN) and isobutamben (BTI) are local anesthetics characterized by a hydrophilic head and a lipophilic aliphatic tail linked by an aminobenzoate group. Previous rotational work on BZ ($H_2N-C_6H_4-COO-Et$)[a] showed that its ethyl aliphatic tail may adopt either in-plane (trans) or out of plane (gauche) conformations, with a low interconversion barrier below 50 cm^{-1}.[b] Here we extend the rotational study to BTN and BTI, isolated in a supersonic jet expansion and vaporized either by heating or UV ps-laser ablation methods. Both molecules share a 14 heavy-atoms skeleton, differing in their butyl (-$(CH_2)_3$-CH_3) or isobutyl (-CH_2-$CH(CH_3)_2$) four-carbon tail. We detected a single conformer for BTN and two conformers for BTI. The two molecules do not adopt an all-trans carbon skeleton. Conversely, the β-ethyl carbon in BTN is gauche. For BTI the β-carbon may be either trans or gauche. The microwave spectrum covered the cm- (BTN, BTI, 6-18 GHz) and mm-wave (BTW, 50-75 GHz) frequency ranges.In all the cases, rotational and centrifugal distortion constants as well as the diagonal elements of the ^{14}N nuclear quadrupole coupling tensor were accurate determined and compared to the theoretical results (ab initio and DFT). No transitions belonging to configurations predicted as higher minima of the PES were found, pointing out that conformational interconversions may take place in the jet.

---

[a] A. Lesarri, S. T. Shipman, G. G. Brown, L. Alvarez-Valtierra, R. D. Suenram, B. H. Pate, Int. Symp. Mol. Spectrosc., 2008, Comm. RH07.
[b] E. Aguado, A. Longarte, E. Alejandro, J. A. Fernández, F. Castaño, J. Phys. Chem. A, 2006, 110, 6010.

MJ03

WE ARE FAMILY: THE CONFORMATIONS OF 1-FLUOROALKANES, $C_nH_{2n+1}F$ (n = 2,3,4,5,6,7,8)

DANIEL A. OBENCHAIN, *Department of Chemistry, Wesleyan University, Middletown, CT, USA*; W. OREL-LANA, S. A. COOKE[a], *Natural and Social Science, Purchase College SUNY, Purchase, NY, USA.*

The pure rotational spectra of the n = 5, 6, 7, and 8 members of the 1-fluoroalkane family have been recorded between 7 GHz and 14 GHz using chirped pulse Fourier transform microwave spectroscopy. The spectra have been analyzed and results will be presented and compared with previous work on the n= 2, 3, and 4 members[b]. The lowest energy conformer for all family members has the common feature that the fluorine is in a gauche position relative to the alkyl tail for which all other heavy atom dihedral angles, where appropriate, are 180 degrees. For the n = 3 and higher family members the second lowest energy conformer has *all* heavy atom dihedral angles equal to 180 degrees. For each family member transitions carried by both low energy conformers were observed in the collected rotational spectra. Quantum chemical calculations were performed and trends in the energy separations between these two common conformers will be presented as a function of chain length. Furthermore, longer chain lengths have been examined using only quantum chemical calculations and results will be presented.

[a]Supported by the Petroleum Research Fund administered by the American Chemical Society, Award No. 53451-UR6

[b]M. Hayashi, M. Fujitake, T. Inagusa, S. Miyazaki, J.Mol.Struct., 216, 9-26, 1990 ; W. Caminati, A. C. Fantoni, F. Manescalchi, F. Scappini, Mol.Phys., 64, 1089 ,1988 ; L. B. Favero, A. Maris, A. Degli Esposti, P. G. Favero, W. Caminati, G. Pawelke, Chem.Eur.J., 6(16), 3018-3025, 2000

MJ04

LARGE MOLECULE STRUCTURES BY BROADBAND FOURIER TRANSFORM MOLECULAR ROTATIONAL SPECTROSCOPY

LUCA EVANGELISTI, *Dipartimento di Chimica G. Ciamician, Università di Bologna, Bologna, Italy*; NATHAN A SEIFERT, *Department of Chemistry, University of Alberta, Edmonton, AB, Canada*; LORENZO SPADA, *Dep. Chemistry 'Giacomo Ciamician', University of Bologna, Bologna, Italy*; BROOKS PATE, *Department of Chemistry, The University of Virginia, Charlottesville, VA, USA.*

Fourier transform molecular rotational resonance spectroscopy (FT-MRR) using pulsed jet molecular beam sources is a high-resolution spectroscopy technique that can be used for chiral analysis of molecules with multiple chiral centers. The sensitivity of the molecular rotational spectrum pattern to small changes in the three dimensional structure makes it possible to identify diastereomers without prior chemical separation. For larger molecules, there is the additional challenge that different conformations of each diastereomer may be present and these need to be differentiated from the diastereomers in the spectral analysis. Broadband rotational spectra of several larger molecules have been measured using a chirped-pulse FT-MRR spectrometer. Measurements of nootkatone ($C_{15}H_{22}O$), cedrol ($C_{15}H_{26}O$), ambroxide ($C_{16}H_{28}O$) and sclareolide ($C_{16}H_{26}O_2$) are presented. These spectra are measured with high sensitivity (signal-to-noise ratio near 1,000:1) and permit structure determination of the most populated isomers using isotopic analysis of the ^{13}C and [18]O isotopologues in natural abundance. The accuracy of quantum chemistry calculations to identify diastereomers and conformers and to predict the dipole moment properties needed for three wave mixing measurements is examined.

MJ05   2:38 – 2:53

CHIRAL ANALYSIS OF ISOPULEGOL BY FOURIER TRANSFORM MOLECULAR ROTATIONAL SPECTROSCOPY

LUCA EVANGELISTI, *Dipartimento di Chimica G. Ciamician, Università di Bologna, Bologna, Italy*; NATHAN A SEIFERT, *Department of Chemistry, University of Alberta, Edmonton, AB, Canada*; LORENZO SPADA, *Dep. Chemistry 'Giacomo Ciamician', University of Bologna, Bologna, Italy*; BROOKS PATE, *Department of Chemistry, The University of Virginia, Charlottesville, VA, USA*.

Chiral analysis on molecules with multiple chiral centers can be performed using pulsed-jet Fourier transform rotational spectroscopy. This analysis includes quantitative measurement of diastereomer products and, with the three wave mixing methods developed by Patterson, Schnell, and Doyle (Nature 497, 475-477 (2013)), quantitative determination of the enantiomeric excess of each diastereomer. The high resolution features enable to perform the analysis directly on complex samples without the need for chromatographic separation. Isopulegol has been chosen to show the capabilities of Fourier transform rotational spectroscopy for chiral analysis. Broadband rotational spectroscopy produces spectra with signal-to-noise ratio exceeding 1000:1. The ability to identify low-abundance (0.1-1%) diastereomers in the sample will be described. Methods to rapidly identify rotational spectra from isotopologues at natural abundance will be shown and the molecular structures obtained from this analysis will be compared to theory. The role that quantum chemistry calculations play in identifying structural minima and estimating their spectroscopic properties to aid spectral analysis will be described. Finally, the implementation of three wave mixing techniques to measure the enantiomeric excess of each diastereomer and determine the absolute configuration of the enantiomer in excess will be described.

MJ06   2:55 – 3:10

ROTATIONAL SPECTROSCOPY OF TETRAHYDRO-2-FUROIC ACID, ITS CHIRAL AGGREGATES AND ITS COMPLEX WITH WATER

JAVIX THOMAS, WOLFGANG JÄGER, YUNJIE XU, *Department of Chemistry, University of Alberta, Edmonton, AB, Canada*.

Rotational spectra of Tetrahydro-2-furoic acid (THA), a chiral acid, and its homo- and heterochiral dimers, and its complex with water have been recorded using a chirped pulse Fourier transform microwave spectrometer. This chiral acid was predicted to have nine conformers, although only the most stable one was detected experimentally and its rotational spectrum assigned. We have analyzed its intramolecular H-bonding pattern in detail. Eleven conformers have been predicted for the 1:1 hydration complex between THA and water and 14 conformers for $(THA)_2$. The assignments of these complexes are currently underway and will be presented.

Intermission

MJ07 3:29 – 3:44

PROBING THE CONFORMATIONAL LANDSCAPE OF POLYETHER BUILDING BLOCKS IN SUPERSONIC JETS

SEBASTIAN BOCKLITZ, *Institute of Physical Chemistry, Georg-August-Universität Göttingen, Göttingen, Germany*; DANIEL M. HEWETT, TIMOTHY S. ZWIER, *Department of Chemistry, Purdue University, West Lafayette, IN, USA*; MARTIN A. SUHM, *Institute of Physical Chemistry, Georg-August-Universität Göttingen, Göttingen, Germany*.

Polyethylene oxides (Polyethylene glycoles) and their phenoxy-capped analogs represent a prominent class of important polymers that are highly used as precursor molecules in supramolecular reactions. After a detailed study on the simplest representative (1,2-dimethoxyethane) [1], we present results on oligoethylene oxides with increasing chain lengths obtained by spontaneous Raman scattering in a supersonic jet.

Through variation of stagnation pressure, carrier gas, nozzle distance and temperature we gain information on the conformational landscape as well as the mutual interconversion of low energy conformers. The obtained results are compared to state-of-the-art quantum chemical calculations.

Additionally, we present UV as well as IR-UV and UV-UV double resonance studies on 1-methoxy-2-phenoxyethane in a supersonic jet. These complementary techniques allow for conformationally selective electronic and vibrational spectra in a closely related conformational landscape.

[1] S. Bocklitz, M. A. Suhm, Constraining the Conformational Landscape of a Polyether Building Block by Raman Jet Spectroscopy, *Z. Phys. Chem.* **2015**, *229*, 1625-1648.

MJ08 3:46 – 4:01

MODELING THE CONFORMATION-SPECIFIC INFRARED SPECTRA OF N-ALKYLBENZENES

DANIEL P. TABOR, EDWIN SIBERT, *Department of Chemistry, University of Wisconsin–Madison, Madison, WI, USA*; DANIEL M. HEWETT, JOSEPH A. KORN, TIMOTHY S. ZWIER, *Department of Chemistry, Purdue University, West Lafayette, IN, USA*.

Conformation-specific UV-IR double resonance spectra are presented for n-alkylbenzenes. With the aid of a local mode Hamiltonian that includes the effects of stretch-bend Fermi coupling, the spectra of ethyl, n-propyl, and n-butylbenzene are assigned to individual conformers. These molecules allow for further development of the work on a first principles method for calculating alkyl stretch spectra. Due to the consistency of the anharmonic couplings from conformer to conformer, construction of the model Hamiltonian for a given conformer only requires a harmonic frequency calculation at the conformer's minimum geometry as an input. The model Hamiltonian can be parameterized with either density functional theory or MP2 electronic structure calculations. The relative strengths and weaknesses of these methods are evaluated, including their predictions of the relative energetics of the conformers. Finally, the IR spectra for conformers that have the alkyl chain bend back and interact with the π cloud of the benzene ring are modeled.

MJ09 4:03 – 4:18

WHERE'S THE BEND? LOCATING THE FIRST FOLDED STRUCTURE IN STRAIGHT CHAIN ALKYLBENZENES IN A SUPERSONIC JET EXPANSION

DANIEL M. HEWETT, *Department of Chemistry, Purdue University, West Lafayette, IN, USA*; SEBASTIAN BOCKLITZ, MARTIN A. SUHM, *Institute of Physical Chemistry, Georg-August-Universität Göttingen, Göttingen, Germany*; TIMOTHY S. ZWIER, *Department of Chemistry, Purdue University, West Lafayette, IN, USA*.

Alkylbenzenes make up 20-30% of petroleum fuels and are important intermediates in combustion. In gasoline, these alkyl chains are relatively short, but extend to 20 or more carbons in length in diesel fuels. While one tends to think of these chains as extending out away from the phenyl ring in an all-trans configuration, dispersive interactions between segments of the alkyl chain and between the alkyl chain and the ring will stabilize more compact geometries in which the alkyl chain folds back on itself and extends over the aromatic π cloud. This talk seeks to answer the following question: How long must the alkyl chain be before it starts to fold back over itself? Studies of the pure n-alkanes by the Suhm group have shown the turn to favorably occur for a chain about 17 carbon atoms in length. The studies presented here focus on the affect the aromatic ring has on when this turn becomes favorable. Jet-cooled laser-induced fluorescence excitation and single-conformation IR spectra have been recorded in the alkyl CH stretch region for a series of alkylbenzenes with chain lengths ranging from two to ten carbon atoms. We show, through a combination of experiment, high level calculation, and theoretical modeling, that conformations begin to form that fold back over the aromatic ring at about n=8.

MJ10 4:20 – 4:35

INTRINSIC OPTICAL ACTIVITY AND ENVIRONMENTAL PERTURBATIONS: SOLVATION EFFECTS IN CHIRAL BUILDING BLOCKS

PAUL M LEMLER, PATRICK VACCARO, *Department of Chemistry, Yale University, New Haven, CT, USA*.

The non-resonant interaction of electromagnetic radiation with an isotropic ensemble of chiral molecules, which causes the incident state of linear polarization to undergo a signed rotation, long has served as a metric for gauging the enantiomeric purity of asymmetric syntheses. While the underlying phenomenon of circular birefringence (CB) typically is probed in the condensed phase, recent advances in ultrasensitive circular-differential detection schemes, as exemplified by the techniques of Cavity Ring-Down Polarimetry (CRDP), have permitted the first quantitative analyses of such processes to be performed in rarefied media. Efforts to extend vapor-phase investigations of CB to new families of chiral substrates will be discussed, with particular emphasis directed towards the elucidation of intrinsic (e.g., solvent-free) properties and their mediation by environmental perturbations (e.g., solvation). Specific species targeted by this work will include the stereoselective building blocks phenylpropylene oxide and α-methylbenzyl amine, both of which exhibit pronounced solvent-dependent changes in measured optical activity. The nature of chiroptical response in different environments will be highlighted, with quantum-chemical calculations serving to unravel the structural and electronic provenance of observed behavior.

MJ11 4:37 – 4:52

DOUBLE-RESONANCE FACILITATED DECOMPOSION OF EMISSION SPECTRA

RYOTA KATO, HARUKI ISHIKAWA, *Department of Chemistry, School of Science, Kitasato University, Sagamihara, Japan.*

Emission spectra provide us with rich information about the excited-state processes such as proton-transfer, charge-transfer and so on. In the cases that more than one excited states are involved, emission spectra from different excited states sometimes overlap and a decomposition of the overlapped spectra is desired. One of the methods to perform a decomposition is a time-resolved fluorescence technique. It uses a difference in time evolutions of components involved. However, in the gas-phase, a concentration of the sample is frequently too small to carry out this method. On the other hand, double-resonance technique is a very powerful tool to discriminate or identify a common species in the spectra in the gas-phase. Thus, in the present study, we applied the double-resonance technique to resolve the overlapped emission spectra. When transient IR absorption spectra of the excited state are available, we can label the population of the certain species by the IR excitation with a proper selection of the IR wavenumbers. Thus, we can obtain the emission spectra of labeled species by subtracting the emission spectra with IR labeling from that without IR.

In the present study, we chose the charge-transfer emission spectra of cyanophenyldisilane (CPDS) as a test system. One of us reported that two charge-transfer (CT) states are involved in the intramolecular charge-transfer (ICT) process of CPDS-water cluster and recorded the transient IR spectra[a]. As expected, we have succeeded in resolving the CT emission spectra of CPDS-water cluster by the double resonance facilitated decomposion technique. In the present paper, we will report the details of the experimental scheme and the results of the decomposition of the emission spectra.

[a]H. Ishikawa, *et al.*, *Chem. Phys. Phys. Chem.*, **9**, 117 (2007).

MJ12 4:54 – 5:09

A CHIRPED PULSE FOURIER TRANSFORM MICROWAVE (CP-FTMW) SPECTROMETER WITH LASER ABLATION SOURCE TO SEARCH FOR ACTINIDE-CONTAINING MOLECULES AND NOBLE METAL CLUSTERS

FRANK E MARSHALL, DAVID JOSEPH GILLCRIST, THOMAS D. PERSINGER, NICOLE MOON, G. S. GRUBBS II, *Department of Chemistry, Missouri University of Science and Technology, Rolla, MO, USA.*

Microwave spectroscopic techniques have traditionally been part of the foundation of molecular structure and this conference. Instrumental developments by Brooks Pate and sourcing developments by Steve Cooke on these instruments have allowed for the dawning of a new era in modern microwave spectroscopic techniques.[ab] With these advances and the growth of powerful computational approaches, microwave spectroscopists can now search for molecules and/or cluster systems of actinide and noble metal-containing species with increasing certainty in molecular assignment even with the difficulties presented with spin-orbit coupling and relativistic effects. Spectrometer and ablation design will be presented along with any preliminary results on actinide-containing molecules or noble metal clusters or interactions.

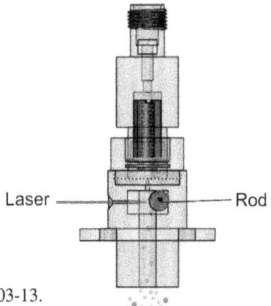

[a]G. G. Brown, B. C. Dian, K. O. Douglass, S. M. Geyer, S. T. Shipman, B. H. Pate, *Rev. Sci. Instrum.* **79** (2008) 053103-1 – 053103-13.
[b]G. S. Grubbs II, C. T. Dewberry, K. C. Etchison, K. E. Kerr, S. A. Cooke, *Rev. Sci. Instrum.* **78** (2007) 096106-1 – 096106-3.

ISOMERIZATION AND FRAGMENTATION OF CYCLOHEXANONE IN A HEATED MICRO-REACTOR

JESSICA P PORTERFIELD, *Department of Chemistry, University of Colorado, Boulder, CO, USA*; THANH LAM NGUYEN, *Department of Chemistry, The University of Texas, Austin, TX, USA*; JOSHUA H BARABAN, *Department of Chemistry, University of Colorado, Boulder, CO, USA*; GRANT BUCKINGHAM, *Department of Chemistry and Biochemistry, University of Colorado, Boulder, CO, USA*; TYLER TROY, OLEG KOSTKO, *Chemical Science Division, Lawrence Berkeley National Laboratory, Berkeley, CA, USA*; MUSAHID AHMED, *UXSL, Chemical Sciences Division, Lawrence Berkeley National Laboratory, Berkeley, CA, USA*; JOHN F. STANTON, *Department of Chemistry, The University of Texas, Austin, TX, USA*; JOHN W DAILY, *Department of Mechanical Engineering, University of Colorado Boulder, Boulder, CO, USA*; BARNEY ELLISON, *Department of Chemistry and Biochemistry, University of Colorado, Boulder, CO, USA*.

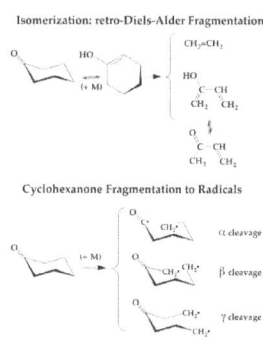

The thermal decomposition of cyclohexanone (C_6H_{10}=O) has been studied in a set of flash-pyrolysis micro-reactors. Samples of C_6H_{10}=O were first observed to decompose at 1200 K. Short residence times of 100 μsec and dilution of samples (<0.1%) isolate unimolecular decomposition. Products were identified by tunable VUV photoionization mass spectroscopy, photoionization appearance thresholds, and complementary matrix infrared absorption spectroscopy. Thermal cracking of cyclohexanone appeared to result from a variety of competing pathways pictured to the right. Isomerization of cyclohexanone to the enol, cyclohexen-1-ol (C_6H_9OH), is followed by retro-Diels-Alder cleavage to CH_2=CH_2 and CH_2=$C(OH)$-CH=CH_2. Further isomerization of CH_2=$C(OH)CH$=CH_2 to methyl vinyl ketone (CH_3COCH=CH_2, MVK) was also observed. Photoionization spectra identified both enols, C_6H_9OH and CH=$C(OH)CH$=CH_2, and the ionization threshold of C_6H_9OH was measured to be 8.2 ± 0.1 eV. At 1200 K, the products of cyclohexanone pyrolysis were found to be: C_6H_9OH, CH_2=$C(OH)CH$=CH_2, MVK, CH_2CHCH_2, CO, CH_2=C=O, CH_3, CH_2=C=CH_2, CH_2=CH-CH=CH_2, CH_2=$CHCH_2CH_3$, CH_2=CH_2, and HCCH.

THEORY OF MICROWAVE 3-WAVE MIXING OF CHIRAL MOLECULES

KEVIN LEHMANN, *Departments of Chemistry and Physics, University of Virginia, Charlottesville, VA, USA*.

The traditional spectroscopic methods to measure enantiomeric excess, based upon optical rotation or circular dichroism arise from an interference of electric and magnetic dipole contributions of an optical transitions. The later is relativisitic and gets smaller with decreasing frequency and thus these effects have not been previously observed in pure rotational spectroscopy. First introduced by the group at Harvard[1], it is possible to use a 3-wave mixing method (with one of the fields potentially a Stark Field) to distinguish enantiomers if the three wave are nonplaner. In the conceptually simplest form of this experiment, a molecule is polarized with X polarization on a $a \rightarrow b$ transition, and then the resulting ρ_{ab} molecular coherence is transferred to a ρ_{ac} coherence by application of a π pulse on the $b \rightarrow c$ transition. For a chiral molecule with nonzero dipole projections on the three inertial axes, this ρ_{ac} coherence can radiate Z polarized emission at the frequency of the $a \rightarrow c$ transition.

In this talk, I will present the full theory of such experiments, including accounting for dirrection cosine matrix elements and M degeneracy. The resulting expressions can be used to calculate the expected size of the signal as a function of the specific transitions used in the $a \rightarrow b \rightarrow c \rightarrow a$ cycle.[2] It will be demonstrated that the maximum size of the ρ_{ac} coherence is nearly that generated by a "$\pi/2$" pulse on the $a \rightarrow c$ transition. However, it is not possible to phase match the emission generated by this polarization due to the requirement that the three fields be orthogonal. Given that in rotational spectroscopy the physical size of the sample produced in a pulsed supersonic jet is comparable to the wavelengths of the microwave fields, the lack of phase matching produces a substantial but not catastrophic loss in the amplitude of the emitted free induction decay field. I will present a proposal to realize an analogy of quasiphase matching to ameliorate the dephasing.

1. D. Patterson, M. Schnell, & JM Doyle, Nature **497**, 475 (2013); D Patterson & JM Doyle, PRL **111**, 023008 (2013)
2. S. Lobsiger *et al*, JCPL **6**, 196 (2015).

MJ15

THEORY OF MICROWAVE 5-WAVE MIXING OF CHIRAL MOLECULES

KEVIN LEHMANN, *Departments of Chemistry and Physics, University of Virginia, Charlottesville, VA, USA.*

Microwave three-wave mixing spectroscopy produces a Free Induction Decay Field that is proportional to the enantiomeric excess (*ee*) of a sample of chiral molecules. However, since there is an unavoidable loss of measured signal strength due to dephasing of the molecular emission, it is not possible to quantitate this *ee* unless one has an enantiomeric pure sample of the same molecule with which to compare the amplitude of the signal of a sample of unknown *ee*.

In this talk, I will demonstrate that it is in principle possible to use a 5 wave mixing experiment, based upon AC Stark shifts produced by nearly resonant fields, to produce a differential splitting of a transition such that one has frequency resolved peaks for the two enantiomers. The peaks corresponding to the two enantiomers can be switched by phase cycling of the fields. This method is promising to allow the quantitative measurement of molecular *ee*'s by microwave spectroscopy. There are experimental issues that make such an experiment difficult. It will likely be required to use of skimmed molecular beam (which will substantially reduce the number of molecular emitters and thus signal level) in order to reduce the field amplitude and phase inhomogeneity of the excited molecules.

MK. Matrix isolation (and droplets)

Monday, June 20, 2016 – 1:30 PM

Room: 140 Burrill Hall

Chair: Paul Raston, James Madison University, Harrisonburg, Virginia, USA

MK01   1:30 – 1:45

IR SPECTROSCOPIC STUDIES ON MICROSOLVATION OF HCl BY WATER

DEVENDRA MANI, RAFFAEL SCHWAN, THEO FISCHER, ARGHYA DEY, MATIN KAUFMANN, *Physikalische Chemie II, Ruhr University Bochum, Bochum, Germany*; BRITTA REDLICH, LEX VAN DER MEER, *Institute for Molecules and Materials (IMM), Radboud University Nijmegen, Nijmegen, Netherlands*; GERHARD SCHWAAB, MARTINA HAVENITH, *Physikalische Chemie II, Ruhr University Bochum, Bochum, Germany.*

Acid dissociation reactions are at the heart of chemistry. These reactions are well understood at the macroscopic level. However, a microscopic level understanding is still in the early stages of development. Questions such as *'how many H_2O molecules are needed to dissociate one HCl molecule?'* have been posed and explored both theoretically and experimentally.[1−5] Most of the theoretical calculations predict that four H_2O molecules are sufficient to dissociate one HCl molecule, resulting in the formation of a solvent separated $H_3O^+(H_2O)_3Cl^-$ cluster.[1−3] IR spectroscopy in helium nanodroplets has earlier been used to study this dissociation process.[3−5] However, these studies were carried out in the region of O-H and H-Cl stretch, which is dominated by the spectral features of undissociated $(HCl)_m$-$(H_2O)_n$ clusters. This contributed to the ambiguity in assigning the spectral features arising from the dissociated cluster.[4,5] Recent predictions from Bowman's group, suggest the presence of a broad spectral feature (1300-1360 cm^{-1}) for the $H_3O^+(H_2O)_3Cl^-$ cluster, corresponding to the umbrella motion of H_3O^+ moiety.[6] This region is expected to be free from the spectral features due to the undissociated clusters. In conjunction with the FELIX laboratory, we have performed experiments on the $(HCl)_m(H_2O)_n$ (m=1-2, n≥4) clusters, aggregated in helium nanodroplets, in the 900-1700 cm^{-1} region. Mass selective measurements on these clusters revealed the presence of a weak-broad feature which spans between 1000-1450 cm^{-1} and depends on both HCl as well as H_2O concentration. Measurements are in progress for the different deuterated species. The details will be presented in the talk.

References: **1)** C.T. Lee et al., *J. Chem. Phys.*, **104**, 7081 (1996). **2)** H. Forbert et al., *J. Am. Chem. Soc.*, **133**, 4062 (2011). **3)** A. Gutberlet et al., *Science*, **324**, 1545 (2009). **4)** S. D. Flynn et al., *J. Phys. Chem. Lett.*, **1**, 2233 (2010). **5)** M. Letzner et al., *J. Chem. Phys.*, **139**, 154304 (2013). **6)** J. M. Bowman et al., *Phys. Chem. Chem. Phys.*, **17**, 6222 (2015).

MK02   1:47 – 2:02

ANETHOLE-WATER: A COMBINED JET, MATRIX, AND COMPUTATIONAL STUDY

JOSH NEWBY, JACKLEEN NESHEIWAT, *Department of Chemistry , Hobart and William Smith Colleges, Geneva, NY, USA.*

Anethole [(E)-1-methoxy-4-(1-propenyl)benzene] is a natural product molecule that is commonly recognized as the flavor component of anise, fennel, and licorice. Previously, we reported the jet-cooled, laser-induced fluorescence (LIF) and single vibronic level fluorescence (SVLF) spectra of anethole.[a] In this work, several weak bands were observed and were tentatively assigned as van der Waals clusters of anethole with water. We have since confirmed this assignment and have conducted a more detailed study to determine the geometry of these clusters. Results from LIF, SVLF, and matrix isolation FTIR[b] spectroscopy, as well as computational results will be presented in this talk.

---

[a] *J. Phys. Chem. A*, **2013**, 117 (48), 12831–12841
[b] Newly built system at Hobart and William Smith Colleges

MK03 *Post-Deadline Abstract* 2:04 – 2:19

VIBRATIONAL SPECTROSCOPY OF CO_2^- RADICAL ANION IN WATER

IRENEUSZ JANIK, G. N. R. TRIPATHI, *Radiation Laboratory, University of Notre Dame, Notre Dame, IN, USA.*

The reductive conversion of CO_2 into industrial products (e.g., oxalic acid, formic acid, and methanol) can occur via aqueous CO_2^- as a transient intermediate. While the formation, structure and reaction pathways of this radical anion have been modelled for decades using various spectroscopic and theoretical approaches, we present here, for the first time, a vibrational spectroscopic investigation in liquid water, using pulse radiolysis time-resolved resonance Raman spectroscopy for its preparation and observation. Excitation of the radical in resonance with its 235 nm absorption displays a transient Raman band at 1298 cm^{-1}, attributed to the symmetric CO stretch, which is at 45 cm^{-1} higher frequency than in inert matrices. Isotopic substitution at C ($^{13}CO_2^-$) shifts the frequency downwards by 22 cm^{-1} which confirms its origin and the assignment. A Raman band of moderate intensity compared to the stronger 1298 cm^{-1} band also appears at 742 cm^{-1}, and is assignable to the OCO bending mode. A reasonable resonance enhancement of this mode is possible only in a bent CO_2^- (C_{2v}/C_s) geometry. These resonance Raman features suggest a strong solute-solvent interaction, the water molecules acting as constituents of the radical structure, rather than exerting a minor solvent perturbation. However, there is no evidence of the non-equivalence (C_s) of the two CO bonds. A surprising resonance Raman feature is the lack of overtones of the symmetric CO stretch, which we interpret due to the detachment of the electron from the CO_2^- moiety towards the solvation shell. Electron detachment occurs at the energies of $0.28^+/_-0.03$ eV or higher with respect to the zero point energy of the ground electronic state. The issue of acid-base equilibrium of the radical which has been in contention for decades, as reflected in a wide variation in the reported pK_a (-0.2 to 3.9), has been resolved. A value of $3.4^+/_-0.2$ measured in this work is consistent with the vibrational properties, bond structure and charge distribution in aqueous CO_2^-.

MK04 2:21 – 2:31

H-π BEATS n-σ IN PHENYLACETYLENE-HCl HYDROGEN BONDED HETERODIMER: A MATRIX ISOLATION INFRARED AND AB INITIO STUDY[a]

GINNY KARIR, K S VISWANATHAN, *Chemical Science, Indian Institute of Science Education and Research, MOHALI, PUNJAB, India.*

Hydrogen bonded complexes of phenylacetylene (PhAc) and HCl were studied using matrix isolation infrared spectroscopy and ab initio computations. An H...π complex was observed in our experiments, which was indicated to be the global minimum by our computations. In this complex, HCl serves as the proton donor to the acetylenic π cloud of PhAc. Computations also located two other minima on the PhAc-HCl potential surface. One was an H...π complex where the proton of HCl interacts with the π cloud of the phenyl ring, which was nearly isoenergetic with the global minimum. The other was an n-σ complex, where the acetylenic hydrogen in PhAc interacted with the chlorine of HCl. The phenylacetylene-HCl system was theoretically investigated, employing MP2 and M06-2X methods, with 6-311++G(d,p) and aug/cc-pVDZ basis sets. AIM, EDA and NBO analysis were also performed to explore the nature, physical origin and the strength of the noncovalent interactions. Experiments with phenylacetylene deuterated at the acetylenic hydrogen (PhAcD) were also performed, to confirm the above observation, through the isotopic effect.

This work is part of a study of the hydrogen bonded interactions of phenylacetylene with various precursors, which provide an interesting interaction landscape ranging from a strong n-σ to a strong H-π interaction. As it turns out, HCl is at one end of this range, displaying a strong H-π interaction. While this presentation will give the details of the phenylacetylene-HCl complex, it will also summarize the landscape mentioned above, putting the present study in perspective.

[a]GK acknowledges fellowship from MHRD, India. Authors thank IISER, Mohali for facilities.

MK05 2:33 – 2:43

WHAT IS DIFFERENT BETWEEN BORAZINE-ACETYLENE AND BENZENE-ACETYLENE?
A MATRIX ISOLATION AND *AB-INITIO* STUDY. [a]

KANUPRIYA VERMA, K S VISWANATHAN, *Chemical Science, Indian Institute of Science Education and Research, MOHALI, PUNJAB, India.*

Borazine ($B_3N_3H_6$)-C_2H_2 system was studied experimentally, using matrix isolation infrared spectroscopy and supported by *ab-initio* computations. $B_3N_3H_6$, also referred to as inorganic benzene, presents an interesting comparison with C_6H_6. While C_6H_6 has a delocalized π system, $B_3N_3H_6$ has electron density centered on the nitrogen atoms, while the boron atoms are electron deficient. In addition, $B_3N_3H_6$ can also serve as a proton donor through N-H group. Similarly, C_2H_2 can act both as a proton donor, using the hydrogen attached to the sp carbon or as a proton acceptor at its π-cloud.

At the MP2/aug-cc-pVDZ level of theory, C_6H_6-C_2H_2 system showed three minima[1]. The global minimum was a structure where the C_2H_2 was the proton donor to the C_6H_6 π system. The next was a local minimum where the C_6H_6 was the proton donor to C_2H_2 and the third was a π stacked structure.

$B_3N_3H_6$-C_2H_2 also shows three minima at the same level of theory mentioned above. One was a structure where C_2H_2 donates a proton to $B_3N_3H_6$, approaching it from above the plane of the ring, much like in C_6H_6-C_2H_2. A second near degenerate structure was also located where the C_2H_2 serves as a proton acceptor towards the N-H group of $B_3N_3H_6$. A similar structure in C_6H_6-C_2H_2 was a local minimum. While in the case of C_6H_6-C_2H_2, the global minimum was the only one observed in the experiments[2], in the case of $B_3N_3H_6$-C_2H_2, both near degenerate minima mentioned above were observed in the matrix. $B_3N_3H_6$-C_2H_2 therefore reveals similarities and differences from the C_6H_6-C_2H_2 system. A π-stacked local minimum was also computationally indicated in the $B_3N_3H_6$-C_2H_2 system, though it was not observed in our experiments. Our earlier work comparing $B_3N_3H_6$-H_2O to C_6H_6-H_2O also yielded a similar behavioral pattern. Details of the experimental data and computational results will be presented.

References: 1. M. Majumder, B. K. Mishra, N. Sathyamurthy Chem. Phys. Lett. 2013,557,59-65
2. K. Sundararajan, K.S. Viswanathan, A.D. Kulkarni and S.R. Gadre. J. Mol. Str. 2002,613,209-222.

---

[a]The authors gratefully acknowledge Dr.Sanjay Singh in the preparation of borazine.

MK06 2:45 – 2:55

INFRARED MATRIX-ISOLATION STUDY OF NEW NOBLE-GAS COMPOUNDS

CHENG ZHU[a], MARKKU RÄSÄNEN, LEONID KHRIACHTCHEV, *Department of Chemistry, University of Helsinki, Helsinki, Finland.*

We identify new noble-gas compounds in solid matrices using IR spectroscopy. The compounds under study belong to two types: HNgY and YNgY' where Ng is a noble-gas atom and Y and Y' are electronegative fragments. The experimental assignments are supported by ab initio calculations at the MP2(full) and CCSD(T) levels of theory with the def2-TZVPPD basis set.

We have prepared and characterized two new HNgY compounds (noble-gas hydrides): HKrCCCl in a Kr matrix and HXeCCCl in a Xe matrix.I The synthesis of these compounds includes two steps: UV photolysis of HCCCl in a noble-gas matrix to form the H + CCCl fragments and annealing of the matrix to mobilize H atoms and to promote the H + Ng + CCCl = HNgCCCl reaction. An interesting observation in the experiments on HXeCCCl in a Xe matrix is the temperature-induced transformation of the three H–Xe stretching bands. This observation is explained by temperature-induced changes of local matrix morphology around the embedded HXeCCCl molecule. In these experiments, we have also obtained the IR spectrum of the CCCl radical, which is produced by photodecomposition of HCCCl.

We have identified three new YNgY' compounds (fluorinated noble-gas cyanides): FKrCN in a Kr matrix and FXeCN and FXeNC in a Xe matrix.II These molecule are formed by photolysis of FCN in a noble-gas matrix due to locality of this process. The amount of these molecules increases upon thermal mobilization of the F atoms in the photolyzed matrix featuring the F + Ng + CN reaction.

---

[a]The author gratefully acknowledges the China Scholarship Council, the Academy of Finland, and the Finnish IT Center for Science.

MK07 2:57 – 3:07

EPR OF CH$_3$ RADICALS IN SIO$_2$ CLATHRATE

<u>YURIJ DMITRIEV</u>, *Division of Plasma Physics, Atomic Physics and Astrophysics, Ioffe Institute, St. Petersburg, Russia*; GIANPIERO BUSCARINO, *Department of Physics and Chemistry, University of Palermo, Palermo, Italy*; NIKOLAS PLOUTARCH BENETIS, *Department of Environmental Engineering and Antipollution Control, Technological Educational Institute of Western Macedonia (TEI), Kozani, Greece.*

EPR lineshape simulations of CH$_3$/SiO$_2$ clathrates reveal the motional conditions of the CH$_3$ radical up to the unusual regime of its stability, the high temperature diffusional regime. This was obvious by the isotropic magnetic interaction at the highest experimental temperatures over 140 K. Special motional and thermodynamics conditions for methyl radical may however prevail for the CH$_3$/SiO$_2$ clathrates system due to the limited space of the host voids, compared to solid gas isolation. The lowest temperature in the experiment was 4.1 K, while the highest one was 300 K. The EPR parameters of the radical revealed non-monotonic temperature dependence. The extremely wide temperature range of the radical stability may be attributed to the solidity of the clathrate voids and the small diameter of their channels that do not allow molecular collisions between the radical species. At the lowest sample temperatures, a portion of the radicals stopped to rotate thus indicating their attachment to specific matrix sites with large radical-host interaction. The unusual increase of the width of the CH$_3$/SiO$_2$ clathrate spectra with the temperature at high sample temperatures indicates resemblance to the spin-rotation interaction relaxation mechanism known only in the case of small species in non-viscous fluids, and is contrasted to the normal difussional decrease of the width in the CH$_3$ hosted in a series of solid. The effect was explained by adopting extremely frequent radical collisions with the clathrate void walls leading to repeated angular momentum alterations, a kind of "reorientation".[a]

[a]Yu. A. D. acknowledges support by the Russian Foundation for Basic Research (RFBR), research project 16-02-00127a.

Intermission

MK08 3:26 – 3:41

OBSERVATION OF TRANS-ETHANOL AND GAUCHE-ETHANOL COMPLEXES WITH BENZENE USING MATRIX ISOLATION INFRARED SPECTROSCOPY

<u>JAY AMICANGELO</u>, MATTHEW J SILBAUGH, *School of Science (Chemistry), Penn State Erie, Erie, PA, USA.*

Ethanol can exist in two conformers, one in which the OH group is trans to the methyl group (trans-ethanol) and the other in which the OH group is gauche to the methyl group (gauche-ethanol). Matrix isolation infrared spectra of ethanol deposited in 20 K argon matrices display distinct infrared peaks that can be assigned to the trans-ethanol and gauche-ethanol conformers, particularly with the O-H stretching vibrations.[a] Given this, matrix isolation experiments were performed in which ethanol (C$_2$H$_5$OH) and benzene (C$_6$H$_6$) were co-deposited in argon matrices at 20 K in order to determine if conformer specific ethanol complexes with benzene could be observed in the infrared spectra. New infrared peaks that can be attributed to the trans-ethanol and gauche-ethanol complexes with benzene have been observed near the O-H stretching vibrations of ethanol. The initial identification of the new infrared peaks as being due to the ethanol-benzene complexes was established by performing a concentration study (1:200 to 1:1600 S/M ratios), by comparing the co-deposition spectra with the spectra of the individual monomers, by matrix annealing experiments (35 K), and by performing experiments using isotopically labeled ethanol (C$_2$D$_5$OD) and benzene (C$_6$D$_6$). Quantum chemical calculations were also performed for the C$_2$H$_5$OH-C$_6$H$_6$ complexes using density functional theory (B3LYP) and ab initio (MP2) methods. Stable minima were found for the both the trans-ethanol and gauche-ethanol complexes with benzene at both levels of theory and were predicted to have similar interaction energies. Both complexes can be characterized as H-π complexes, in which the ethanol is above the benzene ring with the hydroxyl hydrogen interacting with the π cloud of the ring. The theoretical O-H stretching frequencies for the complexes were predicted to be shifted from the monomer frequencies and from each other and these results were used to make the conformer specific infrared peak assignments.

[a]Barnes, A. J.; Hallam, H. E. *Trans. Faraday Soc.*, **1970**, *66*, 1932-1940.

MK09   **3:43 – 3:58**

LOW TEMPERATURE THERMODYNAMIC EQUILIBRIUM OF CO_2 DIMER ANION SPECIES IN CRYOGENIC ARGON AND KRYPTON MATRICES

MICHAEL E. GOODRICH, DAVID T MOORE, *Chemistry Dept., Lehigh University, Bethlehem, PA, USA.*

The separated CO_2 dimer anion, $(CO_2)(CO_2^-)$, is observed by FTIR spectroscopy in matrix isolation experiments at 1652 cm^{-1} upon deposition of high energy argon ions into an argon matrix doped with 0.5% CO_2. It has previously been reported by Andrews that upon annealing the matrix to 25K, the separated species converts to an oxalate-like $C_2O_4^-$ species which appears at 1856 cm^{-1}.[a] We have observed that subsequently holding the matrix at 10K caused the $C_2O_4^-$ species to fully convert back to $(CO_2)(CO_2^-)$. Upon further investigation, we determined that the two species reversibly interconvert between 19K and 23K, suggesting the species are in thermodynamic equilibrium. The associated van't Hoff plot has a linear trend and indicates an endothermic reaction driven by a large increase in entropy. An analogous experiment in a krypton matrix was performed, and the equilibrium was found to occur between 26K and 31K. Interestingly, analysis revealed the reaction in krypton is more endothermic, but has nearly the same entropy value as was observed in the argon experiment.

[a]Zhou, M.; Andrews, L.; J. Chem. Phys. 110, 2414 (1999).

MK10   **4:00 – 4:15**

SIMULTANEOUS DEPOSITION OF MASS SELECTED ANIONS AND CATIONS: IMPROVEMENTS IN ION DELIVERY FOR MATRIX ISOLATION EXPERIMENTS

MICHAEL E. GOODRICH, DAVID T MOORE, *Chemistry Dept., Lehigh University, Bethlehem, PA, USA.*

A focus of the research in our group has been to develop improved methods for ion delivery in matrix isolation experiments. We have previously reported a method to co-deposit low energy, mass selected metal anions and a rare gas counter cation.[a] A modification allowing for mass selection of both the anion and cation will be discussed. Results from preliminary experiments of mass selected, low energy Cu^- and SF_5^+ will also be highlighted. To our knowledge, these experiments are the first time two mass selected beams of ions have been simultaneously deposited into a cryogenic matrix. Co-deposition of the ions into an argon matrix doped with 0.02% CO at 20K resulted in the observation of bands assigned to SF_5^+ and anionic copper carbonyl complexes, $Cu(CO)_n^-$ (n=1-3). Upon irradiation of the matrix with a narrow band, blue LED, the copper carbonyl complexes are converted to the neutral analogues, while the fate of the photodetached electrons can be directly tracked, as a decrease of the SF_5^+ band and a growth of the neutral SF_5 band are observed.

[a]Ludwig, R. M.; Moore, D. T.; J. Chem. Phys. 139, 244202 (2013).

MK11   **4:17 – 4:32**

PUMP AND PROBE SPECTROSCOPY OF CH_3F-(ortho-H_2)$_n$ CLUSTERS IN SOLID PARAHYDROGEN BY USING TWO CW-IR QUANTUM CASCADE LASERS

HIROYUKI KAWASAKI, ASAO MIZOGUCHI, HIDETO KANAMORI, *Department of Physics, Tokyo Institute of Technology, Tokyo, Japan.*

The absorption spectrum of the ν_3 (C-F stretching) mode of CH_3F in solid *para*-H_2 by FTIR showed a series of equal interval peaks[a]. Their interpretation was that the *n*-th peak of this series was due to CH_3F-(ortho-H_2)$_n$ clusters which were formed CH_3F and *n*'s ortho-H_2 in first nearest neighbor sites of the *para*-H_2 crystal with *hcp* structure. In order to understand this system in more detail, we have studied these peaks, especially $n = 0 - 3$ corresponding to 1037 - 1041 cm^{-1}, by using high-resolution and high-sensitive infrared quantum cascade laser (QCL) spectroscopy. Before now, we have observed photochromic phenomena of these peaks by taking an advantage of the high brightness of the laser[b]. However, it has not been revealed what kind of mechanism is undergoing in these processes. In order to solve this problem we introduced two cw-IR QCLs for pump and prove experiment. While the pumping laser is breaching a certain peak with high power, the probing laser is monitoring the increase of other peaks by rapid scan method. The time resolution of this spectroscopy is 5 msec. The new precise kinetic information will help us to understand the molecular interaction in solid *para*-H_2.

[a]K. Yoshioka and D. T. Anderson, J. Chem. Phys. 119 (2003) 4731-4742
[b]A. R. W. McKellar, A. Mizoguchi, and H. Kanamori, Phys. Chem. Chem. Phys. 13 (2011) 11587-11589

MK12

QUANTUM DIFFUSION CONTROLLED CHEMISTRY: THE H + NO REACTION

MORGAN E. BALABANOFF, DAVID T. ANDERSON, *Department of Chemistry, University of Wyoming, Laramie, WY, USA.*

In this study, we present Fourier transform infrared spectroscopic studies of the 193 nm photochemistry of nitric oxide (NO) isolated in a parahydrogen (pH$_2$) matrix over the 1.7 to 4.3 K temperature range. Back in 2003 Fushitani and Momose[a] showed that hydrogen atoms (H atoms) are produced as by-products of the 193 nm photo-initiated reactions of NO trapped in solid pH$_2$. We recently published[b] a further study on the same NO/pH$_2$ system where we showed that H atom reactions with NO produce both HNO and NOH even though the reaction that forms HNO is barrierless and the reaction that forms NOH has a sizeable barrier. Further, we measured the reaction kinetics at 1.8 K and 4.3 K and showed the rate constants follow an Arrhenius-behavior with a small activation energy (E_a=2.39(1) cm^{-1}). In the present studies we are continuing this work using a ^{15}NO enriched sample and are focusing on how we can adjust the experimental conditions to increase the yield of both the HNO/NOH reaction products. We are also performing kinetic experiments at more than just two temperatures to better characterize the temperature dependence of the extracted rate constants. We are conducting these additional experiments to benchmark the reaction kinetics for the H + NO reaction in solid pH$_2$ to better understand what factors influence the rates of these low temperature chemical reactions.

[a]M. Fushitani, T. Momose, *Low Temp. Phys.* **29**, 985-988 (2003).

[b]M. Ruzi, D.T. Anderson, *J. Phys. Chem. A* **119**, 12270-12283 (2015).

MK13

USING INFRARED SPECTROSCOPY TO PROBE THE TEMPERATURE DEPENDENCE OF THE H + N$_2$O REACTION IN PARAHYDROGEN CRYSTALS

FREDRICK M. MUTUNGA, DAVID T. ANDERSON, *Department of Chemistry, University of Wyoming, Laramie, WY, USA.*

In situ photolysis of precursor molecules trapped in a solid parahydrogen matrix has been successfully used in our group to study H atom reactions with other species at temperatures in the range of 1.6 to 4.3 K. At these temperatures, H atoms are known to continuously move through the solid by the H + H$_2$ $\rightarrow$ H$_2$ + H tunneling exchange reaction. We recently studied the reaction of H atoms with ^{15}N$_2$O and in the preliminary communication of this work,[a] we reported a very strange non-Arrhenius temperature dependence to the reaction; the reaction only occurs below 2.4 K and not at higher temperatures. This talk will present our subsequent work on the high-resolution infrared spectroscopy of ^{15}N$_2$O molecules trapped in solid parahydrogen with a focus on the $\nu_1 + \nu_3$ and $2\nu_1$ vibrational bands. For both these vibrational bands we observe multiple peaks and the relative intensities of the observed peaks change with temperature over the measured range similar to the temperature dependence of the ν_3 fundamental reported earlier by Lorenz and Anderson.[b] The temperature dependent changes in intensity imply that there are at least two trapping sites which could potentially explain the observed temperature dependence to the H + ^{15}N$_2$O reaction.

[a]F. M. Mutunga, S. E. Follett, D. T. Anderson, *J. Chem. Phys.* **139**, 151104 (2013).

[b]B. D. Lorenz and D. T. Anderson, *J. Chem. Phys.* **126**, 184506 (2007).

MK14   5:08 – 5:23

INFRARED SPECTROSCOPY OF DEUTERATED ACETYLENE IN SOLID PARAHYDROGEN AND THE HELIUM RECOVERY INITIATIVE

AARON I. STROM, DAVID T. ANDERSON, *Department of Chemistry, University of Wyoming, Laramie, WY, USA.*

The linear tetratomic organic molecule acetylene, HCCH, has been studied extensively throughout the past century via numerous spectroscopic experiments, exploiting wavelengths across the electromagnetic spectrum. Both the mono- and di-deutero acetylene isotopologues have also been widely studied, namely HCCD and DCCD. In this presentation, I will present the Fourier transform infrared (FTIR) spectroscopy of DCCD in solid parahydrogen (pH$_2$) in the low-temperature regime (1.5-5.0 K). We intend to perform UV photochemical studies on DCCD doped solid pH$_2$ and, therefore, the infrared spectroscopy must be characterized prior. The FTIR spectrum of DCCD isolated in solid pH$_2$ exhibits rich fine structure in the ν_3 asymmetric C-D stretch region. Some of the observed peaks may arise from the formation of weakly bound acetylene dimers, or potentially even larger clusters. We can test this hypothesis by varying the DCCD concentration in separate experiments and temperature cycling the matrix to look for irreversible cluster growth. In preliminary experiments we observe trace amounts of the lighter isotopologues (HCCD and HCCH) and so these species can also cluster with DCCD, adding to the complexity of the spectra. We remark that ortho-hydrogen clustering to DCCD may also be occurring and we have ways to check that as well. In order to make better sense of the FTIR spectrum of DCCD doped pH$_2$, a comparison with the simulated low temperature gas-phase spectrum will also be presented. This will allow us to address issues related to the extent of the rotational motion of DCCD in solid pH$_2$.

A liquid helium bath cryostat is used to grow and maintain the DCCD doped pH$_2$ crystals for spectroscopic characterization. Helium is a non-renewable resource and in recent years the Anderson group has been building a helium recovery system. This Helium Recovery Initiative (HRI) will be discussed in an effort to describe how we implemented this new experimental system in our laboratory and to point out the major challenges we faced. One of the main goals of the HRI is to promote sustainable helium use, permitting smaller cryogenics laboratories to continue conducting research with liquid helium.

TA. Mini-symposium: Spectroscopy of Large Amplitude Motions

Tuesday, June 21, 2016 – 8:30 AM

Room: 100 Noyes Laboratory

Chair: Hanno Schmiedt, University of Cologne, Cologne, Germany

---

TA01   *Journal of Molecular Spectroscopy Review Lecture*   8:30 – 9:00

FLOPPY MOLECULES WITH INTERNAL ROTATION AND INVERSION

MAREK KREGLEWSKI, *Faculty of Chemistry, Adam Mickiewicz University, Poznan, Poland.*

There are different ways to analyze rovibrational structure of molecules having several large amplitude motions of different type, like internal rotation and inversion or ring-puckering. In my research group we have developed and used methods starting from potential surfaces for large amplitude motions but also applied purely effective Hamiltonians, where tunneling splittings were key parameters. Whatever is the method the following problems must be solved when addressing a rovibrational problem with large amplitude vibrations: 1) a definition of the permutation-inversion molecular symmetry group, 2) a choice of the internal coordinates and their transformation in the symmetry group, 3) derivation of the Hamiltonian in chosen coordinates, 4) calculation of the Hamiltonian matrix elements in a symmetrized basis set. These points will be discussed. The advantage of methods which start from the geometry and potential surface for large amplitude vibrations give much clearer picture of internal dynamics of molecules but generally the fit to experimental data is much poorer. The fitting procedure is strongly non-linear and the iteration procedure much longer. The effective Hamiltonians the fit is generally much better since almost all optimized parameters are linear but the parameters have no clear physical meaning. This method is very useful in the assignment of experimental spectra. Results of the application of both method to methylamine and hydrazine will be presented.

TA02   9:05 – 9:20

APPLICATION OF THE HYBRID PROGRAM FOR FITTING MICROWAVE AND FAR-INFRARED SPECTRA OF METHYLAMINE

ISABELLE KLEINER, *Laboratoire Interuniversitaire des Systèmes Atmosphériques (LISA), CNRS et Universités Paris Est et Paris Diderot, Créteil, France*; JON T. HOUGEN, *Sensor Science Division, National Institute of Standards and Technology, Gaithersburg, MD, USA.*

Last year we presented a new hybrid-model fitting program for methylamine-like molecules, based on an effective Hamiltonian in which the ammonia-like inversion motion is treated using a tunneling formalism, while the internal-rotation motion is treated using an explicit kinetic energy operator and potential energy function. This new hybrid program was successfully applied to 2-methylmalonaldehyde, for which we refit the already published ground state $v_t = 0$ data. This fit[a], which was of almost the same quality as that obtained using an all-tunneling formalism, removed one of the major puzzles in the isotope-dependence of the internal-rotation tunneling parameters found in the all-tunneling fit. This year we are trying to illustrate a second advantage of the new hybrid formalism, which allows one to carry out global fits of data from two or more torsional states in methylamine-like molecules. We are, in fact, trying to simultaneously fit the $v_t = 0$ and $v_t = 1$ microwave and infrared date on methylamine itself. This data is also in the literature, but the all-tunneling Hamiltonians used could only fit each of the two torsional states separately. At the time of writing this abstract, we have preliminary fits of about 1200 methylamine transitions to 25 or 30 torsion-inversion-rotation parameters, but these hybrid-program fits are not yet at the same level as the all-tunneling-program fits in the literature. We hope to report significant further progress on this work in June.

---

[a] I. Kleiner and J. T. Hougen, J Phys Chem A. 119, 10664-76 (2015)

TA03   9:22 – 9:32

ACCURATE ROVIBRATIONAL ENERGIES FOR THE FIRST EXCITED TORSIONAL STATE OF METHYLAMINE

IWONA GULACZYK, <u>MAREK KREGLEWSKI</u>, *Faculty of Chemistry, Adam Mickiewicz University, Poznan, Poland.*

The first excited torsional band of methylamine, ν_{15}, has been reassigned in a high resolution spectrum in the region from 40 to 360 cm^{-1}. Over 12400 transitions with a resolution of 0.00125 cm^{-1} with K from 0 up to 16 and J from 0 up to 40 have been assigned for all six symmetry species. A global fit of the infrared, pure rotational and microwave data has been carried out and the band centre was determined at 264.5825(60) cm^{-1}. The combined data were fit to a single state model based on the group theoretical formalism of Hougen and Ohashi resulted in the total standard deviation of 0.004 cm^{-1} for the infrared spectrum and 0.40 MHz for microwave spectrum. From the same spectrum the upper state combination differences produced data for the rotational structure of the ground state which could be fitted with the standard deviation of 0.0003 cm^{-1}. The fit to ground state rotational transitions in the MHz region gave the standard deviation of 0.21 MHz. Although the precision of the energies calculated for the excited torsional state is not fully satisfactory it allows us to assign several thousands of lines in the hot bands ν_{15}-$2\nu_{15}$, ν_{15}-$3\nu_{15}$ and ν_{15}-$4\nu_{15}$, which are quite intense in the spectrum recorded at room temperature.

TA04   9:34 – 9:49

ANALYSIS OF THE TORSIONAL SPLITTING IN THE ν_8 BAND OF PROPANE NEAR 870.4 cm^{-1} CAUSED BY FERMI RESONANCE WITH THE $2\nu_{14}+2\nu_{27}$ LEVEL

<u>PETER GRONER</u>, *Department of Chemistry, University of Missouri - Kansas City, Kansas City, MO, USA*; AGNES PERRIN, F. KWABIA TCHANA, *CNRS, Universités Paris Est Créteil et Paris Diderot, LISA, Créteil, France*; LAURENT MANCERON, *Synchrotron SOLEIL, CNRS-MONARIS UMR 8233 and Beamline AILES, Saint Aubin, France.*

Torsional splitting has been observed in the ν_8 and ν_{21} IR bands of propane near 870.4 and 921.4 cm^{-1}, respectively, recorded at the AILES Beamline at the SOLEIL synchrotron.[a] Over 4000 individual transitions of the ν_8 band were assigned and analyzed with an expanded version of the effective rotational Hamiltonian for molecules with two symmetric internal rotors (ERHAM).[b] A least-squares fit approximated a large portion of the assigned transitions with a model of an isolated ν_8 state with acceptable precision. However, this model was unable to reproduce many systematic deviations and local resonances. A torsional analysis of existing experimental data and ab initio predictions allows the conclusion that Fermi resonance between ν_8 and the torsional combination state $2\nu_{14}+2\nu_{27}$ most likely caused the failure of the isolated state model. Additional modifications to ERHAM that include Fermi resonance with another state support the conclusion that most of the observed torsional splitting in ν_8 is caused by the $2\nu_{14}+2\nu_{27}$ state. The continuing detailed analysis is expected to yield more definitive results by the time of this meeting.

[a] A. Perrin et al., J. Mol. Spectrosc. 315 (2015), 55-62; A. Perrin et al., ISMS15, presentation TG04.

[b] P. Groner, J. Chem. Phys. 107 (1997) 4483–4498; P. Groner, J. Mol. Spectrosc. 278 (2012) 52–67.

TA05   9:51 – 10:06

COUPLING OF LARGE AMPLITUDE INVERSION WITH OTHER STATES

<u>JOHN PEARSON</u>, SHANSHAN YU, *Jet Propulsion Laboratory, California Institute of Technology, Pasadena, CA, USA.*

The coupling of a large amplitude motion with a small amplitude vibration remains one of the least well characterized problems in molecular physics. Molecular inversion poses a few unique and not intuitively obvious challenges to the large amplitude motion problem. In spite of several decades of theoretical work numerous challenges in calculation of transition frequencies and more importantly intensities persist. The most challenging aspect of this problem is that the inversion coordinate is a unique function of the overall vibrational state including both the large and small amplitude modes. As a result, the r-axis system and the meaning of the K-quantum number in the rotational basis set are unique to each vibrational state of large or small amplitude motion. This unfortunate reality has profound consequences to calculation of intensities and the coupling of nearly degenerate vibrational states. The case of NH3 inversion and inversion through a plane of symmetry in alcohols will be examined to find a general path forward.

TA06 10:08 – 10:23

FINAL RESULTS ON MODELING THE SPECTRUM OF AMMONIA $2\nu_2$ AND ν_4 STATES

SHANSHAN YU, JOHN PEARSON, TAKAYOSHI AMANO, *Jet Propulsion Laboratory, California Institute of Technology, Pasadena, CA, USA*; OLIVIER PIRALI, *AILES beamline, Synchrotron SOLEIL, Saint Aubin, France.*

At this symposium in 2013, we reported our preliminary results on modeling the spectrum of ammonia $2\nu_2$ and ν_4 states (see Paper TB09 in 2013). This presentation reports the final results on our extensive experimental measurements and data analysis for the $2\nu_2$ and ν_4 inversion-rotation and vibrational transitions. We measured 159 new transition frequencies with microwave precision and assigned 1680 new ones from existing Fourier Transform spectra recorded in Synchrotron SOLEIL. The newly assigned data significantly expand the range of assigned quantum numbers. Combined with all the previously published high-resolution data, the $2\nu_2$ and ν_4 states are reproduced to 1.3σ using a global model. We will discuss the types of transitions included in our global analysis, and fit statistics for date sets from individual experimental work.

Intermission

TA07 10:42 – 10:52

MILLIMETER WAVE SPECTRUM OF NITROMETHANE

V. ILYUSHIN, *Radiospectrometry Department, Institute of Radio Astronomy of NASU, Kharkov, Ukraine.*

A new study of the millimeter wave spectrum of nitromethane CH_3NO_2 is reported. The new measurements covering the frequency range from 49 GHz to 236 GHz have been carried out using spectrometer in IRA NASU (Ukraine). The transitions belonging to the m $\leq$ 8 torsional states have been analyzed using the RAM36 program[a], which has been modified for this study to take into account the quadrupole hyperfine structure due to presence of the nitrogen atom. The dataset consisting of 5838 microwave line frequencies and including transitions with J up to 50 was fit using a model consisting of 93 parameters and weighted root-mean-square deviation of 0.89 has been achieved. In the talk the details of this new study will be discussed.

---

[a]V. Ilyushin, Z. Kisiel, L. Pszczółkowski, H. Mäder, J. T. Hougen J. Mol. Spectrosc. 259 (2010) 26-38.

TA08 10:54 – 11:09

IAM(-LIKE) TUNNELING MATRIX FORMALISM FOR ONE- AND TWO-METHYL-TOP MOLECULES BASED ON THE EXTENDED PERMUTATION-INVERSION GROUP IDEA AND ITS APPLICATION TO THE ANALYSES OF THE METHYL-TORSIONAL ROTATIONAL SPECTRA

NOBUKIMI OHASHI, , *Kanazawa University, Kanazawa, Japan*; KAORI KOBAYASHI, *Department of Physics, University of Toyama, Toyama, Japan*; MASAHARU FUJITAKE, *Division of Mathematical and Physical Sciences, Kanazawa University, Kanazawa, Japan.*

Recently we reanalyzed the microwave absorption spectra of the trans-ethyl methyl ether molecule, state by state, in the ground vibrational, O-methyl torsional, C-methyl torsional and skeletal torsional states with the use of an IAM-like tunneling matrix formalism based on an extended permutation-inversion (PI) group idea, whose results appeared in Journal of Molecular Spectroscopy recently. Since a single rho-axis does not exist in trans-ethyl methyl ether that has two methyl-tops and the IAM formalism is not available as in the case of the one methyl-top molecule, we adopted instead an IAM-like (in other word, partial IAM) formalism. We will show the outline of the present formalism and the results of the spectral analyses briefly. We also would like to review the IAM formalism for the one top molecules based on the extended PI group, and show the result of the application to the spectral analysis.

If possible, we would like to compare the IAM and IAM-like formalisms based on the extended PI group with the ERHAM formalism developed by Groner, especially, in the form of Hamiltonian matrix elements, and discuss about similarity and difference.

TA09 11:11 – 11:26

IS THE COUPLING OF C_{3V} INTERNAL ROTATION AND NORMAL VIBRATIONS A TRACTABLE PROBLEM?

<u>JOHN PEARSON</u>, *Jet Propulsion Laboratory, California Institute of Technology, Pasadena, CA, USA*; PETER GRONER, *Department of Chemistry, University of Missouri - Kansas City, Kansas City, MO, USA*; ADAM M DALY, *Department of Chemistry and Biochemistry, University of Arizona, Tucson, AZ, USA.*

The solution of a C_{3V} internal rotation problem for the torsional manifold of an isolated vibrational state such as the ground state is well established. However, once an interacting small amplitude vibrational state is involved the path to a solution becomes far less clear and there is little guidance in the literature on how to proceed. The fundamental challenge is that the torsional problem and the internal axis system are unique to each torsional manifold of a specific vibrational state. In an asymmetric top molecule vibrational angular momentum can be rotated away, but this sort of rotation changes the angle between the internal rotation axis and the principle axis when there is an internal rotor. This means that there is an angle between the internal axis systems of each torsional manifold of a vibrational state. The net result is that the coupling between the two states must account for the difference in internal axis angle and will have some significant consequences to the selection rules and interactions. Two cases will be discussed, methanol and ethyl cyanide.

TA10 11:28 – 11:43

A HAMILTONIAN TO OBTAIN A GLOBAL FREQUENCY ANALYSIS OF ALL THE VIBRATIONAL BANDS OF ETHANE

<u>NASSER MOAZZEN-AHMADI</u>, JALAL NOROOZ OLIAEE, *Physics and Astronomy/Institute for Quantum Science and Technology, University of Calgary, Calgary, AB, Canada.*

The interest in laboratory spectroscopy of ethane stems from the desire to understand the methane cycle in the atmospheres of planets and their moons and from the importance of ethane as a trace species in the terrestrial atmosphere. Solar decomposition of methane in the upper part of these atmospheres followed by a series of reactions leads to a variety of hydrocarbon compounds among which ethane is often the second most abundant species. Because of its high abundance, ethane spectra have been measured by Voyager and Cassini in the regions around 30, 12, 7, and 3 μm. Therefore, a complete knowledge of line parameters of ethane is crucial for spectroscopic remote sensing of planetary atmospheres. Experimental characterization of torsion-vibration states of ethane lying below 1400 cm^{-1} have been made previously [a], but extension of the Hamiltonian model for treatment of the strongly perturbed ν_8 fundamental and the complex band system of ethane in the 3 micron region requires careful examination of the operators for many new torsionally mediated vibration-rotation interactions. Following the procedures outlined by Hougen [b, c], we have re-examined the transformation properties of the total angular momentum, the translational and vibrational coordinates and momenta of ethane, and for vibration-torsion-rotation interaction terms constructed by taking products of these basic operators. It is found that for certain choices of phase, the doubly degenerate vibrational coordinates with and symmetry can be made to transform under the group elements in such a way as to yield real matrix elements for the torsion-vibration-rotation couplings whereas other choices of phase may require complex algebra. In this talk, I will discuss the construction of a very general torsion-vibration-rotation Hamiltonian for ethane, as well as the prospect for using such a Hamiltonian to obtain a global frequency analysis (based in large part on an extension of earlier programs and ethane fits[a] from our laboratory) of all the vibrational bands of ethane at or below the 3-micron region.

[a]N. Moazzen-Ahmadi and J. Norooz Oliaee, J. Quant. Spectrosc. Radiat. Transfer, submitted.

[b]J.T. Hougen, Can. J. Phys., 42, 1920 (1964)

[c]J. T. Hougen, Can. J. Phys., 43, 935 (1965)

TA11   **11:45 – 12:00**

USING SYMMETRY GROUP CORRELATION TABLES TO EXPLAIN WHY ERHAM (AND OTHER PROGRAMS) CANNOT BE USED TO ANALYZE TORSIONAL SPLITTINGS OF SOME MOLECULES

PETER GRONER, *Department of Chemistry, University of Missouri - Kansas City, Kansas City, MO, USA.*

ERHAM has been used to analyze rotational spectra of many molecules with torsional splitting caused by one or two internal rotors.[a] The gauche form of dimethyl ether-d_1 whose equilibrium structure has C_1 symmetry is an example of a molecule for which ERHAM could not model additional small splittings resolvable for many transitions, whereas the spectrum of the symmetric (anti, trans) form with a C_s equilibrium structure could be analyzed successfully with ERHAM.[b] A more recent example where ERHAM failed is pinacolone $CH_3 - CO - C(CH_3)_3$.[c] In this case, the barriers to internal rotation of the methyl groups within the $-C(CH_3)_3$ unit are too high to produce observable internal rotation splittings, but the splittings due to the $CH_3 - CO$ methyl group could not be modeled correctly with ERHAM nor with any other available program (XIAM, BELGI-C_s, BELGI-C_1, RAM36). In the paper, it was speculated that BELGI-C_s-2tops might be able to the job, but arguments against this possibility have also been put forward. The correlation between irreducible representations of groups and their subgroups according to Watson[d] can be used not only to determine the total number of substates (components) to be expected but also to help decide which particular program has a chance for a successful analysis. As it turns out, the number of components of split lines depends on the molecular symmetry at equilibrium in relation to the highest possible symmetry for a given molecular symmetry group. Therefore, for pinacolone, the vibrational ground state is split into 10 torsional substates.

[a] P. Groner, J. Mol. Spectrosc. 278 (2012) 52–67.

[b] C. Richard et al. A&A 552 (2013), A117.

[c] Y. Zhao et al., J. Mol. Spectrosc. 318 (2015) 91–100, with references to all other programs mentioned in the abstract.

[d] J. K. G. Watson, Can. J. Physics 43 (1965) 1996-2007.

TB. Mini-symposium: Spectroscopy in Atmospheric Chemistry

Tuesday, June 21, 2016 – 8:30 AM

Room: 116 Roger Adams Lab

Chair: Vincent Boudon, CNRS / Université de Bourgogne, Dijon, France

---

TB01   *INVITED TALK*   8:30 – 9:00

RADICALS AND AEROSOLS IN THE TROPOSPHERE AND LOWER STRATOSPHERE

RAINER VOLKAMER, THEODORE KOENIG, BARBARA DIX, *CIRES, University of Colorado, Boulder, CO, United States.*

The remote tropical free troposphere (FT) is one of the most relevant atmospheric environments on Earth. About 75% of the global tropospheric O3 and CH4 loss occurs at tropical latitudes. Tropospheric bromine and iodine catalytically destroy tropospheric O_3, oxidize atmospheric mercury, and modify oxidative capacity, and aerosols. Oxygenated VOCs (OVOC) modify HO_x (= OH + HO_2), NOx (= NO + NO_2), tropospheric O_3, aerosols, and are a sink for BrO_x (= Br + BrO). Until recently, atmospheric models were untested for lack of vertically resolved measurements of BrO and IO radicals in the tropical troposphere. BrO and IO are highly reactive trace gases. Even very low concentrations (parts per trillion; 1 pptv = 10^{-12} volume mixing ratio) can significantly modify the lifetime of climate active gases, and determine (bromine) the rate limiting step of mercury oxidation in air (that is washed out, and subsequently bio-accumulates in fish). Analytical challenges arise when these radicals modify in sampling lines. Sensitive yet robust, portable, and inherently calibrated measurements directly in the open atmosphere have recently been demonstrated by means of limb-measurements of scattered solar photons by the University of Colorado Airborne Multi-AXis DOAS instrument (CU AMAX-DOAS) from research aircraft. The CU AMAX-DOAS instrument is optimized to (1) locate BrO, IO and glyoxal (a short lived OVOC) in the troposphere, (2) decouple stratospheric absorbers, (3) maximize sensitivity at instrument altitude, (4) facilitate altitude control and (5) enable observations over a wide range of solar zenith angles. Further, (6) the filling-in of Fraunhofer lines (Ring-effect) by Raman Scattering offers interesting opportunities for radiative closure studies to assess the effects of aerosols on Climate.

TB02   9:05 – 9:20

APPLICATIONS OF HIGH RESOLUTION MID-INFRARED SPECTROSCOPY FOR ATMOSPHERIC AND ENVIRON-MENTAL MEASUREMENTS

JOSEPH R ROSCIOLI, J BARRY McMANUS, DAVID NELSON, MARK ZAHNISER, SCOTT C HERNDON, JOANNE SHORTER, TARA I YACOVITCH, DYLAN JERVIS, CHRISTOPH DYROFF, CHARLES E KOLB, *Center for Atmospheric and Environmental Chemistry, Aerodyne Research, Inc, Billerica, Massachusetts, United States.*

For the past 20 years, high resolution infrared spectroscopy has served as a valuable tool to measure gas-phase concentrations of ambient gas samples. We review recent advances in atmospheric sampling using direct absorption high resolution mid-infrared spectroscopy from the perspective of light sources, detectors, and optical designs. Developments in diode, quantum cascade and interband cascade laser technology have led to thermoelectrically-cooled single-mode laser sources capable of operation between 800 cm^{-1} and 3100 cm^{-1}, with <10 MHz resolution and >10 mW power. Advances in detector and preamplifier technology have yielded thermoelectriocally-cooled sensors capable of room-temperature operation with extremely high detectivities. Finally, novel spectrometer optical designs have led to robust multipass absorption cells capable of >400 m effective pathlength in a compact package. In combination with accurate spectroscopic databases, these developments have afforded dramatic improvements in measurement sensitivity, accuracy, precision, and selectivity. We will present several examples of the applications of high resolution mid-IR spectrometers in real-world field measurements at sampling towers and aboard mobile platforms such as vehicles and airplanes.

TB03 9:22 – 9:37

A PORTABLE DUAL FREQUENCY COMB SPECTROMETER FOR ATMOSPHERIC APPLICATIONS

KEVIN C COSSEL, ELEANOR WAXMAN, GAR-WING TRUONG, FABRIZIO GIORGETTA, WILLIAM C SWANN, *Applied Physics Division, NIST, Boulder, CO, USA*; SEAN COBURN, ROBERT WRIGHT, GREG B RIEKER, *Department of Mechanical Engineering, University of Colorado Boulder, Boulder, CO, USA*; IAN CODDINGTON, NATHAN R. NEWBURY, *Applied Physics Division, NIST, Boulder, CO, USA.*

Dual frequency comb (DFC) spectroscopy is a new technique that combines broad spectral bandwidth, high spectral resolution, rapid data acquisition, and high sensitivity. In addition, unlike standard Fourier-transform spectroscopy, it has an almost ideal instrument lineshape function, does not require recalibration, and has no moving parts. These features make DFC spectroscopy well suited for accurate measurements of multiple species simultaneously. Because the frequency comb lasers can be well collimated, such a system can be used for long open-path measurements with path lengths ranging from hundreds of meters to several kilometers[a]. This length scale bridges the gap between point measurements and satellite-based measurements and is ideal for providing information about local sources and quantifying emissions.

Here we show a fully portable DFC spectrometer operating over a wide spectral region in the near-infrared (about 1.5-2.1 μm or 6670-4750 cm^{-1} sampled at 0.0067 cm^{-1}) and across several different open-air paths up to a path length of 11.8 km. The current spectrometer fits in about a 500 L volume and has low power consumption. It provides simultaneous measurements of CO_2, CH_4, and water isotopes with a time resolution of seconds to minutes. This system has several potential applications for atmospheric measurements including continuous monitoring city-scale emissions and localizing methane leaks from oil and gas wells.

[a]G. B. Rieker, F. R. Giorgetta, W. C. Swann, J. Kofler, A. M. Zolot, L. C. Sinclair, E. Baumann, C. Cromer, G. Petron, C. Sweeney, P. P. Tans, I. Coddington, and N. R. Newbury, Frequency-comb-based remote sensing of greenhouse gases over kilometer air paths, Optica, 1(5), 290-298 (2014).

TB04 9:39 – 9:54

METHANE DETECTION FOR OIL AND GAS PRODUCTION SITES USING PORTABLE DUAL-COMB SPECTROMETRY

SEAN COBURN, ROBERT WRIGHT, *Department of Mechanical Engineering, University of Colorado Boulder, Boulder, CO, USA*; KEVIN C COSSEL, GAR-WING TRUONG, ESTHER BAUMANN, IAN CODDINGTON, NATHAN R. NEWBURY, *Applied Physics Division, NIST, Boulder, CO, USA*; CAROLINE ALDEN, *Department of Mechanical Engineering, University of Colorado Boulder, Boulder, CO, USA*; SUBHOMOY GHOSH, KULDEEP PRASAD, *Fire Research Division, NIST, Gaithersburg, MD, USA*; GREG B RIEKER, *Department of Mechanical Engineering, University of Colorado Boulder, Boulder, CO, USA.*

Considerable uncertainty exists regarding the contribution of oil and gas operations to anthropogenic emissions of atmospheric methane. Additionally, new proposed EPA regulations on volatile organic compound (VOC) emissions from oil and gas production facilities have been expanded to include methane, making this a topic of growing importance to the oil and gas industry as well as regulators. In order to gain a better understanding of emissions, reliable techniques that enable long-term monitoring of entire production facilities are needed. Recent advances in the development of compact and robust fiber frequency combs are enabling the use of this powerful spectroscopic tool outside of the laboratory. Here we characterize and demonstrate a dual comb spectrometer (DCS) system with the potential to locate and size methane leaks from oil and gas production sites over extended periods of time. The DCS operates over kilometer scale open paths, and the path integrated methane measurements will ultimately be coupled with an atmospheric inversion utilizing local meteorology and a high resolution fluid dynamics simulation to determine leak location and also derive a leak rate. High instrument precision is needed in order to accurately perform the measurement inversion on the highly varying methane background, thus the DCS system has been fully optimized for the detection of atmospheric methane in the methane absorption region around 180-184 THz.

TB05 9:56 – 10:11

TIME-RESOLVED FREQUENCY COMB SPECTROSCOPY FOR STUDYING THE KINETICS AND BRANCHING RATIO OF OD+CO

THINH QUOC BUI, *JILA, National Institute of Standards and Technology and Univ. of Colorado Department of Physics, University of Colorado, Boulder, Boulder, CO, USA*; BRYCE J BJORK, OLIVER H HECKL, BRYAN CHANGALA, BEN SPAUN, *JILA, National Institute of Standards and Technology and Univ. of Colorado Department of Physics, University of Colorado, Boulder, CO, USA*; MITCHIO OKUMURA, *Division of Chemistry and Chemical Engineering, California Institute of Technology, Pasadena, CA, USA*; JUN YE, *JILA, National Institute of Standards and Technology and Univ. of Colorado Department of Physics, University of Colorado, Boulder, Boulder, CO, USA.*

The chemical kinetics of the OH+CO reaction plays important roles in combustion and atmospheric processes. OH+CO has two product channels, $H+CO_2$ and the stabilized HOCO intermediate, with a branching ratio that is highly pressure dependent. Therefore, establishing an accurate kinetic model for this chemical system requires knowledge of the reaction rates and product yields, and the lifetimes of all molecules along a particular reaction pathway. We report the application of time-resolved frequency comb spectroscopy (TRFCS) in the mid-infrared (3.7 μm) spectral region to address the complex reaction kinetics of OD+CO at room temperature. We use the deuterated forms to avoid atmospheric water interference. This technique allows us to detect the lowest energy conformer trans-DOCO intermediate with high time-resolution and sensitivity while also permitting the direct determination of rotational state distributions of all relevant molecules. We simultaneously observe the time-dependent concentrations of trans-DOCO, OD, and D_2O which are used in conjunction with kinetics modeling to obtain the pressure- and collision partner-dependent branching ratio of OD+CO.

Intermission

TB06 10:30 – 10:45

DEMONSTRATION OF A RAPIDLY-SWEPT EXTERNAL CAVITY QUANTUM CASCADE LASER FOR ATMOSPHERIC SENSING APPLICATIONS[a]

BRIAN E BRUMFIELD, MATTHEW S TAUBMAN, MARK C PHILLIPS, *Optical Sensing, Pacific Northwest National Laboratory, Richland, WA, USA*; JONATHAN D SUTER, *Applied Optics, Pacific Northwest National Laboratory, Richland, WA, USA.*

The application of quantum cascade lasers (QCLs) in atmospheric science for trace detection of gases has been demonstrated using sensors in point or remote sensing configurations. Many of these systems utilize single narrowly-tunable ($\sim$10 cm^{-1}) distributed feedback (DFB-) QCLs that limit simultaneous detection to a restricted number of small chemical species like H_2O or N_2O. The narrow wavelength range of DFB-QCLs precludes accurate quantification of large chemical species with broad rotationally-unresolved vibrational spectra, such as volatile organic compounds, that play an important role in the chemistry of the atmosphere. External-cavity (EC-) QCL systems are available that offer tuning ranges greater than 100 cm^{-1}, making them excellent IR sources for measuring multiple small and large chemical species in the atmosphere. While the broad wavelength coverage afforded by an EC system enables measurements of large chemical species, most commercial systems can only be swept over their entire wavelength range at less than 10 Hz. This prohibits broadband simultaneous measurements of multiple chemicals in plumes from natural or industrial sources where turbulence and/or chemical reactivity are resulting in rapid changes in chemical composition on sub-1s timescales.

At Pacific Northwest National Laboratory we have developed rapidly-swept EC-QCL technology that acquires broadband absorption spectra ($\sim$100 cm^{-1}) on ms timescales. The spectral resolution of this system has enabled simultaneous measurement of narrow rotationally-resolved atmospherically-broadened lines from small chemical species, while offering the broad tuning range needed to measure broadband spectral features from multiple large chemical species. In this talk the application of this technology for open-path atmospheric measurements will be discussed based on results from laboratory measurements with simulated plumes of chemicals. The performance offered by the system for simultaneous detection of multiple chemical species will be presented.

[a]The Pacific Northwest National Laboratory is operated for the U.S. Department of Energy (DOE) by the Battelle Memorial Institute under Contract No. DE-AC05-76RL01830.

TB07   **10:47 – 11:02**

DEVELOPMENT OF A QUANTUM CASCADE LASER-BASED SPECTROMETER FOR MEASUREMENTS OF BIO-
GENIC VOLATILE ORGANIC COMPOUNDS

JACOB STEWART, *Department of Chemistry, Connecticut College, New London, CT, USA.*

Biogenic volatile organic compounds (BVOCs) are emitted into Earth's atmosphere by plants and are among the most
abundant reactive organic species in the troposphere. These compounds play an important role in atmospheric chemistry,
including the formation of secondary organic aerosols and production of surface-level ozone, a pollutant which can have
negative health effects. BVOCs are generally measured and monitored using mass spectrometry and gas chromatography, but
infrared spectroscopy is an excellent complementary tool for measuring these species. The development of quantum cascade
lasers (QCLs) has provided robust, coherent light sources which give access to fundamental infrared transitions of BVOCs
that lie in the "infrared window" from 8-14 um. At Connecticut College, we are developing a QCL-based spectrometer for
measuring BVOCs with high resolution and high sensitivity. We will present details on the construction of our spectrometer
and preliminary data for measurements of isoprene (C_5H_8), the most abundant BVOC in the troposphere.

TB08   **11:04 – 11:14**

RAMAN LIDAR PROFILING OF TROPOSPHERIC WATER VAPOR

WATHEQ AL-BASHEER, *Department of Physics, King Fahd University of Petroleum and Minerals, Dhahran,
Saudi Arabia.*

Obtaining vertical profiles of tropospheric water vapor provides critically important information towards understanding
short and long term global climate change. Ground-based Raman lidar technique is a powerful tool to precisely evaluating
Water vapor Mixing Ratio (WVMR) in the troposphere. In this presentation, an overview of the design and basic components
of a Raman water vapor lidar setup employing the third harmonic output (at 355 nm) of a high-powered laser with a telescope
and three detection channels will be presented. Also, detailed discussion of the best method to calibrate and evaluate the
performance of a typical water vapor Raman lidar will be shown and compared with most common calibration methods. By
manipulating the inelastic backscattering lidar signals from the Raman nitrogen channel (386.7 nm) and Raman water vapor
channel (407.5 nm), vertical profiles of water vapor mixing ratio (WVMR) will be deduced, calibrated, and compared against
WVMR profiles obtained from coincident and collocated radiosonde profiles. This presented methodology will be shown to
effectively yield high temporal and spatial resolution measurements of WVMR, with efficient dual detector capability both in
the near-and-far fields.

TB09   **11:16 – 11:31**

PHOTOCHEMICAL FORMATION OF AEROSOL IN PLANETARY ATMOSPHERES: PHOTON AND WATER MEDI-
ATED CHEMISTRY OF SO_2

JAY A KROLL, *Department of Chemistry and Biochemistry, University of Colorado, Boulder, CO, USA*; D.
J. DONALDSON, *Chemistry, University of Toronto, Toronto, Canada*; VERONICA VAIDA, *Department of
Chemistry and Biochemistry, University of Colorado, Boulder, CO, USA.*

Sulfur compounds have been observed in a number of planetary atmospheres throughout our solar system. Our current
understanding of sulfur chemistry explains much of what we observe in Earth's atmosphere. However, several discrepancies
between modeling and observations of the Venusian atmosphere show there are still problems in our fundamental understanding of sulfur chemistry. This is of particular concern due to the important role sulfur compounds play in the formation of
aerosols, which have a direct impact on planetary climates, including Earth's. We investigate the role of water complexes in
the hydration of sulfur oxides and dehydration of sulfur acids and will present spectroscopic studies to document such effects.
I will present recent work investigating mixtures of SO_2 and water that generate large quantities of aerosol when irradiated
with solar UV light, even in the absence of traditional OH chemistry. I will discuss a proposed mechanism for the formation
of sulfurous acid (H_2SO_3) and present recent experimental work that supports this proposed mechanism. Additionally, the
implications that photon-induced hydration of SO_2 has for aerosol formation in the atmosphere of earth as well as other
planetary atmospheres will be discussed.

TB10 **11:33 – 11:48**

GAS PHASE HYDRATION OF METHYL GLYOXAL TO FORM THE GEMDIOL

JAY A KROLL, *Department of Chemistry and Biochemistry, University of Colorado, Boulder, CO, USA*; JESSICA L AXSON, *Department of Environmental Health Sciences, University of Michigan, Ann Arbor, Michigan, USA*; VERONICA VAIDA, *Department of Chemistry and Biochemistry, University of Colorado, Boulder, CO, USA*.

Methylglyoxal is a known oxidation product of volatile organic compounds (VOCs) in Earth's atmosphere. While the gas phase chemistry of methylglyoxal is fairly well understood, its modeled concentration and role in the formation of secondary organic aerosol (SOA) continues to be controversial. The gas phase hydration of methylglyoxal to form a gemdiol has not been widely considered for water-restricted environments such as the atmosphere. However, this process may have important consequences for the atmospheric processing of VOCs. We will report on spectroscopic work done in the Vaida laboratory studying the hydration of methylglyoxal and discuss the implications for understanding the atmospheric processing and fate of methylglyoxal and similar molecules.

TC. Instrument/Technique Demonstration

Tuesday, June 21, 2016 – 8:30 AM

Room: 274 Medical Sciences Building

Chair: David A. Long, National Institute of Standards and Technology, Gaithersburg, MD, USA

TC01   8:30 – 8:40

HIGH HARMONIC GENERATION XUV SPECTROSCOPY FOR STUDYING ULTRAFAST PHOTOPHYSICS OF CO-ORDINATION COMPLEXES

<u>ELIZABETH S RYLAND</u>, MING-FU LIN, MAX A VERKAMP, JOSH VURA-WEIS[a], *Department of Chemistry, University of Illinois at Urbana-Champaign, Urbana, IL, USA.*

Extreme ultraviolet (XUV) spectroscopy is an inner shell technique that probes the $M_{2,3}$-edge excitation of atoms. Absorption of the XUV photon causes a $3p \rightarrow 3d$ transition, the energy and shape of which is directly related to the element and ligand environment. This technique is thus element-, oxidation state-, spin state-, and ligand field specific. A process called high-harmonic generation (HHG) enables the production of ultrashort (~20fs) pulses of collimated XUV photons in a table-top instrument. This allows transient XUV spectroscopy to be conducted as an in-lab experiment, where it was previously only possible at accelerator-based light sources. Additionally, ultrashort pulses provide the capability for unprecedented time resolution (~70fs IRF). This technique has the capacity to serve a pivotal role in the study of electron and energy transfer processes in materials and chemical biology. I will present the XUV transient absorption instrument we have built over the past two years, along with preliminary data and simulations of the $M_{2,3}$-edge absorption data of a battery of small inorganic molecules to demonstrate the high specificity of this ultrafast tabletop technique.

---

[a]PI

TC02   8:42 – 8:57

PHOTOELECTRON VELOCITY MAP IMAGING OF VIBRATIONALLY EXCITED, GAS-PHASE BIOMOLECULES AND THEIR ANIONS

<u>DANIËL BAKKER</u>, SJORS BAKELS, RUTGER VAN DER MADE, ATZE PETERS, ANOUK RIJS, *FELIX Laboratory, Institute for Molecules and Materials (IMM), Radboud University, Nijmegen, Netherlands.*

A powerful method in spectroscopy to characterize the structure of large, gas phase molecules is to probe the ionization yield upon irradiating the molecules with infrared (IR) and/or ultraviolet (UV) radiation. When this spectroscopic technique is employed, the photodetached electrons are usually ignored, although they contain information on, for example, the ionization threshold of the molecule and the excited states of the formed ions.

Here, the novel combination of a molecular beam mass spectrometer equipped with a laser desorption source, the free electron laser FELIX and the powerful velocity map imaging (VMI) technique is presented. With this extended set of tools we can bring large molecules intact into the gas phase and prepare them in specific vibrationally excited states. UV or VUV radiation can subsequently be used to ionize the molecules. The kinetic energy and the radial distribution of the photoelectrons can be measured using VMI combined with ion detection using a time-of-flight mass spectrometer.

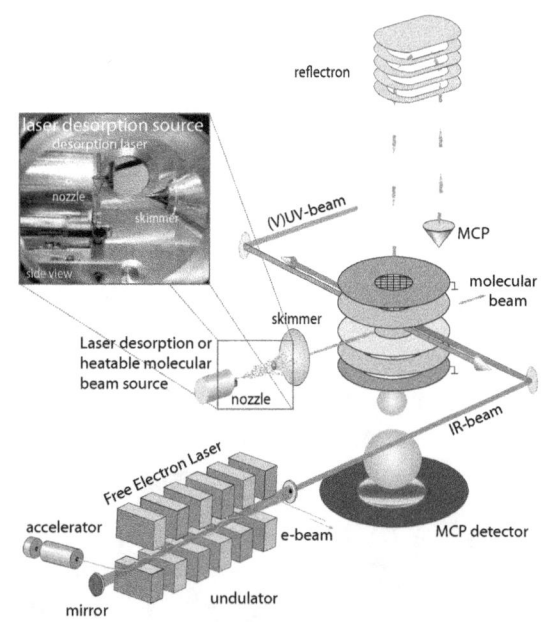

TC03   8:59 – 9:14

AN OPTICALLY ACCESSIBLE PYROLYSIS MICROREACTOR

JOSHUA H BARABAN, *Department of Chemistry, University of Colorado, Boulder, CO, USA*; DONALD E DAVID, *Integrated Instrument Development Facility, CIRES, University of Colorado, Boulder, CO, USA*; BARNEY ELLISON, *Department of Chemistry and Biochemistry, University of Colorado, Boulder, CO, USA*; JOHN W DAILY, *Department of Mechanical Engineering, University of Colorado Boulder, Boulder, CO, USA.*

We report an optically accessible pyrolysis micro-reactor suitable for *in situ* laser spectroscopic measurements. A radiative heating design allows for completely unobstructed views of the micro-reactor along two axes. The maximum temperature demonstrated here is only 1300 K (as opposed to 1700 K for the usual SiC micro-reactor) because of the melting point of fused silica, but alternative transparent materials will allow for higher temperatures. Laser induced fluorescence measurements on nitric oxide are presented as a proof of principle for spectroscopic characterization of pyrolysis conditions. (This work has been published in J. H. Baraban, D. E. David, G. B. Ellison, and J. W. Daily. An Optically Accessible Pyrolysis Micro-Reactor. *Review of Scientific Instruments*, 87(1):014101, 2016.)

TC04   9:16 – 9:31

BOHENDI@FELIX: PROBING THE FAR-INFRARED FINGERPRINT OF SMALL CLUSTERS IN HELIUM NANODROPLETS WITH A FREE ELECTRON LASER

GERHARD SCHWAAB, RAFFAEL SCHWAN, DEVENDRA MANI, ARGHYA DEY, THEO FISCHER, MATIN KAUFMANN, *Physikalische Chemie II, Ruhr University Bochum, Bochum, Germany*; BRITTA REDLICH, LEX VAN DER MEER, *Institute for Molecules and Materials (IMM), Radboud University Nijmegen, Nijmegen, Netherlands*; MARTINA HAVENITH, *Physikalische Chemie II, Ruhr University Bochum, Bochum, Germany.*

Recently, we have installed a helium nanodroplet machine [1,2] at the free electron beamline FELIX in Nijmegen. The current setup allows to study neutral molecules and molecular complexes in the full spectral range from 500–3000 cm^{-1}. First proof of principle experiments using the strong absorber SF_6 were used to verify the overall alignment between helium nanodroplet beam and the FELIX radiation source.

Applications so far included the study of small water clusters and the investigation of microsolvation of small solutes. These results will be presented and compared to recent theoretical predictions of the Bowman group.[3]

[1] K. von Haeften et al., Phys. Rev. B. 73, 054502 (2006)
[2] Choi et al., Int. Rev. Phys. Chem. 25, 15 (2006)
[3] Samantha et al., Acc. Chem. Res. 47, 2700 (2014)

TC05   9:33 – 9:43

SUPERCONTINUUM CAVITY ENHANCED ABSORPTION SPECTROSCOPY FOR H_2O/D_2O SOLUTIONS

MINGYUN LI, KEVIN LEHMANN, *Department of Chemistry, The University of Virginia, Charlottesville, VA, USA.*

Water and heavy water is always a combination of liquids that we would like to know their concentrations in mixtures. In this work, we are trying to make a cavity enhanced absorption spectroscopy (CEAS) setup in liquid phase in the near infrared region. By combining a self-built supercontinuum light source with a fiber loop, we are able to build a setup that has a very broad wavelength coverage to work in the liquid phase. A side-polished-fiber is used as a sensing region on the loop. Some H_2O/D_2O sample pairs are tested first for its properties. The results show that this new setup has the ability in liquid phase detection, and a detection limit of less than 10% H_2O in D_2O solutions can be reached so far.

Intermission

TC06

MID-INFRARED FREQUENCY-AGILE DUAL-COMB SPECTROSCOPY

PEI-LING LUO[a], MING YAN, KANA IWAKUNI, *Laser Spectroscopy Division, Max Planck Institute of Quantum Optics, Garching, Germany*; GUY MILLOT, *Laboratoire ICB, CNRS/Université de Bourgogne, DIJON, France*; THEODOR W. HÄNSCH, NATHALIE PICQUÉ[b], *Laser Spectroscopy Division, Max Planck Institute of Quantum Optics, Garching, Germany.*

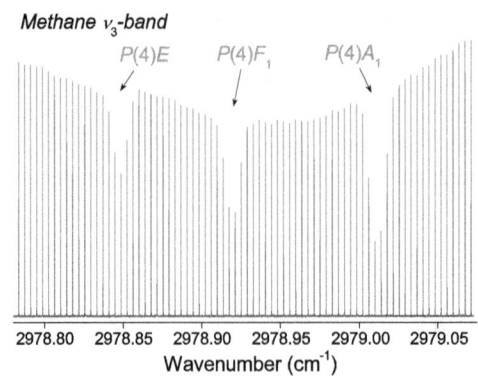

Methane ν_3-band

We demonstrate a new approach to mid-infrared dual-comb spectroscopy. It opens up new opportunities for accurate real-time spectroscopic diagnostics and it significantly simplifies the technique of dual-comb spectroscopy. Two mid-infrared frequency combs of slightly different repetition frequencies and moderate, but rapidly tunable, spectral span are generated in the 2800-3200 cm^{-1} region. The generators rely on electro-optic modulators, nonlinear fibers[c] for spectral broadening and difference frequency generation and do not involve mode-locked lasers. Flat-top frequency combs span up to 10 cm^{-1} with a comb line spacing of 100 MHz (3×10^{-3} cm^{-1}). The performance of the spectrometer without any phase-lock electronics or correction scheme is illustrated with spectra showing resolved comb lines and Doppler-limited spectra of methane. High precision on the spectroscopic parameter (line positions and intensities) determination is demonstrated for spectra measured on a millisecond time scale and it is validated with comparison with literature data.

[a] pei-ling.luo@mpq.mpg.de
[b] nathalie.picque@mpq.mpg.de
[c] G. Millot, S. Pitois, M. Yan, T. Hovannysyan, A. Bendahmane, T.W. Hänsch, N. Picqué, Frequency-agile dual-comb spectroscopy, Nature Photonics 10, 27-30 (2016).

TC07

IMPROVED SPECTROSCOPY OF MOLECULAR IONS IN THE MID-INFRARED WITH UP-CONVERSION DETECTION

CHARLES R. MARKUS, ADAM J. PERRY, JAMES N. HODGES, *Department of Chemistry, University of Illinois at Urbana-Champaign, Urbana, IL, USA*; BENJAMIN J. McCALL, *Departments of Chemistry and Astronomy, University of Illinois at Urbana-Champaign, Urbana, IL, USA.*

Heterodyne detection, velocity modulation, and cavity enhancement are useful tools for observing rovibrational transitions of important molecular ions.[a] We have utilized these methods to investigate a number of molecular ions, such as H_3^+, CH_5^+, HeH^+, and OH^+.[bcde] In the past, parasitic etalons and the lack of fast and sensitive detectors in the mid-infrared have limited the number of transitions we could measure with MHz-level precision. Recently, we have significantly reduced the amplitude of unwanted interference fringes with a Brewster-plate spoiler. We have also developed a detection scheme which up-converts the mid-infrared light with difference frequency generation which allows the use of a faster and more sensitive avalanche photodetector. The higher detection bandwidth allows for optimized heterodyne detection at higher modulation frequencies. The overall gain in signal-to-noise from both improvements will enable extensive high-precision line lists of molecular ions and searches for previously unobserved transitions.

[a] K.N. Crabtree, J.N. Hodges, B.M. Siller, A.J. Perry, J.E. Kelly, P.A. Jenkins II, and B.J. McCall, Chem. Phys. Lett. 551 (2012) 1-6.
[b] A.J. Perry, J.N. Hodges, C.R. Markus, G.S. Kocheril, and B.J. McCall, J. Mol. Spec. 317 (2015) 71-73.
[c] J.N. Hodges, A.J. Perry, P.A. Jenkins II, B.M. Siller, and B.J. McCall, J. Chem. Phys. 139 (2013) 164291.
[d] A.J. Perry, J.N. Hodges, C.R. Markus, G.S. Kocheril, and B.J. McCall. 2014, J. Chem. Phys. 141, 101101
[e] C.R. Markus, J.N. Hodges, A.J. Perry, G.S. Kocheril, H.S.P. Müller, and B.J. McCall, Astrophys. J. 817 (2016) 138.

TC08   **10:36 – 10:51**

HIGH RESOLUTION ROVIBRATIONAL SPECTROSCOPY OF LARGE MOLECULES USING INFRARED FREQUENCY COMBS AND BUFFER GAS COOLING

BRYAN CHANGALA, BEN SPAUN, *JILA, National Institute of Standards and Technology and Univ. of Colorado Department of Physics, University of Colorado, Boulder, CO, USA*; DAVID PATTERSON, *Department of Physics, Harvard University, Cambridge, MA, USA*; BRYCE J BJORK, OLIVER H HECKL, *JILA, National Institute of Standards and Technology and Univ. of Colorado Department of Physics, University of Colorado, Boulder, CO, USA*; JOHN M. DOYLE, *Department of Physics, Harvard University, Cambridge, MA, USA*; JUN YE, *JILA, National Institute of Standards and Technology and Univ. of Colorado Department of Physics, University of Colorado, Boulder, CO, USA.*

We have recently demonstrated the integration of cavity-enhanced direct frequency comb spectroscopy with buffer gas cooling to acquire high resolution infrared spectra of translationally and rotationally cold ($\sim$ 10 K) gas-phase molecules.[a] Here, we extend this method to significantly larger systems, including naphthalene ($C_{10}H_8$), a prototypical polyaromatic hydrocarbon, and adamantane ($C_{10}H_{16}$), the fundamental building block of diamonoids. To the authors' knowledge, the latter molecule represents the largest system for which rotationally resolved spectra in the CH stretch region (3 μm) have been obtained. In addition to the measured spectra, we present several details of our experimental methods. These include introducing non-volatile species into the cold buffer gas cell and obtaining broadband spectra with single comb mode resolution. We also discuss recent modifications to the apparatus to improve its absorption sensitivity and time resolution, which facilitate the study of both larger molecular systems and cold chemical dynamics.

---

[a]B. Spaun, et al. *Probing buffer-gas cooled molecules with direct frequency comb spectroscopy in the mid-infrared*, WF02, 70th International Symposium on Molecular Spectroscopy, Champaign-Urbana, IL, 2015.

TC09   **10:53 – 11:08**

PROGRESS OF A NEW INSTRUMENT TO STUDY MOLECULAR DYNAMICS OF INTERSTELLAR ION-NEUTRAL REACTIONS

KEVIN ROENITZ, *Physical Chemistry, Emory University, Atlanta, GA, USA*; BEN LAMM, *Chemistry, University of South Carolina, Columbus, United States*; LYDIA RUDD, ANDY JUSTL, STEVEN LANDEWEER, DANNY ROADMAN, JUSTYNA KOSCIELNIAK, *Chemistry, Illinois Wesleyan University, Bloomington, United States*; ANDREW SONNENBERGER, *Chemistry, University of Minnesota, Minneapolis, MN, USA*; MANORI PERERA, *Chemistry, Illinois Wesleyan University, Bloomington, United States.*

Astrochemistry, a relatively young field of research, addresses a gap in our understanding of molecular evolution in space. With many space missions gathering data, the number of unresolved spectral lines is growing rapidly. Each year there are about three new molecules that are identified in the interstellar medium (ISM). However, our understanding of molecular processes, branching ratios, and rates are at a beginner level. For instance, we do not yet understand the chemical processes associated with the creation and evolution of even the most basic molecules such as water and methanol in space. One of the important steps toward understanding the chemistry of the ISM is to identify, through laboratory and theoretical work, a list of potential target molecules that are likely to exist in the ISM. This work describes experimental progress towards building a spectrometer that is able to produce complex cold ions that will react with cooled neutral molecules under conditions similar to those in space. I plan to present the astrochemical needs that motivated my research, how the new instrument will meet those needs, and the present status of the instrument and measurements in my lab.

TC10

CAVITY-ENHANCED ULTRAFAST SPECTROSCOPY: ULTRAFAST MEETS ULTRASENSITIVE

THOMAS K ALLISON, *Department of Chemistry, Stony Brook University, Stony Brook, NY, USA*; MELANIE ROBERTS REBER, *Department of Physics and Astronomy, State University of New York, Stony Brook, NY, USA*; YUNING CHEN, *Department of Chemistry, Stony Brook University, Stony Brook, NY, USA.*

Ultrafast optical spectroscopy methods, such as transient absorption spectroscopy and 2D-spectroscopy, are widely used across many disciplines. However, these techniques are typically restricted to optically thick samples, such as solids and liquid solutions. Using a frequency comb laser and optical cavities, we present a new technique for performing ultrafast optical spectroscopy with high sensitivity, enabling work in dilute gas-phase molecular beams. Resonantly enhancing the probe pulses, we demonstrate transient absorption measurements with a detection limit of $\Delta OD = 2 \times 10^{-10}$ ($1 \times 10^{-9}/\sqrt{Hz}$). Resonantly enhancing the pump pulses allows us to produce a high excitation fraction at high repetition-rate, so that signals can be recorded from samples with optical densities as low as $OD \approx 10^{-8}$, or column densities $< 10^{10}$ molecules/cm^2. To our knowledge, this represents a 5,000-fold improvement of the state-of-the-art.

TD. Mini-symposium: Spectroscopy in Traps
Tuesday, June 21, 2016 – 8:30 AM
Room: B102 Chemical and Life Sciences

Chair: Stephan Schlemmer, I. Physikalisches Institut, Köln, Germany

---

TD01   *INVITED TALK*   8:30 – 9:00

BOUND-FREE AND BOUND-BOUND SPECTROSCOPY OF COLD TRAPPED MOLECULAR IONS

ROLAND WESTER, *Institute for Ion Physics and Applied Physics, University of Innsbruck, Innsbruck, Austria.*

Cryogenic radiofrequency ion traps have become a versatile tool to study the spectroscopy and state-selected collision dynamics of molecular ions. Different types of action spectroscopy have been developed to obtain a precise and sensitive spectroscopic signature. In this talk I will give an introduction to molecular ion spectroscopy in multipole traps. Then I will present recent experimental and theoretical investigations from our group on photodetachment spectroscopy and state-selected collisions of cold OH^- anions colliding with helium and hydrogen. Based on these results we performed high resolution terahertz spectroscopy on the two lowest rotational transitions of OD^-. Work is in progress to extend the rotational spectroscopy to polyatomic molecular anions.

TD02   9:05 – 9:20

EQUATION-OF-MOTION COUPLED-CLUSTER CALCULATIONS OF PHOTODETACHMENT CROSS SECTIONS FOR ATOMIC NEGATIVE IONS ACROSS THE PERIODIC TABLE

TAKATOSHI ICHINO, *Department of Chemistry, The University of Texas, Austin, TX, USA*; LAN CHENG, *Department of Chemistry, Johns Hopkins University, Baltimore, MD, USA*; JOHN F. STANTON, *Department of Chemistry, The University of Texas, Austin, TX, USA.*

The innovative application of the ion-trap technique by Wester and coworkers has yielded definitive experimental values of photodetachment cross sections for the atomic oxygen radical anion ($O^{\bullet-}$) [Hlavenka et al., J. Chem. Phys. **130**, 061105 (2009)]. In the present study, equation-of-motion coupled-cluster (EOM-CC) calculations have been performed to derive theoretical values of photodetachment cross sections for the negative ions of atoms in the first two periods of the periodic table as well as of those which belong to the alkali metal and halogen groups. Two methods have been employed to derive the cross sections. One involves the Dyson orbitals obtained from EOM-CC calculations and plane wave functions for the detached electron in the transition dipole moment integrals. The other method utilizes the moment theory following EOM-CC calculations of transition dipole moments for a large number of pseudo-states. The cross sections so evaluated for $O^{\bullet-}$ match the experimental values very well. Generally good agreement has been found between the theoretical and experimental values of the cross sections for the atoms in the first two periods, while the present calculations cast some doubt on reported experimental values for some atoms beyond the second period. Substantial relativistic effects on the cross section have been observed for heavy elements in the alkali metal and halogen groups.

TD03   9:22 – 9:37

THE ELECTRONIC SPECTRUM OF CRYOGENIC RUTHENIUM-TRIS-BIPYRIDINE DICATIONS

SHUANG XU, *JILA and Department of Physics, University of Colorado at Boulder, Boulder, CO, USA*; JAMES E. T. SMITH, J. MATHIAS WEBER, *JILA and the Department of Chemistry and Biochemistry, University of Colorado-Boulder, Boulder, CO, USA.*

We present the electronic spectrum of Ru(II)-tris(2,2'-bipyridine), $Ru(bpy)_3^{2+}$, measured by photodissociation spectroscopy of mass selected $Ru(bpy)_3^{2+} \cdot N_2$ ions prepared in a cryogenic quadrupole ion trap. The spectrum is composed of several metal-to-ligand charge transfer (MLCT) transitions, as well as metal centered bands and ligand centered $\pi\pi^*$ states. We observe several partially resolved electronic transitions in the MLCT band. We discuss the results in the framework of time-dependent density functional theory.

TD04   9:39 – 9:54

ELECTRONIC SPECTRA OF TRIS(2,2'-BIPYRIDINE)-METAL COMPLEX IONS IN GAS PHASE

SHUANG XU, *Department of Physics, JILA - University of Colorado, Boulder, CO, USA*; JAMES E. T. SMITH, *JILA and the Department of Chemistry and Biochemistry, University of Colorado-Boulder, Boulder, CO, USA*; J. MATHIAS WEBER, *Department of Chemistry and Biochemistry, JILA - University of Colorado, Boulder, CO, USA.*

Tris(bpy)-metal complexes (bpy = 2,2'-bipyridine) and their derivatives are important systems in metal-organic chemistry. While tris(bpy)-ruthenium, $Ru(bpy)_3^{2+}$, has been extensively studied, less attention has been paid to analogous complexes involving first row transition metals. Here we report the electronic spectra of a series of dicationic tris(bpy) chelates with different transition metals, measured by photodisscociation spectroscopy of cryogenically prepared ions. We focus our attention on the π-π^* transitions in the UV region of the spectrum.

Intermission

TD05   10:13 – 10:28

ELECTRONIC SPECTROSCOPY OF TRAPPED PAH PHOTOFRAGMENTS

CHRISTINE JOBLIN, ANTHONY BONNAMY, *IRAP, Université de Toulouse 3 - CNRS, Toulouse, France.*

The PIRENEA set-up combines an ion cyclotron resonance cell mass spectrometer with cryogenic cooling in order to study the physical and chemical properties of polycyclic aromatic hydrocarbons (PAHs) of astrophysical interest. In space, PAHs are submitted to UV photons that lead to their dissociation. It is therefore of interest to study fragmentation pathways and search for species that might be good interstellar candidates because of their stability[a].

Electronic spectroscopy can bring major insights into the structure of species formed by photofragmentation. This is also a way to identify new species in space as recently illustrated in the case of C_{60}^+ [b]. In PIRENEA, the trapped ions are not cold enough, and thus we cannot use complexation with rare gas in order to record spectroscopy, as was nicely performed in the work by Campbell et al. on C_{60}^+. We are therefore using the dissociation of the trapped ions themselves instead, which requires in general a multiple photon scheme[c]. This leads to non-linear effects that affect the measured spectrum.

We are working on improving this scheme in the specific case of the photofragment obtained by H-loss from 1-methylpyrene cation (CH_3-$C_{16}H_9^+$). A recent theoretical study has shown that a rearrangement can occur from 1-pyrenemethylium cation (CH_2-$C_{16}H_9^+$) to a system containing a seven membered ring (tropylium like pyrene system)[d]. This study also reports the calculated electronic spectra of both isomers, which are specific enough to distinguish them, and as a function of temperature. We will present experiments that have been performed to study the photophysics of these ions using the PIRENEA set-up and a two-laser scheme for the action spectroscopy. Acknowledgments[e]

[a]J. Montillaud, C. Joblin, D. Toublanc, *Astron. & Astrophys.* **552** (2013), id.A15
[b]E.K. Campbell, M. Holz, D. Gerlich, and J.P. Maier, *Nature* **523** (2015), 322-323
[c]F. Useli-Bacchitta, A. Bonnamy, G. Malloci, et al., *Chem. Phys.* **371** (2010), 16-23; J. Zhen, A. Bonnamy, G. Mulas, C. Joblin, *Mol. Astrophys.* **2** (2015), 12-17
[d]M. Rapacioli, A. Simon, C.C.M. Marshall, et al., *J. Phys. Chem. A* **119** (2015), 12845-12854
[e]European Research Council grant ERC-2013-SyG, Grant Agreement n. 610256 NANOCOSMOS.

TD06   10:30 – 10:45

ELECTRONIC SPECTRA OF BARE AND SOLVATED RUTHENIUM POLYPYRIDINE COMPLEXES

SHUANG XU, *JILA and Department of Physics, University of Colorado at Boulder, Boulder, CO, USA*; JAMES E. T. SMITH, J. MATHIAS WEBER, *JILA and the Department of Chemistry and Biochemistry, University of Colorado-Boulder, Boulder, CO, USA.*

We present work on a prototypical water oxidation catalyst, namely the aqua-complex [(bpy)(tpy)Ru-OH_2]$^{2+}$ (2,2'-bpy = bipyridine, tpy = 2,2':6',2"-terpyridine), and its hydrated clusters [(bpy)(tpy)Ru-OH_2]$^{2+}$ ·$(H2O)_n$, with n = 1 – 4. This complex is the starting species in a catalytic cycle for water oxidation. We couple electrospray ionization mass spectrometry with laser spectroscopy to circumvent challenges that arise in reactive solutions from speciation. Here, we report the electronic spectrum of [(bpy)(tpy)Ru-OH_2]$^{2+}$ by photodissociation spectroscopy of mass selected, cryogenically prepared ions, and we examine effects of its microhydration environment on its electronic structure. In particular, we investigate the solvatochromic shift of the spectral envelope upon sequential addition of water molecules up to the tetrahydrate.

TD07   **10:47 – 11:02**

CONFORMATIONAL SPECIFIC INFRARED AND ULTRAVIOLET SPECTROSCOPY OF COLD YA(D-Pro)AA·H$^+$ IONS: A STEROCHEMICAL "TWIST" ON THE PROLINE EFFECT

CHRISTOPHER P HARRILAL, ANDREW F DeBLASE, NICOLE L BURKE, SCOTT A McLUCKEY, TIMOTHY S. ZWIER, *Department of Chemistry, Purdue University, West Lafayette, IN, USA.*

The "proline effect" is a well-known fragmentation phenomenon in mass spectrometry, in which y-fragments are produced preferentially over b-fragments during the collision induced dissociation of protonated L-proline containing peptide ions. This specific fragmentation channel is favored because of the high basicity of the secondary amine intermediate and the ring instability in alternative b$_n^+$ products [ASMS 2014, 25, 1705]. In contrast, peptides containing the D-Pro stereoisomer have been shown to largely favor the production of b$_4^+$ ions over y$_3^+$ ions. This strongly suggests that differences in the conformational preferences between the D-Pro and L-Pro diastereomers are likely to be responsible but structural evidence has been lacking to date. Using tandem mass spectrometry and IR-UV double resonant action spectroscopy we are able to compare the 3D structures of cold [YA(D-Pro)AA+H]$^+$ to [YA(L-Pro)AA+H]$^+$ ions. The UV action spectra reveals two major conformers in [YA(D-Pro)AA+H]$^+$ and one major conformer in [YA(L-Pro)AA+H]$^+$. Clear differences in the hydrogen bonding patterns are apparent between the two conformers observed in the D-Pro specie which are both distinct from the L-Pro diastereomer. Furthermore, conformer and diastereomer specific photofragmentation patterns are observed. It is also noted that a ten-fold photofragment enhancement unique to one of the D-Pro conformers is observed upon absorption of a resonant IR photon after UV excitation. Differences in the excited state photophysics between the two D-Pro conformers suggest that vibrational excitation of S$_1$ turns on coupling to the dissociative –Tyr channel in one conformer, while this coupling is already present in the vibronic ground state of the other. Calculated harmonic spectra (M052X/6-31+G*) of conformers obtained from Monte Carlo searches to the experimental spectra.

TD08   **11:04 – 11:19**

ALKALI CATION CHELATION IN COLD β-O-4 TETRALIGNOL COMPLEXES

ANDREW F DeBLASE, ERIC T DZIEKONSKI, JOHN R. HOPKINS, NICOLE L BURKE, HILKKA I KENT-TAMAA, SCOTT A McLUCKEY, TIMOTHY S. ZWIER, *Department of Chemistry, Purdue University, West Lafayette, IN, USA.*

Lignins are the second most abundant naturally occurring polymer class, contributing to about 30% of the organic carbon in the biosphere. Their primary function is to provide the structural integrity of plant cell walls and have recently come under consideration as a potential source of biofuels because they have an energy content similar to coal. Herein, we employ cold ion spectroscopy (UV action and IR-UV double resonance) to unravel the spectroscopic signatures of G-type alkali metal cationized (X = Li$^+$, Na$^+$, K$^+$) lignin tetramers connected by β-O-4 linkages. The conformation-specific spectroscopy reveals a variety of conformers, each containing distinct infrared spectra in the OH stretching region building on recent studies on the neutral and alkali metal cationized β-O-4 dimers. Based on comparisons of our infrared spectra to density functional theory [M05-2X/6-31+G*] harmonic level calculations for structures derived from a Monte Carlo conformational search, the alkali metal ion is discovered to engage in M$^+$-OH-O interactions as important motifs that determine the secondary structures of these complexes. This interaction disappears in the major conformer of the K$^+$ adduct, suggesting a reemergence of a neutral dimer segment as the metal binding energy decreases. Chelation of the metal cation by oxygen lone pair(s) of nearby oxygens in the β-O-4 linkage is observed to be the predominant driving force for 3D structure around the charge site, relegating OH-O H-bonds as secondary stabilizing elements.

SINGLE MOLECULAR ION SPECTROSCOPY: TOWARDS PRECISION MEASUREMENTS ON CaH^+

KENNETH R BROWN, NCAMISO B KHANYILE, RENE RUGANGO, *Department of Chemistry and Bio-chemistry, Georgia Institute of Technology, Atlanta, GA, USA*; GANG SHU, AARON CALVIN, *Chemistry and Biochemistry, Georgia Institute of Technology, Atlanta, United States.*

Precision spectroscopy of molecular ions has applications in astrochemistry, quantum state controlled chemical reactions, and measurements of fundamental constants. While spectroscopy of molecular ions is challenging, we present techniques to study molecular ions co-trapped with laser-cooled atomic ions in ion traps. We recently demonstrated the measurement of the $\nu' = 10 \leftarrow \nu = 0$ and $\nu' = 9 \leftarrow \nu = 0$ overtone transitions in CaH^{+}[a]. using resonant two photon dissociation. This technique is extended to the $2^1\Sigma \leftarrow 1^1\Sigma$ electronic transition, which should be rotationally resolvable. This resolution will allow further investigation into the internal state control of CaH^+ by techniques such as optical pumping, cryogenic cooling, and buffer gas cooling.

---

[a]N. B. Khanyile, *et. al.* Nat. Commun. **6** 7825 (2015).

TE. Fundamental interest

Tuesday, June 21, 2016 – 8:30 AM

Room: 217 Noyes Laboratory

Chair: S. A. Cooke, Purchase College SUNY, Purchase, NY, USA

TE01   8:30 – 8:45

HIGH RESOLUTION SPECTROSCOPY OF $A^1B_{1u} \leftarrow X^1A_g 8_0^1 4_0^1$ BAND OF NAPHTHALENE REFERENCED TO AN OPTICAL FREQUENCY COMB

KAZUKI NAKASHIMA, *Applied Physics, Fukuoka University, Fukuoka, Japan*; AKIKO NISHIYAMA, *Department of Engineering Science, The University of Electro-Communications, Chofu, Japan*; MASATOSHI MISONO, *Applied Physics, Fukuoka University, Fukuoka, Japan*.

In the excited vibronic states of naphthalene, there exist various interesting interactions such as intramolecular vibrational energy redistribution (IVR), intersystem crossing (ISC), and internal conversion (IC). More than thirty yeas ago, Beck *et al.* showed that IVR became remarkable when the excess energy exceeded about 2000 cm^{-1}, [a].

In the present study, we observe Doppler-free two-photon absorption spectra of $A^1B_{1u} \leftarrow X^1A_g 8_0^1 4_0^1$ band of naphthalene around 34281 cm^{-1}. The excess energy is 2261 cm^{-1}, which is just above the threshold of IVR. Thus we expect this band is suitable to analyze the dynamics in the excited vibronic states.

In our experiment, the spectral resolution is about 100 kHz, and rovibronic lines are well-resolved. To decide the transition frequencies, frequency shifts, and spectral linewidths with high accuracy and precision, we employed the comb-referenced Doppler-free two-photon absorption spectroscopic system[b]. We proceed to assign the rovibronic lines in ^qQ transition, and to determine molecular constants in the excited vibronic state.

[a] S. M. Beck, J. B. Hopkins, D. E. Powers, and R. E. Smalley, J. Chem. Phys. **74**, 43(1981).

[b] A. Nishiyama, K. Nakashima, A. Matsuba, and M. Misono, J. Mol. Spectrosc. **318**, 40 (2015).

BONDING OF ALKALI-ALKALINE EARTH MOLECULES IN THE LOWEST Σ^+ STATES OF DOUBLET AND QUARTET MULTIPLICITY

JOHANN V. POTOTSCHNIG, ANDREAS W. HAUSER, WOLFGANG E. ERNST, *Institute of Experimental Physics, Graz University of Technology, Graz, Austria.*

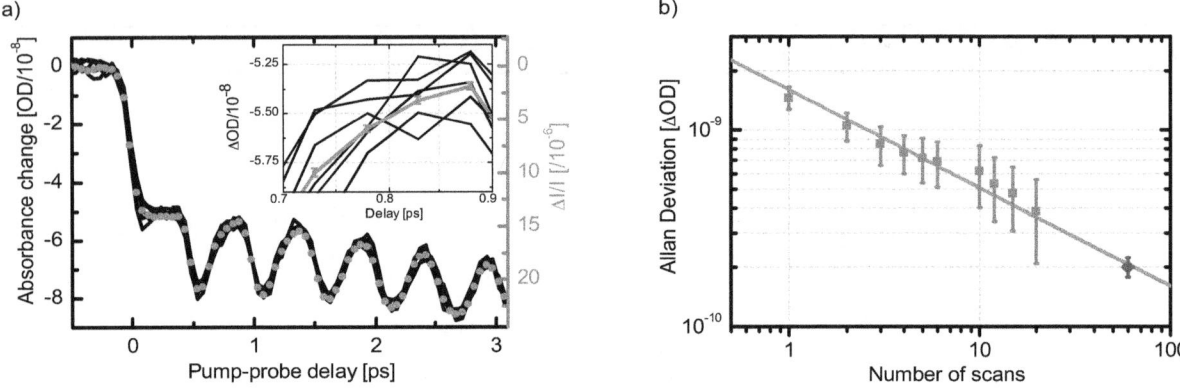

Figure 1: Noise performance of CE-TAS. (a), Transient absorption measurements taken with reduced gas flow and perpendicular polarizations. The red dots represent the average of 60 consecutive scans taken over a 1 hour period. Black curves are every 10th scan from the data set. Inset: Zoom-in around 0.8 ps delay. Error bars represent the uncertainty in the mean. (b), The green squares show the average of the Allan deviations obtained independently for each delay point. Error bars here are the standard deviation (not the uncertainty in the mean) of this ensemble, to represent the spread in the data. The blue diamond is the average of the error bars of (a), along with their standard deviation. The grey line has a slope of -1/2 on the log-log plot, the expected slope for white noise performance

In the present study the ground state as well as the lowest $^4\Sigma^+$ state were determined for 16 AK-AKE molecules.[a] Multireference configuration interaction calculations were carried out in order to understand the bonding of diatomic alkali-alkaline earth (AK-AKE) molecules. The correlations between molecular properties (disociation energy, bond distances, electric dipole moment) and atomic properties (electronegativity, polarizability) will be discussed. A correlation between the dissociation energy and the dipole moment of the lowest $^4\Sigma^+$ state was observed, while the dipole moment of the lowest $^2\Sigma^+$ state does not show such a simple dependency. In this case an empirical relation could be established. The class of AK-AKE molecules was selected for this investigation due to their possible applications in ultracold molecular physics.

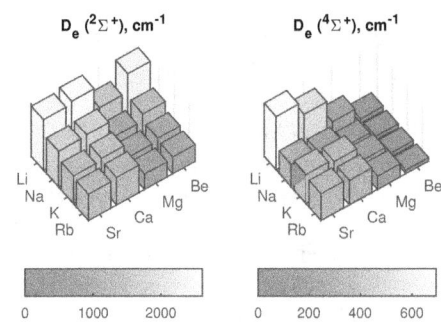

[a] J. V. Pototschnig, A. W. Hauser and W. E. Ernst, Phys. Chem. Chem. Phys., 2016,18, 5964-5973

TE03   9:04 – 9:19

A CANONICAL APPROACH TO MULTI-DIMENSIONAL VAN DER WAALS, HYDROGEN-BONDED, AND HALOGEN-BONDED POTENTIALS

JAY R. WALTON, *Department of Mathematics, Texas A & M University, College Station, TX, USA*; LUIS A. RIVERA-RIVERA, ROBERT R. LUCCHESE, JOHN W. BEVAN, *Department of Chemistry, Texas A & M University, College Station, TX, USA.*

A canonical approach is used to investigate prototypical multi-dimensional intermolecular interaction potentials characteristic of categories in van der Waals, hydrogen-bonded, and halogen-bonded intermolecular potential energy functions. It is demonstrated that well-characterized potentials in Ar-HI, OC-HI, OC-HF, and OC-BrCl, can be canonically transformed to a common dimensionless potential with relative error less than 0.010. The results indicate common intrinsic bonding properties despite other varied characteristics in the systems investigated. The results of these studies are discussed in the context of the previous statement made by J. C. Slater [J. Chem. Phys. 57 (1972) 2389] concerning fundamental bonding properties in the categories of interatomic interactions analyzed.

TE04   9:21 – 9:31

OBSERVATION OF BROADBAND ULTRAVIOLET EMISSION FROM $Hg_3{}^*$

WENTING WENDY CHEN, THOMAS C. GALVIN, J. GARY EDEN, *Department of Electrical and Computer Engineering, University of Illinois at Urbana-Champaign, Urbana, IL, USA.*

A previously-unobserved emission continuum, peaking at ~ 380 nm, has been observed when Hg vapor is photoexcited at 248 nm (KrF laser). Attributed to the mercury trimer, Hg_3, this emission continuum has a spectral breadth (FWHM) increased from ~ 65 to ~ 90 nm and a decay rate growed from $\sim 6 \times 10^3$ to $\sim 7 \times 10^3$ second^{-1}, corresponding to Hg vapor density rising from $\sim 10^{16}$ to $\sim 2 \times 10^{19}$ cm^{-3}. Comparisons of the observed spectra with theory [a] suggest that continuum arises from transitions of the molecule of $D_{\infty h}$ symmetrical linear, D_{3h} equilateral triangular and C_{2h} equilateral triangular configurations.

---

[a] Kitamura, Hikaru. "Theoretical potential energy surfaces for excited mercury trimers." Chemical physics 325.2 (2006): 207-219.

TE05   9:33 – 9:48

QUANTITATIVE DETERMINATION OF LINESHAPE PARAMETERS FROM VELOCITY MODULATION SPECTROSCOPY

JAMES N. HODGES, *Department of Chemistry, University of Illinois at Urbana-Champaign, Urbana, IL, USA*; BENJAMIN J. McCALL, *Departments of Chemistry and Astronomy, University of Illinois at Urbana-Champaign, Urbana, IL, USA.*

Velocity Modulation Spectroscopy (VMS) has stood as the gold standard in molecular ion spectroscopy for 30 years. Whether in a traditional uni-directional experiment or more complicated cavity-enhanced layouts with additional layers of modulation, VMS remains the preferred ion detection scheme and is responsible for the detection and transition frequency determination of around 50 molecules.

Despite its success, VMS still has a great deal of untapped potential. There have only been two other published studies[ab] of VMS lineshapes and both struggle with the highly correlated parameters: linewidth, intensity, and velocity modulation amplitude, *i.e.* the maximum Doppler shift during a period of the discharge. Due to this difficulty, both Gao and Civiš made concessions to achieve a good fit. Careful analysis of the contour of the transition profile allows us to properly disentangle those parameters in order to probe the environment of the positive column. We can extract the precise values for the translational temperature of the ion, the relative transition intensity, the ion mobility, and the electric field strength just from the lineshape of a single transition. A firm understanding of the lineshape will facilitate chemical and physical investigations of positive columns and allow for a better understanding of more complicated detection schemes.

---

[a] H. Gao *et al.*, Acta Phys. Sin. **50**, 1463 (2001).
[b] S. Civiš, Chem. Phys. **186**, 63 (1994).

TE06   9:50 – 10:05

USING NICE-OHVMS LINESHAPES TO STUDY RELAXATION RATES AND TRANSITION DIPOLE MOMENTS

JAMES N. HODGES, *Department of Chemistry, University of Illinois at Urbana-Champaign, Urbana, IL, USA*; BENJAMIN J. McCALL, *Departments of Chemistry and Astronomy, University of Illinois at Urbana-Champaign, Urbana, IL, USA.*

Noise Immune Cavity Enhanced Optical Heterodyne Velocity Modulation Spectroscopy (NICE-OHVMS) is a successful technique that we have developed to sensitively, precisely, and accurately record transitions of molecular ions.[a] It has been used exclusively as a method for precise transition frequency measurement via saturation and fitting of the resultant Lamb dips. NICE-OHVMS has been employed to improve the uncertainties on H_3^+, CH_5^+, HeH^+, and OH^+, reducing the transition frequency uncertainties by two orders of magnitude.[bcde]

Because NICE-OHVMS is a saturation technique, this provides a unique opportunity to access information about the ratio of the transition dipole moment to the relaxation rate of the transition. This can be done in two ways, either through comparison of Lamb dip depth to the transition profile or comparison of the absorption intensity and dispersion intensity. Due to the complexity of the modulation scheme, there are many parameters that affect the apparent intensity of the recorded lineshape. A complete understanding of the lineshape is required to make the measurements of interest.

Here we present a model that accounts for the heterodyne modulation and velocity modulation, assuming that the fundamental lineshape is represented by a Voigt profile. Fits to data are made and interpreted in order to extract the saturation parameter.

---

[a] K.N. Crabtree *et al.*, Chem. Phys. Lett. **551**, 1 (2012).
[b] J.N. Hodges *et al.*, J. Chem. Phys. **139**, 164201 (2013).
[c] A.J. Perry *et al.*, J. Mol. Spectrosc. **317**, 71 (2015).
[d] A.J. Perry *et al.*, J. Chem. Phys. **141**, 101101 (2014).
[e] C.R. Marcus *et al.*, Astrophys. J. **817**, 138 (2016).

TE07   10:07 – 10:12

CO-ASSIGNMENT OF THE MOLECULAR VIBRATIONAL FREQUENCIES IN DIFFERENT ELECTRONIC STATES

YURII PANCHENKO, ALEXANDER ABRAMENKOV, *Department of Chemistry, Lomonosov Moscow State University, Moscow, Russia.*

Ultrafast electron diffraction experimental data for the structural parameters of molecules in excited electronic states are comparatively uncommon, hence these parameters are largely unknown. However, because differences between the molecular geometries of excited and ground electronic states cause differences in their experimental vibrational spectra it is important to establish a correspondence between the molecular vibrational frequencies in the ground state and those of the excited state of interest. The correct co-assignment of the experimental vibrational frequencies between two different electronic states of a molecule may be determined by the analog of the Duschinsky matrix[a] D. This matrix D is defined as $D = (L_I)^{-1} L_{II}$ where L_I and L_{II} are the matrices of the vibrational modes of the two states of the molecule under investigation. They are obtained by solving the vibrational problems in the I and II electronic states, respectively. Choosing the dominant elements in columns of the D matrix and permuting these columns to arrange these elements along the diagonal of the transformed matrix D^* makes it possible to establish the correct co-assignment of the calculated frequencies in the two electronic states. The rows of D^* are for the vibrations in the I electronic state, whereas the columns are for vibrations in the II electronic state. The results obtained may be tested by analogous calculations of D^* for isotopologues. The feasibility of co-assignments of the vibrational frequencies in the ground and T_1 and S_1 excited electronic states are demonstrated for *trans*-$C_2O_2F_2$[b]. The analogs of the Duschinsky matrix D^* were used to juxtapose the vibrational frequencies of this molecule calculated at the CASPT2/cc-pVTZ level in the S_0, T_1 and S_1 states.

---

[a] F. Duschinsky, Acta Physicochim. URSS, 7(4), 551–566 (1937).
[b] Yu. N. Panchenko, Vibrational spectroscopy, 68, 236–240 (2013).

Intermission

TE08   **10:31 – 10:46**

DOPPLER BROADENING THERMOMETRY BASED ON CAVITY RING-DOWN SPECTROSCOPY

JIN WANG, YU ROBERT SUN, CUNFENG CHENG, LEI-GANG TAO, YAN TAN, PENG KANG, AN-WEN LIU, <u>SHUI-MING HU</u>, *Hefei National Laboratory for Physical Science at Microscale, University of Science and Technology of China, Hefei, China.*

A Doppler broadening thermometry (DBT) instrument is implemented based on a laser-locked cavity ring-down spectrometer. [1,2] It can be used to determine the Boltzmann constant by measuring the Doppler width of a molecular ro-vibrational transition in the near infrared. Compared with conventional direct absorption methods, the high-sensitivity of CRDS allows to reach satisfied precision at lower sample pressures, which reduces the influence due to collisions. By measuring the ro-vibrational transition of C_2H_2 at 787 nm, we demonstrate a statistical uncertainty of 6 ppm (part per million) in the determined linewidth by several hours' measurement at a sample pressure of 1.5 Pa. [3] However, the complicity in the spectrum of a polyatomic molecule induces potential systematic influence on the line profile due to nearby "hidden" lines from weak bands or minor isotopologues. Recently, the instrument has been upgraded in both sensitivity and frequency accuracy. A narrow-band fiber laser frequency-locked to a frequency comb is applied, and overtone transitions at 1.56 μm of the $^{12}C^{16}O$ molecule are used in the CRDS-DBT measurements. The simplicity of the spectrum of the diatomic CO molecule eliminates the potential influence from "hidden" lines. Our preliminary measurements and analysis show that it is feasible to pursue a DBT measurement toward the 1 ppm precision.

References

[1] H. Pan, et al., Rev. Sci. Instrum. **82**, 103110 (2011)

[2] Y. R. Sun, et al., Opt. Expr., **19**, 19993 (2011)

[3] C.-F. Cheng, et al., Metrologia, **52**, S385 (2015)

TE09   **10:48 – 11:03**

CHEMICAL SYNTHESIS AND HIGH RESOLUTION SPECTROSCOPIC CHARACTERIZATION OF 1-AZA-ADAMANTANE-4-ONE $C_9H_{13}NO$ FROM THE MICROWAVE TO THE INFRARED

<u>OLIVIER PIRALI</u>[a], *Institut des Sciences Moléculaires d'Orsay, Université Paris-Sud, Orsay, France*; MANUEL GOUBET, *Laboratoire PhLAM, UMR 8523 CNRS - Université Lille 1, Villeneuve d'Ascq, France*; VINCENT BOUDON, *Laboratoire ICB, CNRS/Université de Bourgogne, DIJON, France*; LUCIA D'ACCOLTI, COSIMO ANNESE, CATERINA FUSCO, *Dipartimento di Chimica, Università di Bari A. Moro, Bari, Italy.*

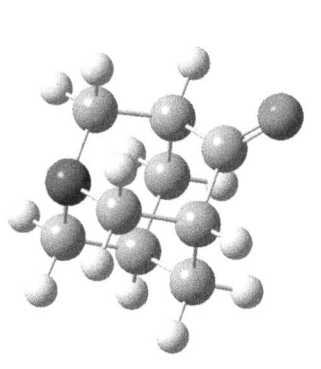

We have synthesized 1-aza-adamantane-4-one ($C_9H_{13}NO$) starting from commercial 1,4-cyclohexanedionemonoethylene acetal and tosylmethylisocianide and following a procedure described in details in the literature.[b] The high degree of sample purity was demonstrated by gas chromatography and mass spectrometric measurements, and its structure evidenced by 1H and 13C NMR spectroscopy. We present a thorough spectroscopic characterization of this molecule by gas phase vibrational and rotational spectroscopy. Accurate vibrational frequencies have been determined by infrared and far-infrared spectra. The pure rotational spectrum of the molecule has been recorded both by cavity-based Fourier-transform microwave spectroscopy in the 2-20 GHz region, by supersonically expanding the vapor pressure of the warm sample, and by room-temperature absorption spectroscopy in the 140-220 GHz range. Quantum-chemical calculations have enabled a fast analysis of the spectra. Accurate sets of rotational and centrifugal distorsion parameters of 1-aza-adamantane-4-one in its ground state and five vibrationally excited states have been derived from these measurements.

[a]also at: AILES beamline, Synchrotron SOLEIL, L'Orme des Merisiers Saint-Aubin, 91192 Gif-sur-Yvette, France
[b]Black, R. M. Synthesis, 1981, 829

TE10   11:05 – 11:20

ASSIGNMENT OF THE PERFLUOROPROPIONIC ACID-FORMIC ACID COMPLEX AND THE DIFFICULTIES OF INCLUDING HIGH K_a TRANSITIONS.

DANIEL A. OBENCHAIN, *Department of Chemistry, Wesleyan University, Middletown, CT, USA*; WEI LIN, *Chemistry, University of Texas Rio Grande Valley, Brownsville, TX, USA*; STEWART E. NOVICK, *Department of Chemistry, Wesleyan University, Middletown, CT, USA*; S. A. COOKE, *Natural and Social Science, Purchase College SUNY, Purchase, NY, USA*.

We recently began an investigation into the perfluoropropionic acid⋯formic acid complex using broadband microwave spectroscopy. This study aims to examine the possible double proton transfer between the two interacting carboxcyclic acid groups. The spectrum presented as a doubled set of lines, with spacing between transitions of $<$ 1 MHz. Transitions appeared to be a-type, R branch transitions for an asymmetric top. Assignment of all $K_a = 1, 0$ transitions yields decent fits to a standard rotational Hamiltonian. Treatment of the doubling as either a two state system (presumably with a double proton transfer) or as two distinct, but nearly identical conformations of the complex produce fits of similar quality. Including higher K_a transitions for the a-type, R-branch lines greatly increases the error of these fits. A previous study involving the trifluoroacetic acid⋯formic acid complex published observed similar high K_a transitions, but did not include them in the published fit.[a] We hope to shed more light on this conundrum. Similarities to other double-well potential minimum systems will be discussed.

[a]Martinache, L.; Kresa, W.; Wegener, M.;, Vonmont, U.; and Bauder, A. *Chem. Phys.* **148** (1990) 129-140.

TE11   11:22 – 11:37

INFRARED SPECTROSCOPIC INVESTIGATION ON CH BOND ACIDITY IN CATIONIC ALKANES

YOSHIYUKI MATSUDA, MIN XIE, ASUKA FUJII, *Department of Chemistry, Tohoku University, Sendai, Japan*.

We have demonstrated large enhancements of CH bond acidities in alcohol, ether, and amine cations through infrared predissociation spectroscopy based on the vacuum ultraviolet photoionization detection. In this study, we investigate for the cationic alkanes (pentane, hexane, and heptane) with different alkyl chain lengths. The σ electrons are ejected in the ionization of alkanes, while nonbonding electrons are ejected in ionization of alcohols, ethers, and amines. Nevertheless, the acidity enhancements of CH in these cationic alkanes have also been demonstrated by infrared spectroscopy. The correlations of their CH bond acidities with the alkyl chain lengths as well as the mechanisms of their acidity enhancements will be discussed by comparison of infrared spectra and theoretical calculations.

TE12   11:39 – 11:54

DIELECTRIC STUDY OF ALCOHOLS USING BROADBAND TERAHERTZ TIME DOMAIN SPECTROSCOPY (THz-TDS).

SOHINI SARKAR, DEBASIS SAHA, SNEHA BANERJEE, ARNAB MUKHERJEE, PANKAJ MANDAL, *Department of Chemistry, Indian Institute of Science Education and Research, Pune, Maharshtra, India*.

Broadband Terahertz-Time Domain Spectroscopy (THz-TDS) (1-10 THz) has been utilized to study the complex dielectric properties of methanol, ethanol, 1-propanol, 2-propanol, 1-butanol, and 1-octanol. Previous reports on dielectric study of alcohols were limited to 5 THz. At THz (1 THz = 33.33 cm^{-1}= 4 meV) frequency range (0.1 to 15 THz), the molecular reorientation and several intermolecular vibrations (local oscillation of dipoles) may coexist and contribute to the overall liquid dynamics. We find that the Debye type relaxations barely contribute beyond 1 THz, rather three harmonic oscillators dominate the entire spectral range. To get insights on the modes responsible for the observed absorption in THz frequency range, we performed all atom molecular dynamics (MD) using OPLS force field and ab initio quantum calculations. Combined experimental and theoretical study reveal that the complex dielectric functions of alcohols have contribution from a) alkyl group oscillation within H-bonded network (1 THz), b) intermolecular H-bond stretching (5 THz) , and c) librational motions in alcohols. The present work, therefore, complements all previous studies on alcohols at lower frequencies and provides a clear picture on them in a broad spectral range from microwave to 10 THz.

TF. Mini-symposium: Spectroscopy of Large Amplitude Motions

Tuesday, June 21, 2016 – 1:30 PM

Room: 100 Noyes Laboratory

Chair: David S. Perry, The University of Akron, Akron, OH, USA

TF01   **1:30 – 1:45**

HIGH-RESOLUTION INFRARED SPECTRSCOPY OF THE HYDROXYMETHYL RADICAL IN SOLID PARAHY-DROGEN

MORGAN E. BALABANOFF, <u>DAVID T. ANDERSON</u>, *Department of Chemistry, University of Wyoming, Laramie, WY, USA.*

Interest in the hydroxymethyl radical, CH_2OH, stems primarily from its importance as a reaction intermediate. However, this radical is also of interest from a spectroscopic point of view with large amplitude COH torsional tunneling and out of plane CH_2 wagging motions. The first IR detection of CH_2OH was accomplished via matrix isolation spectroscopy over 40 years ago by Jacox.[a] Reisler and co-workers[b] detected CH_2OH in the gas-phase using the sensitivity of double resonance ionization detected IR spectroscopy to probe the OH stretch, asymmetric CH stretch, and symmetric CH stretch vibrational modes with partial rotational resolution (0.4 cm^{-1}). Most recently, the Nesbitt group published[c] the first fully rotationally resolved IR spectrum of CH_2OH via the K_a=0←0 band of the symmetric CH stretch. These researchers were able to unambiguously assign the identified transitions to a Watson A-reduced symmetric top Hamiltonian thereby producing improved values for the symmetric CH stretch rotational constants and vibrational band origin. However, in this same work the authors point out a number of remaining unresolved issues. Motivated by these gas-phase observations, we decided to return to the matrix isolation studies of CH_2OH, however utilizing solid parahydrogen as a matrix host to improve upon the sensitivity and resolution of the previous matrix isolation studies. Based on our measurements, while the end-over-end rotation of the CH_2OH radical is quenched, rotational motion around the a-axis is nearly free permitting both A-type and B-type transitions to be resolved. In the case of the OH stretch mode, both A-type and B-type transitions are observed with an energy difference that makes sense based on the gas-phase CH_2OH rotational constants. However, for the symmetric CH stretch mode, the same mode recently assigned by Nesbitt and co-workers, two absorption features are also observed but the energy difference and intensities of the two features do not match predictions based on the rotational constants.

[a]M.E. Jacox, D.E. Milligan, *J. Mol. Spec.* **47**, 148-162 (1973).
[b]L. Feng, J. Wei, H. Reisler, *J. Phys. Chem. A* **108**, 7903-7908 (2004).
[c]M.A. Roberts, E.N. Sharp-Williams, D.J. Nesbitt, *J. Phys. Chem. A* **117**, 7042-7049 (2013).

108

TF02   1:47 – 2:02

INFRARED SPECTROSCOPIC STUDIES OF OCS TRAPPED IN SOLID PARAHYDROGEN: INDIRECT EVIDENCE OF LARGE AMPLITUDE MOTIONS

DAVID T. ANDERSON, *Department of Chemistry, University of Wyoming, Laramie, WY, USA.*

The high-resolution infrared rovibrational spectroscopy of OCS clustered with multiple hydrogen molecules has previously been studied in helium nanodroplets[a] and in the gas-phase[b] in search of another substance other than helium that displays superfluidity. Para-hydrogen (pH$_2$) is one of the most likely candidates because it is a spinless (I=0) composite boson with a light mass similar to helium. However, compared to helium, the pH$_2$-pH$_2$ intermolecular potential is significantly stronger and thus pH$_2$ solidifies at higher temperatures than the predicted superfluid transition temperature thereby blocking access to the superfluid state. Both of these previous studies reveal intriguing results linked to the microscopic details of superfluidity. We were therefore interested to characterize the IR spectrum of OCS in solid pH$_2$. The conventional wisdom is that because pH$_2$ solidifies into a quantum solid, the effects of superfluidity detected in the finite sized clusters should not be manifest in solid pH$_2$. However, the OCS-H$_2$ intermolecular potential strongly favors arranging the first 5 pH$_2$ molecules in a ring around the equator of the OCS (R=3.2 Å). Isolation of OCS in bulk pH$_2$ therefore may result in a solvation site where 6 pH$_2$ molecules in the same basal plane form a ring around the OCS and are pulled inward decoupling them from the bulk. If the periodic barriers to motion around the ring are small, one might expect the 6 equatorial pH$_2$ molecules to become delocalized while still maintaining the permutation symmetry of bosons. These 6 particles-on-a-ring may only show this behavior at low temperatures when thermal excitations are minimized. Analysis of the IR spectroscopy of OCS in solid pH$_2$ indicates 1) the OCS molecule does not freely rotate and 2) there are at least two preferred OCS solvation sites. In principle, the measured OCS peak frequency for these two solvation sites should depend sensitively on the "structure" of the first pH$_2$ solvation shell and therefore provide indirect evidence of this delocalization. We are currently trying to model the effect of pH$_2$ delocalization on the OCS vibrational frequency to compare with experiment and test this hypothesis.

[a]S. Grebenev, B. Sartakov, J.P. Toennies, A.F. Vilesov, *Science* **289**, 1532-1535 (2000).
[b]J. Tang, Y. Xu, A.R.W. McKellar, W. Jäger, *Science* **297**, 2030-2033 (2002).

TF03   2:04 – 2:19

HYPERCONJUGATION IN THE S$_1$ STATE OF SUBSTITUTED TOLUENE PROBED BY INFRARED SPECTROSCOPY

TAKASHI CHIBA, *Department of Chemistry, Tohoku University, Sendai, Japan*; KATSUHIKO OKUYAMA, *Department of Chemical Biology and Applied Chemistry, College of Engineering, Nihon University, Koriyama, Japan*; ASUKA FUJII, *Department of Chemistry, Tohoku University, Sendai, Japan.*

Internal rotation of the methyl group in substituted toluenes is one of prototypes of large amplitude motions in polyatomic molecules. The internal rotation of *o*-fluorotoluene is strongly hindered in the S$_0$ state, but that of *m*-fluorotoluene is almost free. For the S$_1$ state, however, the substantial changes of the internal rotation potentials have been reported; while the potential barrier in the *o*-isomer drastically decreases and the methyl group becomes almost a free rotor, the barrier in the *m*-isomer largely increases[a]. These surprising barrier changes have been attributed to the methyl conformation-dependent stabilization in the S$_1$ state by the π^*-σ^* hyperconjugation[b]. In the present study, to test this interpretation, we observed infrared spectra of *o*- and *m*-fluorotoluenes in the S$_0$ and S$_1$ states. Both the isomers showed decrease of the methyl CH stretch frequencies upon the electronic excitation. We concluded that this frequency decrease is the evidence of the π^*-σ^* hyperconjugation.

[a]K,Okuyama.;N,Mikami.;M,Ito. *J.Phys.Chem.* **1985**, *89*, 5617.
[b]H,Nakai.;M,Kawai. *Chem.Phys.Lett* **1999**, *307*, 272.

TF04

HIGH RESOLUTION DIRECT FREQUENCY COMB SPECTROSCOPY OF VINYL BROMIDE (C_2H_3Br) AND NI-TROMETHANE (CH_3NO_2) IN THE CH STRETCH REGION

BRYAN CHANGALA, BEN SPAUN, *JILA, National Institute of Standards and Technology and Univ. of Colorado Department of Physics, University of Colorado, Boulder, CO, USA*; DAVID PATTERSON, *Department of Physics, Harvard University, Cambridge, MA, USA*; JUN YE, *JILA, National Institute of Standards and Technology and Univ. of Colorado Department of Physics, University of Colorado, Boulder, CO, USA.*

We present high resolution rovibrational spectra of buffer gas cooled vinyl bromide (C_2H_3Br) and nitromethane (CH_3NO_2) in the 3 μm CH stretch region, acquired via cavity-enhanced direct frequency comb absorption spectroscopy. The $\sim$ 10 K translational and rotational temperatures of the molecular gas, as well as the narrow linewidth of the frequency comb, yield well resolved rotational structure, isotope shifts, and nuclear hyperfine splittings. Given the wide bandwidth of the light source and the long path length of the enhancement cavity, we measure entire vibrational bands in a single shot with high signal-to-noise ratios. We discuss spectra of the entire fundamental CH stretch manifolds of both C_2H_3Br and CH_3NO_2, which provide contrasting examples of rovibrational structure of rigid and non-rigid systems. C_2H_3Br is a relatively normal asymmetric top, exhibiting local perturbations to its rotational structure. Conversely, CH_3NO_2 contains an essentially unhindered methyl rotor. Of particular interest are its quasi-degenerate asymmetric CH stretch modes. Here, one must consider multiple couplings between torsional, rotational, and vibrational angular momentum, leading to qualitatively new level patterns and structure.

TF05

FTIR SYNCHROTRON SPECTROSCOPY OF THE ASYMMETRIC C-H STRETCHING BANDS OF METHYL MERCAPTAN (CH_3SH) – A PERPLEXITY OF PERTURBATIONS

RONALD M. LEES, LI-HONG XU, ELIAS M. REID, *Department of Physics, University of New Brunswick, Saint John, NB, Canada*; BISHNU P. THAPALIYA, MAHESH B. DAWADI, DAVID S. PERRY, *Department of Chemistry, The University of Akron, Akron, OH, USA*; SYLVESTRE TWAGIRAYEZU, *Department of Chemistry, Brookhaven National Laboratory, Upton, NY, USA*; BRANT E. BILLINGHURST, *EFD, Canadian Light Source Inc., Saskatoon, Saskatchewan, Canada.*

The infrared Fourier transform spectrum of the asymmetric C-H stretching bands of CH_3SH has been recorded in the 2950-3100 cm^{-1} region at Doppler limited resolution using synchrotron radiation at the FIR beamline of the Canadian Light Source in Saskatoon. Assignment of numerous torsion-rotation sub-bands for the asymmetric stretches has revealed a surprising pseudo-symmetric behavior, in which each band is seen in only one of the two possible ΔK selection rules. The upper states of the two asymmetric stretching vibrational bands thus appear to behave more like $l = \pm 1$ components of a degenerate E state of a symmetric top rather than distinct vibrational states. The two components are separated by about 1.5 cm^{-1} at K = 0, and then diverge linearly at higher K with torsional oscillation amplitude similar to that of the ground state of about 1.3 cm^{-1}. The divergence is consistent with an a-type Coriolis splitting picture with an effective Coriolis constant $\zeta \approx$ 0.075.

TF06   2:55 – 3:10

THE TORSIONAL FUNDAMENTAL BAND AND ROTATIONAL SPECTRA UP TO 940 GHz OF THE GROUND, FIRST AND SECOND EXCITED TORSIONAL STATES OF ACETONE

V. ILYUSHIN, IULIIA ARMIEIEVA, OLGA DOROVSKAYA, E. A. ALEKSEEV, *Radiospectrometry Department, Institute of Radio Astronomy of NASU, Kharkov, Ukraine*; MARCELA TUDORIE, *Service de Chimie Quantique et Photophysique, Université Libre de Bruxelles, Brussels, Belgium*; R. A. MOTIYENKO, L. MARGULÈS, *Laboratoire PhLAM, UMR 8523 CNRS - Université Lille 1, Villeneuve d'Ascq, France*; OLIVIER PIRALI, *AILES beamline, Synchrotron SOLEIL, Saint Aubin, France*; BRIAN DROUIN, *Jet Propulsion Laboratory, California Institute of Technology, Pasadena, CA, USA.*

A new global study of the acetone $(CH_3)_2CO$ spectrum is reported. The new microwave measurements covering the frequency range from 34 GHz to 940 GHz have been carried out using spectrometers in IRA NASU (Ukraine) and PhLAM Lille (France). The far infrared spectrum of acetone has been recorded on the AILES beamline of the synchrotron SOLEIL using a Fourier transform infrared spectrometer coupled to a long path cell. The transitions belonging to the three lowest torsional states as well as to the observed fundamental band associated with the methyl-top torsion mode ($\nu_{17} = 1$) have been analyzed using recently developed model for the molecules with two equivalent methyl rotors and C_{2v} symmetry at equilibrium (PAM_C2v_2tops program)[a]. The dataset consisting of more than 26100 microwave and 1100 FIR line frequencies and including transitions with J up to 89 was fit using a model consisting of 119 parameters and weighted root-mean-square deviation of 0.89 has been achieved. In the talk the details of this new study will be discussed.

---

[a]V. Ilyushin, J.T. Hougen J. Mol. Spectrosc. 289 (2013) 41-49.

TF07   3:12 – 3:22

ASSIGNING THE VIBRATION-ROTATION SPECTRA USING THE LWW PROGRAM PACKAGE

WIESŁAW ŁODYGA, MAREK KREGLEWSKI, *Faculty of Chemistry, Adam Mickiewicz University, Poznan, Poland.*

The LWW program package is based on traditional methods used in assigning rotationally resolved IR molecular spectra. The Loomis-Wood diagrams, which are used to visualize spectral branches and facilitate their identification, are combined with the power of interactive lower state combination difference (LSCD) checking, which provides immediate verification of correct assignments of quantum numbers to spectral lines. The traditional Giessen/Cologne type Loomis-Wood algorithm is also implemented. Predictions of vibration-rotation wavenumbers are calculated from a table of vibration-rotation energies, which can be imported from any external fitting program. Program includes many additional tools like simulation of a spectrum from a catalog file (list of transitions with intensities), build-up of a vibration-rotation band from individual branches and simultaneous displaying of two IR spectra - active one used for assignments and a reference one, both with full link to their peak-list files. Importing energies as well as exporting assigned data for fitting in an external program is made easy and flexible by a user-programmed import/export interface, which facilitates iterative refining of energy levels and gives a possibility of using directly exact vibration-rotation energies. Program is available in tree versions: for symmetric top, asymmetric top and molecules with large amplitude motions. The program is designed for the Windows operating systems and is available with full documentation on www.lww.amu.edu.pl .

Intermission

TF08

FAR-IR SPECTROSCOPY OF NEUTRAL GAS PHASE PEPTIDES: SIGNATURES FROM COMBINED EXPERIMENTS AND SIMULATIONS

JÉRÔME MAHÉ, MARIE-PIERRE GAIGEOT, *Laboratoire Analyse et Modélisation pour la Biologie et l'Environnement, Université d'Evry val d'Essonne, Evry, France*; DANIËL BAKKER, SANDER JAEQX, ANOUK RIJS, *FELIX Laboratory, Radboud University Nijmegen, Nijmegen, The Netherlands.*

Within the past two decades, action vibrational spectroscopy has become an almost routine experimental method to probe the structures of molecules and clusters in the gas phase (neutral and ions). Such experiments are mainly performed in the 1000-4000 cm^{-1} fingerprint regions. Though successful in many respects, these spectral domains can be however restrictive in the information provided, and sometimes reach limitations for unravelling structures without ambiguity. In a collaborative work with the group of Dr A.M. Rijs (FELIX laboratory, Radbout University, The Netherlands) we have launched a new strategy where the far-IR/Tera-Hertz domain (100-800 cm^{-1} domain) is experimentally probed for neutral gas phase molecules.

Our group in Paris apply finite temperature DFT-based molecular dynamics (DFT-MD) simulations in order to unravel the complex signatures arising in the far-IR domain, and provide an unambiguous assignment both of the structural conformation of the gas phase molecules (taking into account the experimental conditions) and an understanding of the spectral signatures/fingerprints.

We will discuss our experimental and theoretical investigations on two neutral peptides in the 100-800 cm^{-1} far-IR spectral domain, i.e. Z-Ala_6 and PheGly dipeptide, that represent two systems which definitive conformational assignment was not possible without the far IR signatures. We will also present our very recent results on the Phe-X peptide series, where X stands for Gly, Ala, Pro, Val, Ser, Cys, combining experiments and DFT-MD simulations, providing a detailed understanding of the vibrational fingerprints in the far-IR domain. In all exemples, we will show how DFT-MD simulations is the proper theoretical tool to account for vibrational anharmonicities and mode couplings, of prime importance in the far-IR domain.

References : J. Mahé, S. Jaeqx, A.M. Rijs, M.P. Gaigeot, Phys. Chem. Chem. Phys., 17 :25905 (2015)
S. Jaeqx, J. Oomens, A. Cimas, M.P. Gaigeot, A.M. Rijs, Angew. Chemie. Int., 53 :3663 (2014)

TF09

INVERSION VIBRATIONAL ENERGY LEVELS OF AsH_3^+ STUDIED BY ZERO-KINETIC-ENERGY PHOTOELECTRON SPECTROSCOPY

YUXIANG MO, *Department of Physics, Tsinghua University, Beijing, China.*

The rotational-resolved vibrational spectra of AsH_3^+ have been measured for the first time with vibrational energies up to 6000 cm^{-1} above the ground state using zero-kinetic energy photoelectron spectroscopic method. The inversion vibrational energy levels (ν_2) and the corresponding rotational constants for the ν_2 =0-16 have been determined. The tunneling splittings of the inversion vibration energy levels have been observed for the ground and the first excited vibrational states. The geometric parameters of AsH_3^+ as a function of inversion vibrational quantum states have been determined, indicating that the geometric structure of the cation changes from near planar structure to a pyramidal structure with more vibrational excitations. In addition to the experimental measurement, a two-dimensional theoretical calculation including the two symmetric vibrational modes was performed to determine the energy levels of the symmetric inversion and As-H stretching vibrations. The calculated vibrational energy levels are in good agreement with the experimental results. The first adiabatic ionization energy (IE) for AsH_3 was also accurately determined. The result of this work will be compared with our published result on the PH_3^+.

TF10 4:15 – 4:30

MICROWAVE MEASUREMENTS OF CYCLOPROPANECARBOXYLIC ACID AND ITS DOUBLY HYDROGEN BONDED DIMER WITH FORMIC ACID*

AARON M PEJLOVAS, *Department of Chemistry and Biochemistry, University of Arizona, Tucson, AZ, USA;* WEI LIN, *Chemistry, University of Texas Rio Grande Valley, Brownsville, TX, USA;* STEPHEN G. KUKOLICH, *Department of Chemistry and Biochemistry, University of Arizona, Tucson, AZ, USA.*

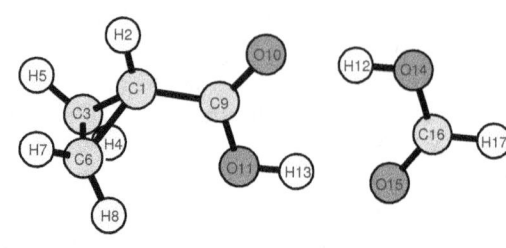

The microwave spectra were measured for cyclopropanecarboxylic acid (CPCA), an excited state conformer of CPCA and a doubly hydrogen bonded dimer formed with formic acid (FA) using a pulsed-beam Fourier transform microwave spectrometer. The rotational constants were determined from the spectra and were used to obtain a best fit gas phase structure of both CPCA and CPCA-FA using a nonlinear least squares fitting program. We obtained the C-C bond lengths in the cyclopropane ring for CPCA and the hydrogen bond distances for the CPCA-FA dimer. For CPCA-FA, there was no evidence of a concerted double proton tunneling motion as singlet b-type transitions were observed. The absence of the tunneling motion is most likely due to the asymmetry of the dimer. The excited stated conformer of the CPCA-FA dimer was also searched for, but was also not observed.

*Supported by the NSF CHE-1057796

TF11 4:32 – 4:47

GAS PHASE MEASUREMENTS OF MONO-FLUORO-BENZOIC ACIDS AND THE DIMER OF 3-FLUORO-BENZOIC ACID[a]

ADAM M DALY, SPENCER J CAREY, *Chemistry and Biochemistry, University of Arizona, Tucson, AZ, USA;* AARON M PEJLOVAS, *Department of Chemistry and Biochemistry, University of Arizona, Tucson, AZ, USA;* KEXIN LI, *Chemistry and Biochemistry, University of Arizona, Tucson, AZ, USA;* LU KANG, *Department of Chemistry and Biochemistry, Kennesaw State University, Kennesaw, GA, USA;* STEPHEN G. KUKOLICH, *Department of Chemistry and Biochemistry, University of Arizona, Tucson, AZ, USA.*

The gas phase homodimer of 3-fluorobenzoic acid was detected and the spectra showed evidence of proton tunneling. Experimental rotational constants are $A(0^+)= 1151.8(5)$, $B(0^+)=100.3(5)$, $C(0^+)= 87.64(3)$ MHz and $A(0^-)=1152.2(5)$, $B(0^-)= 100.7(5)$, $C(0^-)=88.85(3)$ MHz for the two ground vibrational states split by the proton tunneling motion. The tunneling splitting (ΔE) is approximately 560 MHz. This homodimer appears to be the largest carboxylic acid dimer observed with F-T microwave spectroscopy. Additionally, the microwave spectra of the mono-fluoro-benozic acids, (2-fluoro, 3-floro and 4-fluoro) benzoic acid have been measured in the frequency range of 4-14 GHz using a pulsed beam Fourier Transform microwave spectrometer. Measured rotational transition lines were assigned and fit using a rigid rotor Hamiltonian. Assignments were made for 3 conformers of 2-fluorobenzoic acid, 2 conformers of 3-fluorobenzoic acid and 1 conformer of 4-fluorobenzoic acid.

[a]Supported by the NSF CHE-1057796

TF12 **4:49 – 5:04**

MICROWAVE MEASUREMENTS OF THE TROPOLONE-FORMIC ACID DOUBLY HYDROGEN BONDED DIMER*

AARON M PEJLOVAS, *Department of Chemistry and Biochemistry, University of Arizona, Tucson, AZ, USA*; AGAPITO SERRATO III, WEI LIN, *Chemistry, University of Texas Rio Grande Valley, Brownsville, TX, USA*; STEPHEN G. KUKOLICH, *Department of Chemistry and Biochemistry, University of Arizona, Tucson, AZ, USA.*

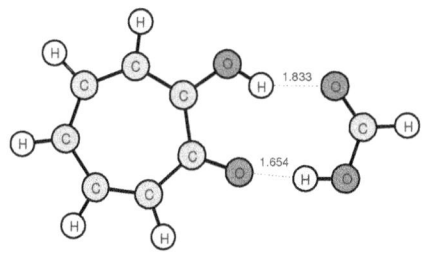

The microwave spectrum was measured for the tropolone-formic acid doubly hydrogen bonded dimer using a pulsed-beam Fourier transform microwave spectrometer in order to search for the concerted double proton tunneling motion. The tunneling motion was expected for the dimer, as the transition state of this motion exhibits C_{2V} symmetry, which has been thought to be a requirement to observe the concerted double proton tunneling. The tunneling motion was not observed for this dimer, as the transitions measured did not show observable splittings into doublets. The barrier height calculated of the dimer using B3LYP/aug-cc-pVTZ was about 15000 cm^{-1}, significantly larger than the value determined for the propiolic acid-formic acid dimer (3800 cm^{-1}),[a] which showed the tunneling motion. The estimated separation of the minima in the potential energy surface is estimated to be very similar to that of propiolic acid-formic acid (about 0.8 Å),[a] so the large barrier height may be why the tunneling process was not observed.

[a]Daly, A. M.; Bunker, P. R.; Kukolich, S. G. Communications: Evidence for Proton Tunneling from the Microwave Spectrum of the Formic Acid-Propiolic Acid Dimer. J. Chem. Phys. 132, 2010, 201101/1.

*Supported by the NSF CHE-1057796

TF13 **5:06 – 5:21**

VIBRATION-ROTATION-TUNNELING SPECTRUM OF FORMIC ACID DIMER IN THE 7.3μm REGION

CHUANXI DUAN, *College of Physical Science and Technology, Central China Normal University, Wuhan, China.*

The vibration-rotation-tunneling spectrum of formic acid dimer, $(HCOOH)_2$, in the spectral region 1369-1375 cm^{-1} has been measured by a multi-step rapid-can method in a slit jet expansion using a distributed-feedback quantum cascade laser. The observed spectrum is assigned to the O-C-H bending fundamental band. The tunneling splitting in the vibrational excited state is determined to be about 0.005 cm^{-1}, which is much smaller than that in the ground state, 0.0165 cm^{-1} (Goroya et al.,J. Chem. Phys. 140, 164311 (2014)). Strong local perturbations involving transitions with J ¿ 9, K = 0 and 1 are identified in the observed spectrum. The deperturbation analysis will be presented.

MICROWAVE MEASUREMENTS OF MALEIMIDE AND ITS DOUBLY HYDROGEN BONDED DIMER WITH FORMIC ACID*

<u>AARON M PEJLOVAS</u>, *Department of Chemistry and Biochemistry, University of Arizona, Tucson, AZ, USA;* LU KANG, *Department of Chemistry and Biochemistry, Kennesaw State University, Kennesaw, GA, USA;* STEPHEN G. KUKOLICH, *Department of Chemistry and Biochemistry, University of Arizona, Tucson, AZ, USA.*

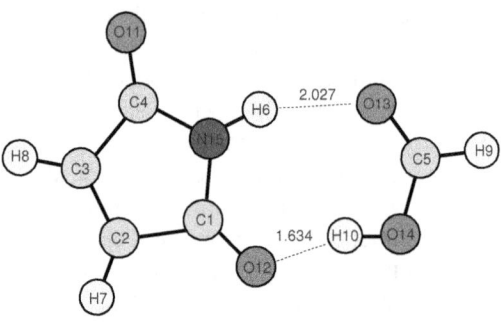

The microwave spectra were measured for the maleimide monomer and the maleimide-formic acid doubly hydrogen bonded dimer using a pulsed-beam Fourier transform microwave spectrometer. Many previously studied doubly hydrogen bonded dimers are formed between oxygen containing species, so it is important to also characterize and study other dimers containing nitrogen, as hydrogen bonding interactions with nitrogen are found in biological systems such as in DNA. The transition state of the dimer does not exhibit C_{2V} symmetry, so the tunneling motion was not expected to be observed based on the symmetry, but it would be very important to also observe the tunneling process for an asymmetric dimer. Single-line b-type transitions were observed, so the tunneling motion was not observed in our microwave spectra. The hydrogen bond lengths were determined using a nonlinear least squares fitting program.

*Supported by the NSF CHE-1057796

TG. Mini-symposium: Spectroscopy in Atmospheric Chemistry
Tuesday, June 21, 2016 – 1:30 PM
Room: 116 Roger Adams Lab

Chair: Rainer Volkamer, University of Colorado Boulder, Boulder, CO, USA

TG01   *INVITED TALK*   1:30 – 2:00

NOVEL IMPLEMENTATIONS OF FARADAY ROTATION SPECTROSCOPY - FROM IN-SITU RADICAL DETECTION TO STUDIES OF ENVIRONMENTAL NITROGEN CYCLING

ERIC ZHANG, JONAS WESTBERG, GERARD WYSOCKI, *Department of Electrical Engineering, Princeton University, Princeton, NJ, USA.*

Radical species play an important role in various chemical processes spanning atmospheric chemistry (e.g. ozone formation), bio-medical science, and combustion. These highly reactive chemicals usually occur at very low concentration levels, and are difficult to quantify in experiments[1]. Generally, laser-based techniques rely on careful selection of the target transition to minimize spectral interference and achieve high selectivity. In case of complex gas mixtures (such as air) a possibility of spectral interference always exists. Since Faraday rotation spectroscopy (FRS) is sensitive only to paramagnetic species (radicals), it can simultaneously provide ultra-high sensitivity and selectivity.

In this talk an overview of novel designs of FRS instrumentation as well as applications of FRS sensing will be provided. Examples will be given for FRS systems that routinely operate at the fundamental limits of optical detection, cavity-enhanced FRS detection schemes for sensitivity enhancement towards sub-pptv detection limits[2], and high-accuracy FRS spectrometers designed specifically for ratiometry of nitrogen isotopes (^{14}N, ^{15}N)[3]. Prospects for the FRS technology to monitor important atmospheric molecules such as HOx radicals (atmospheric "cleansing" agents) will be discussed.

References:
1. Wennberg et al., "Aircraft-borne, laser-induced fluorescence instrument for the in situ detection of hydroxyl and hydroperoxyl radicals," Rev. Sci. Instrum. 65, 1858-1876 (1994).
2. Westberg et al., "Optical feedback cavity-enhanced Faraday rotation spectroscopy for oxygen detection," in CES2015(Boulder, CO, 2015).
3. Zhang, "Nitric Oxide Isotopic Analyzer Based on a Compact Dual-Modulation Faraday Rotation Spectrometer," Sensors 15, 25992 (2015).

TG02   2:05 – 2:20

WITHDRAWN TALK - Check mobile app and last-minute changes for a replacement

TG03   2:22 – 2:37

USING MULTI RESONANCE EFFECTS TOWARDS SINGLE CONFORMER MICROWAVE SPECTROSCOPY

CHAMARA ABEYSEKERA, ALICIA O. HERNANDEZ-CASTILLO, BRIAN M HAYS, TIMOTHY S. ZWIER, *Department of Chemistry, Purdue University, West Lafayette, IN, USA.*

The relationship between the molecular structure and rotational frequencies makes rotational spectroscopy highly structural specific and an ideal tool for complex mixture analysis. The modern developments in broadband microwave techniques have immensely reduced the data acquisition time, while creating a need for high speed data analysis procedures. A new microwave-microwave double resonance method will be introduced, to perform single conformer/isomer microwave spectroscopy in complex chemical mixtures. The method combines the selective excitation schemes possible in chirped pulse microwave spectroscopy with multi-resonance effects observed upon sweeping in the rapid adiabatic passage regime, enabling perturbations to be induced in the intensities of most of the transitions ascribable to a single molecular constituent (e.g. a conformational isomer) in a mixture. Details of the method, experimental implementation and future challenges will be discussed.

TG04   2:39 – 2:54

ISOMER SPECIFIC MICROWAVE SPECTRUM OF (E)- AND (Z)- PHENYLVINYLNITRILE. IMPLEMENTING A NEW MULTI-RESONANT SPECTRAL ANALYSIS TOOL.

ALICIA O. HERNANDEZ-CASTILLO, BRIAN M HAYS, CHAMARA ABEYSEKERA, TIMOTHY S. ZWIER, *Department of Chemistry, Purdue University, West Lafayette, IN, USA.*

There are many circumstances in modern microwave spectroscopy where the observed spectra contain contributions from many distinct sub-populations, creating a complicated spectrum with interleaved transitions due to its components making spectral assignment challenging. A new method, exploiting multi resonance effects with broadband CP-FTMW was developed and implemented to differentiate the structural isomers: (E)- and (Z)-phenylvinylnitrile. This method will output an exclusive set of isomer-specific transitions reducing the spectral assignment time. Details of the method implementation and structural analysis of the two-isomer mixture will be discussed. The application of the method to other circumstances where selective modulation of the transitions due to a single set of connected transitions is vital for complex spectral assignment, will also be considered.

TG05   2:56 – 3:11

PHOTOELECTRON IMAGING OF OXIDE.VOC CLUSTERS

KELLYN M. PATROS, JENNIFER MANN, CAROLINE CHICK JARROLD, *Department of Chemistry, Indiana University, Bloomington, IN, USA.*

Perturbations of the bare O_2- and O_4- electronic structure arising from VOC (VOC = hexane, isoprene, benzene and benzene.D6) interactions are investigated using anion photoelectron imaging at 2.33 and 3.49 eV photon energies. Trends observed from comparing features in the spectra include VOC-identity-dependent electron affinities of the VOC complexes relative to the bare oxide clusters, due to enhance stability in the anion complex relative to the neutral. Autodetachment is observed in all O_4-.VOC spectra and only isoprene with O_2-. In addition, the intensities of transitions to states correlated with the singlet states of O_2 neutral via detachment from the O_2-.VOC anion complexes show dramatic VOC-identity variations. Most notably, benzene as a complex partner significantly enhances these transitions relative to O_2- and O_2-.hexane. A less significant enhancement is also observed in the O_2-.isoprene complex. This enhancement may be due to the presence of low-lying triplet states in the complex partners.

Intermission

TG06

FT-IR MEASUREMENTS OF MID-IR PROPENE (C_3H_6) CROSS SECTIONS FOR TITAN STRATOSPHERE

KEEYOON SUNG, GEOFFREY C. TOON, BRIAN DROUIN, TIMOTHY J. CRAWFORD, *Jet Propulsion Laboratory, California Institute of Technology, Pasadena, CA, USA*; ARLAN MANTZ, *Department of Physics, Astronomy and Geophysics, Connecticut College, New London, CT, USA*; MARY ANN H. SMITH, *Science Directorate, NASA Langley Research Center, Hampton, VA, USA*.

We present temperature dependent cross sections of propene (C_3H_6; CH_2-CH-CH_3, propylene), which was detected in the stratosphere of Titan.[a] For this study, a series of high-resolution (0.0022 cm^{-1}) spectra of pure and N_2-mixture samples were recorded at $150 - 296$ K in the $650 - 1530$ cm^{-1}($6.5 - 15.3$ μm) at the Jet Propulsion Laboratory using a Fourier-transform spectrometer and a custom-designed cold cell[bc]. The observed spectral features cover the strongest band (ν_{19}) with its outstanding Q-branch peak at 912 $cm^{-1}$and three other strong bands: ν_{18}, ν_{16} and ν_7 at 990, 1442, and 1459 cm^{-1}, respectively. In addition, we have generated a HITRAN-format empirical 'pseudoline list' consisting of line positions, intensities, and effective lower state energies, which were determined by fitting all the observed propene spectra simultaneously. A newly derived partition function was used in the analysis. The results are compared with early work from relatively warm temperatures ($278 - 323$ K).[d]

---

[a]C. A. Nixon, et al., Astrophys. J. Lett., 776, L14 (2013).

[b]A.W. Mantz, K. Sung, et al. 65th Symposium on Molecular Spectroscopy, Columbus, OH, 2010.

[c]K. Sung, A.W. Mantz, et al., J. Mol. Spectrosc. 262, 122 – 134 (2010).

[d]Research described in this talk was performed at the Jet Propulsion Laboratory, California Institute of Technology, Connecticut College, and NASA Langley Research Center under contracts and cooperative agreements with the National Aeronautics and Space Administration.

TG07

THE UV SPECTROSCOPY OF JET-COOLED 3-PHENYL-2-PROPYNENITRILE

KHADIJA M. JAWAD, TIMOTHY S. ZWIER, *Department of Chemistry, Purdue University, West Lafayette, IN, USA*.

The atmosphere of Saturn's moon Titan is replete with hydrocarbons and nitriles, but knowledge of the formation and sink processes as well as the identities of molecules on the large end of photochemical models of the atmosphere is very limited. 3-phenyl-2-propynenitrile (Ph-C$\equiv$C-C$\equiv$N) is of potential importance in this atmosphere because it is a likely product of photochemical reaction between cyanoacetylene and benzene, bringing together two of the key functional groups in Titan's atmosphere in a single molecule. We present the UV spectrum of this molecule in the gas phase, under jet-cooled conditions, using 2-color resonant two-photon ionization. The spectrum was recorded from 292-208nm, taking advantage of the wide tunability of a BBO-based OPO as the excitation source. On its long wavelength end, the spectrum has sharp transitions arising from a $\Pi\Pi$* transition characteristic of a phenyl derivative, while deeper into the UV the spectrum is broadened in a manner reminiscent of cyanoacteylene.

118

THEORETICAL STUDIES OF THE RELAXATION MATRIX FOR MOLECULAR SYSTEMS

QIANCHENG MA, *Applied Physis and Applied Mathematics, Columbia University, New York, NY, USA*; C. BOULET, *Institut des Sciences Moléculaires d'Orsay, Université Paris-Sud, Orsay, France.*

The phenomenon of collisional transfer of intensity due to line mixing has an increasing importance for atmospheric monitoring. From a theoretical point of view, all relevant information about the collisional processes is contained in the relaxation matrix where the diagonal elements give half-widths and shifts, and the off-diagonal elements correspond to line interferences. For simple systems such as those consisting of diatom-atom or diatom-diatom, accurate fully quantum calculations based on interaction potentials are feasible. However, fully quantum calculations become unrealistic for more complex systems. On the other hand, the semi-classical Robert-Bonamy formalism, which has been widely used to calculate half-widths and shifts for decades, fails in calculating the off-diagonal matrix elements resulting from applying the isolated line approximation. As a result, in order to simulate atmospheric spectra where the effects from line mixing are important, semi-empirical fitting or scaling laws such as the energy corrected sudden (ECS) and the infinite order sudden (IOS) models are commonly used. Recently, we have found that in developing this semi-classical line shape theory, to rely on the isolated line approximation is not necessary. By eliminating this unjustified assumption, and accurately evaluating matrix elements of the exponential operators, we have developed a more capable formalism that enables one not only to reduce uncertainties for calculated half-widths and shifts, but also to calculate the whole relaxation matrix. This implies that we can address the line mixing with the semi-classical theory based on interaction potentials between molecular absorber and molecular perturber. We have applied this formalism for Raman and infrared spectra of linear and asymmetric-top molecules. Recently, the method has been extended into symmetric-tops with inverse symmetry such as the NH_3 molecule. Our calculated half-widths of NH_3 lines in the ν_1 and the pure rotational bands match measurements very well. Then, the model has been applied to the calculation of the shape of the Q branch and of some R manifolds, for which an obvious signature of line mixing effects has been experimentally demonstrated. Comparisons with measurements show that the present formalism leads to an accurate prediction of the available experimental lineshapes.

CALCULATED VIBRATIONAL STATES OF OZONE UP TO DISSOCIATION

RICHARD DAWES, STEVE ALEXANDRE NDENGUE, *Department of Chemistry, Missouri University of Science and Technology, Rolla, MO, USA*; XIAO-GANG WANG, TUCKER CARRINGTON, *Department of Chemistry, Queen's University, Kingston, ON, Canada*; HUA GUO, *Department of Chemistry and Chemical Biology, University of New Mexico, Albuquerque, NM, USA.*

A new accurate global potential energy surface for the ground electronic state of ozone [J. Chem. Phys. 139, 201103 (2013)] was published fairly recently. The topography near dissociation differs significantly from previous surfaces, without spurious submerged reefs and corresponding van der Waals wells. This has enabled significantly improved descriptions of scattering processes, capturing the negative temperature dependence and large kinetic isotope effects in exchange reaction rates. The exchange reactivity was found to depend on the character of near-threshold resonances and their overlap with reactant and product wavefunctions, which in turn are sensitive to the potential. Here we present global "three-well" calculations of all bound vibrational states of three isotopic combinations of ozone for J = 0 and J = 1 with a focus on the character and density of highly excited states. The calculations were done using a parallel symmetry-adapted Lanczos method with the RTR code, enabling the use of as many as 64.8 million basis functions. Tunneling splittings and the pseudorotation isomerization path will be discussed.

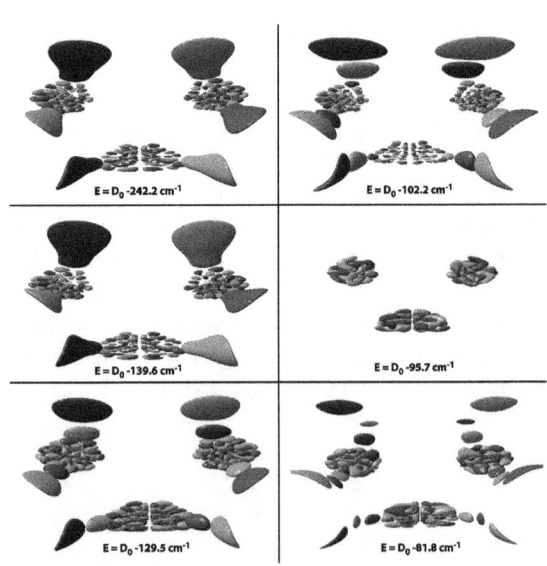

TG10 4:38 – 4:53

ROTATIONAL QUENCHING STUDY IN ISOVALENT H^+ + CO AND H^+ + CS SYSTEMS

RAJWANT KAUR, T. J. DHILIP KUMAR, *Department of Chemistry, Indian Institute of Technology Ropar, Rupnagar, Punjab, India.*

Cooling and trapping of polar molecules has attracted attention at cold and ultracold temperatures. Extended study of molecular inelastic collision processes of polar interstellar species with proton finds an important astrophysical application to model interstellar medium. Present study includes computation of rate coefficient for molecular rotational quenching process in proton collision with isovalent CO and CS molecules using quantum dynamical close-coupling calculations. Full dimensional *ab initio* potential energy surfaces have been computed for the ground state for both the systems using internally contracted multireference configuration interaction method and basis sets. Quantum scattering calculations for rotational quenching of isovalent species are studied in the rigid-rotor approximation with CX (X=O, S) bond length fixed at an experimental equilibrium value of 2.138 and 2.900 a.u., respectively. Asymptotic potentials are computed using the dipole and quadrupole moments, and the dipole polarizability components. The resulting long-range potentials with the short-range ab initio interaction potentials have been fitted to study the anisotropy of the rigid-rotor surface using the multipolar expansion coefficients. Rotational quenching cross-section and corresponding rates from $j=4$ level of CX to lower j' levels have been obtained and found to obey Wigner's threshold law at ultra cold temperatures.

TG11 4:55 – 5:10

MODELING PHOTODETACHMENT FROM HO_2^- USING THE pd CASE OF THE GENERALIZED MIXED CHARACTER MOLECULAR ORBITAL MODEL

CHRISTOPHER C BLACKSTONE, ANDREI SANOV, *Chemistry and Biochemistry, University of Arizona, Tucson, AZ, USA.*

Using the generalized model for photodetachment of electrons from mixed-character molecular orbitals, we gain insight into the nature of the HOMO of HO_2^- by treating it as a coherent superpostion of one p- and one d-type atomic orbital. Fitting the pd model function to the ab initio calculated HOMO of HO_2^- yields a fractional d-character, γ_p, of 0.979. The modeled curve of the anisotropy parameter, β, as a function of electron kinetic energy for a pd-type mixed character orbital is matched to the experimental data.

HITRAN APPLICATION PROGRAMMING INTERFACE (HAPI): EXTENDING HITRAN CAPABILITIES

ROMAN V KOCHANOV[a], IOULI E GORDON, LAURENCE S. ROTHMAN, *Atomic and Molecular Physics, Harvard-Smithsonian Center for Astrophysics, Cambridge, MA, USA*; PIOTR WCISLO, *Institute of Physics, Faculty of Physics, Astronomy and Informatics, Nicolaus Copernicus University, Torun, Poland*; CHRISTIAN HILL, *Department of Physics and Astronomy, University College London, Gower Street, London WC1E 6BT, United Kingdom*; JONAS WILZEWSKI, *Faculty of Physics, Ludwig Maximilians University, Munich, Germany.*

In this talk we present an update on the HITRAN Application Programming Interface (HAPI) [b,c]. HAPI is a free Python library providing a flexible set of tools to work with the most up-to-date spectroscopic data provided by HITRANonline (www.hitran.org) [d,e]. HAPI gives access to the spectroscopic parameters which are continuously being added to HITRANonline. For instance, these include non-Voigt profile parameters [f], foreign broadenings and shifts [g], and line mixing. HAPI enables more accurate spectra calculations for the spectroscopic and astrophysical applications requiring the detailed modeling of the broadener. HAPI implements an expert algorithm for the line profile selection for a single-layer radiative transfer calculation, and can be extended by custom line profiles and algorithms of their calculations, partition sums, instrumental functions, and temperature and pressure dependences. Possible HAPI applications include spectroscopic data validation and analysis [h] as well as radiative-transfer calculations, experiment verification and spectroscopic code benchmarking.

---

[a] Laboratory of Quantum Mechanics of Molecules and Radiative Processes, Tomsk State University, 634050 Tomsk, Russia

[b] Kochanov RV, Gordon IE, et al. Submitted to JQSRT HighRus Special Issue 2016.

[c] Kochanov RV, Hill C, et al. ISMS 2015. http://hdl.handle.net/2142/79241

[d] Rothman LS, Gordon IE, et al. JQSRT 2013;130:4–50.

[e] Hill C, Gordon IE, et al. Accepted to JQSRT HighRus Special Issue 2016.

[f] Wcislo P, Gordon IE, et al. Accepted to JQSRT HighRus Special Issue 2016.

[g] Wilzewski JS, Gordon IE, et al. JQSRT 2016;168:193–206.

[h] Kochanov RV, Gordon IE, et al. Clim Past 2015;11:1097–105.

TH. Astronomy
Tuesday, June 21, 2016 – 1:30 PM
Room: 274 Medical Sciences Building

Chair: Michael C McCarthy, Harvard-Smithsonian CfA, Cambridge, MA, USA

TH01   1:30 – 1:45

LABORATORY MEASUREMENTS AND ASTRONOMICAL SEARCH OF THE HSO RADICAL

GABRIELE CAZZOLI, CRISTINA PUZZARINI, *Dep. Chemistry 'Giacomo Ciamician', University of Bologna, Bologna, Italy*; <u>VALERIO LATTANZI</u>, *The Center for Astrochemical Studies, Max-Planck-Institut für extraterrestrische Physik, Garching, Germany*; BELÉN TERCERO, JOSE CERNICHARO, *Departamento de Astrofísica, Centro de Astrobiología CAB, CSIC-INTA, Madrid, Spain.*

The sulphur chemistry in space is still quite puzzling although several S-bearing species have been detected in the interstellar medium (ISM) in our local system and outside our galaxy. In particular, we observe very large quantities of sulphur harbouring molecules, especially in high-mass star forming regions, that are in perfect accordance with its solar abundance, while in the cold, dense ISM a much lower abundance is observed compared to its solar one. To have a better understanding of the sulphur chemistry in space, it is crucial to derive the broadest picture of the chemical network involving the formation of sulphur species. In this work we report high-resolution spectra of a simple triatomic S-bearing molecule, the HSO radical, with experiments well into the THz region. Thanks to the spectroscopic results of this work, which provide accurate frequency predictions up to the THz, we have also performed a rigorous search for HSO in space. The main outcomes of our work will be briefly presented, showing in particular the synergy between the laboratory and the observations.

TH02   1:47 – 2:02

THE BENDING VIBRATIONS OF THE C_3-ISOTOPOLOGUES IN THE 1.9 TERAHERTZ REGION

A. BREIER, THOMAS BÜCHLING, VOLKER LUTTER, *Institute of Physics, University Kassel, Kassel, Germany*; RICO SCHNIERER, *Institute of Physics, University of Rostock, Rostock, Germany*; <u>GUIDO W FUCHS</u>, *Physics Department, University of Kassel, Kassel, Germany*; THOMAS GIESEN, *Institute of Physics, University Kassel, Kassel, Germany.*

Short carbon chains are fundamental for the chemistry of stellar and interstellar ambiences. The linear carbon chain molecule C_3 has been found in various interstellar and circumstellar environments, encompassing diffuse interstellar clouds, star forming regions, shells of late type stars, as well as cometary tails. Due to the lack of a permanent dipole moment C_3 can only be detected by electronic transitions in the visible spectral range or by vibrational bands in the mid-and far-infrared region. We performed experiments where C_3 was produced via laser-ablation of a graphite rod with a 3 bar He purge and a subsequent adiabatic expansion into a vaccum resulting in a supersonic jet. We report laboratory measurements of the lowest bending mode transitions of six ^{13}C-isotopologues of the linear C_3 molecule. Fifty-eight transitions have been measured between 1.8-1.9 THz with an accuracy of better than 1 MHz. Molecular parameters have been derived to give accurate line frequency positions of all ^{13}C isotopologues to ease their future interstellar detection. A dedicated observation for singly substituted ^{13}CCC is projected within the SOFIA airborne observatory mission.

TH03   2:04 – 2:19

LABORATORY MEASUREMENTS OF SMALL SILICON BEARING MOLECULES OF ASTROPHYSICAL INTEREST

<u>CARL A GOTTLIEB</u>, *Radio and Geoastronomy Division, Harvard-Smithsonian Center for Astrophysics, Cambridge, MA, USA*; MICHAEL C McCARTHY, *Atomic and Molecular Physics, Harvard-Smithsonian Center for Astrophysics, Cambridge, MA, USA.*

We will discuss the status of millimeter-wave laboratory measurements of the rotational spectra in the ground and vibrationally excited levels of small molecules containing two or three silicon atoms that might be the building blocks of dust seeds in carbon- and oxygen-rich AGB stars. The motivation is to provide essential spectroscopic information needed to guide future interferometric observations of the inner envelope of these objects at high angular resolution and sensitivity. The focus will be on a half-dozen species for which there is either no prior high resolution spectroscopy, or only in the centimeter band at best. We will also update the status of the known silicon carbides SiCC and SiCSi.

122

TH04   2:21 – 2:36

MILLIMETER-WAVE SPECTROSCOPY OF METHOXYMETHANOL

R. A. MOTIYENKO, L. MARGULÈS, *Laboratoire PhLAM, UMR 8523 CNRS - Université Lille 1, Villeneuve d'Ascq, France*; J.-C. GUILLEMIN, *Institut des Sciences Chimiques de Rennes, UMR 6226 CNRS - ENSCR, Rennes, France*; DIDIER DESPOIS, *Laboratoire d'Astrophysique de Bordeaux, Université de Bordeaux, Floirac, France*.

Methoxymethanol (CH_3OCH_2OH), a ten atoms molecule, is a very interesting candidate for the detection in the interstellar medium since it can be formed by the reaction between both possible radicals of methanol: CH_3O (already detected in the ISM) and CH_2OH. It could be also formed by addition of CH_3O on formaldehyde (another detected compound in the ISM) followed by the addition or abstraction of a hydrogen radical. According to quantum chemical calculations, methoxymethanol has three stable conformations: *Gg*, *Gg'*, and *Tg*. The most stable *Gg* conformation has a small dipole moment of 0.27 D, whereas two others conformations *Gg'* and *Tg* are considerably less stable (8.5 kJ/mol and 10.9 kJ/mol respectively), but have much greater dipole moments (2.47 D and 2.17 D respectively). Thus, in the room-temperature rotational spectra the three conformations may have approximately the same line intensities. We measured the rotational spectrum of methoxymethanol between 150 and 500 GHz and detected all three conformations. The major difficulties in the spectral analysis consist in the very dense spectrum of methoxymethanol and low signal-to-noise ratio of the lines, and in different large amplitude motions. The methyl top internal rotation splittings were observed for the rotational lines of *Gg* and *Gg'* conformations. The doublet structure of *Tg* conformation rotational lines may be explained by -OH group tunneling between two equivalent gauche configurations. The analysis is in progress, the latest results will be presented. *The support of the "Action sur Projets de l'INSU PCMI, and ANR-13-BS05-0008-02 IMOLABS" is gratefully acknowledged*

TH05   2:38 – 2:53

SUBMILLIMETER SPECTRUM OF THE METHOXY RADICAL TO GUIDE INTERSTELLAR SEARCHES

JACOB LAAS, SUSANNA L. WIDICUS WEAVER, *Department of Chemistry, Emory University, Atlanta, GA, USA*.

The formation of interstellar methanol and its related chemistry have been topics of much discussion and debate within the astrochemical community. This discussion has now also been extended to include methoxy (CH_3O), a closely-related radical, after its initial discovery within a prestellar core was reported by Cernicharo and coworkers (2012). Using a supersonic expansion of methanol diluted in argon and coupled with a plasma discharge, we have collected the rotational spectrum of methoxy at submillimeter wavelengths. By coupling these results with data from a number of other literature reports, we have prepared an updated line catalog that will greatly enhance opportunities to search for interstellar methoxy. We will present these results in the context of related astrochemical processes.

TH06   2:55 – 3:10

THE COMPLETE, TEMPERATURE RESOLVED SPECTRUM OF METHYL CYANIDE BETWEEN 200 AND 277 GHZ

JAMES P. McMILLAN, CHRISTOPHER F. NEESE, FRANK C. DE LUCIA, *Department of Physics, The Ohio State University, Columbus, OH, USA*.

We have studied methyl cyanide, one of the so-called 'astronomical weeds', in the 200–277 GHz band. We have experimentally gathered a set of intensity calibrated, complete, and temperature resolved spectra from across the temperature range of 231–351 K. Using our previously reported method of analysis[a], the point by point method, we are capable of generating the complete spectrum at astronomically significant temperatures. Lines, of nontrivial intensity, which were previously not included in the available astrophysical catalogs have been found. Lower state energies and line strengths have been found for a number of lines which are not currently present in the catalogs. The extent to which this may be useful in making assignments will be discussed.

[a] J. McMillan, S. Fortman, C. Neese, F. DeLucia, ApJ. 795, 56 (2014)

TH07 3:12 – 3:27

COMPREHENSIVE ANALYSIS OF INTERSTELLAR *Iso*-PROPYL CYANIDE UP TO 480 GHZ

LUCIE KOLESNIKOVÁ, E. R. ALONSO, CARLOS CABEZAS, SANTIAGO MATA, JOSÉ L. ALONSO, *Grupo de Espectroscopia Molecular, Lab. de Espectroscopia y Bioespectroscopia, Unidad Asociada CSIC, Universidad de Valladolid, Valladolid, Spain.*

Iso-propyl cyanide, also known as *iso*-butyronitrile, is a branched alkyl molecule recently detected in the interstellar medium.[a] A combination of Stark-modulated microwave spectroscopy and frequency-modulated millimeter and submillimeter wave spectroscopy was used to analyze its rotational spectrum from 26 to 480 GHz. Spectral assignments and analysis include transitions from the ground state, eight excited vibrational states and ^{13}C isotopologues. Results of this work should facilitate astronomers further observations of *iso*-propyl cyanide in the interstellar medium.

[a]A. Belloche, R. T. Garrod, H. S. P. Müller, K. M. Menten, *Science*, **2014**, *345*, 1584.

TH08 3:29 – 3:44

SEARCHING FOR AMINOMETHANOL AMONGST THE REACTION PRODUCTS OF O(^{1}D) INSERTION INTO METHYLAMINE

MORGAN N McCABE, CARSON REED POWERS, BRIAN M HAYS, SAMUEL ZINGA, SUSANNA L. WIDICUS WEAVER, *Department of Chemistry, Emory University, Atlanta, GA, USA.*

Aminomethanol ($HOCH_2NH_2$) is a molecule of astrochemical interest as it is thought to be the precursor to the simplest amino acid, glycine. To date, no laboratory spectrum has been recorded because it is unstable under normal laboratory conditions. As a result, a millimeter spectrometer was developed to study the products of O(^{1}D) insertion into methylamine, with the goal of producing aminomethanol. Here we present the results of this study, including other observed reaction products and a preliminary assignment of aminomethanol.

Intermission

TH09 4:03 – 4:18

SEGMENTED CHIRPED-PULSE MILLIMETER-WAVE SPECTROSCOPY FOR ASTROCHEMISTRY

BENJAMIN E ARENAS, *CoCoMol, Max-Planck-Institut für Struktur und Dynamik der Materie, Hamburg, Germany*; AMANDA STEBER, *The Centre for Ultrafast Imaging (CUI), Universität Hamburg, Hamburg, Germany*; SÉBASTIEN GRUET, MELANIE SCHNELL, *CoCoMol, Max-Planck-Institut für Struktur und Dynamik der Materie, Hamburg, Germany.*

The ability to detect molecules in the interstellar medium (ISM) is afforded to us by the collaboration of state-of-the-art observations, like from the Atacama Large Millimeter/submillimeter Array (ALMA), and high-resolution laboratory spectra. Here, we present our use of a commercial segmented chirped-pulse Fourier transform millimeter-wave rotational spectrometer to study simple oxygen-containing organic molecules. Our spectrometer operates in the region 75 – 110 GHz, providing an overlap with ALMA's Band 3 and allowing direct comparison of our laboratory spectra with observational data. We have measured rotational spectra of 1,2-propanediol[1, 2, 3] and methyl acetate[4, 5] in this spectral range at room temperature – both have been previously studied in the microwave and millimeter-wave regions. The rotational spectrum of the former in the 3 mm region shows eight different conformers to date. Spectral bandwidth overlap with ALMA Band 3 will allow for easier detection of new chemicals in the ISM.

[1] Caminati, W., *J. Mol. Spectrosc.*, 86(1), 193-201, **1981**.
[2] Lovas, F. J., Plusquellic, D. F., Pate, B. H., Neill, J. T., Muckle, M. T. and Remijan, A. J., *J. Mol. Spectrosc.*, 257(1), 82-93, **2009**.
[3] Bossa, J. –B., Ordu, M. H., Müller, H. S. P., Lewen, F. and Schlemmer, S., *Astron. Astrophys.*, 570, A12, **2014**.
[4] Tudorie, M., Kleiner, I., Hougen, J. T., Melandri, S., Sutikdja, L. W. and Stahl, W., *J. Mol. Spectrosc.*, 269, 211-225, **2011**.
[5] Nguyen, H. V. L., Kleiner, I., Shipman, S. T., Mae, Y., Hirose, K., Hatanaka, S. and Kobayashi, K., *J. Mol. Spectrosc.*, 299, 17-21, **2014**.

124

TH10   4:20 – 4:35

THE MILLIMETER-WAVE SPECTROSCOPY OF HYDANTOIN, A POTENTIAL PRECURSOR OF GLYCINE

HIROYUKI OZEKI, RIO MIYAHARA, HIROTO IHARA, SATOSHI TODAKA, *Department of Environmental Science, Toho University, Funabashi, Japan*; KAORI KOBAYASHI, *Department of Physics, University of Toyama, Toyama, Japan*; MASATOSHI OHISHI, *Astronomy Data Center, National Astronomical Observatory of Japan, Mitaka, Japan*.

Hydantoin (Imidazolidine-2,4-dione, $C_3H_4N_2O_2$) is one of five-membered rings with heteroatoms and could be synthesized from carbonyl compounds such as aldehydes or ketones via so-called Bucherer-Bergs reaction. This molecule is also known as a direct precursor of amino acid by hydrolysis, evidently, hydantoin has been found in carbonaceous chondrites with several kinds of amino acids. [a] The aim of this study is to provide spectroscopic information which is useful for the future astronomical search. The hydantoin vapor was obtained by heating hydantoin powder to 150°C, and spectral line survey has been conducted in the frequency range between 138 and 150 GHz. Our DFT calculations suggest that the permanent dipole of this molecule is approximately 2 Debyes and lies mostly along b-molecular axis. Fourty-five spectral lines in the above frequency region can be so far assigned to b-type R-branch transitions, and molecular constants including centrifugal distortion constants up to the 4th-order have been determined. The obtained rotational constants agree well with the calculated values. In addition, some of the unassigned spectral lines were attributed to the hydantoin transitions in the vibrational excited state. We will report the current status of the analysis.

---

[a] *for example,* A. Shimoyama and R. Ogasawara, 2008, *Orig. Life Evol. Biosph.* **32**, 165 (2002).

TH11   4:37 – 4:52

THE MICROWAVE SPECTROSCOPY OF AMINOACETONITRILE IN THE VIBRATIONAL EXCITED STATES 2

CHIHO FUJITA, HARUKA HIGURASHI, HIROYUKI OZEKI, *Department of Environmental Science, Toho University, Funabashi, Japan*; KAORI KOBAYASHI, *Department of Physics, University of Toyama, Toyama, Japan*.

Aminoacetonitrile (NH_2CH_2CN) is a potential precursor of the simplest amino acid, glycine in the interstellar space and was detected toward SgrB2(N). [a] We have extended measurements up to 1.3 THz so that the strongest transitions that may be found in the terahertz region should be covered. [b] Aminoacetonitrile has a few low-lying vibrational excited states [c] and indeed the pure rotational transitions in these vibrational excited states were found. [d] The pure rotational transitions in six vibrational excited states in the 80-180 GHz range have been assigned and centrifugal distortion constants up to the sextic terms were determined. Based on spectral intensities and the vibrational information from Bak et al., They were assigned to the 3 low-lying fundamentals, 1 overtone and 2 combination bands. In the submillimeter wavelength region, perturbations were recognized and some of the lines were off by more than a few MHz. At this moment, these perturbed transitions are not included in our analysis.

---

[a] A. Belloche, K. M. Menten, C. Comito, H. S. P. Müller, P. Schilke, J. Ott, S. Thorwirth, and C. Hieret, 2008, *Astronom. & Astrophys.* **482**, 179 (2008).
[b] Y. Motoki, Y. Tsunoda, H. Ozeki, and K. Kobayashi, *Astrophys. J. Suppl. Ser.* **209**, 23 (2013).
[c] B. Bak, E. L. Hansen, F. M. Nicolaisen, and O. F. Nielsen, *Can. J. Phys.* **53**, 2183 (1975).
[d] C. Fujita, H. Ozeki, and K. Kobayashi, 70th International Symposium on Molecular Spectroscopy (2015), MH14.

TH12 $4:54-5:09$

THE STUDY OF ACENAPHTHENE AND ITS COMPLEXATION WITH WATER

AMANDA STEBER, *The Centre for Ultrafast Imaging (CUI), Universität Hamburg, Hamburg, Germany*; CRISTOBAL PEREZ, *CoCoMol, Max-Planck-Institut für Struktur und Dynamik der Materie, Hamburg, Germany*; ANOUK RIJS, *FELIX Laboratory, Institute for Molecules and Materials (IMM), Radboud University, Nijmegen, Netherlands*; MELANIE SCHNELL, *CoCoMol, Max-Planck-Institut für Struktur und Dynamik der Materie, Hamburg, Germany*.

Acenaphthene (Ace) is a three ring polycyclic aromatic hydrocarbon (PAH), which consists of naphthalene and a non-aromatic five member ring. Ace has been previously been studied by microwave spectroscopy where the rotational constants were reported[1]. New measurements from 2-8 GHz using chirped pulse-Fourier transform microwave spectroscopy (CP-FTMW) will be presented. The high sensitivity achieved enabled us to observe all ^{13}C isotopologues in natural abundance and determine the Kraitchman substitution structure. The spectra of Ace complexed with water and $H_2^{18}O$ were also recorded at this frequency range. From these spectra, we have been able to assign the complexes Ace-$(H_2O)_n$, n=1-3 and $(Ace)_2$-H_2O and experimentally derive the O-atom position of the H_2O. The Ace-$(H_2O)_3$ complex is especially interesting as the water aggregate forms a slightly distorted cyclic water trimer from that observed in the IR[2]. These complexes could give insight about the formation of ice grains in the interstellar medium.

[1] Thorwirth, S., Theulé, P., Gottlieb, C.A., McCarthy, M.C., Thaddeus, P. *Astrophys. J.*, 662, 1309-1314, **2007**.
[2] Keutsch, F.N., Cruzan, J.D., Saykally, R.J. *Chem. Rev.*, 103, 2533-2577, **2003**.

TH13 $5:11-5:26$

THE RADIO SPECTRA AND -VE INERTIAL DEFECTS BEHAVIOR OF PLANAR AROMATIC HETEROCYCLES

DON McNAUGHTON, *School of Chemistry, Monash University, Melbourne, Victoria, Australia*; MICHAELA K JAHN, JENS-UWE GRABOW, *Institut für Physikalische Chemie und Elektrochemie, Gottfried-Wilhelm-Leibniz-Universität, Hannover, Germany*; PETER GODFREY, *School of Chemistry, Monash University, Melbourne, Victoria, Australia*; MICHAEL TRAVERS, DENNIS WACHSMUTH, *Institut für Physikalische Chemie und Elektrochemie, Gottfried-Wilhelm-Leibniz-Universität, Hannover, Germany*.

The simplest tricyclic aromatic nitrogen heterocyclic molecules 5,6 benzoquinoline and 7,8 benzoquinoline are possible candidates for detection of aromatic systems in the interstellar medium. Therefore the pure rotational spectra have been recorded using frequency-scanned Stark modulated, jet-cooled millimetre wave absorption spectroscopy (48-87 GHz) and Fourier Transform Microwave (FT MW) spectroscopy (2-26 GHz) of a supersonic rotationally cold molecular jet. Guided by ab initio molecular orbital predictions, spectral analysis of mm wave spectra, and higher resolution FT MW spectroscopy provided accurate rotational and centrifugal distortion constants together with 14N nuclear quadrupole coupling constants for both species. The determined inertial defects, along with those of similar species are used to develop an empirical formula for calculation of inertial defects of aromatic ring systems. The predictive ability of the formula is shown to be excellent for planar species with a number of pronounced out of plane vibrations. The resultant constants are of sufficient accuracy to be used in potential astrophysical searches.[a]

[a]We gratefully acknowledge support from the Deutsche Forschungsgemeinschaft, the Deutsche Akademische Austauschdienst, as well as the Land Niedersachsen (J.-U.G). DMcN also thanks the Royal Society of Chemistry for their generous travel support.

TH14 $5:28-5:43$

A NEW LABORATORY FOR TERAHERTZ CHARACTERIZATION OF COSMIC ANALOG DUSTS

THUSHARA PERERA, *Physics, Illinois Wesleyan University, Bloomington, Illinois, USA*.

Two efforts have been underway to enable the laboratory study of cosmic analogs dusts in the frequency range 60–2000 GHz. They are: (1) the construction of a novel compact Fourier Transform Spectrometer (FTS) design coupled to a dry 4-K cryostat which houses a cooled sample exchanger (filter wheel) and a bolometer. (2) The production of Mg- and Fe-rich silicate dusts using sol-gel methods; various tests to determine their physical and chemical properties; embedding of samples in LDPE pellets for insertion into the novel FTS. This presentation will focus on the current status of the apparatus and data from its first few months of use.

TH15   **5:45 – 5:55**

MM/SUBMM STUDY OF GAS-PHASE PHOTOPRODUCTS FROM METHANOL INTERSTELLAR ICE ANALOGUES

AJ MESKO, HOUSTON HARTWELL SMITH, *Department of Chemistry, Emory University, Atlanta, GA, USA*; STEFANIE N MILAM, *Astrochemistry, NASA Goddard Space Flight Center, Greenbelt, MD, USA*; SUSANNA L. WIDICUS WEAVER, *Department of Chemistry, Emory University, Atlanta, GA, USA*.

Icy grain reactions have gained quite the popularity in the astrochemistry community to explain the formation of complex organic molecules. Through temperature programmed desorption and photolysis experiments we use rotational spectroscopy to measure the gas-phase products of icy grain reactions. Previous results include testing detection limits of the system by temperature programmed desorption of methanol and water ices, photochemistry of gas-phase methanol, and detection of photodesorbed water from a pure water ice surface. Current work that will be discussed focuses on the detection of gas-phase CO and other photoproducts from an ice surface.

TI. Clusters/Complexes
Tuesday, June 21, 2016 – 1:30 PM
Room: B102 Chemical and Life Sciences

Chair: Zbigniew Kisiel, Institute of Physics, Polish Academy of Sciences, Warszawa, Poland

---

TI01   1:30 – 1:45

INFRARED SPECTROSCOPIC STUDY FOR THE HYDRATED CLUSTERS OF PENTANE CATION

TOMOYA ENDO, YOSHIYUKI MATSUDA, ASUKA FUJII, *Department of Chemistry, Tohoku University, Sendai, Japan.*

We performed infrared predissociation spectroscopy of size-selected pentane-water cluster cations, $[\text{pentane-}(H_2O)_n]^+$, n=1-3, generated through the vacuum-ultraviolet photoionization. In the infrared spectra of the di- and tri-hydrated clusters, there appear broad features which spread to the lower frequency region from 2800 cm^{-1}. These broad features are assigned to vibrations of a proton, which is transferred from CH of the pentane cation to the water molecules. These results indicate that the pentane cation has high proton donor ability. We will discuss these results based on theoretical conputations.

TI02   1:47 – 2:02

INFLUENCE OF AROMATIC MOLECULES ON THE STRUCTURE AND SPECTROSCOPY OF WATER CLUSTERS

DANIEL P. TABOR, EDWIN SIBERT, *Department of Chemistry, University of Wisconsin–Madison, Madison, WI, USA;* PATRICK S. WALSH, TIMOTHY S. ZWIER, *Department of Chemistry, Purdue University, West Lafayette, IN, USA.*

Isomer-specific resonant ion-dip infrared spectra are presented for benzene-(water)$_n$, 1-2-diphenoxyethane-(water)$_n$, and tricyclophane-(water)$_n$ clusters. The IR spectra are modeled with a local mode Hamiltonian that was originally formulated for the analysis of benzene-(water)$_n$ clusters with up to seven waters. The model accounts for stretch-bend Fermi coupling, which can complicate the IR spectra in the 3150-3300 cm^{-1} region. When the water clusters interact with each of the solutes, the hydrogen bond lengths between the water molecules change in a characteristic way, reflecting the strength of the solute-water interaction. These structural effects are also reflected spectroscopically in the shifts of the local mode OH stretch frequencies. When diphenoxyethane is the solute, the water clusters distort more significantly than when bound to benzene. Tricyclophane's structure provides an aromatic-rich binding pocket for the water clusters. The local mode model is used to extract Hamiltonians for individual water molecules. These monomer Hamiltonians divide into groups based on their local H-bonding architecture, allowing for further classification of the wide variety of water environments encountered in this study.

TI03   2:04 – 2:19

VIBRATIONAL COUPLING IN SOLVATED FORM OF EIGEN PROTON: TUNING THE COUPLING VIA ISOTOPO-LOGUES

JHENG-WEI LI, JER-LAI KUO, *Institute of Atomic and Molecular Sciences, Academia Sinica, Taipei, Taiwan.*

Experimental studies have shown that features in the vibrational spectra of H_3O^+ can be modulated not only by the type messengers, but also by the number of messengers. Recently, we compared the experimental $H_3O^+Ar_m$, m=1-3 spectra with accurate theoretical simulations and obtain the peak position and absorption intensity by solving the quantum vibrational Schrodinger equation using the potential and dipole moment obtained ab initio methods.[a] In this work, we studied isotopolgues of this ionic cluster to glean into the details of the vibrational couplings manifested in the spectra region of 1500-3800 cmcm^{-1}.

---

[a] J-W Li, M. Morita, T. Takahashi and J-L Kuo, J. Phys. Chem. A, 119, 10887 (2015)

128

TI04   **2:21 – 2:36**

GAS-PHASE MOLECULAR STRUCTURE OF NOPINONE AND ITS WATER COMPLEXES STUDIED BY MICROWAVE FOURIER TRANSFORM SPECTROSCOPY AND QUANTUM CHEMICAL CALCULATIONS

ELIAS M. NEEMAN, JUAN-RAMON AVILES MORENO, T. R. HUET, *UMR 8523 CNRS - Universités des Sciences et Technologies de Lille, Laboratoire PhLAM, 59655 Villeneuve d'Ascq, France.*

Several monoterpenes and terpenoids are biogenic volatile organic compounds which are emitted in the atmosphere, where they react with OH, O_3 and NO_x etc. to give rise to several oxidation and degradation products.[a] Their decomposition products are a major source of secondray organic aerosol (SOA).[b] Spectroscopic information on these atmospheric species is still very scarce. The rotational spectrum of nopinone ($C_9H_{14}O$) one of the major oxidation products of β-pinene,[c][d] and of its water complexes were recorded in a supersonic jet expansion with a Fourier transform microwave spectrometer over the range 2-20 GHz. The structure of the unique stable conformer of the nopinone was optimized using density functional theory and *ab initio* calculations. Signals from the parent species and from the ^{13}C and ^{18}O isotopomers were observed in natural abundance. A magnetic hyperfine structure associated with the pairs of hydrogen nuclei in the methylene groups was observed and modeled.

The structures of several conformers of the nopinone-water complexes with up to three molecules of water were optimized using density functional theory and *ab initio* calculations. The energetically most stable of calculated conformers were observed and anlyzed. The rotational and centrifugal distortion parameters were fitted to a Watson's Hamiltonian in the A-reduction. The present work provides the first spectroscopic characterization of nopinone and its water complexes in the gas phase.

[a]A. Calogirou, B.R. Larsen, and D. Kotzias, *Atmospheric Environment*, **33**, 1423-1439, (1999).
[b]P. Paasonen et al., *Nat. Geosci.*, **6**, 438-442 (2013)
[c]D. Zhang and R. Zhang *The Journal of Chemical Physics*, **122**, 114308, (2005).
[d]R. Winterhalter et al. *Journal of Atmospheric Chemistry*, **35**, 165-197, (2000).

TI05   **2:38 – 2:53**

CHARACTERIZATION OF MICROSOLVATED CROWN ETHERS FROM BROADBAND ROTATIONAL SPECTROSCOPY

CRISTOBAL PEREZ, MELANIE SCHNELL, *CoCoMol, Max-Planck-Institut für Struktur und Dynamik der Materie, Hamburg, Germany*; SUSANA BLANCO, JUAN CARLOS LOPEZ, *Departamento de Química Física y Química Inorgánica / Grupo de Espectroscopía Molecular, Universidad de Valladolid, Valladolid, Spain.*

Since they were first synthetized, crown ethers have been extensively used in organometallic chemistry due to their unparalleled binding selectivity with alkali metal cations. From a structural point of view, crown ethers are heterocycles containing oxygen and/or other heteroatoms, although the most common ones are formed from ethylene oxide unit. Crown ethers are conventionally seen as being hydrophilic inside and hydrophobic outside when the structures found for the metal cation complexes are considered. However, crown ethers are extremely flexible and in isolation may present a variety of stable conformations[a] so that their structure may be easily adapted in presence of a strong ligand as an alkali metal cation minimize the energy of the resulting complex.

Water can be considered a soft ligand which interacts with crown ethers through moderate hydrogen bonds. It is thus interesting to investigate which conformers are selected by water to form complexes, the preferred interaction sites and the possible conformational changes due to the presence of one or more water molecules. Previous studies identified microsolvated crown ethers but in all cases with a chromophore group attached to the structure.[b] Here we present a broadband rotational spectroscopy study of microsolvated crown ethers produced in a pulsed molecular jet expansion. Several 1:1 and 1:2 crown ether:water aggregates are presented for 12-crown-4, 15-crown-5 and 18-crown-6. Unambiguous identification of the structures has been achieved using isotopic substitution within the water unit. The subtle changes induced in the structures of the crown ether monomer upon complexation and the hydrogen-bonding network that hold them together will be also discussed.

[a]F. Gámez, B. Martínez-Haya, S. Blanco,J. C. Lòpez and J. L. Alonso, *Phys. Chem. Chem. Phys.* **2014**, *14* 12912-12918
[b]V. A. Shubert, C.W. Müller and T. Zwier, *J. Phys. Chem. A* **2009**, *113* 8067-8079

TI06

CONCERTED BREAKING OF TWO HYDROGEN BONDS IN WATER HEXAMER PRISM REVEALED FROM BROADBAND ROTATIONAL SPECTROSCOPY

JEREMY O RICHARDSON, *Department of Chemistry, Durham University, Durham, United Kingdom*; CRISTOBAL PEREZ, *CoCoMol, Max-Planck-Institut für Struktur und Dynamik der Materie, Hamburg, Germany*; SIMON LOBSIGER, *Department of Chemistry, The University of Virginia, Charlottesville, VA, USA*; ADAM A. REID, *Department of Chemistry, University of Cambridge, Cambridge, United Kingdom*; BERHANE TEMELSO, GEORGE C. SHIELDS, *Department of Chemistry, Bucknell University, Lewisburg, PA, USA*; ZBIGNIEW KISIEL, *ON2, Institute of Physics, Polish Academy of Sciences, Warszawa, Poland*; DAVID J. WALES, *Department of Chemistry, University of Cambridge, Cambridge, United Kingdom*; BROOKS PATE, *Department of Chemistry, The University of Virginia, Charlottesville, VA, USA*; STUART C. ALTHORPE, *Department of Chemistry, University of Cambridge, Cambridge, United Kingdom.*

Over the past few years, we have used $H_2^{18}O$ water substitution to determine the structures of water clusters by molecular rotational spectroscopy. In the case of the water hexamer, the energy difference between the cage and prism structures is calculated to be about 0.1 kcal/mol and this energy difference is of the order of the zero-point energy variation between the isomers. Using rotational spectroscopy we provided experimental evidence for three isomers, i.e, cage, prism and book and established their relative energy ordering.[a] In the special case of the prism hexamer, cluster dynamics causes measurable splitting in rotational transitions resulting from tunneling between discernible equivalent minima. Multiple isotopic substitution measurements involving all 64 possible isotopologues of the water hexamer prism $(H_2^{18}O)_n(H_2^{16}O)_{6-n}$ were performed in order to identify the water molecules involved in the tunneling motion. The analysis of these tunneling-rotation spectra suggests that there are two distinct tunneling paths that involve concerted motion of two water molecules, implying a prototype scenario involving the breaking of two hydrogen bonds.[b]

[a]C. Pérez, et al, *Science.* **2012**, *336* 897-901

[b]J. O. Richardson et al, *Science.* **2016**, *in press*

TI07

STRUCTURES OF MICROSOLVATED CAMPHOR FROM BROADBAND ROTATIONAL SPECTROSCOPY

CRISTOBAL PEREZ, ANNA KRIN, AMANDA STEBER, *CoCoMol, Max-Planck-Institut für Struktur und Dynamik der Materie, Hamburg, Germany*; JUAN CARLOS LOPEZ, *Departamento Química Física y Química Inorgánica , Universidad de Valladolid, Valladolid, Spain*; ZBIGNIEW KISIEL, *ON2, Institute of Physics, Polish Academy of Sciences, Warszawa, Poland*; MELANIE SCHNELL, *CoCoMol, Max-Planck-Institut für Struktur und Dynamik der Materie, Hamburg, Germany.*

Using broadband rotational spectroscopy, we will present our results on the microsolvation of camphor ($C_{10}H_{16}O$) complexed with up to three water molecules.[a] Unambiguous assignment was achieved by performing multi $H_2^{18}O$ isotopic substitution of clustered water molecules. The observation of all possible mono- and multi-$H_2^{18}O$ insertions in the cluster structure yielded accurate structural information that is not otherwise achievable with single-substitution experiments. The observed clusters exhibit water chains starting with a strong hydrogen bond to the carbonyl group and terminated by a mainly van der Waals (dispersive) contact to one of the available sites at the monomer moiety. The effect of hydrogen bond cooperativity is noticeable, and will be also discussed.

[a]C. Pérez, A. Krin, A. L. Steber, J. C. López, Z. Kisiel, M. Schnell, *J. Phys. Chem. Lett.* **2016**, *7* 154-160.

Intermission

TI08   **3:46 – 4:01**

ROTATIONAL INVESTIGATION OF THE ADDUCTS OF FORMIC ACID WITH ALCOHOLS, ETHERS AND ESTERS

LUCA EVANGELISTI, LORENZO SPADA, WEIXING LI, <u>WALTHER CAMINATI</u>, *Dep. Chemistry 'Giacomo Ciamician', University of Bologna, Bologna, Italy.*

Mixtures of formic acid with methyl alcohol, with isopropyl alcohol, with *tert*-butyl alcohol, with dimethylether and with isopropylformiate have been supersonically expanded as pulsed jets. The obtained cool plumes have been analyzed by Fourier transform microwave spectroscopy. It has been possible to assign the rotational spectra of the 1:1 adducts of formic acid with *tert*-butyl alcohol, with dimethyl ether and with isopropylformiate. The conformational shapes and geometries of these adducts, as well as the topologies of their itermolecular hydrogen bonds will be presented. An explanation is given of the failure of the assignments of the rotational spectra of the adducts of formic acid with methyl alcohol and isopropyl alcohol.

TI09   **4:03 – 4:18**

MICROWAVE SPECTRUM OF THE ETHANOL-METHANOL DIMER

<u>IAN A FINNERAN</u>, BRANDON CARROLL, GRIFFIN MEAD, GEOFFREY BLAKE, *Division of Chemistry and Chemical Engineering, California Institute of Technology, Pasadena, CA, USA.*

The hydrogen bond donor/acceptor competition in mixed alcohol clusters remains a fundamental question in physical chemistry. Previous theoretical work on the prototype ethanol-methanol dimer has been inconclusive in predicting the energetically preferred structure. Here, we report the microwave spectrum of the ethanol-methanol dimer between 8-18 GHz, using a chirped pulse Fourier transform microwave spectrometer. With the aid of ab initio calculations, 36 transitions have been fit and assigned to a t-ethanol-acceptor, methanol-donor structure in an argon-backed expansion. In a helium-backed expansion, a second excited conformer has been observed, and tentatively assigned to a g-ethanol-acceptor, methanol-donor structure. No ethanol-donor, methanol-acceptor structures have been found, suggesting such structures are energetically disfavored.

TI10   **4:20 – 4:35**

HYDROPEROXIDES AS HYDROGEN BOND DONORS

<u>KRISTIAN H. MØLLER</u>, CAMILLA M. TRAM, ANNE S. HANSEN, HENRIK G. KJAERGAARD, *Department of Chemistry, University of Copenhagen, Copenhagen, Denmark.*

Hydroperoxides are formed in the atmosphere following autooxidation of a wide variety of volatile organics emitted from both natural and anthropogenic sources. This raises the question of whether they can form hydrogen bonds that facilitate aerosol formation and growth.

Using a combination of Fourier transform infrared spectroscopy, FT-IR, and *ab initio* calculations, we have compared the gas phase hydrogen bonding ability of *tert*-butylhydroperoxide (tBuOOH) to that of *tert*-butanol (tBuOH) for a series of bimolecular complexes with different acceptors. The hydrogen bond acceptor atoms studied are nitrogen, oxygen, phosphorus and sulphur. Both in terms of calculated redshifts and binding energies (BE), our results suggest that hydroperoxides are better hydrogen bond donors than the corresponding alcohols. In terms of hydrogen bond acceptor ability, we find that nitrogen is a significantly better acceptor than the other three atoms, which are of similar strength. We observe a similar trend in hydrogen bond acceptor ability with other hydrogen bond donors including methanol and dimethylamine.

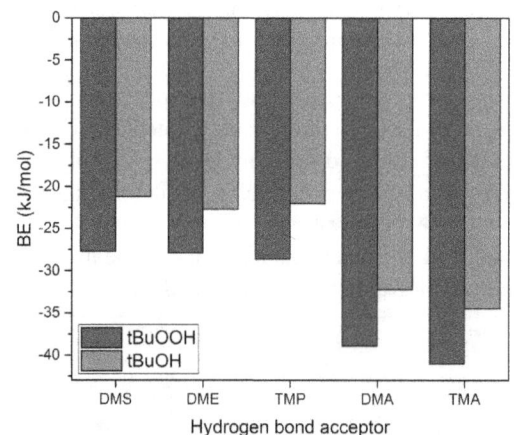

TI11 4:37 – 4:52

GAZ PHASE IR AND UV SPECTROSCOPY OF NEUTRAL CONTACT ION PAIRS

SANA HABKA, VALERIE BRENNER, MICHEL MONS, ERIC GLOAGUEN, *CEA Saclay, LIDYL, Gif-sur-Yvette, France.*

Cations and anions, in solution, tend to pair up forming ion pairs. They play a crucial role in many fundamental processes in ion-concentrated solutions and living organisms. Despite their importance and vast applications in physics, chemistry and biochemistry, they remain difficult to characterize namely because of the coexistence of several types of pairing in solution. However, an interesting alternative consists in applying highly selective gas phase spectroscopy which can offer new insights on these neutral ion pairs. Our study consists in characterizing contact ion pairs (CIPs) in isolated model systems (M^+, Ph-$(CH_2)_n$-COO^- with M=Li, Na, K, Rb, Cs, and n=1-3), to determine their spectral signatures and compare them to ion pairs in solution. We have used laser desorption to vaporize a solid tablet containing the desired salt. Structural information for each system was obtained by mass-selective, UV and IR laser spectroscopy combined with high level quantum chemistry calculations[1]. Evidence of the presence of neutral CIPs was found by scanning the π-π* transition of the phenyl ring using resonant two-photon ionization (R2PI). Then, conformational selective IR/UV double resonance spectra were recorded in the CO_2^- stretch region for each conformation detected. The good agreement between theoretical data obtained at the BSSE-corrected-fullCCSD(T)/dhf-TZVPP//B97-D3/dhf-TZVPP level and experimental IR spectra led us to assign the 3D structure for each ion pair formed. Spectral signatures of (M^+, Ph-CH_2-COO^-) pairs, were assigned to a bidentate CIPs between the alkali cation and the carboxylate group. In the case of (Li^+, Ph-$(CH_2)_3$-COO^-) pairs, the presence of a flexible side chain promotes a cation-π interaction leading to a tridentate O-O-π structure with its unique IR and UV signatures. IR spectra obtained on isolated CIPs were found very much alike the ones published on lithium and sodium acetate in solution[2]. However, in the case of sodium acetate, solution spectra were assigned to solvent shared pairs. Yet, the striking resemblance with our spectral data raises questions about the type assigned, pointing out that CIPs could be more present in these electrolyte solutions than previously thought. The novelty of the gas phase approach to investigate neutral ion pairs, opens the door for various new spectroscopic studies, paving the way to greater knowledge regarding the properties of ion pairs in many scientific fields. 1. Gloaguen, E.; Mons, M.; Topics in Current Chemistry, 2015, Vol 364, 225-270 2. Rudolph, W.W.; Fischer, D.; Irmer, G.; Dalton Transactions 2014, 43, (8), 3174-3185

TI12 *Post-Deadline Abstract* 4:54 – 5:09

DETERMINATION OF STRUCTURAL AND ELECTRONIC PARAMETERS OF ANTIMONY COMPLEX, FROM THEORETICAL CALCULATIONS

BERNA CATIKKAS, ISMAIL KOSAR, *Department of Physics, Mustafa Kemal University, Hatay, Turkey.*

In this study, ground states of antimony (Sb^v) with organic ligands complexes were studied by using density functional theory hybrid methods in order to obtain structural, electronic and vibrational spectral parameters. The mapping molecular electrostatic potential surface of the molecules computed to information about the charge density distribution of the molecules and its chemical reactivity. Frontier molecule orbital properties, HOMO and LUMO energies, global descriptors, and the total density of state diagram analysis were performed by using the time-dependent density functional theory. For the learning nonlinear optical properties, polarizability and hyperpolarizability tensors of the molecule were calculated.

TI13 5:11 – 5:26

EXPERIMENTAL DETERMINATION OF GAS PHASE THERMODYNAMIC PROPERTIES OF BIMOLECULAR COMPLEXES

ANNE S. HANSEN, ZEINA MAROUN, KASPER MACKEPRANG, HENRIK G. KJAERGAARD, *Department of Chemistry, University of Copenhagen, Copenhagen, Denmark.*

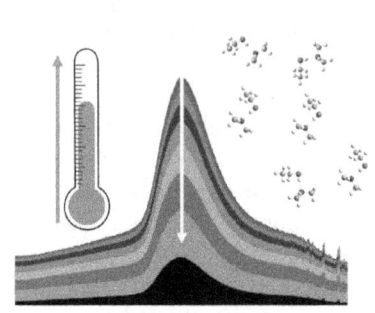

Accurate determination of the atmospheric abundance of hydrogen bound bimolecular complexes is necessary, as hydrogen bonds are partly responsible for the formation and growth of aerosol particles. The abundance of a complex is related to the Gibbs free energy of complex formation (ΔG), which is often obtained from quantum chemical calculations that rely on calculated values of the enthalpy (ΔH) and entropy (ΔS) of complex formation. However, calculations of ΔH and in particular ΔS are associated with large uncertainties, and accurate experimental values are therefore crucial for theoretical benchmarking studies. Infrared measurements of gas phase hydrogen bound complexes were performed in the 300 to 373 K range, and lead to a purely experimental determination of ΔH using the van't Hoff equation. Equilibrium constants were determined by combining an experimental and calculated OH-stretching intensity, from which values of ΔG and hence ΔS could be determined. Thus we can determine ΔG, ΔH and ΔS for a bimolecular complex. We find that in the 300 to 373 K temperature range the determined ΔH and ΔS values are independent of temperature.

TI14 *Post-Deadline Abstract* 5:28 – 5:43

PHOTOELECTRON IMAGING OF TaBO⁻: OBSERVATION OF A BORONYL TRANSITION METAL COMPLEX

JOSEPH CZEKNER, LAI-SHENG WANG, *Department of Chemistry, Brown University, Providence, RI, USA.*

Boronyl (BO) is isoelectronic with CN, but its chemistry is much less known. One cause for the difficulty of the synthesis of boronyl complexes is that BO does not participate in π-backbonding with the d orbitals of transition metals due to an increased energy of the π^* orbital relative to CN or CO. Here we report a velocity map imaging study on TaBO⁻, the first observation of a BO ligand with an early transition metal. We observe transitions from the anion ground state ($^4\Delta$) to two neutral states ($^5\Delta$ and $^3\Delta$). We analyzed the chemical bonding in TaBO and compared it with TaCO. We found that there is a comparable amount of overlap between the d_{xz} and d_{yz} orbitals with the π^* orbitals of BO and CO. Our result suggests Ta may be better suited to bond with BO ligands to allow new transition metal boronyl complexes.

T.J. Vibrational structure/frequencies

Tuesday, June 21, 2016 – 1:30 PM

Room: 217 Noyes Laboratory

Chair: Gary E. Douberly, The University of Georgia, Athens, GA, USA

TJ01   1:30 – 1:45

EXPLORING THE RELATIONSHIPS BETWEEN ANHARMONICITY AND OH BOND LENGTHS IN HYDROGEN BONDED COMPLEXES

ANNE B McCOY, *Department of Chemistry, University of Washington, Seattle, WA, USA*; SOTIRIS XANTH-EAS, *Physical Sciences Division, Pacific Northwest National Laboratory, Richland, WA, USA.*

In this talk we explore the effects of anharmonicity on the zero-point averaged OH bond lengths in hydrogen bonded complexes. Clusters with as many as six HF molecules or water molecules are explored as well as protonated water clusters and complexes of water clusters with F^-, Cl^-, Br^- and OH^-. It is shown that there is a universal correlation between the vibrationally averaged OH or HF bond length and the anharmonc OH or HF stretch frequency. This relationship provides an extension to previously investigated correlations between the equilibrium bond lengths and harmonic frequencies and allows one to anticipate OH or HF bond lengths based on measured frequencies. In addition, differences between the R_z and R_0 structures are discussed within the context of these weakly bound complexes.

TJ02   1:47 – 2:02

SPECTROSCOPIC MANIFESTATION OF VIBRATIONALLY-MEDIATED STRUCTURE CHANGE IN THE ISOLATED FORMATE MONOHYDRATE

JOANNA K. DENTON, CONRAD T. WOLKE, OLGA GORLOVA, HELEN GERARDI, *Department of Chemistry, Yale University, New Haven, CT, USA*; ANNE B McCOY, *Department of Chemistry, University of Washington, Seattle, WA, USA*; MARK JOHNSON, *Department of Chemistry, Yale University, New Haven, CT, USA.*

The breadth of the OH stretching manifold observed in the IR for bulk water is commonly attributed to the thermal population of excited states and the presence of many configurations within the water network. Here, I use carboxylate species as a rigid framework to isolate a single water molecule in the gas phase and cold ion vibrational pre-dissociation spectroscopy to explore excited state contributions to bandwidth. The spectrum of the carboxylate monohydrate exhibits a signature series of peaks in the OH stretching region of this system, providing an archetypal model to study vibrationally adiabatic mode separation. Previous analysis of this behavior accounts for the extensive progression in a Franck-Condon formalism involving displaced vibrationally adiabatic potentials[a,b]. In this talk I will challenge this prediction by using isotopic substation to systematically change the level structure within these potentials. This picture quantitatively accounts for the diffuse spectrum of this complex at elevated temperature providing a convenient spectroscopic reporter for the temperature of ions in a trap.

[a] E. M. Myshakin, K. D. Jordan, E. L. Sibert III, M. A. Johnson J. Chem. Phys. 119, 10138 (2003).
[b] W.H. Robertson, et al. J. Phys Chem. 107, 6527 (2003).

TJ03

THEORETICAL INVESTIGATION OF ANHARMONIC EFFECTS OBSERVED IN THE INFRARED SPECTRA OF THE FORMALDEHYDE CATION AND ITS HYDROXYMETHYLENE ISOMER

LINDSEY R MADISON, *Department of Chemistry, University of Washington, Seattle, WA, USA*; JONATHAN MOSLEY, DANIEL MAUNEY, MICHAEL A DUNCAN, *Department of Chemistry, University of Georgia, Athens, GA, USA*; ANNE B McCOY, *Department of Chemistry, University of Washington, Seattle, WA, USA*.

Formaldehyde is the smallest organic molecule and is a prime candidate for a thorough investigation regarding the anharmonic approximations made in computationally modeling its infrared spectrum. Mass-selected ion spectroscopy was used to detect mass 30 cations which include of $HCOH^+$ and CH_2O^+. In order to elucidate the differences between the structures of these isomers, infrared spectroscopy was performed on the mass 30 cations using Ar predissociation. Interestingly, several additional spectral features are observed that cannot be explained by the fundamental OH and CH stretch vibrations alone. By including anharmonic coupling between OH and CH stretching and various overtones and combination bands involving lower frequency vibrations, we are able to identify how specific modes couple and lead to the experimentally observed spectral features. We combine straight-forward, *ab initio* calculations of the anharmonic frequencies of the mass 30 cations with higher order, adiabatic approximations and Fermi resonance models. By including anharmonic effects we are able to confirm that the isomers of the $CH_2O^+\cdot Ar$ system have substantially different, and thus distinguishable, IR spectra and that many of the features can only be explained with anharmonic treatments.

TJ04

SPECTROSCOPIC SIGNATURES AND STRUCTURAL MOTIFS OF DOPAMINE: A COMPUTATIONAL STUDY

SANTOSH KUMAR SRIVASTAVA, VIPIN BAHADUR SINGH, *Department of Physics, Udai Pratap Autonomous College, Varanasi, India*.

Dopamine (DA) is an essential neurotransmitter in the central nervous system and it plays integral role in numerous brain functions including behaviour, cognition, emotion, working memory and associated learning.[a][b] In the present work the conformational landscapes of neutral and protonated dopamine have been investigated in the gas phase and in aqueous solution by MP2 and DFT (M06-2X, ωB97X-D, B3LYP and B3LYP-D3) methods. Twenty lowest energy structures of neutral DA were subjected to geometry optimization and the gauche conformer, GIa, was found to be the lowest gas phase structure at the each level of theory in agreement with the experimental rotational spectroscopy.[c] All folded gauche conformers (GI) where lone electron pair of the NH2 group is directed towards the π system of the aromatic ring ('non up') are found more stable in the gas phase. While in aqueous solution, all those gauche conformers (GII) where lone electron pair of the NH2 group is directed opposite from the π system of the aromatic ring ('up' structures) are stabilized significantly. Nine lowest energy structures, protonated at the amino group, are optimized at the same MP2/aug-cc-pVDZ level of theory. In the most stable gauche structures, g-1 and g+1, mainly electrostatic cation - π interaction is further stabilized by significant dispersion forces as predicted by the substantial differences between the DFT and dispersion corrected DFT-D3 calculations. In aqueous environment the intra-molecular cation- π distance in g-1 and g+1 isomers, slightly increases compared to the gas phase and the magnitude of the cation- π interaction is reduced relative to the gas phase, because solvation of the cation decreases its interaction energy with the π face of aromatic system. The IR intensity of the bound N-H+ stretching mode provides characteristic 'IR spectroscopic signatures' which can reflect the strength of cation- π interaction energy. The CC2 lowest lying S1 ($1\pi\pi^*$) excited state of neutral dopamine is significantly red shifted upon protonation at amino site.

[a]E. Dragicevic, J. Schiemann and B. Liss, Neuroscience, 2015, 284, 798.
[b]Y. T. Chien et al. Science, 2010, 330, 1091.
[c]Cabezas etal., J. Phys. Chem. Lett. 2013, 4, 486.

TJ05   2:38 – 2:53

SPECTROSCOPIC SIGNATURES AND STRUCTURAL MOTIFS IN ISOLATED AND HYDRATED XANTHINE: A COMPUTATIONAL STUDY

VIPIN BAHADUR SINGH, *Department of Physics, Udai Pratap Autonomous College, Varanasi, India.*

The conformational landscapes of xanthine and its hydrated complex have been investigated by MP2 and DFT methods. The ground state geometry optimization yield five lowest energy conformers of xanth1-(H2O)1 complex at the MP2/6-311++G(d,p) level of theory for the first time. We investigated the low-lying excited states of bare xanthine by means of coupled cluster singles and approximate doubles (CC2) and TDDFT methods and a satisfactory interpretation of the electronic absorption spectra1 is obtained. The difference between the S0-S1 transition energy due to the most stable and the second most stable stable conformation of xanthine was found to be 859 cm^{-1}. One striking feature is the coexistence of the blue and red shift of the vertical excitation energy of the optically bright state S1 of xanthine upon forming complex with a water at C2=O and C6=O carbonyl sites, respectively. The lowest singlet $\pi\pi^*$ excited-state of the xanth1-(H2O)1 complex involving C2=O carbonyl are strongly blue shifted which is in agreement with the result of R2PI spectra of singly hydrated xanthine. While for the most stable and the second most stable xanth1-(H2O)1 complexes involving C6=O carbonyl, the lowest singlet $\pi\pi^*$ excited-state is red shifted. The effect of hydration on S1 excited state due to bulk water environment was mimicked by a combination of polarizable continuum solvent model (PCM) and conductor like screening model (COSMO), which also shows a blue shift in accordance with the result of electronic absorption spectra in aqueous solution.[a] This hypsochromic shift, is expected to be the result of the changes in the π-electron delocalization extent of molecule because of hydrogen bond formation. The optimized structure of xanthine dimer, computed the first time by MP2 and DFT methods. The binding energy of this dimer linked by double N-H...O=C hydrogen bonds was found to be 88 kj/mole at the MP2/6-311++G(d,p) level of theory. Computed IR spectra is found in remarkable agreement with the experiment and the out of phase (C=O)2 stretching mode shows tripling of intensity upon dimerisation. The vertical excitation energy of the optically bright state S1 of xanthine monomer upon forming dimer is shifted towards red as well as blue.

---

[a]J. Chen and B.Kohler, Phys. Chem. Chem. Phys., 2012,14,10677-1068.

TJ06   2:55 – 3:10

VIBRATIONAL SPECTROSCOPY AND THEORY OF $Fe_x^+(CH_4)_n$ (x =2,3) (n = 1–3)

CHRISTOPHER COPELAND, MUHAMMAD AFFAWN ASHRAF, RICARDO B. METZ, *Department of Chemistry, University of Massachusetts, Amherst, MA, USA.*

Vibrational spectra are measured for $Fe_x^+(CH_4)_n$ (x =2,3) (n = 1–3) in the C–H stretching region (2650–3100 cm^{-1}) using photofragment spectroscopy, by monitoring the loss of CH_4. All of the spectra exhibit an intense peak corresponding to the symmetric C–H stretch around 2800 cm^{-1}, which is red shifted by about 100 cm^{-1} from free methane. The presence of a single peak suggests a nearly equivalent interaction between the methane ligands and the iron center. The peak becomes slightly less red shifted as the number of methane ligands increases. Density functional theory calculations, B3LYP and BPW91, are used to identify possible structures and predict the spectra. Results suggest that the methane(s) bind in a terminal configuration and that the Fe_2^+ complexes are in the octet spin state while the Fe_3^+ complexes are in the dectet spin state. Lower C-H stretching frequencies are observed for Fe_3^+ complexes, indicating that the CH_4 interacts more strongly with Fe_3^+ than Fe_2^+.

TJ07

VIBRATIONAL ANALYSIS OF THE SiCN $\tilde{X}\,^2\Pi$ SYSTEM

MASARU FUKUSHIMA, TAKASHI ISHIWATA, *Information Sciences, Hiroshima City University, Hiroshima, Japan.*

The laser induced fluorescence (LIF) spectrum of the $\tilde{A}\,^2\Delta - \tilde{X}\,^2\Pi$ transition was obtained for SiCN generated by laser ablation under supersonic free jet expansion. The vibrational structure of the dispersed fluorescence (DF) spectra from single vibronic levels (SVL's) was analyzed by numerical diagonalization procedure, in which Renner-Teller (R-T), anhamonicity, spin-orbit (SO), Herzberg-Teller (H-T), Fermi, and Sears interactions have been considered, where the Sears resonance is a second-order interaction combined from SO and H-T interactions with $\Delta K = \pm 1$, $\Delta \Sigma = \mp 1$, and $\Delta P = 0$. Four vibronic levels, $(01^10)\ \mu\ \Sigma^{(-)}_{\frac{1}{2}}$, $\kappa\ \Sigma^{(+)}_{\frac{1}{2}}$, $(02^00)\ \mu$ and $\kappa\ \Pi_{\frac{1}{2}}$, are almost closed within the four basis functions by R-T and Sears interactions (i.e. the four-by-four transformation matrix below is close to ortho-normal);

$$
\begin{pmatrix}
|(01^10)\ \mu\ ^2\Sigma^{(-)}\rangle \\
|(01^10)\ \kappa\ ^2\Sigma^{(+)}\rangle \\
|(02^00)\ \mu\ ^2\Pi_{\frac{1}{2}}\rangle \\
|(02^00)\ \kappa\ ^2\Pi_{\frac{1}{2}}\rangle
\end{pmatrix}
=
\begin{pmatrix}
0.9 & -0.4 & 0.0 & 0.0 \\
0.4 & 0.8 & 0.3 & -0.2 \\
0.2 & 0.4 & -0.8 & 0.4 \\
0.0 & 0.0 & -0.5 & -0.8
\end{pmatrix}
\begin{pmatrix}
|-\rangle |+\tfrac{1}{2}\rangle\ |0; 1, +1\rangle \\
|+\rangle |+\tfrac{1}{2}\rangle\ |0; 1, -1\rangle \\
|+\rangle |-\tfrac{1}{2}\rangle\ |+1; 2, 0\rangle \\
|-\rangle |-\tfrac{1}{2}\rangle\ |+1; 2, +2\rangle
\end{pmatrix},
$$

where $|\Lambda\rangle|\Sigma\rangle|K; v_2, l\rangle = |-\rangle|+\tfrac{1}{2}\rangle|0; 1, +1\rangle$ *etc.* are basis functions of the vibronic Hamiltonian for the numerical diagonalization, and $|\Lambda\rangle$, $|\Sigma\rangle$, and $|K; v_2, l\rangle$ are basis functions of electronic, electron spin, and two dimensional harmonic oscillator, respectively. The mixing coefficients of the two vibronic levels agree with those obtained from computational studies[a]. The two levels among the four above, $(01^10)\ \kappa\ \Sigma^{(+)}_{\frac{1}{2}}$ and $(02^00)\ \mu\ \Pi_{\frac{1}{2}}$, with $\Delta K = \pm 1$ and $\Delta P = 0$, show typical example of Sears resonance with an almost one-to-one mixing. Even for levels lying at $\sim 1{,}000$ cm^{-1}, some of them are mixed heavily and widely with several levels, and their vibrational quantum numbers are thus meaningless.

---

[a]V. Brites, A. O. Mitrushchenkov, and C. Léonard, J. Chem. Phys. 138, 104311 (2013); C. Léonard, Private communication.

TJ08

ELECTRONIC STRUCTURE OF SMALL LANTHANIDE CONTAINING MOLECULES

JARED O. KAFADER, MANISHA RAY, JOSEY E TOPOLSKI, CAROLINE CHICK JARROLD, *Department of Chemistry, Indiana University, Bloomington, IN, USA.*

Lanthanide-based materials have unusual electronic properties because of the high number of electronic degrees of freedom arising from partial occupation of 4f orbitals, which make these materials optimal for their utilization in many applications including electronics and catalysis. Electronic spectroscopy of small lanthanide molecules helps us understand the role of these 4f electrons, which are generally considered core-like because of orbital contraction, but are energetically similar to valence electrons. The spectroscopy of small lanthanide-containing molecules is relatively unexplored and to broaden this understanding we have completed the characterization of small cerium, praseodymium, and europium molecules using photoelectron spectroscopy coupled with DFT calculations. The characterization of PrO, EuH, EuO/EuOH, and Ce$_x$O$_y$ molecules have allowed for the determination of their electron affinity, the assignment of numerous anion to neutral state transitions, modeling of anion/neutral structures and electron orbital occupation.

Intermission

TJ09   **4:03 – 4:18**

EXTENSIVE MEASUREMENTS OF VIBRATION-INDUCED PERMANENT ELECTRIC DIPOLE MOMENTS OF
METHANE

SHOKO OKUDA, HIROYUKI SASADA, *Department of Physics, Faculty of Science and Technology, Keio University, Yokohama, Japan.*

A methane molecule (CH_4) has a permanent electric dipole moment (PEDM) in the excited state of the triply-degenerate vibrational modes[a,b]. The rotational dependence of the PEDM was reported in the $2\nu_3$ band[c]. However, in the ν_3 band, it was only determined on the $P(7)E$ transition which fortunately lies in the tunable range of a 3.4 μm He-Ne laser.

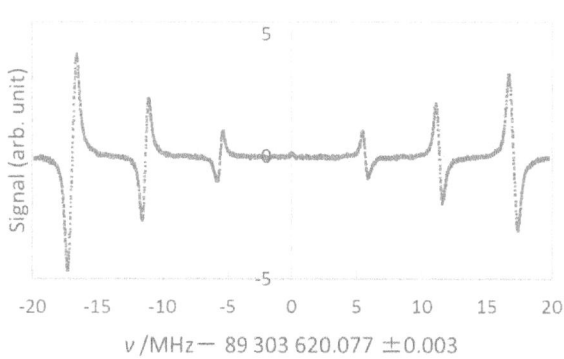

We have developed a mid-infrared broadband sub-Doppler resolution spectrometer consisting of a difference-frequency-generation source and an optical frequency comb linked to International Atomic Time. This spectrometer enables us to measure the Stark effects of 20 transitions in the ν_3 band of methane from 87.7 to 92.8 THz (2927~3095 cm^{-1}). The observed linewidth is 0.5 MHz, and the frequency scale is absolutely calibrated. The figure depicts the Stark modulation spectrum of the $P(4)E$ transition. The applied DC electric field was 3.5 kV/cm. We determined Stark coefficients with a relative uncertainty of 1 %. Our goal is to reveal the rotational dependence of the PEDM. For this end, we yield molecular constants which reproduce the transition frequencies by a least-square method and determine the mixing of the wave functions.

[a]M. Mizushima and P. Venkateswarlu, *J. Chem. Phys.* **21**, 705 (1953).

[b]K. Uehara, K. Sakurai and K. Shimoda, *J. Phys. Soc. Jpn.* **26**, 1018 (1969).

[c]H. Sasada, K. Suzumura and C. Ishibashi, *J. Chem. Phys.* **105**, 9027 (1996).

TJ10   **4:20 – 4:35**

GLOBAL FREQUENCY AND INTENSITY ANALYSIS OF THE $\nu_{10}/\nu_7/\nu_4/\nu_{12}$ BANDS SYSTEM OF $^{12}C_2H_4$ at 10 μm
USING THE D_{2h} TOP DATA SYSTEM

ABDULSAMEE ALKADROU, *Université de Reims/CNRS, Groupe de Spectroscopie Moléculaire et Atmosphérique, Reims, France*; MAUD ROTGER, *Laboratoire GSMA, CNRS / Université de Reims Champagne-Ardenne, REIMS, France*; VINCENT BOUDON, *Laboratoire ICB, CNRS/Université de Bourgogne, DIJON, France*; JEAN VANDER AUWERA, *Service de Chimie Quantique et Photophysique, Université Libre de Bruxelles, Brussels, Belgium.*

A global frequency and intensity analysis of the infrared tetrad located in the 600 − 1500 cm^{-1} region was carried out using the tensorial formalism developed in Dijon for X_2Y_4 asymmetric-top molecules[a] and a program suite called $D_{2h}TDS$ (now part of the XTDS/SPVIEW spectroscopic software)[b]. It relied on spectroscopic information available in the literature and retrieved from absorption spectra recorded in Brussels using a Bruker IFS 120 to 125 HR upgraded Fourier transform spectrometer, in the frame of either the present or previous work[c]. In particular, 645 and 131 lines intensities have been respectively measured for the weak ν_{10} and ν_4 bands. Including the Coriolis interactions affecting the upper vibrational levels 10^1, 7^1, 4^1 and 12^1, a total of 10737 line positions and 1870 line intensities have been assigned and fitted with global root mean square deviations of 2.6×10^{-4} cm^{-1} and 2.4 %, respectively. Relying on the results of the present work and available in the literature, a list of parameters for 65420 lines in the ν_{10}, ν_7, ν_4 and ν_{12} bands of $^{12}C_2H_4$ was generated. The present work provides an obvious improvement over HITRAN and GEISA for the ν_{10} band (see figure), and a marginally better modeling for the ν_7 band (and for the ν_4 band hidden beneath it). To the best of our knowledge, this is the first time that a global intensity analysis is carried out in this range of the ethylene spectrum.

[a]Raballand W, Rotger M, Boudon V, Loëte M. J Mol Spectrosc 2003;217:239–48.

[b]Wenger Ch, Boudon V, Rotger M, Champion JP, Sanzharov M. J Mol Spectrosc 2008;251:102–13.

[c]Rotger M, Boudon V, Vander Auwera J. J Quant Spectrosc Radiat Transf 2008;109:952-62.

138

TJ11 4:37 – 4:52

HIGH-RESOLUTION STIMULATED RAMAN SPECTROSCOPY AND ANALYSIS OF ν_2 AND ν_3 BANDS OF of $^{13}C_2H_4$ USING THE D_{2h} TOP DATA SYSTEM

ABDULSAMEE ALKADROU, *Université de Reims/CNRS, Groupe de Spectroscopie Moléculaire et Atmosphérique, Reims, France*; MAUD ROTGER, *Laboratoire GSMA, CNRS / Université de Reims Champagne-Ardenne, REIMS, France*; DIONISIO BERMEJO, *Inst. Estructura de la Materia, IEM-CSIC, Madrid, Spain*; JOSE LUIS DOMENECH, *Molecular Physics, Instituto de Estructura de la Materia (IEM-CSIC), Madrid, Spain*; VINCENT BOUDON, *Laboratoire ICB, CNRS/Université de Bourgogne, DIJON, France.*

High resolution stimulated Raman spectra of $^{13}C_2H_4$ in the regions of the ν_2 and ν_3 Raman active modes have been recorded at at two temperatures (145 and 296 K) based on the quasi continuous-wave (cw) stimulated Raman spectrometer at Instituto de Estructura de la Materia (CSIC) in Madrid. A tensorial formalism adapted to X_2Y_4 planar asymmetric tops with D_{2h} symmetry has been developed in Dijon[a] and a program suite called $D_{2h}TDS$ (now part of the XTDS/SPVIEW spectroscopic software[b] was proposed to calculate their high-resolution spectra. The effective Hamiltonian operator, involving a polyad structure, and transition moment (dipole moment and polarizability) operators can be systematically expanded to carry out global analyses of many rovibrational bands. A total of 103 and 51 lines corresponding to ν_2 and ν_3 Raman active modes have been assigned and fitted in frequency with a global root mean square deviation of 0.54×10^{-3} cm^{-1} and 0.36×10^{-3} cm^{-1}, respectively. The figures below shows the stimulated Raman spectrum of the ν_2 and ν_3 bands of $^{13}C_2H_4$, compared to the simulation at 296 K.

[a]Raballand W, Rotger M, Boudon V, Loëte M. J Mol Spectrosc 2003;217:239–48.

[b]Wenger Ch, Boudon V, Rotger M, Champion JP, Sanzharov M. J Mol Spectrosc 2008;251:102–13.

TJ12 4:54 – 5:09

OBSERVATION OF THE LOW-LYING $a^3\Delta$ AND $A^1\Delta$ STATES IN JET-COOLED TANTALUM MONONITRIDE

SHEO MUKUND, SOUMEN BHATTACHARYYA, SANJAY G. NAKHATE, *Atomic and Molecular Physics Division, Bhabha Atomic Research Centre, Mumbai-400085, Maharashtra, India.*

The low-lying states of tantalum mononitride (TaN) molecules, produced in the free-jet apparatus have been studied. Laser-induced dispersed fluorescence spectra were recorded by exciting isolated rotational levels of the previously known excited electronic states[a], as well as the newly observed excited states. All the three spin-orbit components along with their vibrational structure of the low-lying $a^3\Delta$ and $A^1\Delta$ states have been observed. The term energy along with harmonic and anharmonic wavenumbers for the ground $X^1\Sigma^+$ and low-lying $a^3\Delta$ states at equilibrium have been determined and are in fairly good agreement with the reported ab initio values[a,b]. A symmetric spin splitting has been observed in the $a^3\Delta$ state in TaN, as opposed to the asymmetric one in the congeneric VN and NbN molecules.

[a]R. S. Ram, J. Liévin, and P. F. Bernath, Journal of Molecular Spectroscopy **215**, 275 (2002)

[b]T. Fleig, M.K. Nayak, M. G. Kozlov, Physical Review A **93**, 012505 (2016)

TJ13   **5:11 – 5:26**

APPLICATION OF MULTIVALUED HIGH ORDER PADE-HERMITE APPROXIMANTS TO RESUMMATION OF PER-TUBATION SERIES. VIBRATIONAL AND ROVIBRATIONAL ENERGY SPECTRUM OF H_2CO MOLECULE.

ANDREY DUCHKO[a], *Molecular Spectroscopy, V.E. Zuev Institute of Atmospheric Optics, Tomsk, Russia*; SERGEI N. YURCHENKO, *Department of Physics and Astronomy, University College London, Gower Street, London WC1E 6BT, United Kingdom*; ALEXANDR BYKOV, *Molecular Spectroscopy, V.E. Zuev Institute of Atmospheric Optics, Tomsk, Russia.*

Perturbation theory is known as the powerful instrument of solving various spectroscopic issues. Nevertheless, direct application of a perturbation theory often lead to divergent series. For instance, it is known that perturbation series diverge in the case of anharmonic resonance coupling between vibrational states [1]. To overcome this problem we use high order Pade-Hermite approximants (degree varies from quadratic up to 10-th degree). The technique was implemented in calculations of vibrational and rovibrational energy levels of H_2CO with potential energy surface taken from [2]. The accuracy of calculated levels is comparable with variational calculations. It is worth noting that using multiple branches of Pade-Hermite approximants allows to get the sum of perturbation series even for highly excited vibrational states with an extremal divergence of corresponding perturbation series. This new approach shows the impact of resonances between different energy levels on the coefficients of the perturbation series, corresponding to each state. Thus, each perturbation series contains additional information about other energy levels, which can be extracted by the use of multivalued approximants.

REFERENCES

1. J. Cizek, V. Spirko, and O. Bludsky, On the use of divergent series in vibrational spectroscopy. Two- and three-dimensional oscillators, J. Chem. Phys. 99, 7331 (1993).
2. A. Yachmenev, S. N. Yurchenko, Per Jensen, and W. Thiel: A new spectroscopic potential energy surface for formaldehyde in its ground electronic state, J. Chem. Phys. 134, 244307/1-11 (2011).

---

[a]National Research Tomsk Polytechnik University, Institute of International Education and Language Communication, Tomsk, Russia

TJ14   **5:28 – 5:43**

CALCULATIONS OF THE ELECTRONIC STRUCTURE AND VIBRATIONAL STATES OF THE EXCITED a^3A" TRIPLET AND A^1A" SINGLET STATES OF FORMALDEHYDE

BRADLEY WELCH, RICHARD DAWES, *Department of Chemistry, Missouri University of Science and Technology, Rolla, MO, USA*; VLADIMIR TYUTEREV, *Groupe de Spectrométrie Moléculaire et Atmosphérique, UMR CNRS 7331, Université de Reims, Reims Cedex 2, France*; LUDOVIC DAUMONT, *Laboratoire GSMA, CNRS / Université de Reims Champagne-Ardenne, REIMS, France.*

Anharmonic forcefields were constructed for the lowest two excited electronic states of formaldehyde (a^3A" triplet and A^1A" singlet) using Davidson corrected explicitly correlated MRCI-F12.

Calculated vibrational levels were used to make comparisons and assignments in some experimental spectra.

Additional calculations for trioxane (a possible contaminant) were performed to aid in the assignments.

TK. Theory and Computation

Tuesday, June 21, 2016 – 1:30 PM

Room: 140 Burrill Hall

Chair: Jonathan Tennyson, University College London, London, United Kingdom

---

TK01   1:30 – 1:45

IS WATSON'S "CHARGE-MODIFIED" REDUCED MASS ALWAYS BEST FOR DIATOMIC IONS ?

ROBERT J. LE ROY, *Department of Chemistry, University of Waterloo, Waterloo, ON, Canada*; NIKESH S. DATTANI, *Graduate School of Science, Kyoto University, Kyoto, Japan.*

Watson's landmark reformulation of the Born-Oppenheimer separation problem has been the basis of most combined-isotopologue analyses of diatomic spectroscopic data since 1980.[a] One noteworthy feature of this work was his proposal that description of the dynamical behaviour of a diatomic ion with ($\pm$) charge Q, formed from atoms with isotopic masses of M_1 and M_2, should use a "charge-modified" reduced mass: $\mu_{\mathrm{Wat}} \equiv M_1 M_2/(M_1 + M_2 - Q m_e)$, in which m_e is the electron mass, and this proposal seems to have been benignly accepted and adopted. The first quantitative test of this proposal was in the pioneering combined-isotopologue direct-potential-fit (CI-DPF) study of HeH$^+$ by Coxon and Hajigeorgiou in 1999,[b] where they compared the quality of fit for analyses that used different choices for the definition of the reduced masses of the various isotopologues, and found that the best choice seemed to be to use conventional two-body reduced masses for $\left(M_1 - \frac{1}{2}m_e\right)$ and $\left(M_2 - \frac{1}{2} m_e\right)$. This question was re-examined recently in the context of a CI-DPF study of CH$^+$, and a rather different conclusion was reached.[c] The present paper combines new CI-DPF studies of HeH$^+$ and BeH$^+$ with our recent CH$^+$ work, and attempts to draw some general conclusions on this matter.

---

[a] J.K.G. Watson, *J. Mol. Spectrosc.* **80**, 411 (1980)

[b] J.A. Coxon and P. Hajigeorgiou, *J. Mol. Spectrosc.* **193**, 306 (1999).

[c] Y.-S. Cho and R.J. Le Roy, *J. Chem. Phys.* **144**, 024311 (2016).

TK02   1:47 – 2:02

INELASTIC SCATTERING OF H+CO: INFLUENCE OF RENNER-TELLER COUPLING

STEVE ALEXANDRE NDENGUE, RICHARD DAWES, *Department of Chemistry, Missouri University of Science and Technology, Rolla, MO, USA.*

Carbon monoxide is after molecular hydrogen the second most abundant molecule in the interstellar medium (ISM) and also an important molecule for processes occurring in the atmosphere, and hydrocarbon combustion. The rate coefficients of CO in collision with dominant species like H, H$_2$, He, etc are common keys to understand the CO emission spectrum or to model combustion chemistry processes. The inelastic scattering of H+CO has indeed been intensively studied in the past decades, using mainly the so-called WKS PES developed by Werner et al or recently a modified version by Song et al. Though the spectroscopic agreement of the WKS surface with experiment is known to be quite good, there is no experimental evidence that the dynamics of the system is correctly reproduced by the surface. We will present in this talk new results on a set of HCO surfaces of the ground and the excited Renner-Teller coupled electronic states with the principal objective of studying the influence of the Renner-Teller coupling on the inelastic scattering of H+CO. Our calculations done using the MCTDH algorithm cover the 0-2 eV energy range and allow one to interpret the effect of the Renner-Teller coupling on the rovibrational inelastic scattering. Additionally, vibrational bound and resonance state calculations on this new PES and comparisons with available experimental data will be presented.

TK03   2:04 – 2:14

ROOM TEMPERATURE LINE LISTS FOR CO$_2$ ISOTOPOLOGUES WITH *AB INITIO* COMPUTED INTENSITIES

EMIL ZAK[a], JONATHAN TENNYSON, OLEG POLYANSKY, LORENZO LODI, *Department of Physics and Astronomy, University College London, Gower Street, London WC1E 6BT, United Kingdom*; NIKOLAY FEDOROVICH ZOBOV, *Microwave Spectroscopy, Institute of Applied Physics, Nizhny Novgorod, Russia*; SERGEY TASHKUN, VALERY PEREVALOV, *Molecular Spectroscopy, V.E. Zuev Institute of Atmospheric Optics, Tomsk, Russia*.

We report 13 room temperature line lists for all major CO$_2$ isotopologues, covering 0-8000 cm^{-1}. These line lists are a response to the need for line intensities of high, preferably sub-percent, accuracy by remote sensing experiments. Our scheme encompasses nuclear motion calculations supported by critical reliability analysis of the generated line intensities. Rotation-vibration wavefunctions and energy levels are computed using DVR3D and a high quality semi-empirical potential energy surface (PES) [1], followed by computation of intensities using a fully *ab initio* dipole moment surface (DMS). Cross comparison of line lists calculated using pairs of high-quality PES's and DMS's is used to assess imperfections in the PES, which lead to unreliable transition intensities between levels involved in resonance interactions. Four line lists are computed for each isotopologue to quantify sensitivity to minor distortions of the PES/DMS. This provides an estimate of the contribution to the overall line intensity error introduced by the underlying PES. Reliable lines are benchmarked against recent state-of-the-art measurements [2] and HITRAN-2012 supporting the claim that the majority of line intensities for strong bands are predicted with sub-percent accuracy [3]. Accurate line positions are generated using an effective Hamiltonian [4]. We recommend use of these line lists for future remote sensing studies and inclusions in databases.

[1] X. Huang, D. W. Schwenke, S. A. Tashkun, T. J. Lee, J. Chem. Phys. 136, 124311, 2012.

[2] O. L. Polyansky, K. Bielska, M. Ghysels, L. Lodi, N. F. Zobov, J. T. Hodges, J. Tennyson, PRL, 114, 243001, 2015.

[3] E. Zak, J. Tennyson, O. L. Polyansky, L. Lodi, S. A. Tashkun, V. I. Perevalov, JQSRT, in press and to be submitted.

[4] S. A. Tashkun, V. I. Perevalov, R. R. Gamache, J. Lamouroux, JQSRT, 152, 45–73, 2015.

[a]Emil Zak Department of Physics and Astronomy, University College London, Gower Street, London, WC1E 6BT Email: emil.zak.14@ucl.ac.uk

---

TK04   2:16 – 2:31

ANALYSIS OF THE VIBRATIONAL SPECTRA TO CALCULATE THE THERMODYNAMIC QUANTITIES CLOSE TO PHASE TRANSITIONS IN NH_4F

HAMIT YURTSEVEN, *Physics Group, Middle East Technical University, Northern Cyprus Campus, Güzelyurt, Turkey*; OZLEM TARI, *Department of Mathematics and Computer Science, Istanbul Arel University, Istanbul, Turkey*.

The pressure dependence of the vibrational frequencies of the Raman modes is analyzed from the literature data and the thermodynamic quantities such as the isothermal compressibility K_T, thermal expansion α_p and the specific heat $C_P - C_V$ are calculated through the mode Gruneisen parameter in NH_4F.

For this calculation, the observed Raman frequencies of the $238 cm^{-1}$ (1 bar) correspondingly $219 cm^{-1}$ (1.9 GPa in phase III) as the translational optic (TO) mode and $603 cm^{-1}$ or $74.7 meV$ (1.9 GPa in phase III) as the librational mode, are used for the phases of I, II and III in NH_4F. We have also calculated the thermodynamic quantities for the phases of V and VI in NH_4F using the observed Raman frequencies of the $238 cm^{-1}$ mode as a function of pressure at T=293 K.

Our calculated thermodynamic quantities (K_T, α_P and $C_P - C_V$) can be compared with the observed data for the phases studied in NH_4F, when available in the literature. This shows that the thermodynamic properties of molecular crystals such as NH_4F can be investigated from the measurements of the vibrational spectra at high pressures.

TK05

THE EFFECT OF INTERMOLECULAR MODES ON THE XH-STRETCHING VIBRATIONS IN HYDROGEN BONDED COMPLEXES

KASPER MACKEPRANG, HENRIK G. KJAERGAARD, *Department of Chemistry, University of Copenhagen, Copenhagen, Denmark.*

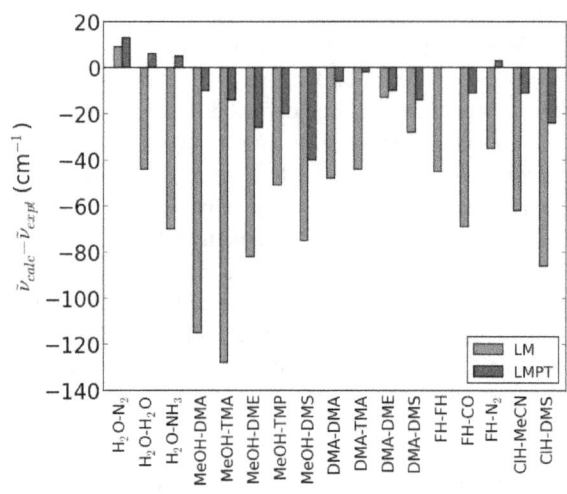

Vibrational spectra of hydrogen bonded bimolecular complexes (XH⋯Y, where X is the hydrogen bond donor atom, and Y is the acceptor atom) have long been a theoretical challenge. Specifically, the XH-stretching motion is difficult to describe due to the effect of the large amplitude intermolecular modes inherent to complexes. We have developed a vibrational model, the Local Mode Perturbation Theory (LMPT) model, to accurately determine the transition wavenumber and oscillator strength of the XH-stretching transition in hydrogen bonded bimolecular complexes. The model is based on a local mode (LM) model of the XH-stretching transition and the effect of the intermolecular modes is included via Rayleigh-Schrödinger perturbation theory. Our model has significantly improved results obtained using the LM model (see Figure). Additionally, our LMPT model does not require a full-dimensional anharmonic calculation, which enables application to large systems and the usage of higher level *ab initio* theory for the required potential energy surfaces. This work was inspired by our recent efforts to accurately determine equilibrium constants of complex formation, which rely on an accurate determination of the oscillator strength of the XH-stretching transition.

TK06

THEORETICAL ANALYSIS OF VCD SPECTRA OF α AND β L-FUCOPYRANOSIDE IN THE CH STRETCHING REGION

SOFIANE MOUSSI, OURIDA OUAMERALI, *Laboratory lctcp, University USTHB, Algiers, Algeria.*

Fucose is a deoxyhexose that is present in a wide variety of organisms. The stereochemical information, namely, glycosidic linkages α or β, gives significant features of the carbohydrate glycosidation position of the glycosylic acceptor [1].

Due to its applicability to all organic molecules and the reliability of ab initio quantum calculation, Vibrational Circular Dichroism VCD has some advantages over conventional electronic spectroscopy. However, for a molecule with many chiral centers such as carbohydrate, determination of the absolute configuration tends to be difficult because the information from each stereochemical center are mixed and averaged over the spectrum.

T. Taniguchi et al [2] reported that in the CH stretching region of carbohydrate, methyl glycosides exhibit a characteristic VCD peak at 2840 cm^{-1}, the sign solely reflects the anomeric carbon absolute configuration.

This work is an investigation of theoretical VCD spectra of α-L-fucopyranoside and β-L-fucopyranoside with an implicit (PCM) and explicit consideration of water molecules using density functional theory DFT and the Potential Energy Distribution's analysis (PED) .

Keywords : VCD, DFT, PED, Fucopyranoside .

[1]. a) C.-H. Wong, Carbohydrate –Based Drug Discovery, Wiley-VCH Weinheinium 2003; b) T. Taniguchi, K. Monde, Wiley VCH Verlag GmbH§ Co. KGaA, Weinhein chem asian J. 2007, 2,1258-1266 [2]. T. Taniguchi et al. Tetrahedron Letters 45 (2004) 8451–8453

Intermission

TK07

FULL CI BENCHMARK POTENTIALS FOR THE $6e^-$ SYSTEM Li_2 WITH A CBS EXTRAPOLATION FROM aug-cc-pCV5Z AND aug-cc-pCV6Z BASIS SETS USING FCIQMC AND DMRG

NIKESH S. DATTANI, *Department of Chemistry, Kyoto University, Kyoto, Japan*; SANDEEP SHARMA, ALI ALAVI, *Max Planck Institute for Solid State Research, Stuttgart, Germany.*

Being the simplest uncharged homonuclear dimer after H_2 that has a stable ground state, Li_2 is one of the most important benchmark systems for theory and experiment. In 1930, Delbrück used Li_2 to test his theory of homopolar binding, and it was used again and again as a prototype to test what have now become some of the most ubiquitous concepts in molecular physics (LCAO, SCF, MO, just to name a few). Experimentally, Roscoe and Schuster studied alkali dimers back in 1874. At the dawn of quantum mechanics, the emerging types of spectroscopic analyses we now use today, were tested on Li_2 in the labs of Wurm (1928), Harvey (1929), Lewis (1931), and many others, independently. Li_2 was at the centre of the development of PFOODR in the 80s, and PAS in the 90s; and Lithium Bose-Einstein condensates were announced only 1 month after the Nobel Prize winning BEC announcement in 1995. Even now in the 2010s, numerous experimental and theoretical studies on Li have tested QED up to the 7th power of the fine structure constant. Li_2 has also been of interest to sub-atomic physicists, as it was spectroscopic measurements on $^7Li_2$ that determined the spin of ^7Li to be 3/2 in 1931; and Li_2 has been proposed in 2014 as a candidate for the first "halo nucleonic molecule".

The lowest triplet state $a(1^3\Sigma_u^+)$ is an excellent benchmark system for all newly emerging *ab initio* techniques because it has only $6e^-$, its potential is only $334\,cm^{-1}$ deep, it avoids harsh complications from spin-orbit coupling, and it is the deepest potential for which *all* predicted vibrational energy levels have been observed with $0.0001\,cm^{-1}$ precision. However the current best *ab initio* potentials do not even yield all vibrational energy spacings correct to within $1\,cm^{-1}$. This could be because the calculation was only done on a cc-pV5Z basis set, or because the QCISD(T,full) method that the authors used, only considered triple excitations while a full CI calculation should include up to hexuple excitations. CCSDTQPH calculations have never yet been reported for anything larger than a DZ basis set, and deterministic FCI calculations for $6e^-$ have not exceeded the level of TZ basis sets. With FCIQMC and DMRG we are able to calculate the potential with all levels of excitation included, and the hardware requirements for an aug-cc-pCV6Z basis set are modest. Energies for aug-cc-pCVQZ have already converged to the full CI limit within $0.3\,cm^{-1}$, and 6Z potentials are underway.

TK08

AN IMPROVED EMPIRICAL POTENTIAL FOR THE HIGHLY MULTI-REFERENCE SEXTUPLY BONDED TRANSITION METAL BENCHAMRK MOLECULE Cr_2

NIKESH S. DATTANI, *Department of Chemistry, Kyoto University, Kyoto, Japan*; MICHAŁ TOMZA, *ICFO - Institut de Ciencies Fotoniques, Barcelona, Spain*; GIOVANNI LI MANNI, *Max Planck Institute for Solid State Research, Stuttgart, Germany.*

The ground electronic state of the chromium dimer dissociates into $Cr(^7S) + Cr(^7S)$ and therefore the fragments are highly open shell systems with a total of 12 singly occupied orbitals among its constituent atoms. It is considered one of the most difficult homonuclear diatomics for *ab initio* methods because of its highly multi-reference character. Therefore, every new multi-reference method must be tested against this benchmark system. However, the best empirical potential to compare with, has its own weaknesses. The photoelectron measurements of $v = 0 - 9$ were fitted to a Morse potential (an old function which has only one parameter controlling the shape from r_e to D_e), and also inverted using a semi-classical theory into a potential after combining these data with measurements from what were hypothesized to be $v = 24 - 43$. This bridging of a $\sim 2000\,cm^{-1}$ gap in data back in 1993 was a valiant spectroscopic analysis. However since 1993, there have been enormous improvements in the field of potentiology. In 2011 a Morse/long-range (MLR) function successfully bridged a gap of more than $5000\,cm^{-1}$ in experimental data[a], and in 2013 an experiment with $\pm 0.000\,02\,cm^{-1}$ resolution confirmed that the 2011 MLR predicted the energy levels in the very center of this gap correctly within $\sim 1\,cm^{-1}$[b]. While *ab initio* methods have very recently been able to predict differences in energy levels correctly to within $1\,cm^{-1}$ for Li_2 [c] and to a lesser extent for BeH[d], *ab initio* methods have still not had this level of success for predicting binding energies.

The MLR function not only has more flexibility than the original Morse function, but it also converges mathematically to the correct long-range limit expected by the state-of-the-art theory. Fitting the data to an MLR potential function in the Schrödinger equation allows for a fully quantum mechanical treatment over the entire range of data. By avoiding a semi-classical treatment, and using this more flexible, more theoretically correct form, we improve the current best empirical potential. This vastly improves the experimental benchmarks against which emerging *ab initio* methods are tested. However, the lack of data for Cr_2 is still a big problem, so further experimental work on Cr_2 is desperately needed.

[a] Dattani & Le Roy (2011) Journal of Molecular Spectroscopy, **268**, 119, [b] Semczuk *et al.* (2013) Physical Review A, **88**, 062510., [c] Dattani (2015) http://arxiv.org/abs/1508.07184, [d] Dattani (2015) Journal of Molecular Spectroscopy **311**, 76.

144

VIBRONIC TRANSITIONS IN THE X-Sr SERIES (X=Li, Na, K, Rb): ON THE ACCURACY OF NUCLEAR WAVE-FUNCTIONS DERIVED FROM QUANTUM CHEMISTRY

RALF MEYER, JOHANN V. POTOTSCHNIG, <u>ANDREAS W. HAUSER</u>, WOLFGANG E. ERNST, *Institute of Experimental Physics, Graz University of Technology, Graz, Austria.*

Research on ultracold molecules has seen a growing interest recently in the context of high-resolution spectroscopy and quantum computation. The preparation of molecules in low vibrational levels of the ground state is experimentally challenging, and typically achieved by population transfer using excited electronic states.

On the theoretical side, highly accurate potential energy surfaces are needed for a correct description of processes such as the coherent de-excitation from the highest and therefore weakly bound vibrational levels in the electronic ground state via couplings to electronically excited states. Particularly problematic is the correct description of potential features at large intermolecular distances. Franck-Condon overlap integrals for nuclear wavefunctions in barely bound vibrational states are extremely sensitive to inaccuracies of the potential at long range. In this study, we compare the predictions of common, wavefunction-based ab initio techniques for a known de-excitation mechanism in alkali-alkaline earth dimers. It is the aim to analyze the predictive power of these methods for a preliminary evaluation of potential cooling mechanisms in heteronuclear open shell systems which offer the experimentalist an electric as well as a magnetic handle for manipulation.

The series of X-Sr molecules, with X = Li, Na, K and Rb, has been chosen for a direct comparison. Quantum degenerate mixtures of Rb and Sr have already been produced,[a] making this combination very promising for the production of ultracold molecules.

[a]B. Pasquiou, A. Bayerle, S. M. Tzanova, S. Stellmer, J. Szczepkowski, M. Parigger, R. Grimm, and F. Schreck, Phys. Rev. A, 2013, 88, 023601

AB INITIO INVESTIGATIONS OF THE EXCITED ELECTRONIC STATES OF CaOCa

<u>WAFAA M FAWZY</u>, *Department of Chemistry, Murray State University, Murray, KY, USA*; MICHAEL HEAVEN, *Department of Chemistry, Emory University, Atlanta, GA, USA.*

Chemical bonding in alkaline earth hypermetalic oxides is of fundamental interest. Previous Ab initio studies of CaOCa predicted a centrosymmetric linear geometry for both the $^1\Sigma_g^+$ ground state and the low lying triplet $^3\Sigma_u^+$ state[a]. However, there have been no reports concerning the higher energy singlet and triplet states. The present work is focused on characterization of the potential energy surface (PES) of the excited $^1\Sigma_u^+$ state (assuming a centrosymmetric linear geometry) and obtaining predictions for the $^1\Sigma_u^+ \leftarrow^1 \Sigma_g^+$ vibronic transitions. We employed the multireference configuration interaction (MRCISD) method with state-averaged, full-valence complete active space self-consistent field (SA-FV-CASSCF) wavefunctions. In these calculations, the active space consisted of ten valence electrons in twelve orbitals, where all the valence electrons were correlated. Contributions of higher excitation and relativistic effects were taken into account using the Davidson correction and the Douglas-Kroll (DK) Hamiltonian, respectively. The correlation-consistent polarized weighed core-valence quadruple zeta basis set (cc-pwCVQZ-DK) was used for all three atoms. The full level of theory is abbreviated as SA-FV-CASSCF (10,12)-MRCISD-Q/cc-pwCVQZ-DK. The calculations were carried out using the MOLPRO2012 suite of programs. For the centrosymmetric linear geometry in all states, initial investigations of one-dimensional radial cuts provided equilibrium bond distances of 2.034 Å, 2.034 Å, and 1.999 Å for the $^1\Sigma_g^+$, $^3\Sigma_u^+$, and $^1\Sigma_u^+$ states, respectively. The vertical excitation frequency of the $^1\Sigma_u^+ \leftarrow^1 \Sigma_g^+$ optical transition was calculated to occur at 14801 cm^{-1}. These predictions were followed by spectroscopic searches by Heaven et al. Indeed, rotationally resolved vibronic progressions were recorded in the vicinity of the predicted electronic band origin. Calculation of the three-dimensional PES showed that the potential minimum in the $^1\Sigma_u^+$ corresponds to a bent equilibrium geometry with a bond angle of 120° (C_{2v} point group, where the electronic symmetry is $^1A_1$). This result suggests that the Ca-O bonds in CaOCa possess covalent character in the $^1A_1$ excited state and ionic character in the $^1\Sigma_g^+$ ground state. The current results, as well as those from ongoing studies will be presented.

[a]B. Ostojicí, P.R. Bunker, P. Schwerdtfeger, Artur Gertych, and Per Jensen, Journal of Molecular Structure 1023 (2012) 101–107.

TK11 4:22 – 4:37
AB INITIO CALCULATION OF NH_3 SPECTRUM

OLEG POLYANSKY, *Department of Physics and Astronomy, University College London, Gower Street, London WC1E 6BT, United Kingdom*; ROMAN I. OVSYANNIKOV, ALEKSANDRA KYUBERIS, *Microwave Spectroscopy, Institute of Applied Physics, Nizhny Novgorod, Russia*; LORENZO LODI, JONATHAN TENNYSON, *Department of Physics and Astronomy, University College London, Gower Street, London WC1E 6BT, United Kingdom*; SERGEI N. YURCHENKO, ANDREY YACHMENEV, *Department of Physics and Astronomy, University College London, Gower Street, London WC1E 6BT, United Kingdom*; NIKOLAY FEDOROVICH ZOBOV, *Microwave Spectroscopy, Institute of Applied Physics, Nizhny Novgorod, Russia.*

An *ab initio* potential energy surface (PES) for NH_3 is computed using the methodology pioneered for water (Polyansky et al. *J. Phys. Chem. A*, **117**, 9633 (2013)). A multireference configuration calculations are performed at 50000 points using quadruple and 5z basis sets to give a complete basis set (CBS) extrapolation. Relativistic and adiabatic surfaces are also computed. The points are fitted to an analytical PES. The rovibrational energy levels are computed using the program TROVE in both linearized and curvilinear coordinates. Better convergence is obtained for the higher energy levels using curvilinear coordinates: an accuracy of about 1 cm^{-1} is achieved for the levels up to 12 000 cm^{-1}. The levels up to 18 000 cm^{-1} are reproduced with the accuracy of a few cm^{-1}. These results are used to assign the visible spectrum of $^{14}NH_3$ recorded by Coy and Lehmann (*J. Chem. Phys.*, **84**, 5239 (1988)). Predicted rovibrational levels for NH_2D, NHD_2, ND_3 and $^{15}NH_3$ are given.

TK12 4:39 – 4:54
AB INITIO EXPLORATION OF THE POTENTIAL ENERGY SURFACE OF THE O_2-SO_2 OPEN-SHELL COMPLEX.

WAFAA M FAWZY, *Department of Chemistry, Murray State University, Murray, KY, USA*; JON T. HOUGEN, *Sensor Science Division, National Institute of Standards and Technology, Gaithersburg, MD, USA.*

The O_2-SO_2 complex is believed to be a precursor to acid rain. The previously observed FTMW spectrum suggested internal motions within the complex, but their nature was not identified. Development of an effective Hamiltonian for an open-shell molecule with tunneling requires knowledge of the potential energy surface (PES) and the intrinsic reaction coordinates (IRC) for the paths between minima. A recent ab initio study reported two different nonplanar minima in the ground electronic state of O_2-SO_2. These predictions were based on geometry optimization calculations at the MP2/aug-cc-pVnZ level of theory, with n = 2 and 3. The current work is focused on a highly correlated ab initio investigation of the global PES (a 9-D problem) in the ground triplet electronic state of O_2-SO_2. Because of the high dimensionality in the complex, the PES calculations are partitioned into several two-dimensional cuts through the PES. We have so far explored only a 3-D part of the global PES to look for stable planar configurations. These calculations included geometry optimization, frequency, and single point energy calculations. Calculations were performed using UCCSD(T)/aug-cc-pV(n+D)Z, where n = 2 and 3, level of theory. We used an axis system that defines the radial and the angular van der Waals coordinates for a planar complex as R_{vW}, θ_1, and θ_2. The bond length (R_{vW}) is the distance between the center of mass of the O_2 unit and the S atom. θ_1 and θ_2 are the angles between the van der Waals bond and the O_2 internuclear axis or one of the SO bonds in the SO_2 moiety, respectively. Full geometry optimization calculations predicted a minimum of C_s symmetry in which both the O_2 and SO_2 units are tilted with respect to the van der Waals bond, and R_{vW} = 3.63 Å. 3-D PES surface calculations, which involve the R_{vW}, θ_1, and θ_2 vdW coordinates, showed that the optimized structure is the global minimum. In addition, a local minimum at R_{vW} = 3.9 Å, which represents a different chemical isomer, was identified. If the four oxygen atoms are labeled, each isomer is a part of four equivalent minima, and three distinguishable transition states between these various minima are identified. These results suggest that PES calculations should consider at least five dimensions. Our progress in exploring possible non-planar coordinates and IRC paths will also be presented.

TK13 4:56–5:11

APPLYING QUANTUM MONTE CARLO TO THE ELECTRONIC STRUCTURE PROBLEM

ANDREW D POWELL, RICHARD DAWES, *Department of Chemistry, Missouri University of Science and Technology, Rolla, MO, USA.*

Two distinct types of Quantum Monte Carlo (QMC) calculations are applied to electronic structure problems such as calculating potential energy curves and producing benchmark values for reaction barriers. First, Variational and Diffusion Monte Carlo (VMC and DMC) methods using a trial wavefunction subject to the fixed node approximation were tested using the CASINO code.[1] Next, Full Configuration Interaction Quantum Monte Carlo (FCIQMC), along with its initiator extension (i-FCIQMC) were tested using the NECI code.[2] FCIQMC seeks the FCI energy for a specific basis set. At a reduced cost, the efficient i-FCIQMC method can be applied to systems in which the standard FCIQMC approach proves to be too costly. Since all of these methods are statistical approaches, uncertainties (error-bars) are introduced for each calculated energy. This study tests the performance of the methods relative to traditional quantum chemistry for some benchmark systems.

References: [1] R. J. Needs et al., J. Phys.: Condensed Matter 22, 023201 (2010). [2] G. H. Booth et al., J. Chem. Phys. 131, 054106 (2009).

TK14 5:13–5:28

CAN WE PREDICT QUANTUM YIELDS USING EXCITED STATE DENSITY FUNCTIONAL THEORY FOR NEW FAMILIES OF FLUORESCENT DYES?

ALEXANDER W. KOHN, ZHOU LIN, JAMES J. SHEPHERD, TROY VAN VOORHIS, *Department of Chemistry, Massachusetts Institute of Technology, Cambridge, MA, USA.*

For a fluorescent dye, the quantum yield characterizes the efficiency of energy transfer from the absorbed light to the emitted fluorescence. In the screening among potential families of dyes, those with higher quantum yields are expected to have more advantages. From the perspective of theoreticians, an efficient prediction of the quantum yield using a universal excited state electronic structure theory is in demand but still challenging. The most representative examples for such excited state theory include time-dependent density functional theory (TDDFT) and restricted open-shell Kohn-Sham (ROKS)[a]. In the present study, we explore the possibility of predicting the quantum yields for conventional and new families of organic dyes using a combination of TDDFT and ROKS. We focus on radiative (k_r) and nonradiative (k_{nr}) rates for the decay of the first singlet excited state (S_1) into the ground state (S_0) in accordance with Kasha's rule.[b] For each dye compound, k_r is calculated with the $S_1 - S_0$ energy gap and transition dipole moment obtained using ROKS and TDDFT respectively at the relaxed S_1 geometry. Our predicted k_r agrees well with the experimental value, so long as the order of energy levels is correctly predicted. Evaluation of k_{nr} is less straightforward as multiple processes are involved. Our study focuses on the $S_1 - T_1$ intersystem crossing (ISC) and the $S_1 - S_0$ internal conversion (IC): we investigate the properties that allow us to model the k_{nr} value using a Marcus-like expression, such as the Stokes shift, the reorganization energy, and the $S_1 - T_1$ and $S_1 - S_0$ energy gaps. Taking these factors into consideration, we compare our results with those obtained using the actual Marcus theory and provide explanation for discrepancy.

---

[a] T. Kowalczyk, T. Tsuchimochi, L. Top, P.-T. Chen, and T. Van Voorhis, *J. Chem. Phys.*, **138**, 164101 (2013).
[b] M. Kasha, *Discuss. Faraday Soc.*, **9**, 14 (1950).

TK15   **5:30 – 5:45**

INCORPORATION OF A ROVIBRATIONAL ANALYSIS OF OC-H$_2$O INTO 6-D MORPHED POTENTIALS OF THE COMPLEX

LUIS A. RIVERA-RIVERA, SEAN D. SPRINGER, BLAKE A. McELMURRY, *Department of Chemistry, Texas A & M University, College Station, TX, USA*; IGOR I LEONOV, *Microwave Spectroscopy, Institute of Applied Physics, Nizhny Novgorod, Russia*; ROBERT R. LUCCHESE, JOHN W. BEVAN, *Department of Chemistry, Texas A & M University, College Station, TX, USA*; L. H. COUDERT, *LISA, CNRS, Universités Paris Est Créteil et Paris Diderot, Créteil, France*.

Rovibrational transitions associated with tunneling states in the water bending vibration in OC-H$_2$O and other available spectroscopic data are included in generation of 6-D morphed potentials of the complex. Six-dimension *ab initio* interaction potentials are initially calculated for the complex to provided the initial functions for the potential morphing. The available spectroscopic data is then used to fit and generate 6-D morphed potentials. Previous prediction of the D_0 of the complex will be incorporated in the analysis. Finally, intermolecular frequencies of the complex will be predicted using the 6-D morphed potentials involving the CO stretching and the H$_2$O bending vibrations.

WA. Mini-symposium: Spectroscopy of Large Amplitude Motions

Wednesday, June 22, 2016 – 8:30 AM

Room: 100 Noyes Laboratory

Chair: Mahesh B. Dawadi, The University of Akron, Akron, OH, USA

WA01   *INVITED TALK*   8:30 – 9:00

TORSION - ROTATION - VIBRATION EFFECTS IN THE GROUND AND FIRST EXCITED STATES OF METHACROLEIN AND METHYL VINYL KETONE

OLENA ZAKHARENKO, R. A. MOTIYENKO, JUAN-RAMON AVILES MORENO, <u>T. R. HUET</u>, *UMR 8523 CNRS - Universités des Sciences et Technologies de Lille, Laboratoire PhLAM, 59655 Villeneuve d'Ascq, France.*

Methacrolein and methyl vinyl ketone are the two major oxidation products of isoprene emitted in the troposphere. New spectroscopic information is provided with the aim to allow unambiguous identification of these molecules, characterized by a large amplitude motion associated with the methyl top. State-of-the-art millimeter-wave spectroscopy experiments coupled to quantum chemical calculations have been performed. Comprehensive sets of molecular parameters have been obtained. The torsion-rotation-vibration effects will be discussed in detail.

From the atmospheric application point of view the results provide precise ground state molecular constants essential as a foundation (by using the Ground State Combination Differences method) for the analysis of high resolution spectrum, recorded from 600 to 1600 cm^{-1}. The infrared range can be then refitted using appropriate Hamiltonian parameters.

The present work is funded by the French ANR through the PIA under contract ANR-11-LABX-0005-01 (Labex CaPPA), by the Regional Council Nord-Pas de Calais and by the European Funds for Regional Economic Development (FEDER).

WA02   9:05 – 9:20

SYNCHROTRON SPECTROSCOPY AND TORSIONAL STRUCTURE OF THE CSH-BENDING AND CH$_3$-ROCKING BANDS OF METHYL MERCAPTAN

<u>RONALD M. LEES</u>, LI-HONG XU, *Department of Physics, University of New Brunswick, Saint John, NB, Canada*; BRANT E. BILLINGHURST, *EFD, Canadian Light Source Inc., Saskatoon, Saskatchewan, Canada.*

The Fourier transform spectra of the CSH-bending and CH$_3$-rocking infrared bands of CH$_3$SH have been investigated at 0.001 cm^{-1} resolution employing synchrotron radiation at the Canadian Light Source in Saskatoon. The relative band strengths and structures are remarkably different from those for the analogous CH$_3$OH relative, with the CSH bend being very weak and both the in-plane and out-of-plane CH$_3$ rocks being strong with comparable intensities. The CSH bend, centered at 801.5 cm^{-1}, has parallel a-type character with no detectable b-type component. The out-of-plane CH$_3$ rock at 957.0 cm^{-1} is a purely c-type perpendicular band, whereas the in-plane rock around 1074 cm^{-1} is of mixed a/b character. The K-reduced $v_t = 0$ sub-state origins for the CSH bend follow the normal oscillatory torsional pattern as a function of K with an amplitude of 0.362 cm^{-1}, as compared to 0.653 cm^{-1} for the ground state and 0.801 cm^{-1} for the C-S stretching mode. The torsional energy curves for the out-of-plane rock are also well-behaved but are inverted, with an amplitude of 1.33 cm^{-1}. In contrast, the sub-state origins for the in-plane rock do not display a clear oscillatory structure but are scattered over a range of about 2 cm^{-1}, with indications of some significant perturbations. The assignments for the three bands all extend up to about $K = 10$ and are well-determined from GSCD relations, particularly for the a/b in-plane rock for which $\Delta K = 0$, +1 and −1 transitions are all observed.

WA03   9:22 – 9:37

VIBRATIONAL CONICAL INTERSECTIONS IN CH$_3$SH: IMPLICATIONS FOR SPECTROSCOPY AND DYNAMICS IN THE CH STRETCH REGION

<u>DAVID S. PERRY</u>, BISHNU P. THAPALIYA, MAHESH B. DAWADI, *Department of Chemistry, The University of Akron, Akron, OH, USA*; RAM BHATTA[a], *Polymer Science, The University of Akron, Akron, OH, USA.*

The adiabatic separation in methyl mercaptan of the high-frequency asymmetric CH stretch vibrations from the low-frequency torsional (γ) and CSH bend (ρ) coordinates yields a set of 7 vibrational conical intersections (CIs). The three CIs in the staggered conformation at $\rho = 79°$ are close to the global minimum energy geometries ($\rho_e = 83.3°$), accounting for the observed near-degeneracy of the two asymmetric CH stretch vibrations. The vibrational frequencies were computed at the CCSD(T)/aug-cc-pVTZ level. A new high-order Exe Jahn-Teller model, which involves a spherical harmonic expansion in ρ and γ, fits the calculated electronic and vibrational energies over the whole range of γ and for ρ between $0°$ and $100°$ to within a standard deviation of 0.2 cm^{-1}. The pattern of the CIs contrasts with that in methanol where the CIs occur only in the eclipsed conformation near the top of the torsional barrier. An examination of three alternative diabatization schemes for the two molecules points to rather different nuclear dynamics. In CH$_3$SH crossings between the upper and lower adiabatic surfaces are predicted to occur predominantly with motion along the CSH bending coordinate; whereas in CH$_3$OH, such crossings are predicted to occur predominantly with torsional motion.

---

[a]present address: College of Pharmacy, The University of North Texas Health Science Center, 3500 Camp Bowie Blvd, Fort Worth, TX 76107

WA04   9:39 – 9:49

THE EQUIVALENCE OF THE METHYL GROUPS IN PUCKERED 3,3-DIMETHYL OXETANE

ALBERTO MACARIO, *Departamento de Química Física y Química Inorgánica, Universidad de Valladolid, Valladolid, Spain*; SUSANA BLANCO, <u>JUAN CARLOS LOPEZ</u>, *Departamento de Química Física y Química Inorgánica / Grupo de Espectroscopía Molecular, Universidad de Valladolid, Valladolid, Spain.*

The spectroscopic study of molecules with large amplitude vibrations have led to reconsider the concept of molecular structure. Sometimes identifying definite bond lengths and angles is not enough to reproduce the experimental data so one must have information on the large amplitude molecular vibration potential energy function and dynamics. 3,3-dimethyloxetane (DMO) has non-planar ring equilibrium configuration and a double minimum potential function for ring-puckering with a barrier of 47 cm^{-1}.[a,b] The observation of endocyclic ^{13}C and ^{18}O monosubstituted isotopologues[c] allow to conclude that the ring is puckered. However an interesting feature was observed for the ^{13}C substitutions at the methyl carbon atoms. While two different axial and equatorial ^{13}C-methyl groups spectra are predicted from a rigid non-planar ring DMO model, only one species was found. The observed rotational transitions appear at a frequency close to the average of the frequencies predicted for each isotopologue. The observed lines have the same intensity as that found for the ^{13}C$_\alpha$ isotopomer and double that that found for the ^{13}C$_\beta$ isotopomer.[c] This behaviour evidences that the two methyl groups of DMO are equivalent as could be expected for a planar ring. In this work we show how consideration of the potential function and the path for ring puckering motion to calculate the proper kinetic energy terms allow to reproduce the experimental results. *Ab initio* computations at the CCSD/6-311++G(d,p) level, tested on related systems, have been done for this purpose.

---

[a]J. A. Duckett, T. L. Smithson, and H. Wieser, *J. Mol. Spectrosc.* **1978**, *69* , 159; *J. Mol. Struct.* **1979**, *56*, 157

[b]J. C. López, A. G. Lesarri, R. M. Villamañán and J. L. Alonso, *J. Mol. Spectrosc.* **1990**, *141*, 231

[c]R. Sánchez, S. Blanco, A. Lesarri, J. C. López and J. L. Alonso, *Phys. Chem. Chem. Phys.* **2005**, *7*, 1157

Intermission

WA05 10:08 – 10:23

SPECTROSCOPIC STUDY OF METHYLGLYOXAL AND ITS HYDRATES : A GASEOUS PRECURSOR OF SECONDARY ORGANIC AEROSOLS.

SABATH BTEICH, MANUEL GOUBET, L. MARGULÈS, R. A. MOTIYENKO, T. R. HUET, *Laboratoire PhLAM, UMR 8523 CNRS - Université Lille 1, Villeneuve d'Ascq, France.*

Secondary organic aerosols (SOA) have a significant effect on climate change. They are mainly produced in the atmosphere by oxidation of gaseous precursors. Fu et al.[a] have suggested trans-methylglyoxal (MG) as a possible precursor of SOA in the cloud for its presence in large quantities in the atmosphere.

The characterization of SOAs precursors by laboratory spectroscopy allows providing elements for the understanding of the process of formation of these aerosols. For this purpose, we completed the existing pure rotational spectrum of MG in the 12-40 GHz range[b] by new records in a supersonic jet in the 4-20 GHz range (FTMW) and at room temperature in the 150-500 GHz range (mm/submm-wave spectrometer).

The analysis was made with the support of quantum chemistry calculations (MP2/CBS and B98/CBS using the Gaussian 09 software). The adjustment of the spectroscopic parameters, taking into account the internal rotation related to the presence of a methyl group, was performed using the RAM36 code. The spectra have been reproduced at the experimental precision up to maximal values of J and K_a equal to 85 and 35, respectively.

The data obtained for the isolated molecule, both experimentally and theoretically, will allow the study of its hydrated complexes and, by comparison, will give access to (micro-) hydration properties. For this purpose, two stable complexes predicted by theoretical calculations will be studied.

[a]T.- M. Fu et al., J. Geophys. Res., 113, (2008).

[b]C.E. Dyltick-Brenzinger and A. Bauder, Chem. Phys. 30, 147 (1978).

WA06 10:25 – 10:40

TORSION - VIBRATION COUPLINGS IN THE $CH_3OO\cdot$ RADICAL

MENG HUANG, TERRY A. MILLER, *Department of Chemistry and Biochemistry, The Ohio State University, Columbus, OH, USA*; ANNE B McCOY, *Department of Chemistry, University of Washington, Seattle, WA, USA*; KUO-HSIANG HSU, YU-HSUAN HUANG, YUAN-PERN LEE, *Applied Chemistry, National Chiao Tung University, Hsinchu, Taiwan.*

A partially rotationally resolved infrared spectrum of $CH_3OO\cdot$ in the CH stretch region has been reported[a]. The rotational contour of the ν_2 CH stretch band in the experimental spectrum can be simulated with an asymmetric rotor model. The simulation shows good agreement with the experimental spectrum except that the broadening of the Q-branch in the experimental spectrum remains unexplained. This broadening is likely due to the sequence band transitions from the torsionally excited levels populated at room temperature to combination levels involving the CH stretch and the same number of torsional quanta. A four dimension model involving three CH stretches and the CH_3 torsion is applied to the $CH_3OO\cdot$ radical to obtain the frequencies and intensities of the vibrational transitions in the CH stretch region. Based on these calculations, the torsional sequence bands are calculated to be slightly shifted from the origin band, because of the couplings between the CH stretches and CH_3 torsion, thereby causing the apparent broadening observed for the ν_2 fundamental. Due to the accidental degeneracy of two different CH stretch and CH_3 torsion combination levels which differ by one quantum in the torsional excitation, the frequencies of the torsional sequence bands will be very sensitive to details of the potential, which makes the shifts difficult to precisely predict with electronic structure calculations. Complementary analyses are now underway for the other two CH stretch vibrational bands, ν_1 and ν_9.

[a]K.-H. Hsu, Y.-H. Huang, Y.-P. Lee, M. Huang, T. A. Miller and A. B. McCoy *J. Phys. Chem. A*, in press, DOI: 10.1021/acs.jpca.5b12334

WA07

WEAK INTRAMOLECULAR INTERACTIONS EFFCTS ON THE STRUCTURE AND THE TORSIONAL SPECTRA OF ETHYLENE GLYCOL, AN ASTROPHYSICAL SPECIES

__MARIA LUISA SENENT__, RAHMA BOUSSESSI, *Inst. Estructura de la Materia, IEM-CSIC, Madrid, Spain.*

A variational procedure of reduced dimensionality based on CCSD(T)-F12 calculations is applied to understand the far infrared spectrum of Ethylene-Glycol. This molecule can be classified in the double molecular symmetry group G8 and displays nine stable conformers, gauche and trans. In the gauche region, the effect of the potential energy surface anisotropy due to the formation of intramolecular hydrogen bonds is relevant. For the primary conformer, the ground vibrational state rotational constants are computed at 6.3 MHz, 7.2 MHz and 3.5 MHz from the experimental parameters.

Ethylene glycol displays very low torsional energy levels whose classification is not straightforward. Given the anisotropy, tunneling splittings are significant and unpredictable. The ground vibrational state splits into 16 sublevels separated approximately 142 cm^{-1}. Transitions corresponding to the three internal rotation modes allow assign previous observed Q branches. Band patterns, calculated between 362.3 cm^{-1} and 375.2 cm^{-1}, between 504 cm^{-1} and 517 cm^{-1} and between 223.3 cm^{-1} and 224.1 cm^{-1}, that correspond to the tunnelling components of the v21 fundamental (ν_{21} = OH-torsional mode), are assigned to the prominent experimental Q branches.

WA08

HIGH-ACCURATE INTERMOLECULAR POTENTIAL ENERGY SURFACE OF HCN − H$_2$ COMPLEX WITH INTRAMOLECULAR VIBRATIONAL MODE OF HCN INCLUDED

__YU ZHAI__, HUI LI, *Institute of Theoretical Chemistry, Jilin University, Changchun, China.*

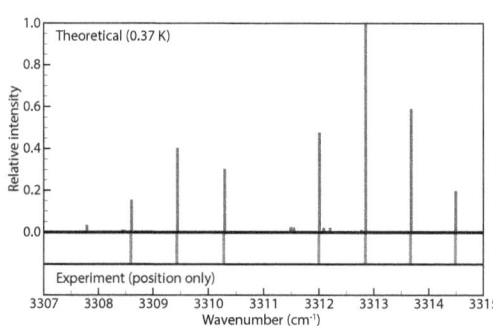

Hydrogen is one of the most abundant interstellar species. Observation of rotational and vibrational spectra of H$_2$ containing complexes is of great importance because they are possible candidates for radioastronomical detection. CO, HCN, HCCH are as isoelectronic molecules of N$_2$, each with a strong triple bond. It had been a big challenge to predict reliable theoretical rovibrational spectra of complexes including such species because the higher order electron correlation energy plays a non-negligible role in improving the accuracy. However, recent works on CO − H$_2$[a] have shown that it is possible to reproduce the experimental spectra quantitatively. In this work, we calculate a five-dimension potential energy surface (PES) of HCN − H$_2$ complex which explicitly include the intramolecular asymmetric stretching vibrational mode(C − H,Q_3) coordinate at CCSD(T)/aug-cc-pVQZ+3s3p2d1f1g level, corrected with electron correlation energy from the triple and quadruple excitation. Vibrational average over intramolecular vibration mode is made with HCN monomer at ground and the first excited vibrational states respectively, and the averaged data are fitted to two four-dimension Morse/Long-Range (MLR) potential energy functions. Based on the MLR PESs, for the first time, we calculated the spectra of HCN − $para$H$_2$ and HCN − $ortho$H$_2$. The results for HCN − $ortho$H$_2$ are in good agreement with the published experimental data[b] with root-mean-square-difference (RMSD) only 0.01cm^{-1}, which validates the accuracy of the PESs.

[a]*J. Chem. Phys.*, **139**, 164315 (2013); *Science*, **336**, 1147 (2012).
[b]*J. Chem. Phys.*, **115**, 5137 (2001).

152

THE ROTATIONAL SPECTRA OF CYANOACETYLENE DIMER, H-C-C-C-N ••• H-C-C-C-N[a]

LU KANG, *Department of Chemistry and Biochemistry, Kennesaw State University, Kennesaw, GA, USA*; PHILIP DAVIS, IAN DORELL, *Department of Physics, Kennesaw State University, Kennesaw, GA, USA*; KEXIN LI, ADAM M. DALY, STEPHEN G. KUKOLICH, *Department of Chemistry and Biochemistry, University of Arizona, Tucson, AZ, USA*; STEWART E. NOVICK, *Department of Chemistry, Wesleyan University, Middletown, CT, USA.*

The rotational spectra of H-C-C-C-N ••• H-C-C-C-N, cyanoacetylene dimer,were recorded using Balle-Flygare type Fourier transform microwave (FTMW) spectrometers at Wesleyan and Arizona. The low-J transitions were measured down to 1.3 GHz at very high resolution, FWHM 1 kHz using a large-cavity spectrometer. The spectral hyperfine structure due to the ^{14}N nuclear quadrupole coupling interactions is well-resolved below 4 GHz using a low frequency spectrometer at the University of Arizona. The experimental spectroscopic constants were fitted as: B_0 = 339.292331(8) MHz, D_J = 32.15(8) Hz, H = -0.0015(2) Hz, eqQ($^{14}N_1$) = -3.990(1) MHz, and eqQ($^{14}N_2$) = -4.171(1) MHz. The vibrationally averaged dimer configuration is H-C-C-C-N_1 ••• H-C-C-C-N_2. Using a simple linear model, the vibrational ground state and the equilibrium hydrogen bond lengths are determined to be: r_0(N ••• H) = 2.2489(3) Åand r_e(N ••• H) = 2.2315 Å. The equilibrium center-of-mass distance between the two HCCCN subunits is = 7.0366 Å. Using the rigid precession model, the vibrational ground state center-of-mass distance and the pivot angles which HCCCN subunits make with the a-axis of H-C-C-C-N_1 ••• H-C-C-C-N_2 are = 7.0603 Å, $\theta_1 = 13°$, and $\theta_2 = 8.7°$, respectively. The calculated hydrogen bond energy of H-C-C-C-N ••• H-C-C-C-N is 1466 cm^{-1} using the MP2/aug-cc-PVTZ method in present work.

[a]Supported by the NSF CHE-1057796 and CHE-1011214

INFRARED SPECTRUM OF CO-O_2, A 'NEW' WEAKLY-BOUND COMPLEX

BOB McKELLAR, *Steacie Laboratory, National Research Council of Canada, Ottawa, ON, Canada*; A. J. BARCLAY, *Department of Physics and Astronomy, University of Calgary, Calgary, AB, Canada*; K. H. MICHAELIAN, *CanmetENERGY, Natural Resources Canada, Edmonton, Alberta, Canada*; NASSER MOAZZEN-AHMADI, *Department of Physics and Astronomy, University of Calgary, Calgary, AB, Canada.*

Only a few weakly-bound complexes containing the O_2 molecule have been characterized by high-resolution spectroscopy, notably N_2O-O_2 [1] and HF-O_2 [2]. This neglect is no doubt due in part to the complications added by the oxygen unpaired electron spin. Here we report an extensive infrared spectrum of CO-O_2, as observed in the CO fundamental band region ($\sim$2150 cm^{-1}) using a tunable quantum cascade laser to probe a pulsed supersonic jet expansion. The derived energy level pattern consists of 'stacks' characterized by K, the projection of the total angular momentum on the intermolecular axis. Five such stacks are observed in the ground vibrational state, and ten in the excited state, v(CO) = 1. They are divided into two groups, with no observed transitions between groups, and we believe these groups correlate with the two lowest rotational states of O_2, namely $(N, J) = (1, 0)$ and $(1, 2)$. In many ways, the spectrum and energy levels are similar to those of CO-N_2 [3], and we use the same approach for analysis, simply fitting each stack with its own origin, B-value, and distortion constants. The rotational constant of the lowest stack in the ground state (with $K = 0$) implies an effective intermolecular separation of 3.82 Å, but this should be interpreted with caution since it ignores possible effects of electron spin.

[1] H.-B. Qian, D. Seccombe, and B.J. Howard, *J. Chem. Phys.* **107**, 7658 (1997).

[2] W.M. Fawzy, C.M. Lovejoy, D.J. Nesbitt, and J.T. Hougen, *J. Chem. Phys.* **117**, 693 (2002); S. Wu, G. Sedo, E.M. Grumstrup, and K.R. Leopold, *J. Chem. Phys.* **127**, 204315 (2007).

[3] M. Rezaei, K.H. Michaelian, N. Moazzen-Ahmadi, and A.R.W. McKellar, *J. Phys. Chem. A* **117**, 13752 (2013), and references therein.

WB. Mini-symposium: Spectroscopy in Atmospheric Chemistry

Wednesday, June 22, 2016 – 8:30 AM

Room: 116 Roger Adams Lab

Chair: Gerard Wysocki, Princeton University, Princeton, NJ, USA

WB01	*INVITED TALK*	8:30 – 9:00

OH WHERE OH WHERE IS OH? MEASURING THE ELUSIVE HYDROXYL RADICAL IN THE ATMOSPHERE USING LASER-INDUCED FLUORESCENCE

PHILIP S. STEVENS, *Department of Chemistry, Indiana University, Bloomington, IN, USA.*

The hydroxyl radical (OH) plays a central role in the chemistry of the atmosphere. In addition to controlling the lifetimes of many trace gases important to issues of global climate change and stratospheric ozone depletion, the OH radical initiates the oxidation of carbon monoxide and volatile organic compounds which in the presence of nitrogen oxides can lead to the production of ground-level ozone and secondary organic aerosols, the primary components of photochemical smog. Accurate measurements of OH radical concentrations in the atmosphere can provide critical tests of our understanding of atmospheric chemistry and ground-level ozone production in urban and rural areas.

Because of its high reactivity, mixing ratios of OH in the atmosphere are extremely low (typically less than 0.1 parts per trillion) and its chemical lifetime very short (less than 1 second). As a result, measurements of OH present a serious analytical challenge, especially on the timescale necessary to test our understanding of the fast photochemistry of the atmosphere. This presentation will describe the Indiana University laser-induced fluorescence instrument for the sensitive detection of OH radicals in the atmosphere, including recent results from several measurement campaigns in both urban and rural environments.

WB02	9:05 – 9:20

EXPERIMENTAL AND THEORETICAL He-BROADENED LINE PARAMETERS OF CARBON MONOXIDE IN THE FUNDAMENTAL BAND

ADRIANA PREDOI-CROSS, HOIMONTI ROSARIO, KOOROSH ESTEKI, SHAMRIA LATIF, HOSSEIN NASERI, *Department of Physics and Astronomy, University of Lethbridge, Lethbridge, Canada*; FRANCK THIBAULT, *Institut de Physique de Rennes, Université de Rennes 1, Rennes, France*; V. MALATHY DEVI, *Department of Physics, College of William and Mary, Williamsburg, VA, USA*; MARY ANN H. SMITH, *Science Directorate, NASA Langley Research Center, Hampton, VA, USA*; ARLAN MANTZ, *Department of Physics, Astronomy and Geophysics, Connecticut College, New London, CT, USA.*

We report experimental measurements and theoretical calculations for He-broadened Lorentz half-width coefficients and He- pressure-shift coefficients of 45 carbon monoxide transitions in the 1-0 band. The high-resolution spectra analyzed in this study were recorded over a range of sample temperatures between 296 and 80 K. The He-broadened line parameters and their temperature dependences were retrieved using a multispectrum nonlinear least squares analysis program. A previous analysis of these spectra[a] used only the Voigt line shape. In the present study four line shape models were compared including Voigt, speed dependent Voigt, Rautian (to take into account confinement narrowing) and Rautian with speed dependence. The line mixing coefficients have been calculated using the Exponential Power Gap scaling law. We were unable to retrieve the temperature dependence of the line mixing coefficients. The current measurements and theoretical results are compared with other published results, where appropriate.

---

[a] A. W. Mantz *et al.*, *J. Molec. Structure* **742** (2005) 99-110.

WB03 9:22–9:37

SELF- AND H$_2$-BROADENED LINE PARAMETERS OF CARBON MONOXIDE IN THE FIRST OVERTONE BAND

ADRIANA PREDOI-CROSS, KOOROSH ESTEKI, HOSSEIN NASERI, *Department of Physics and Astronomy, University of Lethbridge, Lethbridge, Canada*; V. MALATHY DEVI, *Department of Physics, College of William and Mary, Williamsburg, VA, USA*; MARY ANN H. SMITH, *Science Directorate, NASA Langley Research Center, Hampton, VA, USA*; ARLAN MANTZ, *Department of Physics, Astronomy and Geophysics, Connecticut College, New London, CT, USA*; SERGEI V IVANOV, *Institute on Laser and Information Technologies, Russian Academy of Sciences, Troitsk, Moscow, Russia.*

In this study we have re-analyzed high-resolution spectra of pure CO and CO broadened by hydrogen recorded in the spectral range of the first overtone band.[a] We have used four different line shapes in the multispectrum analysis (Voigt, speed dependent Voigt, Rautian, and Rautian with speed dependence) and compared the resulting line shape parameters. The line mixing coefficients have been calculated using the Exponential Power Gap and the Energy Corrected Sudden scaling laws. A classical approach was applied to calculate CO line widths in CO-H$_2$ and CO-CO collisions. The formulas of classical impact theory[b] are used for calculation of dipole absorption half-widths along with exact 3D Hamilton equations for simulation of molecular motion. The calculations utilize Monte Carlo averaging over collision parameters and simple interaction potential (Tipping-Herman + electrostatic).[cd] Molecules are treated as rigid rotors. The dependences of CO half-widths on rotational quantum number $J \leq 24$ are computed and compared with measured data at room temperature.

[a]V. Malathy Devi *et al.*, *J. Mol. Spectrosc.* **228** (2004) 580-592.
[b]R. G. Gordon, *J. Chem. Phys.* **44** (1966) 3083-3089; *ibid.*, **45** (1966) 1649-1655.
[c]J.-P. Bouanich and A. Predoi-Cross, *J. Molec. Structure* **742** (2005) 183-190.
[d]A. Predoi-Cross, J.-P. Bouanich, D. Chris Benner, A. D. May, and J. R. Drummond, *J. Chem. Phys.* **113** (2000) 158-168.

WB04 9:39–9:54

MULTISPECTRUM ANALYSIS OF THE OXYGEN A-BAND

BRIAN DROUIN, LINDA R. BROWN, MATTHEW J. CICH, TIMOTHY J. CRAWFORD, ALEXANDER GUILLAUME, FABIANO OYAFUSO, VIVIENNE H PAYNE, KEEYOON SUNG, SHANSHAN YU, *Jet Propulsion Laboratory, California Institute of Technology, Pasadena, CA, USA*; D. CHRIS BENNER, V. MALATHY DEVI, *Department of Physics, College of William and Mary, Williamsburg, VA, USA*; JOSEPH HODGES, *Material Measurement Laboratory, National Institute of Standards and Technology, Gaithersburg, MD, USA*; ELI J MLAWER, *Atmospheric and Environmental Research, Lexington, MA, USA*; DAVID ROBICHAUD, *Biomass Molecular Science , National Renewable Energy Laboratory , Golden, CO, USA*; EDWARD H WISHNOW, *Space Sciences Laboratory and Department of Physics, University of California, Berkeley, CA, USA.*

Retrievals of atmospheric composition from near-infrared measurements require measurements of airmass to better than the desired precision of the composition. The oxygen bands are obvious choices to quantify airmass since the mixing ratio of oxygen is fixed over the full range of atmospheric conditions. The OCO-2 mission is currently retrieving carbon dioxide concentration using the oxygen A-band for airmass normalization. The 0.25% accuracy desired for the carbon dioxide concentration has pushed the required state-of-the-art for oxygen spectroscopy. To measure O$_2$ A-band cross-sections with such accuracy through the full range of atmospheric pressure requires a sophisticated line-shape model (Rautian or Speed-Dependent Voigt) with line mixing (LM) and collision induced absorption (CIA). Models of each of these phenomena exist, however, this work presents an integrated self-consistent model developed to ensure the best accuracy.

It is also important to consider multiple sources of spectroscopic data for such a study in order to improve the dynamic range of the model and to minimize effects of instrumentation and associated systematic errors. The techniques of Fourier Transform Spectroscopy (FTS) and Cavity Ring-Down Spectroscopy (CRDS) allow complimentary information for such an analysis. We utilize multispectrum fitting software to generate a comprehensive new database with improved accuracy based on these datasets. The extensive information will be made available as a multi-dimensional cross-section (ABSCO) table and the parameterization will be offered for inclusion in the HITRANonline database.

WB05 9:56 – 10:11

HIGH RESOLUTION PHOTOACOUSTIC SPECTROSCOPY OF THE OXYGEN A-BAND

MATTHEW J. CICH, *Jet Propulsion Laboratory, California Institute of Technology, Pasadena, CA, USA*; ELIZ-ABETH M LUNNY, GAUTAM STROSCIO, *Division of Chemistry and Chemical Engineering, California Institute of Technology, Pasadena, CA, USA*; THINH QUOC BUI, *JILA, National Institute of Standards and Technology and Univ. of Colorado Department of Physics, University of Colorado, Boulder, Boulder, CO, USA*; CAITLIN BRAY, *Department of Chemistry, Wesleyan University, Middletown, CT, USA*; DANIEL HOGAN, *Department of Applied Physics, Stanford University, Stanford, CA, USA*; PRIYANKA RUPASINGHE, *Physical Sciences, Cameron University, Lawton, OK, USA*; TIMOTHY J. CRAWFORD, BRIAN DROUIN, CHARLES MILLER, *Jet Propulsion Laboratory, California Institute of Technology, Pasadena, CA, USA*; DAVID A. LONG, JOSEPH HODGES, *Chemical Sciences Division, National Institute of Standards and Technology, Gaithersburg, MD, USA*; MITCHIO OKUMURA, *Division of Chemistry and Chemical Engineering, California Institute of Technology, Pasadena, CA, USA*.

NASA's Orbiting Carbon Observatory missions require spectroscopic parameterization of the Oxygen A-Band absorption (757-775 nm) with unprecedented detail to meet the objective of delivering space-based column CO_2 measurements with an accuracy of better than 1 ppm, and spectroscopic parameters with accuracies at the 0.1% level. To achieve this it is necessary for line shape models to include deviations from the Voigt line shape, including the collisional effects of speed-dependence, line mixing (LM), and collision-induced absorption (CIA). LM and CIA have been difficult to quantify in FTIR and CRDS spectra which have been limited to lower pressure measurements. A photoacoustic spectrometer has been designed to study the pressure-dependence of the spectral line shape up to pressures of 5 atm, where LM and CIA contribute significantly to the A-Band absorption. This spectrometer has a high signal-to-noise (S/N) of about 10,000 and frequency accuracy of 2 MHz. In addition, temperature-dependent effects on the line shape are studied using PID-controlled cooled nitrogen flow/ heater system. The latest acquired spectra and analysis are reported here.

Intermission

WB06 10:30 – 10:45

MEASUREMENTS @ MM-/SUB-MM-WAVE SPECTROSCOPY LABORATORY OF BOLOGNA: ROTATIONAL SPECTROSCOPY APPLIED TO ATMOSPHERIC STUDIES

CRISTINA PUZZARINI, *Dep. Chemistry 'Giacomo Ciamician', University of Bologna, Bologna, Italy.*

The physico-chemistry of the Earth's atmosphere has been one of the main subjects of studies over last years. In particular, the composition of the atmosphere is indeed very important to understand chemical processes linked to depletion of stratospheric ozone and greenhouse effect. The vertical concentration profiles of atmospheric gases can be provided by remote sensing measurements, but they require the accurate knowledge of the parameters involved: line positions, transition intensities, pressure-broadened half-widths, pressure-induced frequency shifts and their temperature dependence. In particular, the collisional broadening parameters have a crucial influence on the accuracy of spectra calculations and on reduction of remote sensing data.

Rotational spectroscopy, thanks to its intrinsic high resolution, is a powerful tool for providing most of the information mentioned above: accurate or even very accurate rotational transition frequencies, accurate spectroscopic as well as hyperfine parameters, accurate pressure-broadening coefficients and their temperature dependence. With respect to collisional phenomena and line shape analysis studies, by applying the source frequency modulation technique it has been found that rotational spectroscopy may provide very good results: not only this technique does not produce uncontrollable instrumental distortions or broadenings, but also, having an high sensitivity, it is particularly suitable for this kind of investigations.

A number of examples will be presented to illustrate the work carried out at the Laboratory of Millimeter/submillimeter-wave Spectroscopy of Bologna in the field of atmospheric studies.

156

WB07   **10:47 – 11:02**

THz AND FT-IR STUDY OF 18-O ISOTOPOLOGUES OF SULFUR DIOXIDE: $^{32}S^{16}O^{18}O$ AND $^{32}S^{18}O_2$

L. MARGULÈS, R. A. MOTIYENKO, *Laboratoire PhLAM, UMR 8523 CNRS - Université Lille 1, Villeneuve d'Ascq, France*; J. DEMAISON, *Laboratoire PhLAM, UMR 8523 CNRS - Université Lille 1, Villeneuve d'Ascq, France*; AGNES PERRIN, F. KWABIA TCHANA, *LISA, CNRS, Universités Paris Est Créteil et Paris Diderot, Créteil, France*; LAURENT MANCERON, *Synchrotron SOLEIL, CNRS-MONARIS UMR 8233 and Beamline AILES, Saint Aubin, France*.

Sulfur dioxide is a molecule that have a great interest in different domains: for atmospheric and planetology chemistry, it is also ubiquitous and abundant in interstellar medium. If the ^{16}O species were extensively studied, this is not the case of the ^{18}O isotopologues. The aim of this study is first to complete the rotational spectra of the ground state with these new measurements up to 1.5 THz, previous measurements are up to 1050 GHz for the $^{32}S^{16}O^{18}O$ species[a], and 145 GHz concerning the $^{32}S^{18}O_2$ species[b]. The second part is making a global fit of the rotational and vibrational transitions for the excited vibrational states. For the v_2 band, we will complete the recent I.R. analysis[c]. About the triad $(v_1, 2v_2, v_3)$: $^{32}S^{18}O_2$ species was studied[d], but not the $^{32}S^{16}O^{18}O$ one.

The FT-IR spectra were recorded on the AILES Beamline at Synchrotron SOLEIL using the Synchrotron light source, coupled to the Bruker IFS125HR Fourier transform spectrometer[e]. The THz spectra were obtained from 150 to 1500 GHz using the Lille's solid state spectrometer[f]. The analysis is in progress, the latest results will be presented.

Support from the French Laboratoire d'Excellence CaPPA (Chemical and Physical Properties of the Atmosphere) through contract ANR-10-LABX-0005 of the Programme d'Investissements d'Avenir is acknowledged

[a]Belov, S. P.; *et al.*, 1998, *J. Mol. Spectrosc.* **191**, 17

[b]Lindermayer, J.; *et al.*, 1985, *J. Mol. Spectrosc.* **110**, 357

[c]Gueye, F.; *et al. Mol. Phys.* **in press**

[d]Ulenikov, O. N.; *et al.*, 2015, *JQSRT* **166**, 13

[e]Brubach, J.; *et al.*, 2010, *AIP Conf. Proc.* **1214**, 81

[f]Zakharenko, O.; *et al.*, 2015, *J. Mol. Spectrosc.* **317**, 41

WB08   **11:04 – 11:19**

IMPROVE THE ABSOLUTE ACCURACY OF OZONE INTENSITIES IN THE 9-11 μm REGION VIA MW/IR MULTI-WAVELENGTH SPECTROSCOPY

SHANSHAN YU, BRIAN DROUIN, *Jet Propulsion Laboratory, California Institute of Technology, Pasadena, CA, USA*.

Ozone (O_3) is crucial for studies of air quality, human and crop health, and radiative forcing. Spectroscopic remote sensing techniques have been extensively employed to investigate ozone globally and regionally. Infrared intensities of $\leq 1\%$ accuracy are desired by the remote sensing community. The accuracy of the current state-of-the-art infrared ozone intensities is on the order of 4-10%, resulting in ad hoc intensity scaling factors for consistent atmospheric retrievals. The large uncertainties on the infrared ozone intensities arise from the fact that pure ozone is very difficult to generate and sustain in the laboratory. Best estimates have employed IR/UV cross beam experiments to determine the accurate O_3 volume mixing ratio of the sample through its standard cross section value at 254 nm.

This presentation reports our effort to improve the absolute accuracy of ozone intensities in the 9-11 μm region via a transfer of the precision of the rotational dipole moment onto the infrared measurement (MW/IR). Our approach was to use MW/IR cross beam experiments and determine the O_3 mixing ratio through alternately measuring pure rotation ozone lines from 692 to 779 GHz. The uncertainty of these pure rotation line intensities is better than 0.1%. The sample cell was a slow flow cross cell and the total pressure inside the sample cell was maintained constant through a proportional–integral–derivative (PID) flow control. Five infrared O_3 spectra were obtained, with a path length of 3.74 m, pressures ranging from 30 to 120 mTorr, and mixing ratio ranging from 0.5 to 0.9. A multi spectrum fitting technique was employed to fit all the FTS spectra simultaneously. The results show that we can determine intensities of the 9.6μm band with absolute accuracy better than 4%.

WB09

FIRST HIGH-RESOLUTION ANALYSIS OF PHOSGENE $^{35}CL_2CO$ AND $^{35}CL^{37}CLCO$ FUNDAMENTALS IN THE 250 - 480 cm^{-1} SPECTRAL REGION

F. KWABIA TCHANA, M. NDAO, *LISA, CNRS, Universités Paris Est Créteil et Paris Diderot, Créteil, France*; LAURENT MANCERON, *Synchrotron SOLEIL, CNRS-MONARIS UMR 8233 and Beamline AILES, Saint Aubin, France*; AGNES PERRIN, JEAN-MARIE FLAUD, *LISA, CNRS, Universités Paris Est Créteil et Paris Diderot, Créteil, France*; WALTER LAFFERTY, *Optical Technology Division, National Institute of Standards and Technology, Gaithersburg, MD, USA.*

Phosgene ($COCl_2$) is relatively more abundant in the stratosphere, but is also present in the troposphere in spite of a shorter lifetime (seventy days). Monitoring its concentration by remote sounding of the upper atmosphere is of importance, since some of its strong infrared absorptions, occurring in the important 8-12 μm atmospheric window, hinder the correct retrieval of Freon-11 concentration profiles[a]. Indeed, the infrared absorptions used to retrieve this ozone depleting compound occur in the same spectral region.

Phosgene, presents two fundamental bands in the 250 - 480 cm^{-1} spectral region, with the lowest (ν_3) near 285 cm^{-1}. These are responsible for hot bands, not yet analysed but of great importance for accurate modeling of the 5.47 μm (ν_1) and 11.75 μm (ν_5) spectral regions and consequently the correct retrieval of Freon-11 atmospheric absorption profiles.

High-resolution absorption spectra of phosgene have been recorded at 0.00102 cm^{-1} resolution in the 250–480 cm^{-1} region by Fourier transform spectroscopy at synchrotron SOLEIL. Due to the spectral congestion, the spectra have been recorded at low temperature (197 K) using a 93.15 m optical path length cryogenic cell[b]. This enables the first detailed far-infrared analyzes of the ν_3 and ν_6 bands of the $^{35}Cl_2CO$ and $^{35}Cl^{37}ClCO$ isotopologues of phosgene. Using a Watson-type Hamiltonian, it was possible to reproduce the upper state rovibrational infrared energy levels to within the experimental accuracy. The results will be presented in this talk.

---

[a]G. Toon, J.F. Blavier, B. Sen and B.J. Drouin, Geophys. Res. Lett., 28/14 (2001) 2835.

[b]F. Kwabia Tchana, F. Willaert, X. Landsheere, J.-M. Flaud, L. Lago, M. Chapuis, P. Roy and L. Manceron, Rev. Sci. Inst., 84 (2013) 093101.

WC. Chirped pulse

Wednesday, June 22, 2016 – 8:30 AM

Room: 274 Medical Sciences Building

Chair: Robert W Field, MIT, Cambridge, MA, USA

WC01   8:30–8:45

CHIRPED PULSE ROTATIONAL SPECTROSCOPY OF A SINGLE THUJONE+WATER SAMPLE

ZBIGNIEW KISIEL, *ON2, Institute of Physics, Polish Academy of Sciences, Warszawa, Poland*; CRISTOBAL PEREZ, MELANIE SCHNELL, *CoCoMol, Max-Planck-Institut für Struktur und Dynamik der Materie, Hamburg, Germany.*

Rotational spectroscopy of natural products dates over 35 years when six different species including thujone were investigated.[a] Nevertheless, the technique of low-resolution microwave spectroscopy employed therein allowed determination of only a single conformational parameter. Advances in sensitivity and resolution possible with supersonic expansion techniques of rotational spectroscopy made possible much more detailed studies such that, for example, the structures of first camphor,[b] and then of multiple clusters of camphor with water[c] were determined.

We revisited the rotational spectrum of the well known thujone molecule by using the chirped pulse spectrometer in Hamburg. The spectrum of a single thujone sample was recorded with an admixture of ^{18}O enriched water and was successively analysed using an array of techniques, including the AUTOFIT program,[d] the AABS package[e] and the STRFIT program.[f] We have, so far, been able to assign rotational transitions of α-thujone, β-thujone, another thujone isomer, fenchone, and several thujone-water clusters in the spectrum of this single sample. Natural abundance molecular populations were sufficient to determine precise heavy atom backbones of thujone and fenchone, and $H_2{}^{18}O$ enrichment delivered water molecule orientations in the hydrated clusters. An overview of these results will be presented.

---

[a]Z.Kisiel, A.C.Legon, *JACS* **100**, 8166 (1978).

[b]Z.Kisiel, O.Desyatnyk, E.Białkowska-Jaworska, L.Pszczółkowski, *PCCP* **5** 820 (2003).

[c]C.Pérez, A.Krin, A.L.Steber, J.C.López, Z.Kisiel, M.Schnell, *J.Phys.Chem.Lett.* **7** 154 (2016).

[d]N.A.Seifert, I.A.Finneran, C.Perez, et al. *J.Mol.Spectrosc.* **312**, 12 (2015).

[e]Z.Kisiel, L.Pszczółkowski, B.J.Drouin, et al. *J.Mol.Spectrosc.* **280**, 134 (2012).

[f]Z.Kisiel, *J.Mol.Spectrosc.* **218**, 58 (2003).

WC02   8:47–9:02

MICROWAVE SPECTRUM AND MOLECULAR STRUCTURE OF THE ARGON-*CIS*-1,2-DICHLOROETHYLENE COMPLEX

MARK D. MARSHALL, HELEN O. LEUNG, CRAIG J. NELSON, LEONARD H. YOON, *Chemistry Department, Amherst College, Amherst, MA, USA.*

The non-planar molecular structure of the complex formed between the argon atom and *cis*-1,2-dichloroethylene is determined via analysis of its microwave spectrum. Spectra of the ^{35}Cl and ^{37}Cl isotopologues are observed in natural abundance and the nuclear quadrupole splitting due to the two chlorine nuclei is fully resolved. In addition, the complete quadrupole coupling tensor for the *cis*-1,2-dichloroethylene molecule, including the single non-zero off-diagonal element, has been determined. Unlike the argon-*cis*-1,2-difluoroethylene and the argon-vinyl chloride complexes, tunneling between the two equivalent non-planar configurations of argon-*cis*-1,2-dichloroethylene is not observed.

WC03

INFLUENCE OF HALOGEN VARIATION ON STRUCTURE AND INTERACTIONS IN VINYL HALIDE $(H_2C=CHX)\cdots CO_2$ (X = F, Cl, Br) COMPLEXES

ASHLEY M. ANDERTON, CORI L. CHRISTENHOLZ, RACHEL E. DORRIS, <u>REBECCA A. PEEBLES</u>, SEAN A. PEEBLES, *Department of Chemistry, Eastern Illinois University, Charleston, IL, USA.*

Chirped-pulse and resonant cavity Fourier-transform microwave spectroscopy have been used to investigate dimers of CO_2 with vinyl fluoride (VF), vinyl chloride (VCl) and vinyl bromide (VBr). For all three complexes, CO_2 is aligned adjacent to the X–C–H end (X = F, Cl, Br) of the ethylene subunit, with C–X$\cdots$C and C–H$\cdots$O contacts. For VF$\cdots CO_2$, a second isomer is also observed, with CO_2 roughly parallel to the H–C=C–F side of VF; however, there is no spectroscopic indication that similar structures are present for VCl$\cdots CO_2$ or VBr$\cdots CO_2$.

For vinyl fluoride$\cdots CO_2$, a full structural analysis has previously been published,[a] while for the Cl- and Br-containing species, insufficient data are presently available for complete structure determinations. However, structural information from ab initio calculations, ^{35}Cl/^{37}Cl and ^{79}Br/^{81}Br isotopic substitution, and analysis of chlorine and bromine nuclear quadrupole coupling constants will be presented. In addition, for this series of dimers containing C–H$\cdots$O contacts, further insight into the nature of the weak interactions may be obtained from Quantum Theory of Atoms in Molecules (QTAIM) and other *ab initio* analyses that are presently in progress.

[a]C. L. Christenholz, R. E. Dorris, R. A. Peebles, S. A. Peebles, *J. Phys. Chem. A*, **118**, (2014), 8765-8772.

WC04

H-BONDING NETWORKS IN SUGAR ALCOHOLS: IDENTIFYING GLUCOPHORES?

<u>E. R. ALONSO</u>, SANTIAGO MATA, CARLOS CABEZAS, ISABEL PEÑA, JOSÉ L. ALONSO, *Grupo de Espectroscopia Molecular, Lab. de Espectroscopia y Bioespectroscopia, Unidad Asociada CSIC, Universidad de Valladolid, Valladolid, Spain.*

The conformational behaviour of sorbitol and dulcitol has been investigated for the first time using a combination of chirped pulse Fourier transform microwave spectroscopy (CP-FTMW) coupled with a laser ablation (LA) source. The observed conformers have been found to be overstabilised by cooperative networks of intramolecular hydrogen bonds between vicinal hydroxyl groups stretching throughout the whole molecule. A common structural signature - involving hydroxyl groups in the H-bond - has been characterized and ascribed to the glucophore's AH and B sites in accordance with Shallenberger's old proposal.[a,b]

[a]R. S. Shallenberger, T. E. Acree, *Nature*, **1967**, *216*, 480-482
[b]R. S. Shallenberger, T. E. Acree, C. Y. Lee, *Nature*, **1969**, *221*, 555-556

WC05

SEVEN CONFORMERS OF PIPECOLIC ACID IDENTIFIED IN THE GAS PHASE

CARLOS CABEZAS, ALCIDES SIMAO, <u>JOSÉ L. ALONSO</u>, *Grupo de Espectroscopia Molecular, Lab. de Espectroscopia y Bioespectroscopia, Unidad Asociada CSIC, Universidad de Valladolid, Valladolid, Spain.*

The multiconformational landscape of the non-proteinogenic cyclic amino acid pipecolic acid has been explored in the gas phase. Solid pipecolic acid (m.p. 280°C) was vaporized by laser ablation (LA) and expanded in a supersonic jet where the rotational spectra of seven conformers were obtained by broadband microwave spectroscopy (CP-FTMW). All conformers were conclusively identified by comparison of the experimental spectroscopic constants with those predicted theoretically. The relative stability of the conformers rests on a delicate balance of the different intramolecular hydrogen bonds established between the carboxylic and the amino groups.

WC06   9:50 – 10:05

PREFERRED CONFORMERS OF NON-PROTEINOGENIC AMINO ACIDS HOMOSERINE AND HOMOCYSTEINE

VERÓNICA DÍEZ, MIGUEL A. RODRÍGUEZ, SANTIAGO MATA, E. R. ALONSO, CARLOS CABEZAS, JOSÉ L. ALONSO, *Grupo de Espectroscopia Molecular, Lab. de Espectroscopia y Bioespectroscopia, Unidad Asociada CSIC, Universidad de Valladolid, Valladolid, Spain.*

Vaporization of solid homoserine and homocysteine by laser ablation in combination with Fourier transform microwave spectroscopy techniques made possible the detection of their most stable structures in a supersonic expansion. All detected conformers have been identified through their rotational and ^{14}N quadrupole coupling constants. They show hydrogen bonds linking the amino and carboxylic group through N–H$\cdots$O=C (type I) or N$\cdots$H–O (type II) interactions. In some of them there are additional hydrogen bonds established between the amino group and the hydroxyl/thiol groups in the gamma position. Entropic effects related to the side chain have been found to be significant in determining the most populated conformations.

WC07   10:07 – 10:22

THE ROTATIONAL SPECTRUM OF THE UREA$\cdots$ISOCYANIC ACID COMPLEX

JOHN C MULLANEY, *School of Chemistry, Newcastle University, Newcastle-upon-Tyne, United Kingdom*; CHRIS MEDCRAFT, *School of Chemistry, Newcastle University, Newcastle upon Tyne, United Kingdom*; NICK WALKER, *School of Chemistry, Newcastle University, Newcastle-upon-Tyne, United Kingdom*; ANTHONY LEGON, *School of Chemistry, University of Bristol, Bristol, United Kingdom*; LUKE LEWIS-BORRELL, BERNARD T GOLDING, *School of Chemistry, Newcastle University, Newcastle-upon-Tyne, United Kingdom.*

A dimer of urea and isocyanic acid has been generated and observed in the gas phase. The complex was generated by laser vaporisation of a rod target containing urea and copper in a 1:1 ratio, then cooled in a supersonic expansion. Six isotopologues of the complex have been characterised using a chirped pulse Fourier-transform microwave spectrometer in the frequency range 6.5-18.5 GHz. The spectra have been fitted to the Hamiltonian for an asymmetric rotor using PGOPHER. Data obtained from the ^{13}C and ^{15}N isotopologues confirms that all nitrogen atoms are close to the a intertial axis while the carbon atoms are not. A tentative structure will be presented.

Intermission

WC08   10:41 – 10:56

GEOMETRY OF AN ISOLATED DIMER OF IMIDAZOLE CHARACTERISED BY ROTATIONAL SPECTROSCOPY AND AB INITIO CALCULATIONS

JOHN C MULLANEY, *School of Chemistry, Newcastle University, Newcastle-upon-Tyne, United Kingdom*; DANIEL P. ZALESKI, *Chemical Sciences and Engineering Division, Argonne National Laboratory, Argonne, IL, USA*; DAVID PETER TEW, *School of Chemistry, University of Bristol, Bristol, United Kingdom*; NICK WALKER, *School of Chemistry, Newcastle University, Newcastle-upon-Tyne, United Kingdom*; ANTHONY LEGON, *School of Chemistry, University of Bristol, Bristol, United Kingdom.*

An isolated, gas-phase dimer of imidazole is generated through laser vaporisation of a solid rod containing a 1:1 mixture of imidazole and copper in the presence of an argon buffer gas undergoing supersonic expansion. The complex is characterised through broadband rotational spectroscopy and is shown to have a twisted, hydrogen-bonded geometry. Calculations at the CCSD(T)(F12*)/cc-pVDZ-F12 level of theory confirm this to be the lowest-energy conformer of the imidazole dimer. The distance between the respective centres of mass of the imidazole monomer subunits is determined to be 5.2751(1) Å, and the twist angle γ describing rotation of one monomer with respect to the other about a line connecting the centres of mass of the monomers is determined to be 87.9(4)^o. Four out of six intermolecular parameters in the model geometry are precisely determined from the experimental rotational constants and are consistent with results calculated ab initio.

WC09   **10:58 – 11:13**

MICROWAVE SPECTRUM OF THE ACETALDEHYDE-WATER DIMER

GRIFFIN MEAD, IAN A FINNERAN, BRANDON CARROLL, GEOFFREY BLAKE, *Division of Chemistry and Chemical Engineering, California Institute of Technology, Pasadena, CA, USA.*

Microwave spectroscopy provides a unique opportunity to study model non-covalent interactions. Of particular interest is the hydrogen bonding of water, which can have both strong and weak interactions. More specifically, measuring the hydrogen bond structure of water-aldehyde dimers investigates both fundamental water-carbonyl interactions and atmospherically relevant molecular clusters. Recently, we have measured the pure rotational spectrum of the acetaldehyde-water dimer using chirped-pulse Fourier transform microwave spectroscopy (CP-FTMW) between 8-18 GHz. Here, we present the spectrum of this dimer and elaborate on the structure's strong and weak hydrogen bonding.

WC10   **11:15 – 11:30**

THE CONFORMATIONAL BEHAVIOUR OF THE ODORANT DIHYDROCARVEOL

DONATELLA LORU, NATASHA JARMAN, M. EUGENIA SANZ, *Department of Chemistry, King's College London, London, United Kingdom.*

The odorant dihydrocarveol ($C_{10}H_{18}O$) has been investigated in the gas phase using a 2-8 GHz chirped-pulse Fourier transform microwave spectrometer. Dihydrocarveol was purchased as a mixture of n-, iso-, neo-, and neoiso- isomers. The sample was placed in a bespoke heating nozzle at about 85°C and seeded in Ne at 5 bar. Three conformers were observed and their rotational constants were determined. By comparing the experimental rotational constants with those calculated ab initio the three conformers were identified as belonging to n-dihydrocarveol. In all three conformers the isopropenyl group is in equatorial position with respect to the six-membered ring, and the OH group maintains the same configuration. The conformers differ in the orientation of the isopropenyl group.

WC11   **11:32 – 11:47**

STRUCTURAL CHARACTERISATION OF FENCHONE AND ITS COMPLEXES WITH ETHANOL BY BROADBAND ROTATIONAL SPECTROSCOPY

DONATELLA LORU, M. EUGENIA SANZ, *Department of Chemistry, King's College London, London, United Kingdom.*

Although significant advances in understanding the human olfactory system have taken place over the last two decades, detailed information on how the interactions between odorants and olfactory receptors occur at the molecular level is still lacking. To achieve a better understanding on the molecular mechanisms involved in olfaction, we are investigating several odorants and their interactions with mimics of amino acid residues in olfactory receptors.

We present here the structural characterisation of fenchone ($C_{10}H_{16}O$) and its complexes with ethanol (to mimic the side chain of serine) using a 2-8 GHz chirped-pulse Fourier transform microwave spectrometer built at King's College London. The rotational spectrum of the parent species and all the ^{13}C and ^{18}O isotopologues of fenchone was observed, and from the experimental rotational constants the substitution (r_0) and effective (r_s) structures of fenchone were determined. The rotational spectrum of fenchone-ethanol was observed by adding ethanol to the carrier gas and passing the mixture through a receptacle with fenchone. Several 1:1 complexes of fenchone-ethanol have been identified in the rotational spectrum. In all the complexes the ethanol molecule binds to the carbonyl group through an O-H$\cdots$O hydrogen bond.

WC12   **11:49 – 12:04**

BROADBAND MICROWAVE SPECTROSCOPY AS A TOOL TO STUDY DISPERSION INTERACTIONS IN CAMPHOR-ALCOHOL SYSTEMS

MARIYAM FATIMA, CRISTOBAL PEREZ, MELANIE SCHNELL, *CoCoMol, Max-Planck-Institut für Struktur und Dynamik der Materie, Hamburg, Germany.*

Many biological processes such as chemical recognition and protein folding are mainly controlled by the interplay between hydrogen bonds and dispersive forces. Broadband rotational spectroscopy studies of weakly bound complexes are able to accurately reveal the structures and internal dynamics of molecular clusters isolated in the gas phase. To investigate the influence of the interplay between different types of weak intermolecular interactions and how it controls the preferred active sites of an amphiphilic molecule, we are using camphor ($C_{10}H_{16}O$, 1,7,7-trimethylbicyclo[2.2.1]hepta-2-one) with different aliphatic alcohol systems. Camphor is a conformationally rigid bicyclic molecule endowed with considerable steric hindrance and has a single polar group (-C=O). The rotational spectrum of camphor and its structure has been previously reported [1] as well as multiple clusters with water [2]. In order to determine the structure of the camphor-alcohol complexes, we targeted low energy rotational transitions in the 2-8 GHz range under the isolated conditions of a molecular jet in the gas phase. The data obtained suggests that camphor forms one complex with methanol and two with ethanol, with differences in the intermolecular interaction in both complexes. With these results, we aim to study the shift in intermolecular interaction from hydrogen bonding to dispersion with the increase in the size of the aliphatic alcohol.

[1] Z. Kisiel, et al., Phys. Chem. Chem. Phys., 5 (2003), 820–826.
[2] C. Pérez, et al, J. Phys. Chem. Lett., 7 (2016), 154–160.

WD. Mini-symposium: Spectroscopy in Traps
Wednesday, June 22, 2016 – 8:30 AM
Room: B102 Chemical and Life Sciences

Chair: Roland Wester, Universität Innsbruck, Innsbruck, Austria

---

WD01　　　　　　　　　　　*INVITED TALK*　　　　　　　　　　　**8:30 – 9:00**

INFRARED ION SPECTROSCOPY AT FELIX: APPLICATIONS IN PEPTIDE DISSOCIATION AND ANALYTICAL CHEMISTRY

JOS OOMENS, *Institute for Molecules and Materials (IMM), Radboud University Nijmegen, Nijmegen, Netherlands.*

Infrared free electron lasers such as those in Paris, Berlin and Nijmegen have been at the forefront of the development of infrared ion spectroscopy. In this contribution, I will give an overview of new developments in IR spectroscopy of stored ions at the FELIX Laboratory. In particular, I will focus on recent developments made possible by the coupling of a new commercial ion trap mass spectrometer to the FELIX beamline.

The possibility to record IR spectra of mass-selected molecular ions and their reaction products has in recent years shed new light on our understanding of collision induced dissociation (CID) reactions of protonated peptides in mass spectrometry (MS). We now show that it is possible to record IR spectra for the products of electron transfer dissociation (ETD) reactions

$$[M + nH]^{n+} + A^- \rightarrow [M + nH]^{(n-1)+} + A \rightarrow \text{dissociation of analyte}$$

These reactions are now widely used in novel MS-based protein sequencing strategies, but involve complex radical chemistry. The spectroscopic results allow stringent verification of computationally predicted product structures and hence reaction mechanisms and H-atom migration.

The sensitivity and high dynamic range of a commercial mass spectrometer also allows us to apply infrared ion spectroscopy to analytes in complex "real-life" mixtures. The ability to record IR spectra with the sensitivity of mass-spectrometric detection is unrivalled in analytical sciences and is particularly useful in the identification of small (biological) molecules, such as in metabolomics. We report preliminary results of a pilot study on the spectroscopic identification of small metabolites in urine and plasma samples.

WD02　　　　　　　　　　　　　　　　　　　　　　　　　　**9:05 – 9:20**

INFRARED PREDISSOCIATION SPECTROSCOPY OF He-TAGGED SMALL MOLECULAR CATIONS OF ASTROCHEMICAL INTEREST

ALEXANDER STOFFELS[a], BRITTA REDLICH, JOS OOMENS, *FELIX Laboratory, Radboud University, Nijmegen, The Netherlands*; OSKAR ASVANY, SANDRA BRÜNKEN, PAVOL JUSKO, SVEN THORWIRTH, STEPHAN SCHLEMMER, *I. Physikalisches Institut, Universität zu Köln, Köln, Germany.*

Small organic hydrocarbon cations containing heteroatoms, e.g. primary alcohol cations, are discussed to play an important role in astrochemical networks, but so far none of these ions have been detected in space. This is mainly due to the lack of reference spectra under astronomically relevant conditions (low T, gas phase). In a cryogenic 22-pole ion trap, held at ca. 4 K, these conditions are approximated. Here we present broadband IR-photodissociation spectra of He-tagged protonated Methanol ($MeOH_2^+$) and Ethanol ($EtOH_2^+$) cations, along with the spectra of their radical analoga ($MeOH^+/EtOH^+$). The covered spectral range is from ca. 500 cm^{-1} to 3800 cm^{-1} at a spectral resolution of a few cm^{-1}. These spectra were recorded by coupling the cryogenic ion trap to FELIX and a commercial mid-IR emitting OPO. The vibrational band positions are assigned using quantum-chemical calculations performed at different levels of theory, with emphasis on anharmonic effects. The computed frequencies correspond favorably to the experimental spectra. For all cations, the (H-)O-H stretching and bending modes are significantly shifted from the position of the neutral analog. In case of $MeOH^+$, theory suggests another isomer, the methylene-oxonium ion ($CH_2OH_2^+$), which can be distinguished by the O-H stretching and bending modes, appearing as a single strong OH stretching band along with a weak, broad feature at ca. 1600 cm^{-1}, which is attributed to an H-O-H bending vibration of $CH_2OH_2^+$. Therefore we conclude that a mixture of $MeOH^+$ isomers is present in the trap experiment. This work provides a good starting point for subsequent high-resolution spectroscopic studies of these cations that are required for unambigous interstellar identification.

---

[a]also at: I. Physikalisches Institut, Universität zu Köln, Köln, Germany.

164

WD03   9:22 – 9:37

PROBING THE VIBRATIONAL SPECTROSCOPY OF THE DEPROTONATED THYMINE RADICAL BY PHOTODE-
TACHMENT AND STATE-SELECTIVE AUTODETACHMENT PHOTOELECTRON SPECTROSCOPY VIA DIPOLE-
BOUND STATES

DAO-LING HUANG, GUO-ZHU ZHU, LAI-SHENG WANG, *Department of Chemistry, Brown University, Providence, RI, USA.*

Deprotonated thymine can exist in two different forms, depending on which of its two N sites is deprotonated: N1[T-H]$^-$ or N3[T-H]$^-$. Here we report a photodetachment study of the N1[T-H]$^-$ isomer cooled in a cryogenic ion trap and the observation of an excited dipole-bound state[a]. Eighteen vibrational levels of the dipole-bound state are observed, and its vibrational ground state is found to be 238 ± 5 cm^{-1} below the detachment threshold of N1[T-H]$^-$. The electron affinity of the deprotonated thymine radical (N1[T-H]$\cdot$) is measured accruately to be $26\,322 \pm 5$ cm^{-1} (3.2635 ± 0.0006 eV). By tuning the detachment laser to the sixteen vibrational levels of the dipole-bound state that are above the detachment threshold, highly non-Franck-Condon resonant-enhanced photoelectron spectra are obtained due to state- and mode-selective vibrational autodetachment. Much richer vibrational information is obtained for the deprotonated thymine radical from the photodetachment and resonant-enhanced photoelectron spectroscopy. Eleven fundamental vibrational frequencies in the low-frequency regime are obtained for the N1[T-H]$\cdot$ radical, including the two lowest-frequency internal rotational modes of the methyl group at 70 ± 8 cm^{-1} and 92 ± 5 cm^{-1}.

[a]D. L. Huang, H. T. Liu, C. G. Ning, G. Z. Zhu and L. S. Wang, *Chem. Sci.*, **6**, 3129-3138 (2015)

WD04   9:39 – 9:54

QUANTIFICATION OF STRUCTURAL ISOMERS VIA MODE-SELECTIVE IRMPD

NICOLAS C POLFER, *Physical Chemistry, University of Florida, Gainesville, FL, USA.*

Mixtures of structural isomers can pose a challenge for vibrational ion spectroscopy. In cases where particular structures display diagnostic vibrations, these structures can be selectively "burned away". In ion traps, the ion population can be subjected to multiple laser shots, in order to fully deplete a particular structure, in effect allowing a quantification of this structure. Protonated para-amino benzoic acid (PABA) serves as an illustrative example. PABA is known to preferentially exist in the N-protonated (N-prot) form in solution, but in the gas phase it is energetically favorable in the O-protonated (O-prot) form. As shown in Figure 1, the N-prot structure can be kinetically trapped in the gas phase when sprayed from non-protic solvent, whereas the O-prot structure is obtained when sprayed from protic solvents, analogous to results by others [1,2].

By parking the light source on the diagnostic 3440 cm^{-1} mode, the percentage of the O-prot structure can be determined, and by default the remainder is assumed to adopt the N-prot structure. It will be shown that the relative percentages of O-prot vs N-prot are highly dependent on the solvent mixture, going from close to 0% O-prot in non-protic solvents, to 99% in protic solvents. Surprisingly, water behaves much more like a non-protic solvent than methanol. It is observed that the capillary temperature, which aids droplet desolvation by black-body radiation in the ESI source, is critical to promote the appearance of O-prot structures. These results are consistent with the picture that a protic bridge mechanism is at play to facilitate proton transfer, and thus allow conversion from N-prot to O-prot, but that this mechanism is subject to appreciable kinetic barriers on the timescale of solvent evaporation. 1. J. Phys. Chem. A 2011, 115, 7625. 2. Anal. Chem. 2012, 84, 7857.

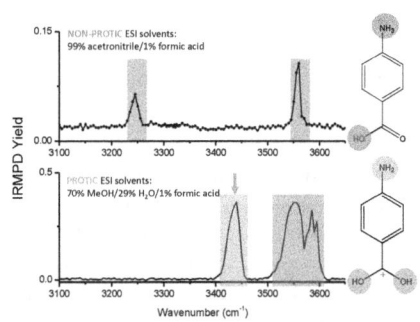

WD05

URIDINE NUCLEOSIDE THIATION: GAS-PHASE STRUCTURES AND ENERGETICS

LUCAS HAMLOW, JUSTIN LEE, M T RODGERS, *Department of Chemistry, Wayne State University, Detroit, MI, USA*; GIEL BERDEN, JOS OOMENS, *Institute for Molecules and Materials (IMM), Radboud University Nijmegen, Nijmegen, Netherlands.*

The naturally occurring thiated uridine nucleosides, 4-thiouridine (s^4Urd) and 2-thiouridine (s^2Urd), play important roles in the function and analysis of a variety of RNAs. 2-Thiouridine and its C5 modified analogues are commonly found in tRNAs and are believed to play an important role in codon recognition possibly due to their different structure, which has been shown by NMR to be predominantly C3′-endo. 2-Thiouridine may also play an important role in facilitating nonenzymatic RNA replication and transcription. 4-Thiouridine is a commonly used photoactivatable crosslinker that is often used to study RNA-RNA and RNA-protein cross-linking behavior. Differences in the base pairing between uracil and 4-thiouracil with adenine and guanine are an important factor in their role as a cross linker. The photoactivity of s^4Urd may also aid in preventing near-UV lethality in cells. An understanding of their intrinsic structure in the gas-phase may help further elucidate the roles these modified nucleosides play in the regulation of RNAs.

In this work, infrared multiple photon dissociation (IRMPD) action spectra of the protonated forms of s^2Urd and s^4Urd were collected in the IR fingerprint region. Structural information is determined by comparison with theoretical linear IR spectra generated from density functional theory calculations using molecular modeling to generate low-energy candidate structures. Present results are compared with analogous results for the protonated forms of uridine and 2′-deoxyuridine as well as solution phase NMR data and crystal structures.

Intermission

WD06

STRUCTURE DETERMINATION OF ORNITHINE-LINKED CISPLATIN BY INFRARED MULTIPLE PHOTON DIS-SOCIATION ACTION SPECTROSCOPY

CHENCHEN HE, BETT KIMUTAI, LUCAS HAMLOW, HARRISON ROY, Y-W NEI, XUN BAO, *Department of Chemistry, Wayne State University, Detroit, MI, USA*; JUEHAN GAO, JONATHAN K MARTENS, GIEL BERDEN, JOS OOMENS, *Institute for Molecules and Materials (IMM), Radboud University Nijmegen, Nijmegen, Netherlands*; PHILIPPE MAITRE, VINCENT STEINMETZ, *Institut des Sciences Moléculaires d'Orsay, Université Paris-Sud, Orsay, France*; CHRISTOPHER P McNARY, PETER B ARMENTROUT, *Department of Chemistry, University of Utah, Salt Lake City, UT, USA*; C S CHOW, M T RODGERS, *Department of Chemistry, Wayne State University, Detroit, MI, USA.*

Cisplatin [$(NH_3)_2PtCl_2$], the first FDA-approved platinum-based anticancer drug, has been widely used in cancer chemotherapy. Its pharmacological mechanism has been identified as its ability to coordinate to genomic DNA with guanine as its major target. Amino acid-linked cisplatin derivatives are being investigated as alternatives for cisplatin that may exhibit altered binding selectivity such as that found for ornithine-linked cisplatin (Ornplatin, [$(Orn)PtCl_2$]), which exhibits a preference for adenine over guanine in RNA. Infrared multiple photon dissociation (IRMPD) action spectroscopy experiments and complementary electronic structure calculations are performed on a series of Ornplatin complexes to elucidate the nature of binding of the Orn amino acid to the Pt center and how that binding is influenced by the local environment. The complexes examined in the work include: [$(Orn - H)PtCl_2$]$^-$, [$(Orn)PtCl$]$^+$, [$(Orn)Pt(H_2O)Cl$]$^+$, and [$(Orn)PtCl_2 + Na$]$^+$. In contrast to that found previously for the glycine-linked cisplatin complex (Glyplatin), which binds via the backbone amino and carboxylate groups, binding of Orn in these complexes is found to involve both the backbone and sidechain amino groups. Extensive broadening of the IRMPD spectrum for the [$(Orn)Pt(H_2O)Cl$]$^+$ complex suggests that either multiple structures are contributing to the measured spectrum or strong intra-molecular hydrogen-binding interactions are present. The results for Ornplatin lead to an interesting discussion about the differences in selectivity and reactivity versus cisplatin.

166

WD07   10:47 – 11:02

CONTROLLED FORMATION AND VIBRATIONAL CHARACTERIZATION OF LARGE SOLVATED IONIC CLUS-
TERS IN CRYOGENIC ION TRAPS

ETIENNE GARAND, BRETT MARSH, JONATHAN VOSS, ERIN M. DUFFY, *Department of Chemistry, University of Wisconsin, Madison, WI, USA.*

An experimental approach for the formation of solvated ionic clusters and their vibrational spectroscopy will be presented. This recently developed apparatus combines an electrospray ionization source, two temperature controlled cryogenic ion traps and a time-of-flight infrared photofragmentation spectrometer, to allow for a universal and controlled formation and characterization of solvent clusters around ionic core as well as product of ion-molecule reaction.

Recent results on the spectroscopy of such solvated ions, will be presented and discussed. In particular, this talk will present the structural evolution of glycylglycine as a function of stepwise solvation, and show how the presence of just a few water can modify the geometry of this model peptide. I will also present results solvation of ion that do not form hydrogen bond or strongly interactions with the solvent.

WD08   11:04 – 11:19

VIBRATIONAL CHARACTERIZATION OF CATALYTIC INTERMEDIATES IN A DUAL CRYOGENIC ION TRAP
SPECTROMETER

ERIN M. DUFFY, JONATHAN VOSS, BRETT MARSH, ETIENNE GARAND, *Department of Chemistry, University of Wisconsin, Madison, WI, USA.*

Decades of research on water oxidation catalysis has yielded much progress in making water splitting a viable option for alternative energy. However, precise molecular-level understanding of the catalytic mechanism remains elusive due to the difficulty of studying reaction intermediates by traditional methods. In this talk, vibrational characterization of a ruthenium water oxidation catalyst and catalytic intermediates will be presented. In particular, infrared spectra acquired using a recently developed approach that employs two cryogenic ion traps, which enable the isolation of the chemical species discussed here, will be the focus of this talk.

WD09   11:21 – 11:36

MODELING AND OPTIMIZING RF MULTIPOLE ION TRAPS

SVEN FANGHAENEL, OSKAR ASVANY, *I. Physikalisches Institut, Universität zu Köln, Köln, Germany*; STEPHAN SCHLEMMER, *I. Physikalisches Institut, University of Cologne, Cologne, Germany.*

Radio frequency (rf) ion traps are very well suited for spectroscopy experiments thanks to the long time storage of the species of interest in a well defined volume. The electrical potential of the ion trap is determined by the geometry of its electrodes and the applied voltages. In order to understand the behavior of trapped ions in realistic multipole traps it is necessary to characterize these trapping potentials. Commercial programs like SIMION or COMSOL, employing the finite difference and/or finite element method, are often used to model the electrical fields of the trap in order to design traps for various purposes, e.g. introducing light from a laser into the trap volume. For a controlled trapping of ions, e.g. for low temperature trapping, the time dependent electrical fields need to be known to high accuracy especially at the minimum of the effective (mechanical) potential. The commercial programs are not optimized for these applications and suffer from a number of limitations. Therefore, in our approach the boundary element method (BEM) has been employed in home-built programs to generate numerical solutions of real trap geometries, e.g. from CAD drawings. In addition the resulting fields are described by appropriate multipole expansions. As a consequence, the quality of a trap can be characterized by a small set of multipole parameters which are used to optimize the trap design. In this presentation a few example calculations will be discussed. In particular the accuracy of the method and the benefits of describing the trapping potentials via multipole expansions will be illustrated. As one important application heating effects of cold ions arising from non-ideal multipole fields can now be understood as a consequence of imperfect field configurations.

WE. (Hyper)fine structure, tunneling
Wednesday, June 22, 2016 – 8:30 AM
Room: 217 Noyes Laboratory

Chair: Trevor Sears, Brookhaven National Laboratory, Upton, NY, USA

WE01   8:30 – 8:45

IODINE: MANY ELECTRONS AND MUCH TO DISCUSS... THE NUCLEAR QUADRUPOLE COUPLING, NUCLEAR SPIN-ROTATION, CONFORMATIONAL ANALYSIS, AND STRUCTURAL DETERMINATION OF 2-IODOBUTANE

ERIC A. ARSENAULT, YOON JEONG CHOI, DANIEL A. OBENCHAIN, *Department of Chemistry, Wesleyan University, Middletown, CT, USA*; S. A. COOKE, *Natural and Social Science, Purchase College SUNY, Purchase, NY, USA*; THOMAS A. BLAKE, *Chemical Physics, Pacific Northwest National Laboratory, Richland, WA, USA*; STEWART E. NOVICK, *Department of Chemistry, Wesleyan University, Middletown, CT, USA.*

The rotational spectrum of 2-iodobutane (sec-butyl-iodide) has been collected from 5.5-16.5 GHz using jet-pulsed Fourier transform microwave spectroscopy on both broadband[a] and Balle-Flygare cavity[b] instruments. Transitions belonging to three unique conformers were observed, namley the gauche-, anti-, and gauche'- species. All four ^{13}C isotopologues of the gauche-2-iodobutane were observed. The complete nuclear quadrupole coupling tensor of iodine has been determined for all conformers and ^{13}C isotopologues. A comparison between these nuclear quadrupole coupling tensors and those of similar iodine-containing molecules will be presented. Changes in the quadrupole coupling of iodine upon isotopic substitution will also be discussed. Additionally, isotopic substitution in conjunction with *ab initio* calculations allowed for both an r_s and r_0 structural analysis of gauche-2-iodobutane.

---

[a]Brown, G. G.; Dian, B. C.; Douglass, K. O.; Geyer, S. M.; Shipman, S. T.; Pate, B. H. *Review of Scientific Instruments* **2008**, *79*, 053103.
[b]Balle, T.; Flygare, W. *Review of Scientific Instruments* **1981**, *52*, 33-45.

WE02   8:47 – 9:02

A STUDY OF THE CONFORMATIONAL ISOMERISM OF 1-IODOBUTANE BY MICROWAVE SPECTROSCOPY

ERIC A. ARSENAULT, DANIEL A. OBENCHAIN, *Department of Chemistry, Wesleyan University, Middletown, CT, USA*; S. A. COOKE, *Natural and Social Science, Purchase College SUNY, Purchase, NY, USA*; THOMAS A. BLAKE, *Chemical Physics, Pacific Northwest National Laboratory, Richland, WA, USA*; STEWART E. NOVICK, *Department of Chemistry, Wesleyan University, Middletown, CT, USA.*

The rotational spectrum of 1-iodobutane was measured in a frequency range of 7-13 GHz, revealing a dense set of rotational transitions. Over 400 of the observed transitions were assigned to three different low energy conformational isomers. A previous low resolution microwave study[a] of 1-haloalkanes, including 1-iodobutane, confirmed that the three conformers present are the anti-anti, gauche-anti, and gauche-gauche species. From this high resolution study, the complete nuclear quadrupole coupling tensor of iodine was determined for each conformer. Rotational, centrifugal distortion, nuclear spin-rotation coupling constants will be discussed. Nuclear quadrupole coupling constants will also be presented and compared to other iodoalkane species.

gauche-gauche   gauche-anti   anti-anti

---

[a]Steinmetz, W. E.; Hickernell, F.; Mun, I. K.; Scharpen, L. H. *J. Mol. Spectrosc.* **1977**, *68*, 173-182.

WE03   9:04 – 9:19

^{14}N QUADRUPOLE COUPLING IN THE MICROWAVE SPECTRA OF N-VINYLFORMAMIDE

RAPHAELA KANNENGIEßER, WOLFGANG STAHL, *Institute for Physical Chemistry, RWTH Aachen University, Aachen, Germany*; <u>HA VINH LAM NGUYEN</u>, *Laboratoire Interuniversitaire des Systèmes Atmosphériques (LISA), CNRS et Universités Paris Est et Paris Diderot, Créteil, France*; WILLIAM C. BAILEY, *Department of Chemistry-Physics, Kean University (Retired), Union, NJ, USA.*

The microwave spectra of two conformers, *trans* and *cis*, of the title compound were recorded using two molecular beam Fourier transform microwave spectrometers operating in the frequency range 2 GHz to 40 GHz, and aimed at analysis of their ^{14}N quadrupole hyperfine structures. Rotational constants, centrifugal distortion constants, and nuclear quadrupole coupling constants (NQCCs) χ_{aa} and χ_{bb} - χ_{cc}, were all determined with very high accuracy. Two fits including 176 and 117 hyperfine transitions were performed for the *trans* and *cis* conformers, respectively. Standard deviations of both fits are close to the measurement accuracy of 2 kHz. The NQCCs of the two conformers are almost exactly the same, and are compared with values found for other saturated and unsaturated formamides.

Complementary quantum chemical calculations - MP2/6-311++G(d,p) rotational constants, MP2/cc-pVTZ centrifugal distortion constants, and B3PW91/6-311+G(d,p)//MP2/6-311++G(d,p) nuclear quadrupole coupling constants - give spectroscopic parameters in excellent agreement with the experimental parameters. B3PW91/6-311+G(d,p) calculated electric field gradients, in conjunction with eQ/h = 4.599(12) MHz/a.u., yields more reliable NQCCs for formamides possessing conjugated π-electron systems than does the B3PW91/6-311+G(df,pd) model recommended in Ref. [a], whereas this latter performs better for aliphatic formamides.[b] We conclude from this that f-polarization functions on heavy atoms hinder rather than help with modeling of conjugated π-electron systems.

[a] W. C. Bailey, *Chem. Phys.*, **2000**, *252*, 57.

[b] W. C. Bailey, Calculation of Nuclear Quadrupole Coupling Constants in Gaseous State Molecules, http://nqcc.wcbailey.net/index.html.

WE04   9:21 – 9:36

SOLVING THE TAUTOMERIC EQUILIBRIUM OF PURINE THROUGH THE ANALYSIS OF THE COMPLEX HYPERFINE STRUCTURE OF THE FOUR ^{14}N NUCLEI

<u>EMILIO J. COCINERO</u>, ICIAR URIARTE, PATRICIA ECIJA, *Physical Chemistry Department, Universidad del País Vasco (UPV/EHU), Bilbao, Spain*; LAURA B. FAVERO, *Istituto per lo Studio dei Materiali Nanostrutturati, Consiglio Nazionale delle Ricerche (ISMN-CNR), Bologna, Italy*; LORENZO SPADA, CAMILLA CALABRESE, WALTHER CAMINATI, *Dep. Chemistry 'Giacomo Ciamician', University of Bologna, Bologna, Italy.*

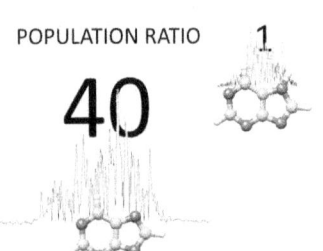

POPULATION RATIO

Microwave spectroscopy has been restricted to the investigation of small molecules in the last years. However, with the advent of FTMW[a,b] and CP-FTMW[c] spectroscopies coupled with laser vaporization techniques[d] it has turned into a very competitive methodology in the studies of moderate-size biomolecules.

Here, we present the study of purine, characterized by two aromatic rings, one six- and one five-membered, fused together to give a planar aromatic bicycle. Biologically, it is the mainframe of two of the five nucleobases of DNA and RNA. Two tautomers were observed by FTMW spectroscopy coupled to UV ultrafast laser vaporization system. The population ratio of the two main tautomers [N(7)H]/[N(9)H] is about 1/40 in the gas phase. It contrasts with the solid state where only the N(7)H species is present, or in solution where a mixture of both tautomers is observed. For both species, a full quadrupolar hyperfine analysis has been performed. This has led to the determination of the full sets of diagonal quadrupole coupling constants of the four ^{14}N atoms, which have provided crucial information for the unambiguous identification of both species.

[a] T. J. Balle and W. H. Flygare *Rev. Sci. Instrum.* **52**, 33-45, 1981.

[b] J.-U. Grabow, W. Stahl and H. Dreizler *Rev. Sci. Instrum.* **67**, 4072-4084, 1996.

[c] G. G. Brown, B. D. Dian, K. O. Douglass, S. M. Geyer, S. T. Shipman and B. H. Pate *Rev. Sci. Instrum.* **79**, 0531031/1-053103/13, 2008.

[d] E. J. Cocinero, A. Lesarri, P. Écija, F. J. Basterretxea, J. U. Grabow, J. A. Fernández and F. Castaño *Angew. Chem. Int. Ed.* **51**, 3119-3124, 2012.

WE05

CP-FTMW SPECTRUM OF BROMOPERFLUOROACETONE

FRANK E MARSHALL, NICOLE MOON, THOMAS D. PERSINGER, DAVID JOSEPH GILLCRIST, G. S. GRUBBS II, *Department of Chemistry, Missouri University of Science and Technology, Rolla, MO, USA.*

The microwave spectrum of the molecule bromoperfluoroacetone has been measured on a CP-FTMW spectrometer in the 6-18. The spectra is dense with approximately one transition every 5 MHz on average. Rotational constants, centrifugal distortion parameters, and nuclear quadrupole coupling constants will be discussed. Comparisons to the previously studied halogen analogues perfluoroacetone[a] and chloroperfluoroacetone[b] along with a family of previously studied halogenated acetone species will be discussed.

[a]J.-U. Grabow, N. Heineking, and W. Stahl, *Z. Naturforsch.* **46a** (1991) 229.
[b]G. Kadiwar, C. T. Dewberry, G. S. Grubbs II and S. A. Cooke, Talk **RH11**, 65th International Symposium on Molecular Spectroscopy (2010).

Intermission

WE06

ROTATIONAL SPECTROSCOPY OF CF_2ClCCl_3 AND ANALYSIS OF HYPERFINE STRUCTURE FROM FOUR QUADRUPOLAR NUCLEI

ZBIGNIEW KISIEL, EWA BIAŁKOWSKA-JAWORSKA, ON2, *Institute of Physics, Polish Academy of Sciences, Warszawa, Poland*; ICIAR URIARTE, FRANCISCO J. BASTERRETXEA, EMILIO J. COCINERO, *Departamento de Química Física, Universidad del País Vasco (UPV-EHU), Bilbao, Spain.*

CF_2ClCCl_3 has recently been identified among several new ozone- depleting substances in the atmosphere.[a] There are no literature reports concerning rotational spectroscopy of this molecule, although we were recently able to report its first chirped pulse, supersonic expansion spectrum.[b] CF_2ClCCl_3 has a rather small dipole moment so that the spectrum is weak and each transition displays very complex nuclear quadrupole hyperfine structure resulting from the presence of four chlorine nuclei.

We have presently been able to carry out a complete analysis of the hyperfine structure by combining the information from chirped pulse spectra with dedicated higher resolution measurements made with a cavity supersonic expansion instrument. The hyperfine analysis was carried out with Pickett's SPFIT/SPCAT package and the sizes of Hamiltonian matrices are sufficiently large to require the use of 64-bit compilation of these programs (made available for both Windows and Linux systems on the PROSPE website). The resulting fit is to within experimental accuracy and is supported by *ab initio* calculations. The precise values of off-diagonal hyperfine constants for all nuclei lead to useful angular information that is complementary to direct structural information from moments of inertia.[c]

[a]J.C.Laube, M.J.Newland, C.Hogan, et al., *Nature Geoscience* **7**, 266 (2014).
[b]Z.Kisiel, E.Białkowska-Jaworska, L.Pszczółkowski, I.Uriarte, P.Ejica, F.J.Basterretxea, E.J.Cocinero, 70th ISMS, Champaign-Urbana, Illinois, RF-11 (2015).
[c]Z.Kisiel, E.Białkowska-Jaworska, L.Pszczółkowski, *J.Chem.Phys.* **109**, 10263 (1998).

WE07

AB INITIO CALCULATIONS OF SPIN-ORBIT COUPLING FOR HEAVY-METAL CONTAINING RADICALS

LAN CHENG, *Department of Chemistry, Johns Hopkins University, Baltimore, MD, USA.*

The perturbative treatment of spin-orbit coupling (SOC) on top of scalar-relativistic calculations is a cost-effective alternative to rigorous fully relativistic calculations. In this work the applicability of the perturbative scheme in the framework of spin-free exact two-component theory is demonstrated with calculations of SO splittings and SOC contributions to molecular properties in small heavy-metal containing radicals, including AuO, AuS, and ThO^+. The equation of motion coupled cluster techniques have been used to accurately account for the electron-correlation effects in these radicals, and basis-set effects are carefully analyzed. The computed results are compared with experimental measurements for SO splittings and dipole moments when available.

170

WE08 10:46 – 11:01

MOLECULAR BEAM OPTICAL ZEEMAN SPECTROSCOPY OF VANADIUM MONOXIDE, VO[a]

TRUNG NGUYEN, RUOHAN ZHANG, TIMOTHY STEIMLE, *School of Molecular Sciences, Arizona State University, Tempe, AZ, USA.*

Like almost all astronomical studies, exoplanet investigations are observational endeavors that rely primarily on remote spectroscopic sensing to infer the physical properties of planets. Most exoplanet related information is inferred from to temporal variation of luminosity of the parent star. An effective method of monitoring this variation is via Magnetic Doppler Imaging (MDI)[b], which uses optical polarimetry[c] of paramagnetic molecules or atoms. One promising paramagnetic stellar absorption is the near infrared spectrum of VO[d]. With this in mind, we have begun a project to record and analyze the field-free and Zeeman spectrum of the band. A cold (approx. 20 K) beam of VO was probed with a single frequency laser and detected using laser induced fluorescence. The determined spectral parameters will be discussed and compared to those extracted from the analysis of a hot spectrum[e].

[a]Supported by the National Science Foundation under the Grant No. CHE-1265885.
[b]O. Kochukhov, N. Rusomarov, J. A. Valenti, H. C. Stempels, F. Snik, M. Rodenhuis, N. Piskunov, V. Makaganiuk, C. U. Keller and C. M. Johns-Krull, Astron. Astrophys. 574 (Pt. 2), A79/71-A79/12 (2015).
[c]S. V. Berdyugina, Astron. Soc. Pac. Conf. Ser. 437 (Solar Polarization 6), 219-235 (2011).
[d]S. V. Berdyugina, P. A. Braun, D. M. Fluri and S. K. Solanki, Astron. Astrophys. 444 (3), 947-960 (2005).
[e]A. S. C. Cheung, P. G. Hajigeorgiou, G. Huang, S. Z. Huang and A. J. Merer, J. Mol. Spectrosc. 163 (2), 443-458 (1994).

WE09 11:03 – 11:18

MOLECULAR BEAM OPTICAL STUDY OF GOLD SULFIDE AND GOLD OXIDE[a]

RUOHAN ZHANG, YUANQIN YU, TIMOTHY STEIMLE, *School of Molecular Sciences, Arizona State University, Tempe, AZ, USA.*

Gold-sulfur and gold-oxygen bonds are key components to numerous established and emerging technologies that have applications as far ranging as medical imaging, catalysis, electronics, and material science. A major theoretical challenge for describing this bonding is correctly accounting for the large relativistic and electron correlation effects. Such effects are best studied in diatomic, AuX, molecules. Recently, the observed AuS electronic state energy ordering was measured and compared to a simple molecular orbital diagram prediction[b]. Here we more thoroughly investigate the nature of the electronic states of both AuS and AuO from the analysis of high-resolution (FWHM$\cong$ 35MHz) optical Zeeman spectroscopy of the $(0,0)B^2\Sigma^- - X^2\Pi_{3/2}$ bands. The determined fine and hyperfine parameters for the $B^2\Sigma^-$ state of AuO differ from those extracted from the analysis of a hot, Doppler-limited, spectrum[c]. It is demonstrated that the nature of the $B^2\Sigma^-$ states of AuO and AuS are radically different. The magnetic tuning of AuO and AuS indicates that the $B^2\Sigma^-$ states are heavily contaminated.

[a]Supported by the National Science Foundation under Grant No.1265885.
[b]D. L. Kokkin, R. Zhang, T. C. Steimle, I. A. Wyse, B. W. Pearlman and T. D. Varberg, *J. Phys. Chem. A.*, **119(48)**, 4412, 2015.
[c]L. C. O'Brien, B. A. Borchert, A. Farquhar, S. Shaji, J. J. O'Brien and R. W. Field, *J. Mol. Spectrosc.*, **252(2)**, 136, 2008.

WE10 **11:20 – 11:35**

HYPERFINE SPLITTINGS IN THE NEAR-INFRARED SPECTRUM OF $^{14}NH_3$

SYLVESTRE TWAGIRAYEZU, TREVOR SEARS[a], GREGORY HALL, *Division of Chemistry, Department of Energy and Photon Sciences, Brookhaven National Laboratory, Upton, NY, USA.*

Sub-Doppler, saturation dip, measurements of transitions in the $\nu_1 + \nu_3$ band of $^{14}NH_3$ have been made by frequency comb-referenced diode laser absorption spectroscopy. The observed spectra exhibit either resolved or partially-resolved hyperfine splittings that are primarily determined by the ^{14}N quadrupole coupling in the molecule. Modeling of the line shapes based on the known hyperfine level structure of the ground state of the molecule shows that, in nearly all cases, the upper state level has splittings similar to that of the same rotational level in the ground state. The data provide accurate frequencies for the line positions and the observed hyperfine splittings can be used to make or confirm rotational assignments. Of all the measurements, one transition, $^pP(5,4)_a$ at 195 994.73457 GHz, exhibits hyperfine structure which does not conform to that expected based on extrapolation from the known lower state hyperfine splittings. Examination of the known vibration-rotation level structure near the upper state energy shows that there exists a near degeneracy between this level and one in the $\nu_1 + 2\nu_4$ manifold which is of the appropriate symmetry to be mixed by magnetic hyperfine terms that couple ortho- and para- modifications of the molecule. It is possible that the unusual hyperfine splittings are a consequence of ortho-paro mixing, which has been predicted, but not previously seen in ammonia and further experimental measurements to investigate this possibility are ongoing.

Acknowledgments: Work at Brookhaven National Laboratory was carried out under Contract No. DE-SC0012704 with the U.S. Department of Energy, Office of Science, and supported by its Division of Chemical Sciences, Geosciences and Biosciences within the Office of Basic Energy Sciences.

---

[a]Also at: *Chemistry Department, Stony Brook University, Stony Brook, NY 11794, USA*

WE11 **11:37 – 11:52**

HYPERFINE STRUCTURE IN ROTATIONAL SPECTRA OF DEUTERATED MOLECULES: THE HDS AND ND_3 CASE STUDIES

GABRIELE CAZZOLI, CRISTINA PUZZARINI, *Dep. Chemistry 'Giacomo Ciamician', University of Bologna, Bologna, Italy.*

The determination of hyperfine parameters (quadrupole-coupling, spin-spin coupling, and spin-rotation constants) is one of the aims of high-resolution rotational spectroscopy. These parameters are relevant not only from a spectroscopic point of view, but also from a physical and/or chemical viewpoint, as they might provide detailed information on the chemical bond, structure, etc. In addition, the hyperfine structure of rotational spectra is so characteristic that its analysis may help in assigning the spectra of unknown species. In astronomical observations, hyperfine structures of rotational spectra would allow us to gain information on column densities and kinematics, and the omission of taking them into account can lead to a misinterpretation of the line width of the molecular emission lines. Nevertheless, the experimental determination of hyperfine constants can be a challenge not only for actual problems in resolving hyperfine structures themselves, but also due to the lack of reliable estimates or the complexity of the hyperfine structure itself. It is thus important to be able to rely on good predictions for such parameters, which can nowadays be provided by quantum-chemical calculations. In fact, the fruitful interplay of experiment and theory will be demonstrated by means of two study cases: the hyperfine structure of the rotational spectra of HDS and ND_3.

From an experimental point of view, the Lamb-dip technique has been employed to improve the resolving power in themillimeter- and submillimeterwave frequency range by at least one order of magnitude, thus making it possible to perform sub-Doppler measurements as well as to resolve narrow hyperfine structures. Concerning theory, it will be demonstrated that high-level calculations can provide quantitative estimates for hyperfine parameters (quadrupole coupling constants, spin-rotation tensors, spin-spin couplings, etc.) and shown how theoretical predictions are often essential for a detailed analysis of the hyperfine structure of the recorded spectra.

WF. Mini-symposium: Spectroscopy of Large Amplitude Motions

Wednesday, June 22, 2016 – 1:30 PM

Room: 100 Noyes Laboratory

Chair: Jon T. Hougen, National Institute of Standards and Technology, Gaithersburg, MD, USA

WF01   1:30 – 1:45

THE ROLE OF SYMMETRIC-STRETCH VIBRATION IN ASYMMETRIC-STRETCH VIBRATIONAL FREQUENCY SHIFT: THE CASE OF 2CH EXCITATION INFRARED SPECTRA OF ACETYLENE-HYDROGEN VAN DER WAALS COMPLEX

DAN HOU, YONG-TAO MA, XIAO-LONG ZHANG, YU ZHAI, <u>HUI LI</u>, *Institute of Theoretical Chemistry, Jilin University, Changchun, China.*

Direct infrared spectra predictions for van der Waals (vdW) complexes rely on accurate intra-molecular vibrationally excited inter-molecular potential. Due to computational cost increasing with number of freedom, constructing an effective reduced-dimension potential energy surface, which only includes direct relevant intra- molecular modes, is the most feasible way and widely used in the recent potential studies. However, because of strong intra-molecular vibrational coupling, some indirect relevant modes are also play important roles in simulating infrared spectra of vdW complexes. The questions are how many intra-molecular modes are needed, and which modes are most important in determining the effective potential and direct infrared spectra simulations. Here, we explore these issues using a simple, flexible and efficient vibration-averaged approach, and apply the method to vdW complex $C_2H_2 - H_2$. With initial examination of the intra-molecular vibrational coupling, an effective seven-dimensional *ab initio* potential energy surface(PES) for $C_2H_2 - H_2$, which explicitly takes into account the Q_1, Q_2 symmetric-stretch and Q_3 asymmetric-stretch normal modes of the C_2H_2 monomer, has been generated. Analytic four-dimensional PESs are obtained by least-squares fitting vibrationally averaged interaction energies for $\nu_3(C_2H_2)=0$ and 1 to the Morse/long-range(MLR) potential function form. We provide the first prediction of the infrared spectra and band origin shifts for $C_2H_2 - H_2$ dimer. We particularly examine the dependence of the symmetric-stretch normal mode on asymmetric-stretch frequency shift for the complex.

WF02   1:47 – 2:02

HALOGEN BONDING VS HYDROGEN BONDING IN CHF_2I COMPLEXES WITH NH_3 AND $N(CH_3)_3$

<u>CHRIS MEDCRAFT</u>, *School of Chemistry, Newcastle University, Newcastle-upon-Tyne, United Kingdom;* YANNICK GEBOES, *Department of Chemistry, University of Antwerpen, Antwerpen, Belgium;* ANTHONY LEGON, *School of Chemistry, University of Bristol, Bristol, United Kingdom;* NICK WALKER, *School of Chemistry, Newcastle University, Newcastle-upon-Tyne, United Kingdom.*

Ammonia and trimethylamine (TMA) were used to probe preference of hydrogen over halogen bonding in molecular complexes containing CHF_2I via chirped pulse Fourier transform microwave spectroscopy. The halogen bonded complex of TMA is ≈ 2 kJ/mol more energetically favourable (extrapolation to CCSD(T)/CBS level) than the hydrogen bonded complex. The reverse is true for the ammonia complex where the hydrogen bonded complex is ≈ 3kJ/mol more favourable. Although the spectra of both complexes were perturbed by large amplitude motions around the intermolecular bond effective fits of the lower rotational energy levels appear to confirm that TMA prefers to bind to the iodine whilst ammonia prefers the hydrogen.

WF03 2:04 – 2:19

THE CURIOUS CASE OF PYRIDINE - WATER

BECCA MACKENZIE, CHRIS DEWBERRY, CJ SMITH, *Chemistry Department, University of Minnesota, Minneapolis, MN, USA*; RYAN D. CORNELIUS, *Chemistry Department, St. Cloud State University , St. Cloud, MN, USA*; KEN LEOPOLD, *Chemistry Department, University of Minnesota, Minneapolis, MN, USA.*

The rotational spectrum of the pyridine$\cdots$water complex has been observed in the 2-18 GHz region using chirped-pulse and cavity Fourier transform microwave spectroscopy. The water is hydrogen bonded to the nitrogen, as expected, but the hydrogen bond is bent, with the oxygen tilted toward either of the ortho hydrogens of the pyridine. This gives rise to a pair of equivalent configurations and the possibility of a tunneling motion involving an in-plane rocking of the water. DFT calculations support this view. Experimentally, a pair of states with severely perturbed rotational structure has been identified and the spectra assigned. Analysis of the perturbations in the a-type (pure rotation) spectra has enabled an accurate determination of the tunneling splitting, which has been confirmed by direct observation of b-type (rotation-tunneling) transitions. A simultaneous fit of the a- and b- type transitions gives the most accurate value of the tunneling splitting. Results for the H_2O, D_2O, and D-bound HOD complexes will be presented. The tunneling splittings are as follows: H_2O-pyridine (10402.9 MHz), HOD-pyridine (12513.4 MHz, determined only from perturbation analysis), and D_2O-pyridine (13582.3 MHz). Curiously, the tunneling splitting increases with increased deuteration. Additional small splittings have been observed in some transitions, indicating the possibility of further internal dynamics. This system offers an interesting test case for theoretical treatments of large amplitude motion.

WF04 2:21 – 2:36

LABORATORY ROTATIONAL SPECTRUM AND ASTRONOMICAL SEARCH OF S-METHYL THIOFORMATE

ATEF JABRI, ISABELLE KLEINER, *Laboratoire Interuniversitaire des Systèmes Atmosphériques (LISA), CNRS et Universités Paris Est et Paris Diderot, Créteil, France*; R. A. MOTIYENKO, L. MARGULÈS, *Laboratoire PhLAM, UMR 8523 CNRS - Université Lille 1, Villeneuve d'Ascq, France*; J.-C. GUILLEMIN, *UMR 6226 CNRS - ENSCR, Institut des Sciences Chimiques de Rennes, Rennes, France*; E. A. ALEKSEEV, *Radiospectrometry Department, Institute of Radio Astronomy of NASU, Kharkov, Ukraine*; BELÉN TERCERO, JOSE CERNICHARO, *Departamento de Astrofísica, Centro de Astrobiología CAB, CSIC-INTA, Madrid, Spain.*

Methyl thioformate $CH_3SC(O)H$, is a monosulfur derivative of methyl formate, a relatively abundant component of the interstellar medium (ISM)[a]. Methyl thioformate being the thermodynamically most stable isomer with a C_2H_4OS formula, it can be reasonably proposed for detection in the ISM. Theoretical investigations on this molecule have been done recently by Senent et al.[b]. Previous experimental studies on this molecule have been performed by Jones et al.[c] and Caminati et al.[d] and its microwave spectrum was recorded between 10 and 41 GHz.
In this study, S-methyl thioformate has been synthesized by reaction of methyl mercaptan with formic-acetic anhydride. The millimeter wave spectrum was then recorded for the first time from 150 to 660 GHz with the Lille's spectrometer based on solid-state sources [e]. Around 2300 lines were assigned up to $J = 70$ and $K = 15$ and a fit for the ground torsional state $\nu_t = 0$ performed with the *BELGI-C$_s$* code[f] will be presented and discussed. Our aim is to provide a line list for an astrophysical research.

[a]E. Chruchwell, G. Winnewisser, A&A, 45, 229 (1975)

[b]M. L. Senent, C. Puzzarini, M. Hochlaf, R. Domínguez-Gómez and M. Carvajal, J. Chem. Phys., 141, 104303 (2014)

[c]G. I. L. Jones, D. G. Lister and N. L. Owen, J. Mol. Spectrosc., 60, 348 (1976)

[d]W. Caminati, B. P. V. Eijck and D. G. Lister, J. Mol. Spectrosc., 90, 15 (1981)

[e]O. Zakharenko, R. A. Motiyenko, L. Margulès, T. R. Huet., J. Mol. Spectrosc., 317, 41 (2015)

[f]J. T. Hougen, I. Kleiner, and M. Godefroid, J. Mol. Spectrosc. 163, 559 (1994)

WF05 2:38 – 2:53

LOW BARRIER METHYL ROTATION IN 3-PENTYN-1-OL AS OBSERVED BY MICROWAVE SPECTROSCOPY

KONRAD EIBL, RAPHAELA KANNENGIEßER, WOLFGANG STAHL, *Institute for Physical Chemistry, RWTH Aachen University, Aachen, Germany*; HA VINH LAM NGUYEN, ISABELLE KLEINER, *Laboratoire Interuniversitaire des Systèmes Atmosphériques (LISA), CNRS et Universités Paris Est et Paris Diderot, Créteil, France.*

It is known that the barrier to internal rotation of the methyl groups in ethane (**1**) is about 1000 cm^{-1}.[a] If a C-C-triple bond is inserted between the methyl groups as a spacer (**2**), the torsional barrier is assumed to be dramatically lower, which is a common feature of ethinyl groups in general.

To study this effect of almost free internal rotation, we measured the rotational spectrum of 3-pentyn-1-ol (**3**) by pulsed jet Fourier transform microwave spectroscopy in the frequency range from 2 to 26.5 GHz. Quantum chemical calculations at the MP2/6-311++G(d,p) level of theory yielded five stable conformers on the potential energy surface. The most stable conformer, which possesses C$_1$ symmetry, was assigned and fitted using two theoretical approaches treating internal rotations, the rho axis method (*BELGI-C$_1$*) and the combined axis method (*XIAM*). The molecular parameters as well as the internal rotation parameters were determined. A very low barrier to internal rotation of the methyl group of only 9.4545(95) cm^{-1} was observed.

[a]R. M. Pitzer, *Acc. Chem. Res.*, **1983**, *16*, 207–210.

$$H_3C—CH_3 \qquad H_3C—C≡C—CH_3 \qquad H_3C—C≡C—CH\begin{smallmatrix}OH\\CH_3\end{smallmatrix}$$

(**1**) (**2**) (**3**)

WF06 2:55 – 3:10

MILLIMETER WAVE SPECTRA OF METHYL CYANATE, METHOXYAMINE AND N-METHYLHYDROXYLAMINE: LABORATORY STUDIES AND ASTRONOMICAL SEARCH IN SPACE

LUCIE KOLESNIKOVÁ, JOSÉ L. ALONSO, CELINA BERMÚDEZ, E. R. ALONSO, *Grupo de Espectroscopia Molecular, Lab. de Espectroscopia y Bioespectroscopia, Unidad Asociada CSIC, Universidad de Valladolid, Valladolid, Spain*; BELÉN TERCERO, JOSE CERNICHARO, *Molecular Astrophysics, ICMM, Madrid, Spain*; J.-C. GUILLEMIN, *Institut des Sciences Chimiques de Rennes, UMR 6226 CNRS - ENSCR, Rennes, France.*

Recent discovery of methyl isocyanate (CH$_3$NCO) in Sgr B2(N) and Orion KL[a,b] makes methyl cyanate (CH$_3$OCN) a potential molecule in the interstellar medium. Methoxyamine (CH$_3$ONH$_2$) and its isomeric form N-methylhydroxylamine (CH$_3$NHOH) may be considered as a potential interstellar amines.[c] Pure rotational transitions belonging to the ground state and several excited vibrational states were measured and analyzed up to 400 GHz. Rotational transitions revealed *A-E* splitting due to the methyl internal rotation and were globally analyzed in order to provide a precise set of the spectroscopic constants. Results of this work were used to search for the spectral features of methyl cyanate in Orion KL, Sgr B2(N), B1-b and TMC-1 molecular clouds.[d]

[a]D. T. Halfen, V. V. Ilyushin, L. Ziurys, *Astrophys. J. Lett.*, **2015**, *812*, L5.
[b]J. Cernicharo, Z. Kisiel, B. Tercero, L. Kolesniková, I. R. Medvedev et. al., *Astron. & Astrophys.*, **2016**, *587*, L4.
[c]R. T. Garrod, S. L. Widicus Weaver, E. Herbst, *Astrophys. J.*, **2008**, *682*, 283.
[d]L. Kolesniková, J. L. Alonso, C. Bermúdez, E. R. Alonso, J. Cernicharo et. al., *Astron. & Astrophys.*, **2016**, accepted.

WF07 3:12 – 3:27

FURTHER ANALYSIS OF THE LABORATORY ROTATIONAL SPECTRUM OF CH_3NCO

ZBIGNIEW KISIEL, *ON2, Institute of Physics, Polish Academy of Sciences, Warszawa, Poland*; LUCIE KOLESNIKOVÁ, E. R. ALONSO, JOSÉ L. ALONSO, *Grupo de Espectroscopia Molecular, Lab. de Espectroscopia y Bioespectroscopia, Unidad Asociada CSIC, Universidad de Valladolid, Valladolid, Spain*; MANFRED WINNEWISSER, FRANK C. DE LUCIA, *Department of Physics, The Ohio State University, Columbus, OH, USA*; IVAN MEDVEDEV, *Department of Physics, Wright State University, Dayton, OH, USA*; BELÉN TERCERO, JOSE CERNICHARO, *Departamento de Astrofísica, Centro de Astrobiología CAB, CSIC-INTA, Madrid, Spain*; J.-C. GUILLEMIN, *Institut des Sciences Chimiques de Rennes, UMR 6226 CNRS - ENSCR, Rennes, France.*

Identification by the Rosetta mission that CH_3NCO is among the more plentiful molecules on the surface of the comet Churyumov-Gerasimenko stimulated rapid detection of this molecule in the interstellar medium.[a,b] In particular, we have been successful in detecting almost 400 lines of CH_3NCO in Orion[b] by extending the Koput[c] cm-wave assignment to frequencies relevant to mm-wave radio-telescopes through measurement of the complete laboratory spectrum up to 363 GHz.[b,d]

Presently, we describe further progress in understanding the laboratory rotational spectrum of CH_3NCO. Assignment has been extended to transitions with $K > 3$ by analysis of Stark and hyperfine patterns of the corresponding lowest-J transitions. Broadband spectra of synthezised pure $^{13}CH_3NCO$ and $CH_3N^{13}CO$ isotopic species have also been recorded and assigned. Furthermore, the progress in fitting this very low barrier and highly perturbed internal rotation spectrum is described.

[a]D.T.Halfen, V.V.Ilyushin, L.Ziurys, *ApJ* **812**, L5 (1915).

[b]J.Cernicharo, Z.Kisiel, B.Tercero, et al., *A&A* **587**, L4 (2016).

[c]J.Koput, *J. Mol. Spectrosc.* **115**, 131 (1986).

[d]Z.Kisiel et al., 65th ISMS, Columbus, Ohio, RC-13 (2010); 70th ISMS, Champaign-Urbana, Illinois, TG-08 (2015).

Intermission

WF08 3:46 – 3:56

MOLECULAR ELECTRONIC ENVIRONMENT FROM METHYL TORSION AND ^{14}N QUADRUPOLE COUPLING

RANIL M. GURUSINGHE, MICHAEL TUBERGEN, *Department of Chemistry and Biochemistry, Kent State University, Kent, OH, USA.*

Methyl rotors are sensitive indicators of the local and non-local electronic environment, so a study of methyl torsional barriers at different sites around the perimeter of a conjugated ring system may reveal electronic environment of the indole ring. Previously reported results[a] are limited to microwave spectroscopic identification of the 1-methylindole, 2-methylindole, 3-methylindole, and 5-methylindole. The new measurements of 4-methylindole, 6-methylindole, and 7-methylindole complete the series.

Over one thousand transitions were assigned in the 10.5 – 21 GHz frequency range, resolving both nuclear quadrupole coupling and methyl internal rotation for all seven species. Electronic structure calculations at MP2/6-311++G(d,p) level, torsional barrier calculations and molecular electrostatic potential maps at ωB97XD/6-311++G(d,p) level are used along with the experimental data. ^{14}N nuclear quadrupole coupling tensor elements were used to determine the ionic character of the NH sigma bond, π bonding character of the nitrogen p_z orbital, and the amount of negative charge on the nitrogen of each methylindole. Fitted torsional barriers were compared with theoretical investigations of the origin of methyl torsional barriers to confirm that overall barrier arises from both hyperconjugative interactions and structural interactions of bonding and anti-bonding orbitals.

[a]R. M. Gurusinghe, M. J. Tubergen. 69th International Symposium of Molecular Spectroscopy, Urbana-Champaign, IL, 2014, RJ03.

WF09 3:58 – 4:13

THz SPECTROSCOPY OF EXCITED TORSIONAL STATES OF MONODEUTERATED METHYL FORMATE ($DCOOCH_3$)

MIGUEL CARVAJAL, *Dpto. Fisica Aplicada, Unidad Asociada CSIC, Facultad de Ciencias Experimentales, Universidad de Huelva, Huelva, Spain*; CHUANXI DUAN, *College of Physical Science and Technology, Central China Normal University, Wuhan, China*; SHANSHAN YU, JOHN PEARSON, BRIAN DROUIN, *Jet Propulsion Laboratory, California Institute of Technology, Pasadena, CA, USA*; ISABELLE KLEINER, *Universités Paris Est et Paris Diderot , Laboratoire Interuniversitaire des systèmes atmosphériques (LISA), CNRS, Creteil, France.*

Recently, a measurement of the rotational spectrum of $DCOOCH_3$ has been carried out in the frequency range from 0.85 to 1.5 THz at Jet Propulsion Laboratory (JPL) (Duan et al. 2015). These JPL data and the available spectroscopic millimeter- and submillimeter-wave data (Margulès et al. 2010 and references therein) of the ground state were analyzed using the Rho Axis Method (RAM) (Kleiner 2010).

At present, a new analysis of JPL lines of $DCOOCH_3$ in the first excited $v_t=1$ torsional states is undertaken. This analysis may help the future identification of $v_t=1$ lines in the interstellar and circumstellar media as was carried out for the $v_t=0$ lines in Orion KL (Margulès et al. 2010). In this communication, the progress of this study is presented as well as a short outline of the spectral analyses accomplished for other methyl formate isotopologues. [a]

C. Duan, M. Carvajal, S. Yu, J.C. Pearson, B.J. Drouin, I. Kleiner 2015, A&A, 576, A39
I. Kleiner 2010, J. Mol. Spectroc., 260, 1
L. Margulès, T.R. Huet, J. Demaison, M. Carvajal, I. Kleiner, H. Møllendal, B. Tercero, N. Marcelino, J. Cernicharo 2010, ApJ, 714, 1120

[a]This research is supported by the FIS2014-53448-C2-2-P project (MINECO, Spain), the French PCMI (Programme National de Physique Chimie du Milieu Interstellaire), and the National Natural Science Foundation of China (Grant No. 11174098). Portions of this work is carried out at the Jet Propulsion Laboratory, California Institute of Technology, under contract with the National Aeronautics and Space Administration. Government sponsorship is acknowledged.

WF10 4:15 – 4:30

TWO EQUIVALENT METHYL INTERNAL ROTATIONS IN 2,5-DIMETHYLTHIOPHENE INVESTIGATED BY MICROWAVE SPECTROSCOPY

VINH VAN, WOLFGANG STAHL, *Institute for Physical Chemistry, RWTH Aachen University, Aachen, Germany*; HA VINH LAM NGUYEN, *Laboratoire Interuniversitaire des Systèmes Atmosphériques (LISA), CNRS et Universités Paris Est et Paris Diderot, Créteil, France.*

The microwave spectrum of 2,5-dimethylthiophene, a sulfur-containing five-membered heterocyclic molecule with two conjugated double bonds, was recorded using two molecular beam Fourier transform microwave spectrometers operating in the frequency range from 2 to 40 GHz. Highly accurate molecular parameters were determined. The rotational constants obtained by geometry optimizations at different levels of theory are in good agreement with the experimental values. A C_{2v} equilibrium structure was calculated, where one hydrogen atom of each methyl group is antiperiplanar to the sulfur atom, and the two methyl groups are thus equivalent.

Transition states were optimized at different levels of theory using the Berny algorithm to calculate the barrier height of the two equivalent methyl rotors. The fitted experimental torsional barrier of $247.95594(30)$ cm^{-1} is in reasonable agreement with the calculated barriers. Similar barriers to internal rotation were found for the monomethyl derivatives 2-methylthiophene (194.1 cm^{-1}) and 3-methylthiophene (258.8 cm^{-1}). A labeling scheme for the group G_{36} written as the semi-direct product $(C_3{}^I \times C_3{}^I) (\times C_{2v}$ was introduced.

WF11

PROBING THE METHYL TORSIONAL BARRIERS OF THE E AND Z ISOMERS OF BUTADIENYL ACETATE BY MICROWAVE SPECTROSCOPY

ATEF JABRI, HA VINH LAM NGUYEN, ISABELLE KLEINER, *Laboratoire Interuniversitaire des Systèmes Atmosphériques (LISA), CNRS et Universités Paris Est et Paris Diderot, Créteil, France*; VINH VAN, WOLF-GANG STAHL, *Institute for Physical Chemistry, RWTH Aachen University, Aachen, Germany.*

The Fourier transform microwave spectra of the E and the Z isomer of butadienyl acetate have been measured in the frequency range from 2 to 26.5 GHz under molecular beam conditions. The most stable conformer of each isomer, in which all heavy atoms are located in a symmetry plane, was identified after analyzing the spectrum by comparison with results from quantum chemical calculations. The barrier to internal rotation of the acetyl methyl group was found to be 149.1822(20) cm^{-1} and 150.2128(48) cm^{-1} for the E and the Z isomer, respectively, which are similar to that of vinyl acetate [a,b]. A comparison between two theoretical approaches treating internal rotations, the rho axis method (using the program *BELGI-C_s*) and combined axis method (using the program *XIAM*), is also performed.

Since several years we study the barriers to internal rotation of the acetyl methyl group in acetates, CH_3-COOR. Currently, we assume that all acetates can be divided into three classes. Class I contains α,β saturated acetates, where the torsional barrier is always close to 100 cm^{-1}. Examples are a series of alkyl acetates such as methyl acetate and ethyl acetate. Class II contains α,β-unsaturated acetates where the C=C double bond is located in the COO plane. This is the case of vinyl acetate and butadienyl acetate. Finally, in class III with isopropenyl acetate and phenyl acetate as two representatives, α,β-unsaturated acetates, in which the double bond is not located in the COO plane, are collected. There, we observed a barrier height around 135 cm^{-1}. This observation will be discussed in details.

[a] B. Velino, A. Maris, S. Melandri, W. Caminati, J. Mol. Spectrosc. 2009, 256, 228
[b] H. V. L. Nguyen, A. Jabri, V. Van, and W. Stahl, J. Phys. Chem. A, 2014, 118, 12130

WF12

PROGRESS IN THE ROTATIONAL ANALYSIS OF THE GROUND AND LOW-LYING VIBRATIONALLY EXCITED STATES OF MALONALDEHYDE

E. S. GOUDREAU, DENNIS W. TOKARYK, STEPHEN CARY ROSS, *Department of Physics, University of New Brunswick, Fredericton, NB, Canada*; BRANT E. BILLINGHURST, *EFD, Canadian Light Source Inc., Saskatoon, Saskatchewan, Canada.*

Despite being an important prototype molecule for intramolecular proton tunnelling, the far-IR spectrum of the internally hydrogen-bonded species malonaldehyde ($C_3O_2H_4$) is not yet well understood. In the talk I gave at the ISMS meeting in 2015 I discussed the high-resolution spectra we obtained at the Canadian Light Source synchrotron in Saskatoon, Saskatchewan. These spectra include a number of fundamental vibrational bands in the 100-2000 cm^{-1} region. In our efforts to analyze these bands we have noticed that our ground state combination differences show a large drift (up to an order of magnitude larger than our experimental error) away from those calculated using constants established by Baba *et al.*,[a] particularly in regions of high J (above 30) and low K_a (below 5). An examination of the previous microwave and far-IR studies[bc] reveals that this region of J-K_a space was not represented in the lines that Baba *et al.* used to generate the values for their fitting parameters. By including our own measurements in the fitting, we were able to improve the characterization of the ground state so that it is now consistent with all of the existing data. This characterization now covers a much larger range of J-K_a space and has enabled us to make significant progress in analyzing our far-IR synchrotron spectra. These include an excited vibrational state at 241 cm^{-1} as well as several states split by the tunnelling effect at higher wavenumber.

[a] T. Baba, T. Tanaka, I. Morino, K. M. T. Yamada, K. Tanaka. *Detection of the tunneling-rotation transitions of malonaldehyde in the submillimeter-wave region.* J. Chem. Phys., **110**. 4131-4133 (1999)
[b] P. Turner, S. L. Baughcum, S. L. Coy, Z. Smith. *Microwave Spectroscopic Study of Malonaldehyde. 4. Vibration-Rotation Interaction in Parent Species.* J. Am. Chem. Soc., **106**. 2265-2267 (1984)
[c] D. W. Firth, K. Beyer, M. A. Dvorak, S. W. Reeve, A. Grushow, K. R. Leopold. *Tunable far-infrared spectroscopy of malonaldehyde.* J. Chem. Phys., **94**. 1812-1819 (1991)

WF13

THE ORIGINS OF INTRA- AND INTER-MOLECULAR VIBRATIONAL COUPLINGS: A CASE STUDY OF $H_2O - Ar$ ON FULL AND REDUCED-DIMENSIONAL POTENTIAL ENERGY SURFACE

DAN HOU, YONG-TAO MA, XIAO-LONG ZHANG, HUI LI, *Institute of Theoretical Chemistry, Jilin University, Changchun, China.*

The origin and strength of intra- and inter-molecular vibrational coupling is difficult to probe by direct experimental observations. However, explicitly including or not including some specific intramolecular vibrational modes to study intermolecular interaction provides a precise theoretical way to examine the effects of anharmonic coupling between modes. In this work, a full-dimension intra- and inter-molecular *ab initio* potential energy surface (PES) for $H_2O - Ar$, which explicitly incorporates interdependence on the intramolecular normal-mode coordinates of the H_2O monomer, has been calculated. In addition, four analytic vibrational-quantum-state-specific PESs are obtained by least-squares fitting vibrationally averaged interaction energies for the (ν_1,ν_2,ν_3)=(0,0,0),(0,0,1),(1,0,0),(0,1,0) states of H_2O to the three-dimensional Morse/long-range potential function. The resulting vibrationally averaged PESs provide good representations of the experimental infrared data, with RMS discrepancies smaller than 0.02 cm^{-1} for all three rotational branches of the asymmetric stretch fundamental transitions. The infrared band origin shifts associated with three fundamental bands of H_2O in $H_2O - Ar$ complex are predicted for the first time and are found to be in good agreement with the (extrapolated) experimental values. Upon introduction of additional intramolecular degrees of freedom into the intermolecular potential energy surface, there is clear spectroscopic evidence of intra- and intermolecular vibrational couplings.[a]

---

[a] *J. Chem. Phys.*, **144**, 014301 (2016)

WF14

ROTATIONAL SPECTRA OF T-SHAPED CYANOACETYLENE – CARBON DIOXIDE COMPLEX, HCCCN — CO_2

LU KANG, *Department of Chemistry and Biochemistry, Kennesaw State University, Kennesaw, GA, USA;* IAN DORELL, PHILIP DAVIS, *Department of Physics, Kennesaw State University, Kennesaw, GA, USA;* ONUR ONCER, STEPHEN G. KUKOLICH, *Department of Chemistry and Biochemistry, University of Arizona, Tucson, AZ, USA;* STEWART E. NOVICK, *Department of Chemistry, Wesleyan University, Middletown, CT, USA.*

The rotational spectra of T-shaped cyanoacetylene carbon dioxide complex, HCCCN — CO_2, were measured using two Balle-Flygare type Fourier transform microwave (FTMW) spectrometers between 1.4 GHz and 22 GHz. The low J transitions were recorded using the low frequency FTMW spectrometer at the University of Arizona with a state-of-the-art resolution of "full width at half maximum" (FWHM) 1 kHz. The spectra above 4 GHz were recorded at Wesleyan University. Spectral hyperfine structures due to the ^{14}N nuclear quadrupole coupling interactions can be fully resolved in low frequency bands. Since all $K_a = 1$ branches were not observed, this implies that HCCCN — CO_2 possesses a rigorous T-shaped structure. Assuming that A_0 is the same as that of HCN — CO_2, 11824 MHz, the spectroscopic constants of HCCCN — CO_2 are: $B_0 = 794.59686(63)$ MHz, $C_0 = 715.74488(60)$ MHz, $\Delta_J = 0.50067(18)$ kHz, $\Delta_{JK} = 120.892(12)$ kHz, $\delta_J = 0.04253(31)$ kHz, $\delta_K = 65.32(12)$ kHz, $H_J = -0.00117(33)$ Hz, $H_{JK} = 0.034876(21)$ kHz, $H_{KJ} = -0.68254(73)$ kHz, $\chi_{aa}(^{14}N) = -4.12873(78)$ MHz, $\chi_{bb}(^{14}N) = 2.110(25)$ MHz, and $\chi_{cc}(^{14}N) = 2.019(25)$ MHz.

WG. Clusters/Complexes
Wednesday, June 22, 2016 – 1:30 PM
Room: 116 Roger Adams Lab

Chair: J. Mathias Weber, University of Colorado, Boulder, CO, USA

WG01   1:30 – 1:45

SOLVENT-INDUCED REDUCTIVE ACTIVATION IN GAS PHASE $[Bi(CO_2)_n]^-$ CLUSTERS

MICHAEL C THOMPSON, *Department of Chemistry and Biochemistry, JILA - University of Colorado, Boulder, CO, USA*; JACOB SONDERGAARD RAMSAY, *Department of Chemistry, University of Aarhus, Aarhus, Denmark*; J. MATHIAS WEBER, *Department of Chemistry and Biochemistry, JILA - University of Colorado, Boulder, CO, USA.*

We report infrared photodissociation spectra of $[Bi(CO_2)_n]^-$ ($n = 2-9$) cluster anions. We determine the charge carrier geometry by comparing calculated vibrational frequencies based on density functional theory to the experimental spectra. The vibrational frequencies and the charge carrier geometry depend strongly on the solvation environment present in the cluster. We discuss the interaction of bismuth and CO_2 in the presence of an excess electron in the context of heterogeneous catalytic reduction of CO_2.

WG02   1:47 – 2:02

STRUCTURES AND SOLVATION EFFECTS OF $[Fe(CO_2)_n]^-$ CLUSTER ANIONS

MICHAEL C THOMPSON, J. MATHIAS WEBER, *Department of Chemistry and Biochemistry, JILA - University of Colorado, Boulder, CO, USA.*

We present infrared photodissociation spectra of $[Fe(CO_2)_n]^-$ ($n = 3 - 7$) cluster anions. We use density functional theory to compare calculated vibrational frequencies to our experimental spectra to determine plausible structures for the molecular charge carriers. The spectra display similar characteristics to those of other complexes of first-row transition metals with CO_2 ligands, and show signatures of several structural motifs.

WG03   2:04 – 2:19

NITROGEN MOLECULE-ETHYLENE SULFIDE COMPLEX INVESTIGATED BY FOURIER TRANSFORM MICROWAVE SPECTROSCOPY AND AB INITIO CALCULATION

SAKAE IWANO, YOSHIYUKI KAWASHIMA, *Applied Chemistry, Kanagawa Institute of Technology, Atsugi, Japan*; EIZI HIROTA, *The Central Office, The Graduate University for Advanced Studies, Hayama, Kanagawa, Japan.*

We have systematically investigated the van der Waals complexes consisting of the one from each of the two groups: (Rg, CO, N_2 or CO_2) and (dimethyl ether, dimethyl sulfide, ethylene oxide or ethylene sulfide), by using Fourier transform microwave spectroscopy supplemented by ab initio MO calculations, in order to understand the dynamical behavior of van der Waals complexes and to obtain information on the potential function to internal motions in complexes.[a] Two examples of the N_2 complex were investigated: N_2-DME (dimethyl ether), for which we reported a preliminary result[b] and N_2-EO (ethylene oxide).[c] In the present study we focused attention to the N_2-ES (ethylene sulfide) complex. We have detected two sets of the b-type transitions for the $^{15}N_2$-ES in ortho and para states, and have analyzed them by using the asymmetric-rotor program of A-reduction. In contrast with the N_2-EO, for which each of the ortho and para states were found split into a strong/weak pair, only some transitions of the $^{15}N_2$-ES were accompanied by two or three components. The observed spectra of the $^{14}N_2$-ES were complicated because of hyperfine splittings due to the nuclear quadrupole coupling of the two nitrogen atoms. We concluded that the N_2 moiety was located in the plane perpendicular to the C-S-C plane and bisecting the CSC angle of the ES. Two isomers were expected to exist for ^{15}NN-ES, one with ^{15}N in the inner and the other in the outer position, and in fact two sets of the spectra were detected. We have carried out ab initio molecular orbital calculations at the level of MP2 with basis sets 6-311++G(d, p), aug-cc-pVDZ, and aug-cc-pVTZ, to complement the information on the intracomplex motions obtained from the observed rotational spectra.

[a]Y. Kawashima, A. Sato, Y. Orita, and E. Hirota, *J.Phys.Chem.A* **2012** *116* 1224

[b]Y. Kawashima, Y. Tatamitani, Y. Morita, and E. Hirota, *61stInternationalSymposiumonMolecularSpectroscopy*, TE10 (2006)

[c]Y. Kawashima and E. Hirota, *J.Phys.Chem.A* **2013** *117* 13855

WG04   2:21 – 2:36

NITROGEN MOLECULE-DIMETHYL SULFIDE COMPLEX INVESTIGATED BY FOURIER TRANSFORM MICROWAVE SPECTROSCOPY AND AB INITIO CALCULATION

YOSHIYUKI KAWASHIMA, SAKAE IWANO, *Applied Chemistry, Kanagawa Institute of Technology, Atsugi, Japan*; EIZI HIROTA, *The Central Office, The Graduate University for Advanced Studies, Hayama, Kanagawa, Japan.*

This paper presents an extension of the preceding talk on the FTMW spectroscopy of N_2-ES (ethylene sulfide), namely the results on N_2-DMS (dimethyl sulfide). We have previously investigated two N_2 complexes: N_2-DME (dimethyl ether), for which we reported a prelimanary result,[a] and N_2-EO (ethylene oxide).[b] We have observed the ground-state rotational spectrum of the N_2-DMS complex, i.e. c-type transitions in the frequency region from 5 to 24 GHz, which we assigned to the normal, $^{15}N_2$-DMS, and ^{15}NN-DMS species of the N_2-DMS. We have found both the ortho and para states for the $^{14}N_2$-DMS and $^{15}N_2$-DMS species. In the case of the $^{15}N_2$-DMS, some transitions with $K_a = 2$ and 3 were observed slightly split by the internal rotation of the two methyl tops of the DMS. The observed spectra of the $^{15}N_2$-DMS were analyzed by using the XIAM program. In the case of the para state of the $^{15}N_2$-DMS, three rotational and five centrifugal distortion constants with the V_3 barrier to the methyl group internal rotation, whereas, in the case of the ortho state of the $^{15}N_2$-DMS, two more centrifugal distortion constants, Φ_{JK} and Φ_{KJ}, were needed to reproduce the observed spectra. For the N_2-DMS complex, we concluded that the N_2 moiety was located in a plane perpendicular to the C-S-C plane and bisecting the CSC angle of the DMS.

We have carried out ab initio molecular orbital calculations at the level of MP2 with basis sets 6-311++G(d, p), aug-cc-pVDZ, and aug-cc-pVTZ, to complement the information on the intracomplex motions obtained from the observed rotational spectra. We have applied a natural bond orbital (NBO) analysis to the N_2-DMS and N_2-ES to calculate the stabilization energy CT ($=\Delta E_{\sigma\sigma*}$), which was closely correlated with the binding energy E_B, as found for other related complexes.

[a]Y. Kawashima, Y. Tatamitani, Y. Morita, and E. Hirota, *61st International Symposium on Molecular Spectroscopy*, TE10 (2006)
[b]Y. Kawashima and E. Hirota, *J. Phys. Chem. A* **2013** *117* 13855

WG05   2:38 – 2:53

PLANAR CoB$_{18}^-$ CLUSTER: A NEW MOTIF FOR HETERO- AND METALLO-BOROPHENES

TENG-TENG CHEN, TIAN JIAN, GARY LOPEZ, *Department of Chemistry, Brown University, Providence, RI, USA*; WAN-LU LI, XIN CHEN, JUN LI, *Department of Chemistry, Tsinghua University, Beijing, China*; LAI-SHENG WANG, *Department of Chemistry, Brown University, Providence, RI, USA.*

Combined Photoelectron Spectroscopy (PES) and theoretical calculations have found that anion boron clusters (B_n^-) are planar and quasi-planar up to B_{25}^-. Recent works show that anion pure boron clusters continued to be planar at B_{27}^-,B_{30}^-,B_{35}^- and B_{36}^-. B_{35}^- and B_{36}^- provide the first experimental evidence for the viability of the two-dimensional (2D) boron sheets (Borophene). The 2D to three-dimensional (3D) transitions are shown to happen at B_{40}^-,B_{39}^- and B_{28}^-, which possess cage-like structures. These fullerene-like boron cage clusters are named as Borospherene. Recently, borophenes or similar structures are claimed to be synthesized by several groups.

Following an electronic design principle, a series of transition-metal-doped boron clusters (M©B_n^-, n=8-10) are found to possess the monocyclic wheel structures. Meanwhile, CoB$_{12}^-$ and RhB$_{12}^-$ are revealed to adopt half-sandwich-type structures with the quasi-planar B_{12} moiety similar to the B_{12}^- cluster. Very lately, we show that the CoB$_{16}^-$ cluster possesses a highly symmetric Cobalt-centered drum-like structure, with a new record of coordination number at 16.

Here we report the CoB$_{18}^-$ cluster to possess a unique planar structure, in which the Co atom is doped into the network of a planar boron cluster. PES reveals that the CoB$_{18}^-$ cluster is a highly stable electronic system with the first adiabatic detachment energy (ADE) at 4.0 eV. Global minimum searches along with high-level quantum calculations show the global minimum for CoB$_{18}^-$ is perfectly planar and closed shell ($^1A_1$) with C_{2v} symmetry. The Co atom is bonded with 7 boron atoms in the closest coordination shell and the other 11 boron atoms in the outer coordination shell. The calculated vertical detachment energy (VDE) values match quite well with our experimental results. Chemical bonding analysis by the Adaptive Natural Density Partitioning (AdNDP) method shows the CoB$_{18}^-$ cluster is π-aromatic with four 4-centered-2-electron (4c-2e) π bonds and one 19-centered-2-electron (19c-2e) π bond, 10 π electrons in total. This perfectly planar structure reveals the viability of creating a new class of hetero-borophenes and metallo-borophenes by doping metal atoms into the plane of monolayer boron atoms. This gives a new approach to design perspective hetero-borophenes and metallo-borophenes materials with tunable chemical, magnetic and optical properties.

WG06 $\qquad$ **2:55 – 3:10**

CO_2 DIMER: FOUR INTERMOLECULAR MODES OBSERVED VIA INFRARED COMBINATION BANDS

JALAL NOROOZ OLIAEE, *Department of Physics and Astronomy, University of Calgary, Calgary, AB, Canada*; MEHDI DEHGHANY, *Department of Mathematics, Physics and Engineering, Mount Royal University, Calgary, Canada*; MOJTABA REZAEI, *Department of Physics and Astronomy, University of Calgary, Calgary, AB, Canada*; BOB McKELLAR, *Steacie Laboratory, National Research Council of Canada, Ottawa, ON, Canada*; NASSER MOAZZEN-AHMADI, *Physics and Astronomy/Institute for Quantum Science and Technology, University of Calgary, Calgary, AB, Canada.*

Study of the carbon dioxide dimer has a long history, but there is only one previous observation of an intermolecular vibration [1]. Here we analyze four new combination bands of $(CO_2)_2$ in the CO_2 ν_3 region ($\sim$2350 cm^{-1}), observed using tunable infrared lasers and a pulsed slit-jet supersonic expansion. The previous combination band at 2382.2 cm^{-1}was simple to assign [1]. A much more complicated band ($\sim$2370 cm^{-1}) turns out to involve *two* upper states, one at 2369.0 cm^{-1}(B_u symmetry), and the other at 2370.0 cm^{-1}(A_u). The spectrum can be nicely fit by including the Coriolis interactions between these states. Another complicated band around 2443 cm^{-1}also involves two nearby upper states which are highly perturbed in so-far unexplained ways (possibly related to tunneling shifts).

With the help of new *ab initio* calculations [2], we assign the results as follows. The 2369.0 cm^{-1}band is the combination of the forbidden A_g intramolecular fundamental (probably [1] at about 2346.76 cm^{-1}) and the intermolecular geared bend (B_u). The 2370.0 cm^{-1}band is the combination of the same A_g fundamental and the intermolecular torsion (A_u). This gives about 22.3 and 23.2 cm^{-1}for the geared bend and torsion. The previous 2382.2 cm^{-1}band [1] is the allowed B_u fundamental (2350.771 cm^{-1}) plus two quanta of the geared bend (B_u), giving 31.509 cm^{-1}for this overtone. The highly perturbed 2442.7 cm^{-1}band is the B_u fundamental plus the antigeared bend (A_g), giving about 91.9 cm^{-1}for the antigeared bend. Finally, the perturbed 2442.1 cm^{-1}band is due to an unknown combination of modes which gains intensity from the antigeared bend by a Fermi-type interaction. Calculated values [2] are: 20.64 (geared bend), 24.44 (torsion), 32.34 (geared bend overtone), and 92.30 cm^{-1}(antigeared bend), in good agreement with experiment.

[1] M. Dehghany, A.R.W. McKellar, Mahin Afshari, and N. Moazzen-Ahmadi, *Mol. Phys.* **108**, 2195 (2010).

[2] X.-G. Wang, T. Carrington, Jr., and R. Dawes, private communication.

WG07 $\qquad$ **3:12 – 3:27**

JET-COOLED HIGH RESOLUTION INFRARED SPECTROSCOPY OF SMALL VAN DER WAALS SF_6 CLUSTERS

PIERRE ASSELIN, *MONARIS UMR 8233, CNRS UPMC, Paris, France*; VINCENT BOUDON, *Laboratoire ICB, CNRS/Université de Bourgogne, DIJON, France*; ALEXEY POTAPOV, *I. Physikalisches Institut, Universität zu Köln, Köln, Germany*; LAURENT BRUEL, *CEA Marcoule, DEN, Bagnols-sur-Cèze, FRANCE*; MARC-ANDRÉ GAVEAU, MICHEL MONS, *CEA Saclay, LIDYL, Gif-sur-Yvette, France.*

Using a pulsed slit nozzle multipass absorption spectrometer with a tunable quantum cascade laser we investigated van der Waals clusters involving sulfur hexafluoride in the spectral range near the ν_3 stretching vibration. Different sized homo-complexes were generated in a planar supersonic expansion with typically 0,5 % SF_6 diluted in 6 bar He. Firstly, several rotationally resolved parallel and perpendicular bands of $(SF_6)_2$, at 934,0 and 956,1 cm^{-1}(#1 structure) in agreement with Takami et al.[a] but also one band at 933,6 cm^{-1}(#2 structure) never observed previously, were analyzed in light of a recent theoretical study predicting three nearly isoenergetic isomers of D_{2d}, C_{2h} and C_2 symmetry for the dimer. [b] Furthermore, some broader bands were detected around 938 and 964 cm^{-1}and assigned to $(SF_6)_3$ and $(SF_6)_4$ clusters on the grounds of concentration effects and/or ab initio calculations. Lastly, with 0,5 % rare gas Rg (Rg = Ne, Ar, Kr and Xe) added to the SF_6:He gas mixture, a series of van der Waals $(SF_6)_2$-Rg hetero-trimers were observed, which display a remarkable linear dependence of the vibrational shift with the polarizability of the rare gas atom provided that the initial SF_6 dimer structure is #2 . In the same time no transitions belonging to the binary complexes SF_6-Rg were found near the ν_3 monomer band. This result suggests a complex thermodynamics within the pulsed supersonic expansion leading to the preponderance of $(SF_6)_2$-Rg clusters over SF_6-Rg binary systems.

[a]R. D. Urban and M. Takami, J. Chem. Phys. 103, 9132 (1995).

[b]T. Vazhappily, A. Marjolin and K. D. Jordan, J. Phys. Chem. B, DOI: 10.1021 / acs.jpcb.5b09419 (2015).

Intermission

WG08   3:46 – 4:01

DOES A SECOND HALOGEN ATOM AFFECT THE NATURE OF INTERMOLECULAR INTERACTIONS IN PROTIC ACID-HALOETHYLENE COMPLEXES? IN (*E*)-1-CHLORO-2-FLUOROETHYLENE-HYDROGEN CHLORIDE IT DEPENDS ON HOW YOU LOOK AT IT

HELEN O. LEUNG, MARK D. MARSHALL, *Chemistry Department, Amherst College, Amherst, MA, USA.*

As part of a systematic study of the effect of chlorine substitution on the structures of protic acid haloethylene complexes, the structure of the (*E*)-1-chloro-2-fluoroethylene-hydrogen chloride complex has been investigated using *ab initio* quantum chemistry calculations and microwave spectroscopy. Although theory predicts a non-planar equilibrium structure for this species, it is only 7 cm^{-1} lower in energy than the planar geometry connecting the two equivalent minima on either side of the haloethylene plane, and the observed spectrum is consistent with a planar, average structure, likely the result of zero-point averaging. The geometry is very similar to the fluorine binding, vinyl fluoride-hydrogen chloride complex, suggesting that the substitution of chlorine for a hydrogen *trans* to the fluorine atom has very little effect on intermolecular interactions in this case. On the other hand, vinyl chloride-hydrogen chloride adopts a non-planar, chlorine binding configuration so that alternatively one could say that the presence of fluorine has a large effect on protic acid-chlorine interactions.

WG09   4:03 – 4:18

DOES A SECOND HALOGEN ATOM AFFECT THE NATURE OF INTERMOLECULAR INTERACTIONS IN PROTIC ACID-HALOETHYLENE COMPLEXES? IN (*Z*)-1-CHLORO-2-FLUOROETHYLENE-HYDROGEN CHLORIDE IT MOST CERTAINLY DOES!

HANNAH K. TANDON, HELEN O. LEUNG, MARK D. MARSHALL, *Chemistry Department, Amherst College, Amherst, MA, USA.*

As part of a systematic study of the effect of chlorine substitution on the structures of protic acid-haloethylene complexes, the structure of the (*Z*)-1-chloro-2-fluoroethylene-hydrogen chloride complex has been investigated using *ab initio* quantum chemistry calculations and microwave spectroscopy. Although theory predicts a non-planar equilibrium structure for this species, it is only 6 cm^{-1} lower in energy than the planar geometry connecting the two equivalent minima on either side of the haloethylene plane, and the observed spectrum is consistent with a planar, average structure, likely the result of zero-point averaging. The geometry is unlike that of any previously characterized protic acid-haloethylene complex with a bifurcated primary interaction in which the hydrogen of the acid interacts with both the fluorine and the chlorine atoms on the haloethylene and there is no evidence for a secondary interaction involving the electron rich region of the acid. This structure can be contrasted to those of vinyl fluoride-hydrogen chloride (fluorine bound, planar "top-binding," across the double bond), vinyl chloride-hydrogen chloride (chlorine bound, non-planar) and (*Z*)-1-chloro-2-fluoroethylene-acetylene (chlorine bound, planar "side-binding," at one end of the double bond).

WG10   4:20 – 4:35

FIRST OBSERVATION OF THE N$_2$O-OC VAN DER WAALS COMPLEX AND NEW SET OF EXPERIMENTAL MEASUREMENTS ON THE N$_2$O-CO COMPLEX.

CLÉMENT LAUZIN, *Institue of Condensed Matter and Nanosciences (IMCN), Université catholique de Louvain, Louvain-la-Neuve, Belgium*; A. J. BARCLAY, S. SHEYBANI-DELOUI, NASSER MOAZZEN-AHMADI, *Department of Physics and Astronomy, University of Calgary, Calgary, AB, Canada.*

Jet cooled infrared spectrum of the N$_2$O-CO van der Waals complex was observed in the region of the ν_1 fundamental band of the N$_2$O monomer (2224 cm^{-1}) and in the CO stretch region (2143 cm^{-1}). These new measurements allowed the predicted less stable isomer, N$_2$O-OC, to be observed for the first time in both spectral regions. In addition, four combination bands were observed in the CO region. Two of these were assigned to N$_2$O-CO and the other two to N$_2$O-OC. Finally, a combination band in the N$_2$O region was assigned to the most stable isomer. In this talk I will discuss our results for the intermolecular vibrational frequencies and compare these to the recently published experimental values on similar systems CO$_2$-CO and CO$_2$-OC [a] and to ab initio predictions on this complex [b].

[a] S . Sheybani-Deloui, A. J. Barclay, K. H. Michaelian, A. R. W. McKellar, and N. Moazzen-Ahmadi, J. Chem. Phys 143, 121101 (2015).
[b] M. Venayagamoorthy, T. A. Ford, THEOCHEM 717,111 (2005)

WG11   4:37 – 4:52

LASER-INDUCED FLUORESCENCE SPECTRA OF C_3Ar NEAR 25400-25600 cm^{-1}

YI-JEN WANG, <u>YEN-CHU HSU</u>, *Institute of Atomic and Molecular Sciences, Academia Sinica, Taipei, Taiwan.*

About 14 bands of C_3Ar near the 0 4^- 0-000 and 0 2^+ 0-000 bands of the $\tilde{A}$-$\tilde{X}$ system of C_3 have been recorded by laser-induced fluoresence with a laser resolution of 0.035 cm^{-1}. Bands at 25428 amd 25515 cm^{-1} are found to be type A, and those at 25431, 25496, and 25519 cm^{-1} are type C. Bands at 25504 and 25507 cm^{-1} are too diffuse for rotational analysis. The bands near 25500 cm^{-1} form part of two progressions with about 10 cm^{-1} separations, which appear to represent the van der Waals in-plane-bending vibration. A third diffuse feature was observed near the R(3) line of the 25519 cm^{-1} band. Possible dissociation processes will be discussed.

WG12   4:54 – 5:09

A MICROWAVE STUDY OF 3,5 DIFLUOROPYRIDINE$\cdots CO_2$: THE EFFECT OF META-FLUORINATION ON INTER-MOLECULAR INTERACTIONS OF PYRIDINE

CHRIS DEWBERRY, *Chemistry Department, University of Minnesota, Minneapolis, MN, USA*; <u>RYAN D. CORNELIUS</u>, *Chemistry Department, St. Cloud State University , St. Cloud, MN, USA*; BECCA MACKENZIE, CJ SMITH, *Chemistry Department, University of Minnesota, Minneapolis, MN, USA*; MICHAEL A. DVORAK, *Chemistry Department, St. Cloud State University , St. Cloud, MN, USA*; KEN LEOPOLD, *Chemistry Department, University of Minnesota, Minneapolis, MN, USA.*

The rotational spectrum of the weakly bound complex 3,5-difluoropyridine$\cdots CO_2$ has been observed using pulsed-nozzle Fourier transform microwave spectroscopy. Spectroscopic constants are reported for the parent and the $^{13}CO_2$ isotopologues. The data indicate a structure in which the nitrogen approaches the carbon of the CO_2 with the C_2 axis of the difluoropyridine perpendicular to the CO_2. The N$\cdots$C van der Waals bond distance is 2.827(17) Å and the oxygen$\cdots$ortho-hydrogen distance is 3.045(3) Å. The amplitude of the zero point bending vibrational motion of the difluoropyridine moiety away from the C_2 axis of the complex is estimated from ^{14}N nuclear hyperfine structure to be 8.8°. The N$\cdots$C van der Waals bond length is 0.029 Å longer than that previously determined for pyridine-CO_2, but is still considerably shorter than the 2.997(1) Å distance in HCN$\cdots CO_2$. Density functional theory calculations place the binding energy of the complex at 3.2 kcal/mol.

WG13   5:11 – 5:26

HIGH RESOLUTION INFRARED SPECTROSCOPY OF THE CO_2-CO DIMERS AND $(CO_2)_2$-CO TRIMER

<u>A. J. BARCLAY</u>, S. SHEYBANI-DELOUI, *Department of Physics and Astronomy, University of Calgary, Calgary, AB, Canada*; K. H. MICHAELIAN, *CanmetENERGY, Natural Resources Canada, Edmonton, Alberta, Canada*; BOB McKELLAR, *Steacie Laboratory, National Research Council of Canada, Ottawa, ON, Canada*; NASSER MOAZZEN-AHMADI, *Department of Physics and Astronomy, University of Calgary, Calgary, AB, Canada.*

Infrared spectra in the carbon monoxide CO stretch region (≈ 2150 cm^{-1}) are assigned to the previously unobserved O-bonded form of the CO_2-CO dimer ("isomer 2"), which has a planar T-shaped structure like that of the previously observed C-bonded form ("isomer 1"). Results will also be reported for both isomers of the $^{12}C^{18}O_2$-substituted form of the dimer. In addition, we have observed two combination bands for each isomer yielding the first experimental determinations of intermolecular frequencies for the planar T-shaped structures. Within both of the fundamental bands, weak "satellite bands" are observed. These are tentatively assigned to the trimer He-CO_2-CO. To the higher side of the fundamental for "isomer 1", we have observed a weaker b-type band which we have assigned to $(CO_2)_2$-CO trimer. This trimer has a "pin wheel" structure with C2 symmetry and the derived experimental structural parameters match well with those obtained from *ab initio* calculations.

WG14   **5:28 – 5:43**

MICROWAVE OBSERVATION OF THE VAN DER WAALS COMPLEX O_2-CO

FRANK E MARSHALL, THOMAS D. PERSINGER, DAVID JOSEPH GILLCRIST, NICOLE MOON, STEVE ALEXANDRE NDENGUE, RICHARD DAWES, <u>G. S. GRUBBS II</u>, *Department of Chemistry, Missouri University of Science and Technology, Rolla, MO, USA.*

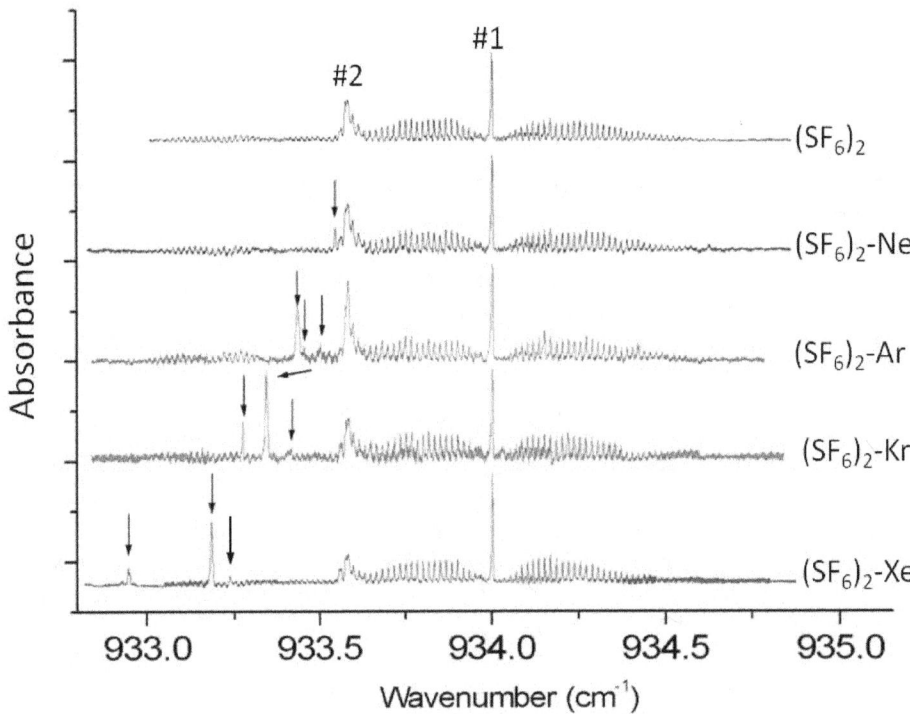

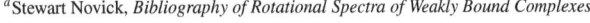

FTMW spectroscopy has long been known to be a powerful tool in characterizing van der Waals complexes.[a] Along with this, advances in microwave technology and computing have made complicated spin-interaction systems much easier to observe and characterize. One such system, O_2-CO has been observed for the first time on a CP-FTMW spectrometer operational in the 6-18 GHz region. Preliminary observations and calculations indicate a slipped-parallel structure. High level calculations are ongoing, including the construction of a 4D potential energy surface. Rotational assignments, along with any observed fine structure due to the $^3\Sigma$ ground state of O_2 will be discussed.

[a]Stewart Novick, *Bibliography of Rotational Spectra of Weakly Bound Complexes*

WH. Astronomy

Wednesday, June 22, 2016 – 1:30 PM

Room: 274 Medical Sciences Building

Chair: Brett A. McGuire, National Radio Astronomy Observatory, Charlottesville, VA, USA

WH01   1:30 – 1:45

SPECTROSCOPIC STUDY AND ASTRONOMICAL DETECTION OF VIBRATIONALLY EXCITED *n*-PROPYL CYANIDE

HOLGER S. P. MÜLLER, NADINE WEHRES, OLIVIA H. WILKINS, FRANK LEWEN, STEPHAN SCHLEMMER, *I. Physikalisches Institut, Universität zu Köln, Köln, Germany*; ADAM WALTERS, RÉMI VICENTE, DELONG LIU, *IRAP, Université de Toulouse 3 - CNRS - OMP, Toulouse, France*; ROBIN T. GARROD, *Departments of Chemistry and Astronomy, The University of Virginia, Charlottesville, VA, USA*; ARNAUD BELLOCHE, KARL M. MENTEN, *Millimeter- und Submillimeter-Astronomie, Max-Planck-Institut für Radioastronomie, Bonn, NRW, Germany*.

We have obtained ALMA data of Sagittarius (Sgr for short) B2(N) between 84.0 and 114.4 GHz in its Early Science Cycles 0 and 1. We have focused our analyses on the northern, secondary hot molecular core Sgr B2(N2) because of the smaller line widths. The survey led to the first detection of a branched alkyl compound, *iso*-propyl cyanide, *i*-C_3H_7CN, in space[a] besides the $\sim$2.5 times more abundant straight chain isomer *n*-propyl cyanide, a molecule which we had detected in our IRAM 30 m survey.[b] We suspected to be able to detect *n*-propyl cyanide in vibrationally excited states in our ALMA data.

We have recorded laboratory rotational spectra of this molecule in three large frequency regions and identified several excited vibrational states. The analyses of these spectra have focused on the 36 to 70 GHz and 89 to 127 GHz regions and on the four lowest excited vibrational states of both the lower lying *gauche*- and the slightly higher lying *anti*-conformer for which rotational constants had been published.[c] We will present results of our laboratory spectroscopic investigations and will report on the detection of these states toward Sgr B2(N2).

---

[a] A. Belloche et al., *Science* **345** (2014) 1584.
[b] A. Belloche et al., *A&A* **499** (2009) 215.
[c] E. Hirota, *J. Chem. Phys.* **37** (1962) 2918.

WH02   1:47 – 2:02

THE INTERSTELLAR DETECTION OF CH_3NCO IN Sgr B2(N)

DeWAYNE T HALFEN, *Steward Observatory, University of Arizona, Tucson, AZ, USA*; V. ILYUSHIN, *Radiospectrometry Department, Institute of Radio Astronomy of NASU, Kharkov, Ukraine*; LUCY M. ZIURYS, *Department of Chemistry and Biochemistry; Department of Astronomy, Arizona Radio Observatory, University of Arizona, Tucson, AZ, USA*.

A new interstellar molecule, CH_3NCO (methyl isocyanate), has been detected for the first time using the 12 m telescope of the Arizona Radio Observatory (ARO). CH_3NCO was identified in spectra covering 68 - 116 GHz in the 3 mm segment of a broadband survey of Sgr B2(N). This study was based on previous laboratory work by Koput (1986) and new Fourier transform millimeter-wave (FTmmW) measurements performed at Arizona in the 60 - 88 GHz range. Spectroscopic constants were determined for CH_3NCO in a combined fit, and were used to predict other transitions at 3 mm.

Thirty very favorable rotational lines ($K_a = 0$ and $K_a = 1$ only; $E_u < 60$ K) originating in five consecutive transitions of CH_3NCO in both the A and E internal rotation states were found to be present in the Sgr B2(N) survey from 68 - 105 GHz. Emission was observed at all of the predicted frequencies, with 17 lines appearing as distinct, uncontaminated spectral features, clearly showing the classic a-type, asymmetric top pattern. The CH_3NCO spectra also appear to exhibit two velocity components near $V_{LSR} \approx 62$ and 73 km s^{-1}, both with $\Delta V_{1/2} \approx 10$ km s^{-1}, typical of most molecules in Sgr B2(N). The fractional abundances were determined to be $f = 7.6$ x 10^{-12} and 5.0 x 10^{-12} for the 62 and 73 km s^{-1} components, relative to H_2, respectively. CH_3NCO was recently detected in volatized material from comet 67P/Churyumov-Gerasimenko by Rosetta's Philae lander.

WH03   2:04 – 2:19

A SURVEY OF HNCO AND CH$_3$NCO IN MOLECULAR CLOUDS

<u>DeWAYNE T HALFEN</u>, *Steward Observatory, University of Arizona, Tucson, AZ, USA*; LUCY M. ZIURYS, *Department of Chemistry and Biochemistry; Department of Astronomy, Arizona Radio Observatory, University of Arizona, Tuscon, AZ, USA.*

Following the first interstellar detection of CH$_3$NCO in Sgr B2(N) by Halfen et al. (2015), a survey of this species and its likely precursor HNCO has been conducted towards several dense molecular clouds. Three transitions of CH$_3$NCO in its $K_a = 0$ ladder for both A and E internal rotation states were searched for at 3 mm, using the new ARO ALMA Prototype 12 m telescope. In addition, two transitions of HNCO in its $K_a = 0$ and 2 ladders were observed near 88 and 110 GHz. Emission from CH$_3$NCO was detected towards Orion-KL, G34.3, W51M, Sgr B2(2N), and DR-21(OH) with intensities of $T_A{}^* \approx$ 10-40 mK. HNCO was also found in each source observed. The ratio of HNCO/CH$_3$NCO estimated from these data is around 25 - 45, consistent with that derived from Sgr B2(N). These results suggest that HNCO is most likely the chemical precursor to CH$_3$NCO.

WH04   2:21 – 2:36

MODELLING STUDY OF INTERSTELLAR ETHANIMINE ISOMERS

<u>DONGHUI QUAN</u>, *Department of Chemistry, Eastern Kentucky University, Richmond, KY, USA*; ERIC HERBST, *Department of Chemistry, The University of Virginia, Charlottesville, VA, USA*; JOANNA F. CORBY, *Department of Astronomy, The University of Virginia, Charlottesville, VA, USA*; ALLISON DURR, *Department of Chemistry NSB, Eastern Kentucky University, Richmond, Kentucky, United States*; GEORGE HASSEL, *Department of Physics and Astronomy, Siena College, Loudonville, NY, USA.*

Ethanimine (CH3CHNH) , including both the E- and Z- isomers, were detected towards the star-forming region Sgr B2(N) using the GBT PRIMOS data (Loomis et al 2013), and were recently imaged by the ACTA (Corby et al. 2015). These aldimines can serve as precursors of biological molecules such as amino acids thus are considered prebiotic molecules in interstellar medium. In this study, we present chemical simulations of ethanimine with various physical conditions. From models for Sgr B2(N) and environs, calculated ethanimine abundances show reasonable agreement with observed values, while the translucent cloud models yield much lower abundances. These results agree with locations suggested by observations that ethanimine isomers were detected in the foreground of the shells of the hot core.

WH05   2:38 – 2:53

COMPLETE RESULTS FROM A SPECTRAL-LINE SURVEY OF Sgr B2(N)

<u>DeWAYNE T HALFEN</u>, *Steward Observatory, University of Arizona, Tucson, AZ, USA*; LUCY M. ZIURYS, *Department of Chemistry and Biochemistry; Department of Astronomy, Arizona Radio Observatory, University of Arizona, Tuscon, AZ, USA.*

A confusion-limited spectral line survey of the Galactic center molecular cloud Sgr B2(N) at 3, 2, and 1 mm (68 - 116, 130 - 172, and 210 - 280 GHz) using the Kitt Peak 12 m and the Submillimeter Telescope (SMT) of the Arizona Radio Observatory was recently completed. About 15,000 spectral lines were observed in the survey range. The data have been analyzed using two techniques. First, the rotational temperature diagram methods was used for each individual species. Second, an LTE code was used to model and ultimately fit the data to a set of parameters for each species, using a least squares approach. Seventy-four molecules are identified in the data, along with 81 isotopologue species. In addition, 26 excited vibrational states of the identified molecules were detected, as well as H and He recombination lines. Source and Galactic absorption lines are seen in several abundant species, as well as multiple maser lines of methanol and possibly SO$_2$.

WH06   2:55 – 3:10

DISCOVERY OF THE FIRST INTERSTELLAR CHIRAL MOLECULE: PROPYLENE OXIDE

BRANDON CARROLL, *Division of Chemistry and Chemical Engineering, California Institute of Technology, Pasadena, CA, USA*; BRETT A. McGUIRE, *NAASC, National Radio Astronomy Observatory, Charlottesville, VA, USA*; RYAN LOOMIS, *Department of Astronomy, Harvard University, Cambridge, MA, USA*; IAN A FINNERAN, *Division of Chemistry and Chemical Engineering, California Institute of Technology, Pasadena, CA, USA*; PHILIP JEWELL, ANTHONY REMIJAN, *ALMA, National Radio Astronomy Observatory, Charlottesville, VA, USA*; GEOFFREY BLAKE, *Division of Chemistry and Chemical Engineering, California Institute of Technology, Pasadena, CA, USA*.

Life on Earth relies on chiral molecules, that is, species not superimposable on their mirror images. This manifests itself as a reliance on a single molecular handedness, or homochirality that is characteristic of life and perhaps most readily apparent in the large enhancement in biological activity of particular amino acid and sugar enantiomers. Yet, the ancestral origin of biological homochirality remains a mystery. The non-racemic ratios in some organics isolated from primitive meteorites hint at a primordial chiral seed but even these samples have experienced substantial processing during planetary assembly, obscuring their complete histories. To determine the underlying origin of any enantiomeric excess, it is critical to understand the molecular gas from which these molecules originated. Here, we present the first extra-solar, astronomical detection of a chiral molecule, propylene oxide (CH_3CHCH_2O), in absorption toward the Galactic Center. With the detection of propylene oxide, we at last have a target for broad-ranging searches for the possible cosmic origin of the homochirality of life.

WH07   3:12 – 3:27

THE CO TRANSITION FROM DIFFUSE MOLECULAR GAS TO DENSE CLOUDS: PRELIMINARY RESULTS

JOHNATHAN S RICE, STEVEN FEDERMAN, *Physics and Astronomy, University of Toledo, Toledo, OH, USA*.

The atomic to molecular transitions occurring in diffuse interstellar gas surrounding molecular clouds are affected by the local physical conditions (density and temperature) and the radiation field penetrating the material. The material is closely connected to CO-dark gas, which is not associated with emission from H I at 21 cm or from CO at 2.6 mm. Using optical observations of CH, CH^+ and CN absorption from McDonald Observatory and the European Southern Observatory in conjunction with UV observations of CO and H_2 absorption from FUSE, we explore the changing environment between diffuse and dense gas, emphasizing trends in column density, excitation temperature, gas density, and velocity structure. This presentation will focus on the completed analysis involving H_2 and on the preliminary results of CO for our sample.

WH08 3:29 – 3:44

CENTRAL 300 PC OF THE GALAXY PROBED BY THE INFRARED SPECTRA OF H_3^+ AND CO: I. PREDOMINANCE OF WARM AND DIFFUSE GAS AND HIGH H_2 IONIZATION RATE

<u>TAKESHI OKA</u>, *Department of Astronomy and Astrophysics, Chemistry, The University of Chicago, Chicago, IL, USA*; THOMAS R. GEBALLE, *Gemini Observatory, Hilo, HI, USA*; MIWA GOTO, *The Center for Astrochemical Studies, Max-Planck-Institut für extraterrestrische Physik, Garching, Germany*; TOMONORI USUDA, *National Astronomical Observatory of Japan, Tokyo, Japan*; NICK INDRIOLO, *Department of Astronomy, University of Michigan, Ann Arbor, MI, USA.*

A low-resolution 2.0-2.5 μm survey of $\sim$500 very red point-like objects in the Central Molecular Zone (CMZ) of our Galaxy, initiated in 2008, has revealed many new bright objects with featureless spectra that are suitable for high resolution absorption spectroscopy of H_3^+ and CO.[a] We now have altogether 48 objects mostly close to the Galactic plane located from 142 pc to the west of Sgr A* to 120 pc east allowing us to probe dense and diffuse gas by H_3^+ and dense gas by CO. Our observations demonstrate that the warm ($\sim$250 K) and diffuse ($\leq$100 cm^{-3}) gas with a large column length ($\geq$30 pc) initially observed toward the brightest star in the CMZ, GCS3-2 of the Quintuplet Cluster,[b] exists throughout the CMZ with the surface filling factor of $\sim$ 100% dominating the region.

The column densities of CO in the CMZ are found to be much less than those in the three foreground spiral arms except in the directions of Sgr B and Sgr E complexes and indicate that the volume filling factor of dense clouds of 10% previously estimated is a gross overestimate for the front half of the CMZ. Nevertheless the predominance of the newly found diffuse molecular gas makes the term "Central Molecular Zone" even more appropriate. The ultra-hot X-rays emitting plasma which some thought to dominate the region must be non existent except near the stars and SNRs.

Recently the H_2 fraction $f(H_2)$ in diffuse gas of the CMZ has been reported to be $\sim$0.6[c]. If we use this value, the cosmic ray H_2 ionization rate ζ of a few times 10^{-15} s^{-1} reported earlier[b] on the assumption of $f(H_2)$=1 needs to be increased by a factor of $\sim$3 since the value is approximately inversely proportional to $f(H_2)^2$.

---

[a]Geballe, T. R., Oka, T., Lambridges, E., Yeh, S. C. C., Schlegelmilch, B., Goto, M., Westrick, C. W., WI07 at the 70th ISMS, Urbana, IL, USA,2015

[b]Oka, T., Geballe, T. R., Goto, M., Usuda, T., McCall, B. J. 2005, ApJ, 632, 882

[c]Le Petit, F., Ruaud, M., Bron, E., Godard, B., Roueff, E., Languignon, D., Le Bourlot, J. 2016, A&A, 585, A105

Intermission

WH09 4:03 – 4:18

THE PRECISE RADIO OBSERVATION OF THE ^{13}C ISOTOPIC FRACTIONATION FOR CARBON CHAIN MOLECULE HC_3N IN THE LOW-MASS STAR FORMING REGION L1527

<u>MITSUNORI ARAKI</u>, *IR Free Electron Laser Research Center, Tokyo University of Science, Tokyo, Japan*; SHURO TAKANO, *College of Engineering, Nihon University, Fukushima, Japan*; NAMI SAKAI, *RIKEN Center for Advanced Photonics, RIKEN, Wako, Japan*; SATOSHI YAMAMOTO, *Department of Physics and Research Center for the Early Universe, The University of Tokyo, Tokyo, Japan*; TAKAHIRO OYAMA, *Faculty of Science Division I, Tokyo University of Science, Tokyo , Japan*; NOBUHIKO KUZE, *Faculty of Science and Technology, Sophia University, Tokyo, Japan*; KOICHI TSUKIYAMA, *Faculty of Science Division I, Tokyo University of Science, Shinjuku-ku, Tokyo, Japan.*

We observed the three ^{13}C isotopic species of HC_3N with the high signal-to-noise ratios in L1527 using Green Bank 100 m telescope and Nobeyama 45 m telescope to explore the production scheme of HC_3N, where L1527 is the low-mass star forming region in the phase of a warm carbon chain chemistry region.[a] The spectral lines of the J = 5–4, 9–8, 10–9, and 12–11 transitions in the 44–109 GHz region were used to measure isotopic ratios. The abundance of HCCCN was determined from the line intensities of the two weak hyperfine components of the J = 5–4 transition. The isotopic ratios were precisely determined to be 1.00 : 1.01 : 1.35 : 86.4 for [H^{13}CCCN] : [HC^{13}CCN] : [HCC^{13}CN] : [HCCCN]. It was found that the abundance of H^{13}CCCN is equal to that of HC^{13}CCN, and it was implied that HC_3N is mainly formed by the reaction schemes via C_2H_2 and $C_2H_2^+$ in L1527. This would suggest a universality of dicarbide chemistry producing HC_3N irrespective of evolutional phases from a starless dark cloud[b] to a warm carbon chain chemistry region.

---

[a]Sakai, N., Sakai, T., Hirota, T., & Yamamoto, S. 2008, ApJ, 672, 371

[b]Takano, S., Masuda, A., Hirahara, Y., et al. 1998, A&A, 329, 1156

WH10 4:20 – 4:35

CARMA OBSERVATIONS OF L1157: CHEMICAL COMPLEXITY IN THE SHOCKED OUTFLOW

ANDREW M BURKHARDT, NIKLAUS M DOLLHOPF, JOANNA F. CORBY, *Department of Astronomy, The University of Virginia, Charlottesville, VA, USA*; BRANDON CARROLL, *Division of Chemistry and Chemical Engineering, California Institute of Technology, Pasadena, CA, USA*; CHRISTOPHER N SHINGLEDECKER, *Department of Chemistry, The University of Virginia, Charlottesville, VA, USA*; RYAN LOOMIS, *Department of Astronomy, Harvard University, Cambridge, MA, USA*; S. TOM BOOTH, *Department of Astronomy, The University of Virginia, Charlottesville, VA, USA*; GEOFFREY BLAKE, *Division of Chemistry and Chemical Engineering, California Institute of Technology, Pasadena, CA, USA*; ANTHONY REMIJAN, *ALMA, National Radio Astronomy Observatory, Charlottesville, VA, USA*; BRETT A. McGUIRE, *NAASC, National Radio Astronomy Observatory, Charlottesville, VA, USA.*

L1157, a molecular dark cloud with an embedded Class 0 protostar possessing a bipolar outflow, is an excellent source for studying shock chemistry, including grain-surface chemistry prior to shocks, and post-shock, gas-phase processing. Prior to shock events an estimated $\sim$2000 and 4000 years ago, temperatures were too low for most complex organic molecules to undergo thermal desorption. Thus, the shocks should have liberated these molecules from the ice grain-surfaces *en masse*. Here, we present high spatial resolution ($\sim$3") maps of CH_3OH, HNCO, HCN, and HCO^+ in the southern portion of the outflow containing B1 and B2, as observed with CARMA. The HNCO maps are the first interferometric observations of this species in L1157. The maps show distinct differences in the chemistry within the various shocked regions in L1157B. This is further supported through constraints of the molecular abundances using the non-LTE code RADEX. We find the east/west chemical differentiation in C2 may be explained by the contrast of the shock's interaction with either cold, pristine material or warm, previously-shocked gas, as seen in enhanced HCN abundances. In addition, the enhancement of HNCO abundance toward the the older shock, B2, suggests the importance of high-temperature O-chemistry in shocked regions.

WH11 4:37 – 4:52

MODELING THE AFTER-EFFECTS OF SHOCKS TOWARD L1157

ANDREW M BURKHARDT, *Department of Astronomy, The University of Virginia, Charlottesville, VA, USA*; BRETT A. McGUIRE, *NAASC, National Radio Astronomy Observatory, Charlottesville, VA, USA*; NIKLAUS M DOLLHOPF, *Department of Astronomy, The University of Virginia, Charlottesville, VA, USA*; ERIC HERBST, *Department of Chemistry, The University of Virginia, Charlottesville, VA, USA.*

Shocks have been found to be ubiquitous throughout the interstellar medium and in star forming regions. How these phenomena affect the chemistry, especially the interplay between gas-phase and grain-surface processes, in these regions has yet to be fully understood. In the prototypical shocked-outflow of L1157, we can study the effects that recent shocks ($\sim$$10^3$-$10^4$ years ago) can have on previously cold, quiescent gas, where many of the complex molecules are thought to be locked within grains. Toward a single shock event, C2, a significant chemical differentiation is observed between the previously shocked gas along the eastern wall and the newly shocked gas along the western wall. In addition, substantial enhancement of HNCO towards the younger shock, C1, may imply high-temperature O-chemistry is important soon after the passage of a shock. Here, we utilize the gas-grain chemical network model NAUTILUS in order to investigate the prominence of these effects.

WH12

FILAMENTARY STRUCTURE OF SERPENS MAIN AND SERPENS SOUTH SEEN IN N_2H^+, HCO^+, AND HCN

ERIN GUILFOIL COX, *Department of Astronomy, University of Illinois at Urbana-Champaign, Urbana, IL, USA*; MANUEL FERNANDEZ-LOPEZ, *Instituto Argentino de Radioastronomía, Centro Científico Tecnológico La Plata, Villa Elisa, Argentina*; LESLIE LOONEY, *Department of Astronomy, University of Illinois at Urbana-Champaign, Urbana, IL, USA*; HÉCTOR ARCE, *Astronomy Department, Yale University, New Haven, CT, USA*; LEE MUNDY, SHAYE STORM, *Department of Astronomy, University of Maryland, College Park, MD, USA*; ROBERT J HARRIS, *Department of Astronomy, University of Illinois at Urbana-Champaign, Urbana, IL, USA*; PETER J. TEUBEN, *Department of Astronomy, University of Maryland, College Park, MD, USA*.

We present the N_2H^+ ($J = 1 \rightarrow 0$) map of the Serpens Main and Serpens South molecular cloud obtained as part of the CARMA Large Area Star Formation Survey (CLASSy). The observations cover 150 arcmin^2 and 250 arcmin^2, respectively, and fully sample structures from 3000 AU to 3 pc with a velocity resolution of 0.16 km s^{-1}. They can be used to constrain the origin and evolution of molecular cloud filaments. The spatial distribution of the N_2H^+ emission is characterized by long filaments that resemble those observed in the dust continuum emission by Herschel. However, the gas filaments are typically narrower such that, in some cases, two or three quasi-parallel N_2H^+ filaments comprise a single observed dust continuum filament. Our results suggest that single filaments seen in Serpens South by Herschel may in fact be comprised of multiple narrower filaments. Some molecular filaments show velocity gradients along their major axis, and two are characterized by a steep velocity gradient in the direction perpendicular to the filament axis. The observed velocity gradient along one of these filaments was previously postulated as evidence for mass infall toward the central cluster, but these kind of gradients can be interpreted as projection of large-scale turbulence. Finally we compare the morphologies of these N_2H^+ filaments with those detected in HCO^+ and HCN. In Serpens South we find that the N_2H^+ and dust maps are well correlated, whereas HCO^+ and HCN do not have regularly have N_2H^+ counterparts. We postulate that this difference is due to large-scale shocks creating the HCO^+ and HCN emission.

WH13

SULFUR CHEMISTRY IN VY CANIS MAJORIS REVEALED BY ALMA

ANDREW M BURKHARDT, *Department of Astronomy, The University of Virginia, Charlottesville, VA, USA*; BRETT A. McGUIRE, *NAASC, National Radio Astronomy Observatory, Charlottesville, VA, USA*; GILLES ADANDE, *Astrochemistry, NASA Goddard Space Flight Center, Greenbelt, MD, USA*; LUCY M. ZIURYS, *Department of Astronomy, University of Arizona, Tucson, AZ, USA*; ANTHONY REMIJAN, *ALMA, National Radio Astronomy Observatory, Charlottesville, VA, USA*.

In O-rich circumstellar envelopes, such as the red hypergiant VY Canis Majoris (VY CMa), SO_2 and SO are particularly important species for understanding the sulfur chemistry in the region, especially with respect to H_2S. Unlike the prototypical circumstellar environments, which can be treated as spherically expanding shells, VY CMa has a rich and complex spatial and velocity structure due to numerous outflows punching through the spherical shells. This produces a dynamic, and potentially shock-rich, environment, resulting in competing processes such as photodissociation and shock-induced chemistry. From previous studies (Fu et al., 2010, Andande et al., 2013), sulfur-bearing species, such as SO and SO_2, are believed to trace these numerous outflows. Through the high spatial resolution of ALMA, we are able to resolve this structure and confirm this. Here, we present further analysis of the ALMA Band 7 and 9 Science Verification Survey of VY CMa and what it implies for the sulfur-chemistry in the wide variety of conditions presents here. We will discuss the difference in formation processes near the photosphere and far out into the asymmetric outflows.

WH14 5:28 – 5:43

TRACING THE ORIGINS OF NITROGEN BEARING ORGANICS TOWARD ORION KL WITH ALMA

BRANDON CARROLL, *Division of Chemistry and Chemical Engineering, California Institute of Technology, Pasadena, CA, USA*; NATHAN CROCKETT, *Geological and Planetary Sciences , California Institute of Techonolgy, Pasadena, CA, USA*; EDWIN BERGIN, *Department of Astronomy, University of Michigan, Ann Arbor, MI, USA*; GEOFFREY BLAKE, *Division of Chemistry and Chemical Engineering, California Institute of Technology, Pasadena, CA, USA*.

A comprehensive analysis of a broadband 1.2 THz wide spectral survey of the Orion Kleinmann-Low nebula (Orion KL) from the Herschel Space Telescope has shown that nitrogen bearing complex organics trace systematically hotter gas than O-bearing organics toward this source. The origin of this O/N dichotomy remains a mystery. If complex molecules originate from grain surfaces, N-bearing species may be more difficult to remove from grain surfaces than O-bearing organics. Theoretical studies, however, have shown that hot (T=300 K) gas phase chemistry can produce high abundances of N-bearing organics while suppressing the formation of O-bearing complex molecules. In order to distinguish these distinct formation pathways we have obtained extremely high angular resolution observations of methyl cyanide (CH_3CN) using the Atacama Large Millimeter/Submillimeter Array (ALMA) toward Orion KL. By simultaneously imaging $^{13}CH_3CN$ and CH_2DCN we map the temperature structure and D/H ratio of CH_3CN. We will present the initial results of these observations and discuss their implications for the formation of N-bearing organics in the interstellar medium.

WH15 5:45 – 6:00

SPECTROSCOPIC FITS TO THE ALMA SCIENCE VERIFICATION BAND 6 SURVEY OF THE ORION HOT CORE AND COMPACT RIDGE

SATYAKUMAR NAGARAJAN, JAMES P. McMILLAN, *Department of Physics, The Ohio State University, Columbus, OH, USA*; ANDREW M BURKHARDT, *Department of Astronomy, The University of Virginia, Charlottesville, VA, USA*; CHRISTOPHER F. NEESE, FRANK C. DE LUCIA, *Department of Physics, The Ohio State University, Columbus, OH, USA*; ANTHONY REMIJAN, *ALMA, National Radio Astronomy Observatory, Charlottesville, VA, USA*.

We have studied methyl cyanide, one of the so-called 'astronomical weeds', in the 200–277 GHz band. We have experimentally gathered a set of intensity calibrated, complete, and temperature resolved spectra from across the temperature range of 231–351 K. Using our previously reported method of analysis[a], the point by point method, we are capable of generating the complete spectrum at astronomically significant temperatures. Lines, of nontrivial intensity, which were previously not included in the available astrophysical catalogs have been found. Lower state energies and line strengths have been found for a number of lines which are not currently present in the catalogs. The extent to which this may be useful in making assignments will be discussed.

---

[a]J. McMillan, S. Fortman, C. Neese, F. DeLucia, ApJ. 795, 56 (2014)

WI. Mini-symposium: Spectroscopy in Traps

Wednesday, June 22, 2016 – 1:30 PM

Room: B102 Chemical and Life Sciences

Chair: Jos Oomens, Radboud University, Nijmegen, The Netherlands

---

WI01   ***INVITED TALK***   1:30 – 2:00

ISOLATING SITE-SPECIFIC SPECTRAL SIGNATURES OF INDIVIDUAL WATER MOLECULES IN H-BONDED NETWORKS WITH ISOTOPOMER-SELECTIVE, IR-IR DOUBLE RESONANCE VIBRATIONAL PREDISSOCIATION SPECTROSCOPY

CONRAD T. WOLKE, <u>MARK JOHNSON</u>, *Department of Chemistry, Yale University, New Haven, CT, USA.*

We will discuss an experimental method that directly yields the embedded correlations between the two OH stretches and the intramolecular bending modes associated with a single H_2O water molecule embedded in an otherwise all-D isotopologue. This is accomplished using isotopomer-selective IR-IR hole-burning on the $Cs^+(D_2O)_5(H_2O)$ clusters formed by gas-phase exchange of a single, intact H_2O molecule for D_2O in the $Cs^+(D_2O)_6$ ion. The OH stretching pattern of the $Cs^+(H_2O)_6$ isotopologue is accurately recovered by superposition of the isotopomer spectra, thus establishing that the H_2O incorporation is random and that the OH stretching manifold is largely due to contributions from decoupled water molecules. This behavior enables a powerful new way to extract structural information from vibrational spectra of size-selected clusters by explicitly identifying the local environments responsible for specific infrared features. Extension of this method to address the degree to which OH stretches are decoupled in the protonated water clusters will also be discussed.

WI02   2:05 – 2:20

IR-UV DOUBLE RESONANCE SPECTROSCOPY OF A COLD PROTONATED FIBRIL-FORMING PEPTIDE: NNQQNY·H^+

<u>ANDREW F DeBLASE</u>, CHRISTOPHER P HARRILAL, PATRICK S. WALSH, SCOTT A McLUCKEY, TIMOTHY S. ZWIER, *Department of Chemistry, Purdue University, West Lafayette, IN, USA.*

Protein aggregation to form amyloid-like fibrils is a purported molecular manifestation that leads to Alzheimer's, Huntington's, and other neurodegenerative diseases. The propensity for a protein to aggregate is often driven by the presence of glutamine (Q) and asparagine (N) rich tracts within the primary sequence. For example, Eisenberg and coworkers [Nature 2006, 435, 773] have shown by X-ray crystallography that the peptides NNQQNY and GNNQQNY aggregate into a parallel β-sheet configuration with side chains that intercalate into a "steric zipper". These sequences are commonly found at the N-terminus of the prion-determining domain in the yeast protein Sup35, a typical fibril-forming protein. Herein, we invoke recent advances in cold ion spectroscopy to explore the nascent conformational preferences of the protonated peptides that are generated by electrospray ionization. Towards this aim, we have used UV and IR spectroscopy to record conformation-specific photofragment action spectra of the NNQQNY monomer cryogenically cooled in an octopole ion trap. This short peptide contains 20 hydride stretch oscillators, leading to a rich infrared spectrum with at least 18 resolved transitions in the 2800-3800 cm^{-1} region. The infrared spectrum suggests the presence of both a free acid OH moiety and an H-bonded tyrosine OH group. We compare our results with resonant ion dip infrared spectra (RIDIRS) of the acyl/NH-benzyl capped neutral glutamine amino acid and its corresponding dipeptide: Ac-Q-NHBn and Ac-QQ-NHBn, respectively. These comparisons bring empirical insight to the NH stretching region of the spectrum, which contains contributions from free and singly H-bonded NH_2 side-chain groups, and from peptide backbone amide NH groups. We further compare our spectrum to harmonic calculations at the M05-2X/6-31+G* level of theory, which were performed on low energy structures obtained from Monte Carlo conformational searches using the Amber* and OPLS force fields to assess the presence of sidechain-sidechain and sidechain-backbone interactions.

WI03 2:22 – 2:37

VIBRATIONAL AND ROTATIONAL SPECTROSCOPY OF CD_2H^+

OSKAR ASVANY, PAVOL JUSKO, SANDRA BRÜNKEN, STEPHAN SCHLEMMER, *I. Physikalisches Institut, Universität zu Köln, Köln, Germany.*

The lowest rotational levels (J=0-5) of the CD_2H^+ ground state have been probed by high-resolution rovibrational and pure rotational spectroscopy in a cryogenic 22-pole ion trap. For this, the ν_1 rovibrational band has been revisited[a], detecting 107 transitions, among which 35 are new. The use of a frequency comb system allowed to measure the rovibrational transitions with high precision and accuracy, typically better than 1 MHz. The high precision has been confirmed by comparing combination differences in the ground and vibrationally excited state. For the ground state, this allowed for equally precise predictions of pure rotational transitions, 24 of which have been measured directly by a novel IR - mm-wave double resonance method[b].

[a] M.-F. Jagod et al, J. Molec. Spectrosc. 153, 666, 1992
[b] S. Gärtner et al, J. Phys. Chem. A 117, 9975, 2013

WI04 2:39 – 2:54

INFRARED PREDISSOCIATION SPECTROSCOPY OF THE HYDROCARBON CATIONS C_3H^+, C_2H^+, and $C_3H_2^+$

SANDRA BRÜNKEN, *I. Physikalisches Institut, Universität zu Köln, Köln, Germany*; FILIPPO LIPPARINI, JÜRGEN GAUSS, *Institut für Physikalische Chemie, Universität Mainz, Mainz, Germany*; ALEXANDER STOFFELS, BRITTA REDLICH, LEX VAN DER MEER, GIEL BERDEN, JOS OOMENS, *Institute for Molecules and Materials (IMM), Radboud University Nijmegen, Nijmegen, Netherlands*; STEPHAN SCHLEMMER, *I. Physikalisches Institut, Universität zu Köln, Köln, Germany.*

Reactive hydrocarbon cations play an important role in the astrochemistry of the interstellar medium, but spectroscopic data, needed for their identification in astronomical observations, is sparse. Here we report the first gas-phase vibrational spectra of the linear C_3H^+ ($^1\Sigma$), the radical cation C_2H^+ ($^3\Pi$), and the linear-/cyclic-$C_3H_2^+$ ($^2\Pi$ /$^2A_1$, resp.). Broadband spectra were recorded by Ne- and He-messenger infrared-predissociation (IR-PD) action spectroscopy in a cryogenic (4 − 11 K) ion trap instrument (FELion) in the $250 - 3500$ cm^{-1} range using a free electron laser and a MIR-OPO at the FELIX (Free-Electron Laser for Infrared eXperiments) laboratory. The band positions (determined with a precision of $1 - 2$ cm^{-1}) covering the C-H and C-C stretching as well as several bending modes are compared to high-level (CCSD(T) with large basis sets) quantum-chemical calculations with an emphasis on anharmonic effects and on the influence of the rare-gas messenger atom. The experimental and theoretical data provide a solid basis for subsequent IR high-resolution studies, with the ultimate goal to predict and measure accurate rotational spectra for a radio-astronomical search of these molecular ions in space.

WI05 2:56 – 3:11

FREQUENCY COMB ASSISTED IR MEASUREMENTS OF H_3^+, H_2D^+ AND D_2H^+ TRANSITIONS

PAVOL JUSKO, OSKAR ASVANY, STEPHAN SCHLEMMER, *I. Physikalisches Institut, Universität zu Köln, Köln, Germany.*

We present recent measurements of the fundamental transitions of H_3^+, H_2D^+ and D_2H^{+a} in a 4 K 22-pole trap[b] by action spectroscopic techniques. Either Laser Induced Inhibition of Cluster Growth (He attachment at T$\approx$4 K), endothermic reaction of H_3^+ with O_2, or deuterium exchange has been used as measurement scheme. We used a 3 μm optical parametric oscillator coupled to a frequency comb[c] in order to achieve accuracy generally below 1 MHz. Five transitions of H_3^+, eleven of H_2D^+ and ten of D_2H^+ were recorder in our spectral range. We compare our H_3^+ results with two previous frequency comb assisted works[de]. Moreover, accurate determination of the frequency allows us to predict pure rotational transitions for H_2D^+ and D_2H^+ in the THz range.

[a] P. Jusko, C. Konietzko, S. Schlemmer, O. Asvany, *J. Mol. Spec.* 319 (2016) 55
[b] O. Asvany, S. Brünken, L. Kluge, S. Schlemmer, *Appl. Phys. B* 114 (2014) 203
[c] O. Asvany, J. Krieg, S. Schlemmer, *Rev. Sci. Instr.* 83 (2012) 093110
[d] J.N. Hodges, A.J. Perry, P.A. Jenkins, B.M. Siller, B.J. McCall, *J. Chem. Phys.* 139 (2013) 164201
[e] H.-C. Chen, C.-Y. Hsiao, J.-L. Peng, T. Amano, J.-T. Shy, *Phys. Rev. Lett.* 109 (2012) 263002

Intermission

WI06 3:30–3:45

INFRARED SPECTROSCOPY OF IONS IN SELECTED ROTATIONAL AND SPIN-ORBIT STATES

UGO JACOVELLA, JOSEF A. AGNER, HANSJÜRG SCHMUTZ, FREDERIC MERKT, *Laboratorium für Physikalische Chemie, ETH Zurich, Zurich, Switzerland.*

First results are presented obtained using an experimental setup developed to record IR spectra of rotationally state-selected ions. The method we use is a state-selective version of a method developed by Schlemmer *et al.* [a] to record IR spectra of ions.

Ions are produced in specific rotational levels using mass-analysed threshold ionisation (MATI) spectroscopy combined with single-photon excitation of neutral molecules in supersonic expansions with a vacuum-ultraviolet laser. The ions generated by pulsed-field ionisation of Rydberg states of high principal quantum number ($n \approx 200$) are extracted toward an octupole ion guide containing a neutral target gas. Prior to entering the octupole the ions are excited by an IR laser. The target gas is chosen so that only excited ions react to form product ions. These product ions are detected mass selectively as function of the IR laser wavenumber.

To illustrate this method, we present IR spectra of $C_2H_2^+$ in selected rotational levels of the $^2\Pi_{3/2}$ and $^2\Pi_{1/2}$ spin-orbit components of the electronic ground state.

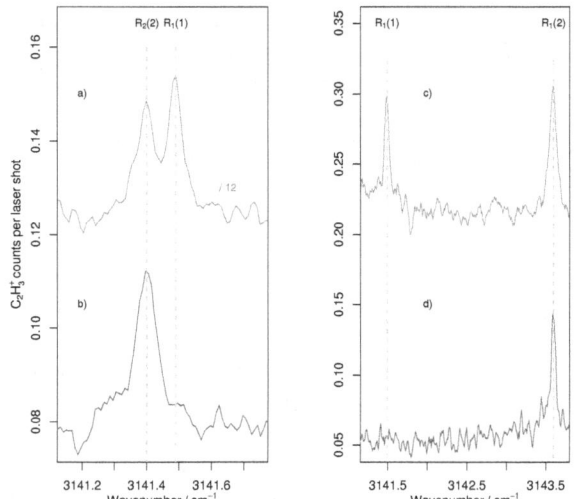

IR spectra of $C_2H_2^+$ with(b, and d) and without (a, and c) selection of the rotational state by MATI.

---

[a]Schlemmer *et al.*, J. Chem. Phys. **117**, 2068 (2002)

WI07 3:47–4:02

PRECISION SPECTROSCOPY ON SINGLE COLD TRAPPED MOLECULAR NITROGEN IONS

GREGOR HEGI, KAVEH NAJAFIAN, MATTHIAS GERMANN, ILIA SERGACHEV, STEFAN WILLITSCH, *Department of Chemistry, University of Basel, Basel, Switzerland.*

The ability to precisely control and manipulate single cold trapped particles has enabled spectroscopic studies on narrow transitions of ions at unprecedented levels of precision[a]. This has opened up a wide range of applications, from tests of fundamental physical concepts, e.g., possible time-variations of fundamental constants, to new and improved frequency standards. So far most of these experiments have concentrated on atomic ions. Recently, however, attention has also been focused on molecular species, and molecular nitrogen ions have been identified as promising candidates for testing a possible time-variation of the proton/electron mass ratio[b]. Here, we report progress towards precision-spectroscopic studies on dipole-forbidden vibrational transitions in single trapped N_2^+ ions[c]. Our approach relies on the state-selective generation of single N_2^+ ions[d], subsequent infrared excitation using high intensity, narrow-band quantum-cascade lasers and a quantum-logic scheme for non-destructive state readout. We also characterize processes limiting the state lifetimes in our experiment, which impair the measurement fidelity.

---

[a]P. O. Schmidt *et. al.*, *Science* **309** (2005), 749.
[b]M. Kajita *et. al.*, *Phys. Rev. A* **89** (2014), 032509.
[c]M. Germann , X. Tong, S. Willitsch, *Nature Physics* **10** (2014), 820.
[d]X. Tong, A. Winney, S. Willitsch, *Phys. Rev. Lett.* **105** (2010), 143001.

WI08 4:04 – 4:19

UV PHOTODISSOCIATION SPECTROSCOPY OF TEMPERATURE CONTROLLED HYDRATED PHENOL CLUSTER CATION

ITARU KURUSU, REONA YAGI, YASUTOSHI KASAHARA, HARUKI ISHIKAWA, *Department of Chemistry, School of Science, Kitasato University, Sagamihara, Japan.*

Owing to various developments of spectroscopic techniques, microscopic hydration structures of various clusters in the gas phase have been determined so far. The next step for further understanding of the microscopic hydration is to reveal the temperature effect, such as a fluctuation of the hydration structure. Thus, we have been carrying out photodissociation spectroscopy on the hydrated phenol cation clusters, $[PhOH(H_2O)_n]^+$, trapped in our temperature-variable ion trap[a].

After the last symposium[b], we succeeded in improving our experimental condition and recorded the UV photodissociation spectra of $[PhOH(H_2O)_5]^+$ at the trap temperatures of 20, 50, and 100 K. We identified three groups of bands by their temperature dependence in the spectra. Based on the results of the DFT calculations, we estimated the temperature dependence of the relative populations among the isomers. As a results, the isomers were grouped into three groups having different motifs of the hydrogen-bond structures. Comparing the experimental with the theoretical results, we assigned the relation between the band carriers and the hydrogen-bond structure motifs. Details of the discussion will be presented in the paper.

[a]H. Ishikawa, T. Nakano, T. Eguchi, T. Shibukawa, K. Fuke, *Chem. Phys. Lett.* **514**, 234 (2011).

[b]R. Yagi, Y. Kasahara, H. Ishikawa, WH12, the 70th International Symposium on Molecular Spectroscopy (2015)

WI09 4:21 – 4:36

STRUCTURAL INFORMATION INFERENCE FROM LANTHANOID COMPLEXING SYSTEMS: PHOTOLUMINESCENCE STUDIES ON ISOLATED IONS

JEAN FRANCOIS GREISCH, MICHAEL E. HARDING, *Institute of Nanotechnology, Karlsruhe Institute of Technology, Karlsruhe, Germany*; JIRI CHMELA, WILLEM M. KLOPPER, *Institute of Physical Chemistry, Karlsruhe Institute of Technology, Karlsruhe, Germany*; DETLEF SCHOOSS, *Institute of Nanotechnology, Karlsruhe Institute of Technology, Karlsruhe, Germany*; MANFRED M KAPPES, *Institute of Physical Chemistry, Karlsruhe Institute of Technology, Karlsruhe, Germany.*

The application of lanthanoid complexes ranges from photovoltaics and light-emitting diodes to quantum memories and biological assays. Rationalization of their design requires a thorough understanding of intramolecular processes such as energy transfer, charge transfer, and non-radiative decay involving their subunits. Characterization of the excited states of such complexes considerably benefits from mass spectrometric methods since the associated optical transitions and processes are strongly affected by stoichiometry, symmetry, and overall charge state. We report herein spectroscopic measurements on ensembles of ions trapped in the gas phase and soft-landed in neon matrices. Their interpretation is considerably facilitated by direct comparison with computations.

The combination of energy- and time-resolved measurements on isolated species with density functional as well as ligand-field and Franck-Condon computations enables us to infer structural as well as dynamical information about the species studied. The approach is first illustrated for sets of model lanthanoid complexes whose structure and electronic properties are systematically varied via the substitution of one component (lanthanoid or alkali,alkali-earth ion): (i) systematic dependence of ligand-centered phosphorescence on the lanthanoid(III) promotion energy and its impact on sensitization, and (ii) structural changes induced by the substitution of alkali or alkali-earth ions in relation with structures inferred using ion mobility spectroscopy. The temperature dependence of sensitization is briefly discussed. The focus is then shifted to measurements involving europium complexes with doxycycline an antibiotic of the tetracycline family. Besides discussing the complexes' structural and electronic features, we report on their use to monitor enzymatic processes involving hydrogen peroxide or biologically relevant molecules such as adenosine triphosphate (ATP).

WI10 4:38 – 4:53

PHOTO-IONIZATION AND PHOTO-DISSOCIATION OF TRAPPED PAH CATIONS

CHRISTINE JOBLIN, JUNFENG ZHEN, SARAH RODRIGUEZ CASTILLO, GIACOMO MULAS, HASSAN SABBAH, *IRAP, Université de Toulouse 3 - CNRS, Toulouse, France*; AUDE SIMON, *LCPQ, Université de Toulouse 3 - CNRS, Toulouse, France*; ALEXANDRE GIULIANI, LAURENT NAHON, *DESIRS beamline, Synchrotron SOLEIL, Gif-sur-Yvette, France*; SERGE MARTIN, *UMR 5306, ILM University Lyon 1 and CNRS, Villeurbanne, France*; JEAN-PHILIPPE CHAMPEAUX, *LCAR, Université de Toulouse 3 - CNRS, Toulouse, France*; PAUL M. MAYER, *Department of Chemistry and Biomolecular Sciences,, University of Ottawa, Ottawa, Canada.*

In astrophysical environments, polycyclic aromatic hydrocarbons (PAHs) are submitted to VUV photons of energy up to $\sim$20 eV. In the laboratory, photoelectron-photoion spectroscopy is usually performed using VUV synchrotron radiation, in which the same photon (15-25 eV), is used to ionize and dissociate PAHs[a]. These experiments explore specific conditions and complementary studies in ion traps are required for a wider investigation of interstellar conditions.

We have used the LTQ linear ion trap available on the DESIRS VUV beamline at SOLEIL to study the interaction of PAH cations with photons in the 7-20 eV range. We recorded by action spectroscopy the relative intensities of photo-fragmentation and photo-ionization of eight PAH cations ranging in size from 14 to 24 carbon atoms and with different structures. We found that, at photon energies below $\sim$13.6 eV, fragmentation dominates for the smaller species, while for larger species ionization is competitive immediately above the second ionization potential. At higher photon energies, all species behave similarly, the ionization yield gradually increases, levelling off between 0.8 and 0.9 at $\sim$18 eV [b].

We have also recorded the competition between the different dissociation channels as a function of the VUV photon energy, such as the C_2H_2 versus H/H_2 loss. We will discuss how these data can be compared to results of photoelectron spectroscopy performed on neutral PAHs at the VUV beamline at the Swiss Light Source.[c]. Acknowledgments[d]

[a]H.W. Jochims et al., *Astron. & Astrophys.* **420** (1994), 307-317; P. M. Mayer et al., *J. Chem. Phys.* **134** (2011), 244312-244312-8
[b]J. Zhen et al., *Astron. & Astrophys.* (2016), in press
[c]B. West et al., *J. Phys. Chem. A* **118** (2014), 7824-7831; B. West et al., *J. Phys. Chem. A* **118** (2014), 9870-9878
[d]European Research Council grant ERC-2013-SyG, Grant Agreement n. 610256 NANOCOSMOS.

WI11 4:55 – 5:10

PROGRESS OF THE JILA ELECTRON EDM EXPERIMENT

YAN ZHOU, WILLIAM CAIRNCROSS, MATT GRAU, DAN GRESH, KIA BOON NG, YIQI NI, JUN YE, ERIC CORNELL, *JILA, National Institute of Standards and Technology and Univ. of Colorado Department of Physics, University of Colorado, Boulder, Boulder, CO, USA.*

A more accurate value for the permanent electric dipole moment of the electron (eEDM), whether consistent with zero or otherwise, will have important implications for extensions to the Standard Model. The JILA eEDM experiment uses trapped HfF^+ ions to attain a large effective electric field and a long measurement coherence time. We prepare HfF^+ ions in a low-lying, metastable $^3\Delta_1$ state and perform Ramsey spectroscopy with free-evolution times of > 500 ms between two Zeeman sub-levels in the presence of rotating bias electric and magnetic fields. In this talk, we present initial results of an eEDM-sensitive 100-hour data run and a thorough investigation of various systematic errors. Several of our leading systematics have been suppressed to the 10^{-30} e.cm level. In addition, we report two ongoing experiments aimed towards increasing the statistical sensitivity: (1) applying STImulated Raman Adiabatic Passage (STIRAP) for increased coherent population transfer to the eEDM-sensitive $^3\Delta_1$ state; and (2) implementing a new ion counting detector for increased detection efficiency and for suppressing shot-to-shot noise.

WJ. Small molecules
Wednesday, June 22, 2016 – 1:30 PM
Room: 217 Noyes Laboratory

Chair: Colan Linton, University of New Brunswick, Fredericton, NB, Canada

WJ01 1:30 – 1:45

SPECTRAL LINE SHAPE PARAMETERS FOR THE ν_1, ν_2, and ν_3 BANDS OF HDO: SELF AND CO_2 BROADENED

V. MALATHY DEVI, D. CHRIS BENNER, *Department of Physics, College of William and Mary, Williamsburg, VA, USA*; KEEYOON SUNG, TIMOTHY J. CRAWFORD, *Jet Propulsion Laboratory, California Institute of Technology, Pasadena, CA, USA*; ROBERT R. GAMACHE, CANDICE L. RENAUD, *Department of Environmental, Earth, and Atmospheric Sciences, University of Massachusetts Lowell, Lowell, MA, USA*; ARLAN MANTZ, *Department of Physics, Astronomy and Geophysics, Connecticut College, New London, CT, USA*; MARY ANN H. SMITH, *Science Directorate, NASA Langley Research Center, Hampton, VA, USA*; GERONIMO L. VILLANUEVA, *Astrochemistry, NASA Goddard Space Flight Center, Greenbelt, MD, USA*.

To provide precise information relevant to Martian atmospheric remote sensing, high resolution high signal-to-noise ratio spectra of HDO in mixture with CO_2 were recorded in the ν_1, ν_2, and ν_3 fundamental bands between 2.7 and 7 μm regions. The spectra were obtained with the Bruker IFS-125HR Fourier transform spectrometer at the Jet Propulsion Laboratory along with two specially built coolable absorption cells with path lengths of 0.2038 m[a] and 20.941 m[b] at various sample gas temperatures ($\sim$220 – 296 K), total sample pressures and volume mixing ratios. A multispectrum nonlinear least squares technique[c] was applied to fit simultaneously all the spectra obtained. The measured line parameters include accurate line positions, intensities, self- and CO_2-broadened Lorentz halfwidth and pressure-shift coefficients, and temperature dependences of CO_2 broadened HDO halfwidth and pressure-shift coefficients. Line mixing coefficients using the relaxation matrix formalism and quadratic speed dependence parameters were also measured where appropriate. Example results for select transitions in each band will be presented and comparisons made to other measured/calculated values[d].

[a]K. Sung, A.W. Mantz, M.A.H. Smith, L.R. Brown, T.J. Crawford, V.M. Devi, D.C. Benner. J. Mol. Spectrosc. 162 (2010) 124-134.

[b]A.W. Mantz, K. Sung, T.J. Crawford, L.R. Brown, M.A.H. Smith, V.M. Devi, D.C. Benner, J. Mol. Spectrosc. 304 (2014) 12-24.

[c]D.C. Benner, C.P. Rinsland, V. Malathy Devi, M.A. H. Smith, and D. Atkins. JQSRT 53 (1995) 705-721.

[d]Research described in this paper are performed at the College of William and Mary, Jet Propulsion Laboratory, California Institute of Technology, Connecticut College, and NASA Langley Research Center under contracts and cooperative agreements with the National Aeronautics and Space Administration. RRG and CLR were supported by the National Science Foundation through Grant # AGS-1156862.

WJ02 1:47 – 2:02

SPECTRAL LINE SHAPES IN THE $2\nu_3$ Q BRANCH OF $^{12}CH_4$

V. MALATHY DEVI, D. CHRIS BENNER, *Department of Physics, College of William and Mary, Williamsburg, VA, USA*; KEEYOON SUNG, LINDA R. BROWN, TIMOTHY J. CRAWFORD, SHANSHAN YU, *Jet Propulsion Laboratory, California Institute of Technology, Pasadena, CA, USA*; MARY ANN H. SMITH, SYED ISMAIL, *Science Directorate, NASA Langley Research Center, Hampton, VA, USA*; ARLAN MANTZ, *Department of Physics, Astronomy and Geophysics, Connecticut College, New London, CT, USA*; VINCENT BOUDON, *Laboratoire ICB, CNRS/Université de Bourgogne, DIJON, France*.

We will present the first experimental measurements of spectral line shapes (self- and air-broadened half width, pressure shift, and line mixing (via off-diagonal relaxation matrix element) coefficients and their temperature dependences, where appropriate, for transitions in the $2\nu_3$ Q branch manifolds of $^{12}CH_4$ in the 1.6 μm region. Employing a multispectrum nonlinear least squares technique[a], we simultaneously fitted 23 high-resolution spectra of $^{12}CH_4$ and mixtures of $^{12}CH_4$ in air, recorded at different pressure-temperature combinations between 130 and 296 K. These data were recorded using the Bruker IFS 125 HR Fourier transform spectrometer at the Jet Propulsion Laboratory together with two coolable sample cells[b,c]. By applying a set of constraints to the parameters of severely blended transitions, a self-consistent set of broadening, shift and line mixing parameters for CH_4-CH_4 and CH_4-air collisions were retrieved. A quadratic speed dependence parameter common for all transitions in each Q(J) manifold was determined. In addition to line shape parameters, line positions and line intensities were also measured for over 100 transitions in the whole Q branch region (5996.5 - 6007.7 cm^{-1}). Comparisons of present results with values in HITRAN2012 will be provided[d]

[a]D.C. Benner, C.P. Rinsland, V. Malathy Devi, M.A. H. Smith, and D. Atkins. JQSRT 53 (1995) 705-721

[b]K. Sung, A.W. Mantz, M.A.H. Smith, L.R. Brown, T.J. Crawford, V.M. Devi, D.C. Benner. J.Mol. Spectrosc. 162 (2010)124-134.

[c]A.W. Mantz, K. Sung, T.J. Crawford, L.R. Brown, M.A.H. Smith, V.M. Devi, D.C. Benner, J. Mol. Spectrosc. 304 (2014) 12-24.

[d]Research described in this paper are performed at the College of William and Mary, Jet Propulsion Laboratory, California Institute of Technology, Connecticut College, and NASA Langley Research Center under contracts and cooperative agreements with the National Aeronautics and Space Administration.

WJ03 2:04 – 2:19

ABSORPTION CROSS SECTIONS OF HOT HYDROCARBONS IN THE 3 μm REGION

CHRISTOPHER A. BEALE, *Department of Ocean, Earth and Atmospheric Sciences, Old Dominion University, Norfolk, VA, USA*; ROBERT J. HARGREAVES, ERIC M. BUZAN, PETER F. BERNATH, *Department of Chemistry and Biochemistry, Old Dominion University, Norfolk, VA, USA.*

The 3 μm region of ethane, propane and propene contains a number of vibrational modes dominated by various C-H stretches. Transmission spectra for these hydrocarbons have been obtained at high resolution (0.005 cm^{-1}) and at elevated temperatures (up to 773 K). The integrated absorption cross sections over the isolated 3 μm region are calibrated to those from the Pacific Northwest National Laboratory (PNNL) in order to obtain an effective pressure for the sample. With the Jovian Infrared Auroral Mapper (JIRAM on Juno) due to arrive at Jupiter in July 2016, these cross sections will find use in the study of hot emission from hydrocarbons in the auroral regions of Jupiter. They are also appropriate for modeling atmospheres of hot Jupiter exoplanets and brown dwarfs.

WJ04 2:21 – 2:36

LINE LISTS AND ASSIGNMENTS OF HOT AMMONIA IN THE INFRARED

CHRISTOPHER A. BEALE, *Department of Ocean, Earth and Atmospheric Sciences, Old Dominion University, Norfolk, VA, USA*; ROBERT J. HARGREAVES, ANDY WONG, PETER F. BERNATH, *Department of Chemistry and Biochemistry, Old Dominion University, Norfolk, VA, USA.*

Transmission spectra for hot ammonia (23-700 °C) have been recorded in the region 5500-8800 cm^{-1} and line lists have been produced from these spectra that include line positions, intensities and empirical lower state energies. Transmission spectra were obtained at high resolution (0.02 cm^{-1}) by recording absorption, emission and background spectra using a Bruker IFS 125HR with a hot quartz cell heated by a tube furnace. Temperature-appropriate line lists for the 2500-5500 cm^{-1} region have been completed and assignment of both regions is undwer way. These empirical line lists will be compared to the theoretical line lists for ammonia available from the ExoMol project and can be used in efforts to model the atmospheres of exoplanets and brown dwarfs.

WJ05 2:38 – 2:53

THE EFFECT OF TERMINAL SUBSTITUTION ON THE HELICAL CARBON STRUCTURE OF FLUORO-ALKANE CHAINS: A PURE ROTATIONAL STUDY OF CH_2OH-$C_{n-1}F_{2n-1}$ (n = 4, 5,& 6)

AARON Z. A. SCHWARTZ, *Natural and Social Science, Purchase College SUNY, Purchase, NY, USA*; MARK P. MATURO, DANIEL A. OBENCHAIN, *Department of Chemistry, Wesleyan University, Middletown, CT, USA*; S. A. COOKE, *Natural and Social Science, Purchase College SUNY, Purchase, NY, USA.*

Continuing a series of studies to investigate the change in structure of hydrocarbons as the amount of fluorination is increased to varying degrees of substitution, we present a survey on the change in the helical nature of the fluorinated carbon backbone when a $^-CH_2OH$ group is substituted for a terminal $^-CF_3$ group. Spectra for 1H,1H-heptafluorobutan-1-ol, 1H,1H-nonafluoropentan-1-ol, and 1H,1H-undecafluorohexan-1-ol were collected separately using a chirped-pulse FTMW spectrometer in the range of 7-13 GHz. Only one conformation was observed for each molecule. Additional measurements of the 1H,1H-heptafluorobutan-1-ol were completed using a Balle-Flygare cavity instrument. Assignments of the singly-substituted ^{13}C isotopologues of the 1H,1H-heptafluorobutan-1-ol were also measured. A comparison of both *ab initio* and experimental structures will be presented.

WJ06 2:55 – 3:05

DEVELOPMENT OF A NEW DETECTION SCHEME TO PROBE PREDISSOCIATED LEVELS OF THE S_1 STATE OF ACETYLENE

JUN JIANG, TREVOR J. ERICKSON, *Department of Chemistry, MIT, Cambridge, MA, USA*; ANTHONY MERER, *Department of Chemistry, University of British Columbia, Vancouver, BC, Canada*; ROBERT W FIELD, *Department of Chemistry, MIT, Cambridge, MA, USA*.

A new spectroscopic scheme has been developed to probe the predissociated levels of the S_1 state of acetylene. Our new scheme is based on detection of visible fluorescence that is a result of multi-photon excitation of acetylene (resonantly through single rovibronic S_1 levels). The new detection scheme is not subject to decreases in fluorescence quantum yield of S_1 levels that lie above the predissociation limit, and laser scatter-light can be easily eliminated by a long-pass filter with a cutoff in the visible range. For the S_1 predissociated levels, the new detection scheme offers much improved signal-to-noise ratio compared to the conventional laser-induced fluorescence technique, based on detection of UV fluorescence from the S_1 levels. The new method is also easier to implement than various H-atom detection schemes, which involve one additional laser of different wavelength than the excitation wavelength. Based on the power dependence and lifetime of the fluorescence signals, electronically excited C_2H and/or C_2 fragments are the likely emitters of the detected visible fluorescence. The new method is currently being used to extend the vibrational and rotational assignments of both gerade and ungerade levels of the S_1 state of acetylene in the region of the *cis-trans* isomerization barrier, >1000 cm^{-1} above the onset of S_1 predissociation.

WJ07 3:07 – 3:22

FOURIER TRANSFORM MICROWAVE SPECTROSCOPY OF Sc^{13}C$_2$ AND Sc^{12}C^{13}C: ESTABLISHING AN ACCURATE STRUCTURE OF ScC$_2$ ($\tilde{X}^2A_1$)

MARK BURTON, *Department of Chemistry and Biochemistry, University of Arizona, Tucson, AZ, USA*; DeWAYNE T HALFEN, *Steward Observatory, University of Arizona, Tucson, AZ, USA*; JIE MIN, *Department of Chemistry and Biochemistry, University of Arizona, Tucson, AZ, USA*; LUCY M. ZIURYS, *Department of Chemistry and Biochemistry; Department of Astronomy, Arizona Radio Observatory, University of Arizona, Tuscon, AZ, USA*.

Pure rotational spectra of Sc^{13}C$_2$ and Sc^{12}C^{13}C ($\tilde{X}^2A_1$) have been obtained using Fourier Transform Microwave methods. These molecules were created from scandium vapor in combination with ^{13}CH$_4$ and/or ^{12}CH$_4$, diluted in argon, using a Discharge Assisted Laser Ablation Source (DALAS). Transitions in the frequency range of 14-30 GHz were observed for both species including hyperfine splitting due to the nuclear spin of Sc ($I = 7/2$) and ^{13}C ($I = 1/2$). Rotational, spin-rotational, and hyperfine constants have been determined for Sc^{13}C$_2$ and Sc^{12}C^{13}C, as well as a refined structure for ScC$_2$. In agreement with theoretical calculations and previous Sc^{12}C$_2$ results, these data confirm a cyclic (or T-shaped) structure for this molecule.

Scandium carbides have been shown to form endohedral-doped fullerenes, which have unique electrical and magnetic properties due to electron transfer between the metal and the carbon-cage. Spectroscopy of ScC$_2$ provides data on model systems for comparison with theory.

WJ08 3:24 – 3:39

ANALYSIS OF QUARTET AND DOUBLET STATES OF NO MOLECULE EXCITED BY GLOW DISCHARGE

MOHAMMED A GONDAL, *Department of Physics, King Fahd University of Petroleum and Minerals, Dhahran, Saudi Arabia*.

In this work, we report the fluorescence emission spectra of NO molecules excited using a low power glow discharge under different experimental conditions such as different gas pressure, buffer gases, NO concentration, discharge voltage and time evolution of Ar/NO density ratio as well. This glow discharge electronic excitation populated different high lying energy states like quartet and doublet states of NO in its proximity such as the A $^2\Sigma$ ($\nu = 2$), b $^4\Sigma$ - ($\nu = 3$), B $^2\Pi$ ($\nu = 4$) and X $^2\Pi$ ($\nu = 33 - 32$) states. Due to intersystem crossing, emission lines originating from these levels to lower lying states were recorded and spectral line assignments were performed. Observed systems included b b $^4\Sigma$– a $^4\Pi$, B $^2\Pi$ - a $^4\Pi$, a $^4\Pi$ - X $^2\Pi$, A $^2\Pi$ -X $^2\Pi$ and X $^2\Pi$ -X $^2\Pi$. This investigation could assist in understanding the interesting features of NO molecule such as collision processes, population dynamics and energy transfer within molecules.

Intermission

WJ09

SPECTROSCOPY OF THE $X^1\Sigma^+$, $A^1\Pi$ and $B^1\Sigma^+$ ELECTRONIC STATES OF MgS

NICHOLAS CARON, <u>DENNIS W. TOKARYK</u>, *Department of Physics, University of New Brunswick, Fredericton, NB, Canada*; ALLAN G. ADAM, *Department of Chemistry, University of New Brunswick, Fredericton, NB, Canada*; COLAN LINTON, *Department of Physics, University of New Brunswick, Fredericton, NB, Canada*.

The spectra of some astrophysical sources contain signatures from molecules containing magnesium or sulphur atoms. Therefore, we have extended previous studies of the diatomic molecule MgS, which is a possible candidate for astrophysical detection. Microwave spectra of $X^1\Sigma^+$, the ground electronic state, were reported in 1989[a] and 1997[b], and the $B^1\Sigma^+$–$X^1\Sigma^+$ electronic absorption spectrum in the blue was last studied in 1970[c]. We have investigated the $B^1\Sigma^+$–$X^1\Sigma^+$ 0-0 spectrum of MgS at high resolution under jet-cooled conditions in a laser-ablation molecular source, and have obtained laser-induced fluorescence spectra from four isotopologues. Dispersed fluorescence from this source identified the low-lying $A^1\Pi$ state near 4520 cm^{-1}. We also created MgS in a Broida oven, with the help of a stream of activated nitrogen, and took rotationally resolved dispersed fluorescence spectra of the $B^1\Sigma^+$–$A^1\Pi$ transition with a grating spectrometer by laser excitation of individual rotational levels of the $B^1\Sigma^+$ state via the $B^1\Sigma^+$–$X^1\Sigma^+$ transition. These spectra provide a first observation and analysis of the $A^1\Pi$ state.

[a] S. Takano, S. Yamamoto and S. Saito, Chem. Phys. Lett. **159**, 563-566 (1989).
[b] K. A. Walker and M. C. L. Gerry, J. Mol. Spectrosc **182**, 178-183 (1997).
[c] M. Marcano and R. F. Barrow, Trans. Faraday Soc. **66**, 2936-2938 (1970).

WJ10

IDENTIFICATION OF TWO NEW ELECTRONIC STATES OF NiCl USING INTRACAVITY LASER SPECTROSCOPY AND THE CORRELATION BETWEEN THEORETICAL PREDICTIONS AND EXPERIMENTAL OBSERVATIONS

<u>JACK C HARMS</u>, ETHAN M GRAMES, SHU HAN, *Chemistry and Biochemistry, University of Missouri, St. Louis, MO, USA*; LEAH C O'BRIEN, *Department of Chemistry, Southern Illinois University, Edwardsville, IL, USA*; JAMES J O'BRIEN, *Chemistry and Biochemistry, University of Missouri, St. Louis, MO, USA*.

The near-infrared spectrum of NiCl has been recorded in high resolution in the 13,200-13,500 cm^{-1} and 13,600-13,750 cm^{-1} regions using Intracavity Laser Spectroscopy (ILS). The NiCl Molecules were produced in the plasma discharge of a Ni-lined copper hollow cathode with 0.3-0.6 torr of argon as the sputter gas, and a trace amount of CCl_4. The hollow cathode was located within the laser cavity of a Verdi V-10 pumped Ti:sapphire system. A generation of 90 μsec resulted in an effective pathlength of approximately 700 m for the absorption measurements. Several transitions were observed, including 3 transitions involving 2 previously unreported electronic states. The (0,0) and (1,0) bands of the [13.5] $^2\Phi_{7/2}$-[0.16] A $^2\Delta_{5/2}$ transition were observed near 13,709 cm^{-1} and 13,318 cm^{-1}, respectively. The (0,0) band of the [13.8] $^2\Pi_{1/2}$ - [0.38] X $^2\Pi_{1/2}$ transition was observed near 13,480 cm^{-1}. With the analysis of these transitions, molecular constants have been obtained for 9 of the 12 doublet states of NiCl predicted by Zou and Lou in 2006. Analysis of these transitions and a comparison between the experimentally observed transitions and the theoretically predicted states of NiCl will be presented.

WJ11 4:32 – 4:47

NEW EMPIRICAL POTENTIAL ENERGY FUNCTIONS FOR THE HEAVIER HOMONUCLEAR RARE GAS PAIRS: Ne_2, Ar_2, Kr_2, and Xe_2

PHILIP THOMAS MYATT, MATTHEW T. BAKER, JU-HEE KANG, <u>ANDRES ESCOBAR MOYA</u>, FREDERICK R. W. McCOURT, ROBERT J. LE ROY, *Department of Chemistry, University of Waterloo, Waterloo, ON, Canada.*

The many decades of work on determining accurate analytic pair potentials for rare gas dimers from experimental data focussed largely on the use of bulk non-ideal gas and collisional properties, with the use of spectroscopic data being somewhat of an afterthought, for testing the resulting functions. This was a natural result of experimental challenges, as the very weak binding of ground-state rare gas pairs made high resolution spectroscopy a relatively late arrival as a practical tool in this area. However, we believe that it is now time for a comprehensive reassessment. Following up on a preliminary report at this meeting five years ago,[a] this paper describes work to determine a new generation of empirical potential energy functions for the four heavier (i.e., not involving He) homonuclear rare gas pairs from direct fits to all available spectroscopic, pressure virial, and acoustic virial coefficient data, with the resulting functions being 'tuned' by comparisons with available thermal transport property data: viscosity, mass diffusion and thermal diffusion, and thermal conductivity data, and tested against the best available *ab initio* potentials. The resulting functions are everywhere smooth and differentiable to all orders, incorporate the correct (damped) theoretical inverse-power long-range behaviour, and have sensible short-range extrapolation behaviour.

---

[a]R.J. Le Roy, C.J.W. Mackie, P. Chandrasekhar and K.M. Sentjens, *"Accurate New Potential Energy Functions From Spectroscopic and Virial Coefficient Data for the Ten Rare Gas Pairs formed from Ne, Ar, Kr and Xe*, paper MF03 at the 66th Ohio State University International Symposium on Molecular Spectroscopy, Columbus, Ohio, June 13-17 (2011).

WJ12 4:49 – 5:04

SPECTROSCOPIC LINE PARAMETERS OF HELIUM- AND HYDROGEN-BROADENED $^{12}C^{16}O$ TRANSITIONS IN THE 3–0 BAND FROM 6270 cm^{-1} TO 6402 cm^{-1}.

<u>ZACHARY REED</u>, JOSEPH HODGES, *Chemical Sciences Division, National Institute of Standards and Technology, Gaithersburg, MD, USA.*

We present helium- and hydrogen-broadened linewidths, pressure-induced shifts, and collisional narrowing coefficients for selected lines in the P- and R- branch of the second overtone (3–0) band of CO, spanning from 6270 cm^{-1} to 6402 cm^{-1}. The contribution of speed dependent effects and partial correlation between velocity-changing and dephasing collisions on the foreign broadened line shapes are also discussed. The data were obtained using the frequency-stabilized cavity ring-down spectroscopy technique. Spectra were collected at room temperature over a pressure range from 13.3 kPa to 100 kPa. The spectrum frequency axis is referenced via an optical frequency comb to a Cs clock, which provides pressure shifting values with uncertainties as low as 100 kHz/atm. The spectra exhibited signal-to-noise ratios as high as 20,000:1, which enables rigorous tests of theoretical line profiles through multi-spectrum least squares data analysis. The partially correlated, quadratic-speed-dependent Nelkin Ghatak profile gives a quality of fit mostly commensurate with the high spectrum signal-to-noise and minimizes structural residuals.

WJ13

AN *AB INITIO* STUDY OF SbH_2 AND BiH_2: THE RENNER EFFECT, SPIN-ORBIT COUPLING, LOCAL MODE VIBRATIONS AND ROVIBRONIC ENERGY LEVEL CLUSTERING IN SbH_2

BOJANA OSTOJIC, *Institute of Chemistry, Technology and Metallurgy, University of Belgrade, Belgrade, Serbia*; PETER SCHWERDTFEGER, *The New Zealand Institute for Advanced Study, Massey University, Auckland, New Zealand*; PHIL BUNKER, *Steacie Laboratory, National Research Council of Canada, Ottawa, ON, Canada*; PER JENSEN, *Faculty of Mathematics and Natural Sciences, University of Wuppertal, Wuppertal, Germany*.

We present the results of *ab initio* calculations for the lower electronic states of the Group 15 (pnictogen) dihydrides, SbH_2 and BiH_2. For each of these molecules the two lowest electronic states become degenerate at linearity and are therefore subject to the Renner effect. Spin-orbit coupling is also strong in these two heavy-element containing molecules. For the lowest two electronic states of SbH_2, we construct the three dimensional potential energy surfaces and corresponding dipole moment and transition moment surfaces by multi-reference configuration interaction techniques. Including both the Renner effect and spin-orbit coupling, we calculate term values and simulate the rovibrational and rovibronic spectra of SbH_2. Excellent agreement is obtained with the results of matrix isolation infrared spectroscopic studies and with gas phase electronic spectroscopic studies in absorption [1,2]. For the heavier dihydride BiH_2 we calculate bending potential curves and the spin-orbit coupling constant for comparison. For SbH_2 we further study the local mode vibrational behavior and the formation of rovibronic energy level clusters in high angular momentum states.

[1] X. Wang, P. F. Souter and L. Andrews, J. Phys. Chem. A 107, 4244-4249 (2003)
[2] N. Basco and K. K. Lee, Spectroscopy Letters 1, 13-15 (1968)

WJ14

THE BICHROMATIC FORCE ON SMALL MOLECULES

LELAND M. ALDRIDGE, SCOTT E. GALICA, DONAL SHEETS, EDWARD E. EYLER, *Department of Physics, University of Connecticut, Storrs, CT, USA*.

The bichromatic force is a coherent optical force that has been demonstrated to exceed the saturated radiative force from a monochromatic cw laser by orders of magnitude in atomic systems. By stimulating photon emission between two states, the bichromatic force allows us to increase the photon scattering rate beyond the spontaneous emission rate while also suppressing decays into dark states. We present studies of the efficacy of the bichromatic force on molecular systems using the test cases of $B\text{-}X$ $(0,0)$, $P_{11}(1.5)/^{P}Q_{12}(0.5)$ in CaF and $\tilde{A}(000) - \tilde{X}(000)$, $P_{11}(1.5)/^{P}Q_{12}(0.5)$ in the linear triatomic molecule SrOH. Computational results from detailed multilevel models indicate that both of these molecular systems are suitable for the use of the bichromatic force, with neither repumping nor magnetic destabilization of dark states interrupting the coherent cycling at the heart of the force. We comment on the applicability of the bichromatic force to arbitrary polyatomic molecules, and present our experimental progress in demonstrating the bichromatic force on CaF and possibly on SrOH.[a]

---

[a]Supported by the National Science Foundation.

WK. Spectroscopy as an analytical tool
Wednesday, June 22, 2016 – 1:30 PM
Room: 140 Burrill Hall

Chair: Kyle N. Crabtree, University of California, Davis, CA, USA

WK01   1:30 – 1:45

IDENTIFICATION AND CHARACTERIZATION OF 1,2-BN CYCLOHEXENE USING MICROWAVE SPECTROSCOPY[a]

STEPHEN G. KUKOLICH, MING SUN, <u>ADAM M. DALY</u>, *Department of Chemistry and Biochemistry, University of Arizona, Tucson, AZ, USA*; JACOB S. A. ISHIBASHI, SHIH-YUAN LIU, *Chemistry, Boston College, Chesnut Hill, MA, USA.*

1,2-BN Cyclohexene was produced from 1,2-BN Cyclohexane through the loss of H_2 and characterized and identified using a pulsed-beam Fourier-transform microwave spectrometer. The first microwave spectra for 1,2-^{10}BN Cyclohexene 1,2-^{11}BN Cyclohexene have been measured in the frequency range of 5.5-12.5 GHz, providing accurate rotational constants and nitrogen and boron quadrupole coupling strengths for two isotopologues. High-level ab initio calculations provided rotational constants and quadrupole coupling strengths for the precursor 1,2-BN Cyclohexane ($C_4H_{12}BN$) and 1,2-BN Cyclohexene($C_4H_{10}BN$). Calculated molecular properties for 1,2-BN Cyclohexene are in very good agreement with measured parameters. Calculated parameters for the starting material, 1,2-BN Cyclohexane do not agree with the experimental data. Rotational constants for 1,2-^{11}BN Cyclohexene are A = 4702.058(2) MHz, B = 4360.334(1) MHz and C = 2494.407(1) MHz. The inertial defect is Δ_0 = -20.78 amu-Å^2 clearly indicating a nonplanar structure. These microwave experiments show that heating the initial compound, 1,2-BN Cyclohexane, to 60 C in a 1 atm neon stream results in the loss of H_2 and conversion to 1,2-BN Cyclohexene. This appears to be the first characterization of the 1,2-BN Cyclohexene monomer.

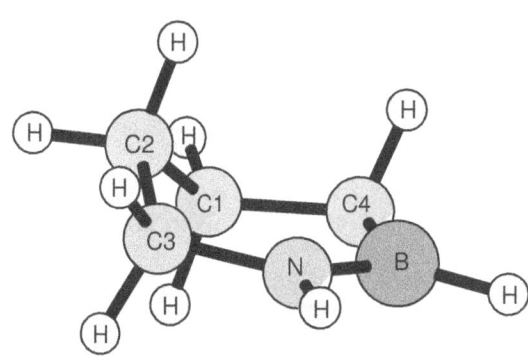

[a]Supported by the NSF CHE-1057796 and DOE DE-EE-0005658

WK02   1:47 – 2:02

AUTOMATED MICROWAVE DOUBLE RESONANCE SPECTROSCOPY: A TOOL TO IDENTIFY AND CHARACTERIZE CHEMICAL COMPOUNDS

<u>MARIE-ALINE MARTIN-DRUMEL</u>, MICHAEL C McCARTHY, *Atomic and Molecular Physics, Harvard-Smithsonian Center for Astrophysics, Cambridge, MA, USA*; DAVID PATTERSON, *Department of Physics, Harvard University, Cambridge, MA, USA*; BRETT A. McGUIRE, *NAASC, National Radio Astronomy Observatory, Charlottesville, VA, USA*; KYLE N. CRABTREE, *Department of Chemistry, The University of California, Davis, CA, USA.*

Owing to its unparalleled structural specificity, rotational spectroscopy is a powerful technique to unambiguously identify and characterize volatile, polar molecules. We present here a new experimental approach, automated microwave double resonance (AMDOR) spectroscopy, to rapidly determine the rotational constants of these compounds without any *a priori* knowledge of elemental composition or molecular structure. This task is achieved by rapidly acquiring the classical (frequency vs. intensity) broadband spectrum of a molecule using chirped-pulse Fourier transform microwave (FTMW) spectroscopy, and subsequently analyzing it in near-real time using complementary cavity FTMW detection and double resonance. AMDOR measurements provide a unique "barcode" for each compound from which rotational constants can be extracted. To illustrate the power of this approach, AMDOR spectra of three aroma compounds — *trans*-cinnamaldehyde, α- and β-ionone — have been recorded and analyzed. The prospects to extend this approach to mixture characterization and purity assessment are described.

WK03   2:04 – 2:19

UTILIZATION OF MICROWAVE SPECTROSCOPY TO IDENTIFY AND PROBE REACTION DYNAMICS OF HSNO, A CRUCIAL BIOLOGICAL SIGNALING MOLECULE

MATTHEW NAVA, *Department of Chemistry, MIT, Cambridge, MA, USA*; MARIE-ALINE MARTIN-DRUMEL, *Atomic and Molecular Physics, Harvard-Smithsonian Center for Astrophysics, Cambridge, MA, USA*; JOHN F. STANTON, *Department of Chemistry, The University of Texas, Austin, TX, USA*; CHRISTOPHER CUMMINS, *Department of Chemistry, MIT, Cambridge, MA, USA*; MICHAEL C McCARTHY, *Atomic and Molecular Physics, Harvard-Smithsonian Center for Astrophysics, Cambridge, MA, USA*.

Thionitrous acid (HSNO), a potential key intermediate in biological signaling pathways, has been proposed to link NO and H_2S biochemistries. Its existence and stability in vivo, however, remain controversial. By means of Fourier-transform microwave spectroscopy, we establish that HSNO is spontaneously formed in high concentration when NO and H_2S gases are simply mixed at room temperature in the presence of metallic surfaces. Our measurements reveal that HSNO is formed with high efficiency by the reaction H_2S and N_2O_3 to produce HSNO and HNO_2, where N_2O_3 is a product of NO disproportionation. These studies also suggest that further reaction of HSNO with H_2S may form HNO and HSSH. The length of the S–N bond has been derived to high precision from isotopic studies, and is found to be unusually long, 1.84 Å – the longest S–N bond reported to date for an SNO compound. The present structural and reactivity investigations of this elusive molecule provide a firm fundation to better understand its physiological chemistry and propensity to undergo S–N bond homolysis in vivo.

WK04   2:21 – 2:36

ROTATIONAL SPECTRUM AND CARBON ATOM STRUCTURE OF DIHYDROARTEMISINIC ACID

LUCA EVANGELISTI, *Dipartimento di Chimica G. Ciamician, Università di Bologna, Bologna, Italy*; NATHAN A SEIFERT, *Department of Chemistry, University of Alberta, Edmonton, AB, Canada*; LORENZO SPADA, *Dep. Chemistry 'Giacomo Ciamician', University of Bologna, Bologna, Italy*; BROOKS PATE, *Department of Chemistry, The University of Virginia, Charlottesville, VA, USA*.

Dihydroartemisinic acid (DHAA, $C_{15}H_{24}O_2$, five chiral centers) is a precursor in proposed low-cost synthetic routes to the antimalarial drug artemisinin. In one reaction process being considered in pharmaceutical production, DHAA is formed from an enantiopure sample of artemisinic acid through hydrogenation of the alkene. This reaction needs to properly set the stereochemistry of the asymmetric carbon for the synthesis to produce artemisinin. A recrystallization process can purify the diastereomer mixture of the hydrogenation reaction if the unwanted epimer is produced in less than 10% abundance. There is a need in the process analytical chemistry to rapidly (less than 1 min) measure the diastereomer excess and current solutions, such a HPLC, lack the needed measurement speed. The rotational spectrum of DHAA has been measured at 300:1 signal-to-noise ratio in a chirped-pulsed Fourier transform microwave spectrometer operating from 2-8 GHz using simple heating of the compound. The ^{13}C isotope analysis provides a carbon atom structure that confirms the diastereomer. This structure is in excellent agreement with quantum chemistry calculations at the B2PLYPD3/ 6-311++G** level of theory. The DHAA spectrum is expected to be fully resolved from the unwanted diastereomer raising the potential for fast diastereomer excess measurement by rotational spectroscopy in the pharmaceutical production process.

WK05   2:38 – 2:53

PROBING THE CH_3SH + N_2O_3 REACTION BY AUTOMATED MICROWAVE DOUBLE RESONANCE SPECTROSCOPY

MICHAEL C McCARTHY, MARIE-ALINE MARTIN-DRUMEL, *Atomic and Molecular Physics, Harvard-Smithsonian Center for Astrophysics, Cambridge, MA, USA*; MATTHEW NAVA, *Department of Chemistry, MIT, Cambridge, MA, USA*; SVEN THORWIRTH, *I. Physikalisches Institut, Universität zu Köln, Köln, Germany*.

Because HSNO is formed abundantly and selectively from H_2S and N_2O_3 in the presence of metallic surfaces, it may be feasible to synthesize larger RSNOs in analogous reactions using RSH precursors. To critically explore this possibility, products of the CH_3SH + N_2O_3 reaction have been studied using a combination of chirped-pulse microwave spectroscopy and automated double resonance techniques. As with HSNO, we find that *anti*-CH_3SNO is formed in high abundance under similar experimental conditions, suggesting that this production method might be extended to study still larger *S*-nitrothiols in the gas-phase. This talk will provide a status report of our analysis, high-level quantum chemical calculations of minima on the CH_3SNO potential energy surface, and searches for secondary products.

WK06 2:55 – 3:10

MILLIMETER-WAVE SPECTROSCOPY FOR ANALYTICAL CHEMISTRY: THERMAL EVOLUTION OF LOW VOLATILITY IMPURITIES AND DETECTION WITH A FOURIER TRANSFORM MOLECULAR ROTATIONAL RESONANCE SPECTROMETER (TEV FT-MRR)

BRENT HARRIS, SHELBY S. FIELDS, JUSTIN L. NEILL, ROBIN PULLIAM, MATT MUCKLE, *BrightSpec Labs, BrightSpec, Inc., Charlottesville, VA, USA*; BROOKS PATE, *Department of Chemistry, The University of Virginia, Charlottesville, VA, USA.*

Recent advances in Fourier transform millimeter-wave spectroscopy techniques have renewed the application reach of molecular rotational spectroscopy for analytical chemistry. We present a sampling method for sub ppm analysis of low volatility impurities by thermal evolution from solid powders using a millimeter-wave Fourier transform molecular rotational resonance (FT-MRR) spectrometer for detection. This application of FT-MRR is relevant to the manufacturing of safe oral pharmaceuticals. Low volatility impurities can be challenging to detect at 1 ppm levels with chromatographic techniques. One such example of a potentially mutagenic impurity is acetamide (v.p. 1 Torr at 40 C, m.p. 80 C). We measured the pure reference spectrum of acetamide by flowing the sublimated vapor pressure of acetamide crystals through the FT-MRR spectrometer. The spectrometer lower detection level (LDL) for a broadband ($¿$ 20 GHz, 10 min.) spectrum is 300 nTorr, 30 pmol, or 2 ng. For a 50 mg powder, perfect sample transfer efficiency can yield a w/w % detection limit of 35 ppb. We extended the sampling method for the acetamide reference measurement to an acetaminophen sample spiked with 5000 ppm acetamide in order to test the sample transfer efficiency when liberated from an pharmaceutical powder. A spectral reference matching algorithm detected the presence of several impurities including acetaldehyde, acetic acid, and acetonitrile that evolved at the melting point of acetaminophen, demonstrating the capability of FT-MRR for identification without a routine chemical standard. The method detection limit (MDL) without further development is less than 10 ppm w/w %. Resolved FT-MRR mixture spectra will be presented with a description of sampling methods.

Intermission

WK07 3:29 – 3:44

DETECTION OF *in vitro* S-NITROSYLATED COMPOUNDS WITH CAVITY RING-DOWN SPECTROSCOPY

MARY LYNN RAD, *Department of Chemistry, The University of Virginia, Charlottesville, VA, USA*; MONIQUE MICHELE MEZHER, *School of Engineering and Applied Sciences, University of Virginia, Charlottesville, Virginia, USA*; BENJAMIN M GASTON, *Department of Pediatrics , Case Western Reserve University, Cleveland, Ohio, USA*; KEVIN LEHMANN, *Department of Chemistry and Physics, The University of Virginia, Charlottesville, VA, USA.*

Nitric oxide has been of strong biological interest for nearly 40 years due to its role in cardiovascular and nervous signaling. It has been shown that S-nitrosocompounds are the main carrier molecule for nitric oxide in biological systems. These compounds are also of interest due to their relationship to several diseases including muscular dystrophy, stroke, myocardial infarction, Alzheimer's disease, Parkinson's disease, cystic fibrosis, asthma, and pulmonary arterial hypertension. Understanding the role of these S-nitrosocompounds in these diseases requires concentration studies in healthy and diseased tissues as well as metabolic studies using isotopically labeled S-nitroso precursors such at ^{15}N-arginine. The current widely used techniques for these studies include chemiluminescence, which is blind to isotopic substitution, and mass spectrometry, which is known to artificially create and break S-NO bonds in the sample preparation stages. To this end we have designed and constructed a mid-IR cavity ring-down spectrometer for the detection of nitric oxide released from the target S-nitrosocompounds. Progress toward measuring S-NO groups in biological samples using the CRDS instrument will be presented.

206

WK08 3:46 – 4:01

CAVITY RING DOWN ABSORPTION OF OXYGEN IN AIR AS A TEMPERATURE SENSOR

CARLOS MANZANARES, PARASHU R NYAUPANE, *Department of Chemistry and Biochemistry, Baylor University, Waco, TX, USA.*

The A-band of oxygen has been measured at low resolution at temperatures between 90 K and 373 K using the phase shift cavity ring down (PS-CRD) technique. For temperatures between 90 K and 295 K, the PS-CRD technique presented here involves an optical cavity attached to a cryostat. The static cell and mirrors of the optical cavity are all inside a vacuum chamber at the same temperature of the cryostat. The temperature of the cell can be changed between 77 K and 295 K. For temperatures above 295 K, a hollow glass cylindrical tube without windows has been inserted inside an optical cavity to measure the temperature of air flowing through the tube. The cavity consists of two highly reflective mirrors which are mounted parallel to each other and separated by a distance of 93 cm. In this experiment, air is passed through a heated tube. The temperature of the air flowing through the tube is determined by measuring the intensity of the oxygen absorption as a function of the wavenumber. The A-band of oxygen is measured between 298 K and 373 K, with several air flow rates. Accuracy of the temperature measurement is determined by comparing the calculated temperature from the spectra with the temperature obtained from a calibrated thermocouple inserted at the center of the tube.

WK09 4:03 – 4:18

ATMOSPHERIC REMOTE SENSING VIA INFRARED-SUBMILLIMETER DOUBLE RESONANCE

SREE SRIKANTAIAH, JENNIFER HOLT, CHRISTOPHER F. NEESE, *Department of Physics, The Ohio State University, Columbus, OH, USA*; DANE PHILLIPS, *IERUS Technologies, Huntsville, AL, USA*; HENRY O. EVERITT, *Army Aviation and Missile Research Development and Engineering Center, Redstone Arsenal, AL, USA*; FRANK C. DE LUCIA, *Department of Physics, The Ohio State University, Columbus, OH, USA.*

Specificity and sensitivity in atmospheric pressure remote sensing have always been big challenges. This is especially true for approaches that involve the submillimeter/terahertz (smm/THz) spectral region because atmospheric pressure broadening precludes taking advantage of the small Doppler broadening in the region. The Infrared-submillimeter (IR-smm) double resonance spectroscopic technique allows us to obtain a more specific two-dimensional signature as well as a means of modulating the molecular signal to enhance its separation from background and system variation. Applying this technique at atmospheric pressure presents a unique bandwidth requirement on the IR pump laser, and the smm/THz receiver. We will discuss the pump system comprising of a CO2 TEA laser, plasma switch and a free induction decay hot cell designed to produce fast IR pulses on the time scale of atmospheric pressure relaxation and a high bandwidth fast pulse smm/THz receiver. System diagnostics will also be discussed. Results as a function of pressure and pump pulse width will be presented.

WK10 4:20 – 4:35

MOLECULAR STRUCTURE AND REACTIVITY IN THE PYROLYSIS OF ALDEHYDES

ERIC SIAS, SARAH COLE, JOHN SOWARDS, BRIAN WARNER, EMILY WRIGHT, LAURA R. McCUNN, *Department of Chemistry, Marshall University, Huntington, WV, USA.*

The effect of alkyl chain structure on pyrolysis mechanisms has been investigated in a series of aldehydes. Isovaleraldehyde, $CH_3CH(CH_3)CH_2CHO$, and pivaldehyde, $(CH_3)_3CCHO$, were subject to thermal decomposition in a resistively heated SiC tubular reactor at $800 - 1200°C$. Matrix-isolation FTIR spectroscopy was used to identify pyrolysis products. Carbon monoxide and isobutene were major products from each of the aldehydes, which is consistent with what is known from previous studies of unbranched alkyl-chain aldehydes. Other products observed include vinyl alcohol, propene, acetylene, and ethylene, revealing complexities to be considered in the pyrolysis of large, branched-chain aldehydes.

WK11 4:37 – 4:52

VUV FLUORESCENCE OF JET-COOLED WATER AS A VEHICLE FOR SATELLITE THRUSTER PLUME CHARACTERIZATION.

JUSTIN W. YOUNG, JAIME A. STEARNS, *Space Vehicles Directorate, Air Force Research Lab, Kirtland AFB, NM, USA.*

A quantified characterization of a spacecraft's thruster plume is obtainable through measurements of fluorescence from the plume. Fluorescence in a plume is due to electronic excitation of a plume's molecular species, such as water and ammonia, from solar photons. For instance, electronic excitation of water with Lyman-alpha (121.6 nm) causes photodissociation to OH radical by following one of several possible pathways. One pathway leads to an electronically excited OH radical which fluoresces at 310 nm. Here, the emission spectra of H_2O excited at wavelengths ranging from 128-121 nm are presented and the role of temperature in fluorescence is discussed.

WK12 4:54 – 5:04

INFLUENCE OF BIODEGRADATION ON THE ORGANIC COMPOUNDS COMPOSITION OF PEAT.

OLGA SEREBRENNIKOVA, *Institute of Natural Resources, National Research Tomsk Polytechnic University, Tomsk, Russia*; LIDIYA SVAROVSKAYA, *Laboratory of Colloidal Oil Chemistry, Institute of Petroleum Chemistry, Tomsk, Russia*; MARIA DUCHKO, *Institute of Natural Resources, National Research Tomsk Polytechnic University, Tomsk, Russia*; EVGENIYA STRELNIKOVA, IRINA RUSSKIKH, *Laboratory of Naphthides Geochemistry, Institute of Petroleum Chemistry, Tomsk, Russia.*

Largest wetland systems are situated on the territory of the Tomsk region. They are characterized by the high content of organic matter (OM), which undergoes transformation as a result of physical, chemical and biological processes. The composition of peat OM is determined by the nature of initial peat-forming plants, their transformation products and bacteria. An experiment in stimulated microbial impact was carried out for estimating the influence of biodegradation on the composition of peat lipids. The composition of the functional groups in the bacterial biomass, initial peat and peat after biodegradation was determined by IR-spectroscopy using the spectrometer NICOLET 5700. The IR spectra of peat and bacteria organic matter are characterized by the presence of absorption bands in ranges: 3400-3200 cm^{-1}, which refers to the stretching vibrations of OH-group of carboxylic acids and various types of hydrogen bonds; 1738-1671 cm^{-1} – characteristic stretching vibrations of the C = O group of carboxylic acids and ketones; 1262 cm^{-1} – stretching vibrations of C-O of carboxylic acids. Group and individual composition of organic compounds in studied samples was determined by gas chromatography-mass-spectrometry.

WK13 5:06 – 5:21

NON-LINEAR THERMAL LENS SIGNAL OF THE ($\Delta v = 6$) C-H VIBRATIONAL OVERTONE OF BENZENE IN LIQUID SOLUTIONS OF HEXANE

PARASHU R NYAUPANE, CARLOS MANZANARES, *Department of Chemistry and Biochemistry, Baylor University, Waco, TX, USA.*

The thermal lens technique is applied to vibrational overtone spectroscopy of solutions of benzene. The pump and probe thermal lens technique has been found to be very sensitive for detecting samples of low concentration in transparent solvents. The C-H fifth vibrational ($\Delta v = 6$) overtone spectrum of benzene is detected at room temperature for compositions per volume in the range (1 to 1×10^{-4}) using n-C_6H_{14} as the solvent. By detecting the absorption band in a 100 ppm solution, the peak absorption of the signal is approximately $(2.2 \pm 0.3) \times 10^{-7}$ cm^{-1}. The parameters that determine the magnitude of the thermal lens signal such as the pump laser power and the thermodynamic properties of the solvent and solute are discussed. A plot of normalized integrated intensity as a function of composition of benzene in solution reveals a non-linear behavior. The non-linearity cannot be explained assuming solvent enhancement at low concentrations. A two color absorption model that includes the simultaneous absorption of the pump and probe lasers could explain the enhanced magnitude and the non-linear behavior of the thermal lens signal for solutions of composition below 0.01.

208

WK14 5:23 – 5:38

SCREENING OF POLY CYSTIC OVARIAN SYNDROME BY MID INFRARED SPECTROMETRY

MOHAMMADREZA KHANMOHAMMADI, FATEMEH GOLPOUR, AMIR BAGHERI GARMARUDI, *Department of Chemistry, Imam Khomeini International University, Qazvin, Iran*; FAHIMEH RAMEZANI TEHRANI, *Reproductive Endocrinology Research Center, Research Institute for Endocrine Sciences, Shahid Beheshti University of Medical Sciences, Tehran, Iran.*

Poly Cystic Ovarian Syndrome (PCOS) is the most common gynecological endocrinopathy with various reproductive manifestations including infertility, endometrial cancer, late menopause and also metabolic aberrations, including insulin resistance, type 2 diabetes mellitus, dyslipidemia and cardiovascular diseases [1,2]. These women mainly have enlarges ovaries with thickened sclerotic capsules and abnormal high number of follicles that may contemporaneously exist in varying states of growth, maturation, or atresia [3]. Hormonal profiles of affected women may be useful for diagnosis, however they may be at normal range in some PCOS women [4]. The objective of this work was to propose a non-invasive diagnostic method for Poly cystic ovarian syndrome based on Infrared spectroscopy. It was attempted to evaluate the spectral variations in human blood serum samples obtained from healthy and PCOS women using infrared spectroscopy, to be employed for discriminative diagnosis being compared with the outcome of Eliza kit assay. A total number of 93 samples of PCOs and 46 ones were obtained being analyzed in mid infrared spectral region. After preprocessing on initial data set, principal component analysis was employed to determine general relationships in data collection. In the next step, soft independent modeling of class analogy and partial least-squares discriminant analysis were used to classify normal and PCOs samples .In conclusion, PLS-DA function was more better than SIMCA model because SIMCA had 10 percent of not assigned data in its model despite same classification parameters for both models.

WK15 5:40 – 5:55

OBSERVATION OF ORTHO-PARA DEPENDENCE OF PRESSURE BROADENING COEFFICIENT IN ACETYLENE $\nu_1+\nu_3$ VIBRATION BAND USING DUAL-COMB SPECTROSCOPY

KANA IWAKUNI, *Department of Physics, Faculty of Science and Technology, Keio University, Yokohama, Japan*; SHO OKUBO, HAJIME INABA, ATSUSHI ONAE, FENG-LEI HONG, *National Metrology Institute of Japan (NMIJ), Ntional Institute of Advanced Industrial Science and Technology (AIST), Tsukuba, Japan*; HIROYUKI SASADA, *Department of Physics, Faculty of Science and Technology, Keio University, Yokohama, Japan*; KOICHI MT YAMADA, *National Metrology Institute of Japan (NMIJ), Ntional Institute of Advanced Industrial Science and Technology (AIST), Tsukuba, Japan.*

We observe that the pressure-broadening coefficients depend on the ortho-para levels. The spectrum is taken with a dual-comb spectrometer which has the resolution of 48 MHz and the frequency accuracy of 8 digit when the signal-to-noise ratio is more than 20[a].

In this study, about 4.4-Tz wide spectra of the $P(31)$ to $R(31)$ transitions in the $\nu_1+\nu_3$ vibration band of $^{12}C_2H_2$ are observed at the pressure of 25, 60, 396, 1047, 1962 and 2654 Pa. Each rotation-vibration absorption line is fitted to Voight function and we determined pressure-broadening coefficients for each rotation-vibration transition. The Figure shows pressure broadening coefficient as a function of m. Here m is $J"+1$ for R and $-J"$ for P-branch. The graph shows obvious dependence on ortho and para. We fit it to Pade function considering the population ratio of three-to-one for the ortho and para levels. This would lead to detailed understanding of the pressure boarding mechanism.

[a]S. Okubo *et al.*, Applied Physics Express 8, 082402 (2015)

RA. Plenary
Thursday, June 23, 2016 – 8:30 AM
Room: Foellinger Auditorium

Chair: Brian D. Fields, The University of Illinois, Urbana, IL, USA

RAO AWARDS   **8:30**
Presentation of Awards by Gary Douberly, University of Georgia

2015 Rao Award Winners
Daniel Bakker, Radboud University
Scott Dubowsky, University of Illinois at Urbana-Champaign
James McMillan, The Ohio State University

MILLER PRIZE   **8:40**
Introduction by Jinjun Liu, University of Louisville

RA01   *Miller Prize Lecture*   **8:55 – 9:10**

HIGH-RESOLUTION SPECTROSCOPY OF He_2^+ USING RYDBERG-SERIES EXTRAPOLATION AND ZEEMAN-DECELERATED SUPERSONIC BEAMS OF METASTABLE He_2

PAUL JANSEN, LUCA SEMERIA, FREDERIC MERKT, *Laboratorium für Physikalische Chemie, ETH Zurich, Zurich, Switzerland.*

Having only three electrons, He_2^+ represents a system for which highly accurate *ab initio* calculations are possible. The latest calculations of rovibrational energies in He_2^+ do not include relativistic or QED corrections but claim an accuracy of 120 MHz[a]. We have performed high-resolution Rydberg spectroscopy of metastable He_2 molecules[b] and employed multichannel-quantum-defect-theory extrapolation techniques[c] to determine the rotational energy-level structure in the He_2^+ ion. To this end, we have produced samples of metastable helium molecules in supersonic beams with velocities tunable down to 100 m/s by combining a cryogenic supersonic-beam source with a multistage Zeeman decelerator[d]. The metastable He_2 molecules are excited to np Rydberg states using the frequency-doubled output of a pulse-amplified ring dye laser. Although the bandwidth of the laser system is too large to observe the reduction of the Doppler width resulting from deceleration, the deceleration greatly simplifies the spectral assignments because of its spin-rotational state selectivity. Our approach enabled us to determine the rotational structure of He_2 with an unprecedented accuracy of 18 MHz, to quantify the size of the relativistic and QED corrections by comparison with the results of Tung *et al.* and to precisely measure the rotational structure of the metastable state for comparison with the results of Focsa *et al.*[e] Here, we present an extension of these measurements in which we have measured higher rotational intervals of He_2^+. In addition, we have replaced the pulsed UV laser by a cw UV laser and improved the resolution of the spectra by a factor of more than five[f].

[a] W.-C. Tung, M. Pavanello and L. Adamowicz, J. Chem. Phys. **136**, 104309 (2012).
[b] P. Jansen, L. Semeria, L. Esteban Hofer, S. Scheidegger, J.A. Agner, H. Schmutz, and F. Merkt, Phys. Rev. Lett. **115**, 133202 (2015).
[c] D. Sprecher, J. Liu, T. Krähenmann, M. Schäfer, and F. Merkt, J. Chem. Phys. **140**, 064304 (2014).
[d] M. Motsch, P. Jansen, J. A. Agner, H. Schmutz, and F. Merkt, Phys. Rev. A **89**, 043420 (2014).
[e] C. Focsa, P. F. Bernath, and R. Colin, J. Mol. Spectrosc. **191**, 209 (1998).
[f] P. Jansen, L. Semeria, and F. Merkt, J. Mol. Spectrosc. **322**, 9 (2016).

COBLENTZ AWARD   **9:15**
Presentation of Award by Linda Kidder, Coblentz Society

EXAMINING THE NANOWORLD USING A MOLECULAR SPECTROSCOPIST'S TOOLBOX

KENNETH L. KNAPPENBERGER, JR., *Chemistry and Biochemistry, Florida State University , Tallahassee, FL, USA.*

I will describe recent advances in understanding the influence of nanoscale structure on plasmon-mediated electron dynamics. Steady-state extinction spectra of plasmonic nanoparticle networks are accurately described using hybridization models reminiscent of molecular orbitals. We have extended these molecular-based descriptions to account for nanoparticle electron dynamics by quantifying the coherence dephasing times of collective inter-particle plasmon modes of single nanostructures. In particular, we demonstrate that interference between plasmon modes of different angular momenta leads to increased coherence times. These observations are consistent with a model based on superpositions of molecular-like electronic states. These fundamental studies are important for understanding the structure-photonic-function relationship of plasmonic nanoparticles. This is because the spectroscopically determined coherence times reflect mode quality factors, which determine achievable amplification factors of optical signals. These new insights are made possible by recent advances in single-nanoparticle/molecule spectroscopy based on interferometric nonlinear optical detection. I will describe how the generation of sequences of phase-locked femtosecond laser pulses (33mrad phase stability) and their integration to an optical microscope were critical for this research.

Intermission

TWO DECADES OF ADVANCES IN HIGH-RESOLUTION SPECTROSCOPY OF LARGE-AMPLITUDE MOTIONS IN N-FOLD POTENTIAL WELLS, AS ILLUSTRATED BY METHANOL

LI-HONG XU, *Department of Physics, University of New Brunswick, Saint John, NB, Canada.*

Methanol is a simple and intensively studied organic molecule possessing one large-amplitude torsional motion. It has, for nearly a century, been a favorite of researchers in many fields, e.g., instrument builders, for whom methanol is often the first molecule chosen for testing an improved or a newly built instrument (including HIFI, the Heterodyne Instrument for the Far Infrared on board the Herschel space mission); theorists and/or dynamicists studying the challenging effects of a large-amplitude motion coupling with small-amplitude motions to enhance intramolecular vibrational energy redistribution; astronomers who have elevated methanol to their #1 interstellar weed because of its rich and omnipresent spectrum in the interstellar garden, where it serves as a unique probe for diagnosing conditions in star-forming regions; astrochemists studying isotopic ratios as clues to the chemical evolution of the universe; and fundamentalists seeking possible time variation of the proton/electron mass ratio in the standard model; just to name a few.

From high-resolution to high-precision spectroscopy, the large-amplitude internal rotation of the methyl top against its OH framework in methanol has never failed to produce new surprises in spectral regions from the microwave all the way to the near IR. The very recent observation of completely unexpected large methanol hyperfine splittings is a vivid testimonial that the large-amplitude torsional motion can still lead us to unexplored landscapes. This talk will focus on the complicated vibration-torsion-rotation energy networks and interactions deduced from high resolution spectra; our efforts to understand some of them using ab-initio-assisted approaches and the modeling of torsion-rotation and torsionally mediated spin-rotation hyperfine splittings in methanol. These topics represent one part of the much larger fascinating world inhabited by methanolics.

RA04   **11:15 – 11:55**

MOLECULAR SPECTROSCOPY IN SPACE: DISCOVERING NEW MOLECULES FROM LINE SURVEYS AND LABORATORY SPECTROSCOPY

JOSE CERNICHARO, *Molecular Astrophysics, ICMM, Madrid, Spain.*

The increasing sensitivity offered by the new generation of radio astronomical receivers and radio telescopes (single dishes and radio interferometers) has provided an enormous impact in our capacity to study the molecular content of interstellar and circumstellar clouds. Astronomers face now the challenging problem of interpreting the thousands of lines detected in hot cores which arise from isotopologues and vibrationally excited states of most known molecules. Although all strong features have been already assigned to abundant species, many of the lines still pending to be assigned could arise from very abundant molecular species having low dipole moment and/or very large partition functions.

The only way to address this problem in astrophysics is through a close collaboration between astrophysicists and laboratory spectroscopists. In this talk I am going to present the results obtained over the last 10 years in interpreting the line surveys of Orion gathered with the 30m IRAM radio telescope and with ALMA. The most recent molecule found in this cloud is methyl isocyanate, CH_3NCO, for which near 400 lines have been found in Orion[a] in the 80-280 GHz domain. This molecule has an abundance only a factor 5-20 below that of the well-known species HNCO and CH_3CN. The molecule has been also found towards the giant cloud SgrB2[b] in the galactic center.

Finally, I will present the case of the submillimeter spectrum of the carbon-rich evolved star IRC+10216 in which we have recently found Si_2C with an abundance similar to SiC_2. Our recent ALMA observations in a narrow band of 20 GHz around 265 GHz show near 200 features corresponding to the J=3-2 transition of hot HCN (vibrational levels up to 11000 cm^{-1}). In addition to HCN lines, a forest of several hundreds of U lines dominates the spectrum. Most of these lines arise from molecules that condensate very quickly into dust grains[c].

----

[a]J. Cernicharo, Z.Kisiel, B.Tercero, et al., A&A 587, L4 (2016).

[b]D.T. Halfen, V.V.Ilyushin, L.Ziurys, ApJ 812, L5 (2015).

[c]J. Cernicharo, F. Daniel, A. Castro-Carrizo, et al., ApJ, 778, L25 (2015).

RF. Radicals
Thursday, June 23, 2016 – 1:30 PM
Room: 100 Noyes Laboratory

Chair: Terry A. Miller, The Ohio State University, Columbus, OH, USA

RF01   *Post-Deadline Abstract*   1:30 – 1:45

2C-R4WM SPECTROSCOPY OF JET COOLED NO_3

MASARU FUKUSHIMA, TAKASHI ISHIWATA, *Information Sciences, Hiroshima City University, Hiroshima, Japan*; EIZI HIROTA, *The Central Office, The Graduate University for Advanced Studies, Hayama, Kanagawa, Japan.*

We have generated NO_3 from pyrolysis of N_2O_5 following supersonic free jet expansion, and carried out two color resonant four wave mixing (2C-R4WM) spectroscopy of the $\tilde{B}\ ^2E' - \tilde{X}\ ^2A_2'$ electronic transition. One laser was fixed to pump NO_3 to a ro-vibronic level of the $\tilde{B}$ state, and the other laser (probe) was scanned across two levels of the $\tilde{X}$ $^2A_2'$ state lying at 1051 and 1492 cm^{-1}, the ν_1 (a_1') and ν_3 (e') fundamentals, respectively. The 2C-R4WM spectra have unexpected back-ground signal of NO_3 (stray signal due to experimental set-up is also detected) similar to laser induced fluorescence (LIF) excitation spectrum of the 0-0 band, although the back-ground signal was not expected in considering the 2C-R4WM scheme. Despite the back-ground interference, we have observed two peaks at 1051.61 and 1055.29 cm^{-1} in the ν_1 region of the spectrum, and the frequencies agree with the two bands, 1051.2 and 1055.3 cm^{-1}, of our relatively higher resolution dispersed fluorescence spectrum, the former of which has been assigned to the ν_1 fundamental. Band width of both peaks, ~ 0.2 cm^{-1}, is broader than twice the experimental spectral-resolution, 0.04 cm^{-1} (because this experiment is double resonance spectroscopy), and the 1051.61 cm^{-1} peak is attributed to a Q branch band head (a line-like Q branch) of the ν_1 fundamental. The other branches are suspected to be hidden in noise of the back-ground signal. The 1055.29 cm^{-1} peak is also attributed to a Q band head. The $\tilde{B}\ ^2E'_{\frac{1}{2}}$ ($J' = \frac{3}{2}$, $K' = 1$) $- \tilde{X}\ ^2A_2'$ ($N'' = 1$, $K'' = 0$) ro-vibronic transition was used as the pump transition. The dump (probe) transition to both a_1' and e' vibronic levels are then allowed as perpendicular transition. Accordingly, it cannot be determined from present results whether the 1055.29 cm^{-1} band is attributed to a_1' or e' (ν_3), unfortunately. The 2C-R4WM spectrum of the 1492 cm^{-1} band region shows one Q head at 1499.79 cm^{-1}, which is consistent with our dispersed fluorescence spectrum. By considering with the $\nu_3 + \nu_4 - \nu_4$ hot band[a], the present results suggest that both 1055.29 and 1499.79 cm^{-1} levels are a_1' level.

---

[a]K. Kawaguchi *et al.*, *J. Phys. Chem. A* **117**, 13732 (2013) and E. Hirota, *J. Mol. Spectrosco.* **310**, 99 (2015).

RF02   1:47 – 2:02

HIGH-RESOLUTION LASER SPECTROSCOPY OF THE $\tilde{B} \leftarrow \tilde{X}$ TRANSITION OF $^{14}NO_3$ RADICAL: VIBRATIONALLY EXCITED STATES OF THE $\tilde{B}$ STATE

SHUNJI KASAHARA, KOHEI TADA, *Molecular Photoscience Research Center, Kobe University, Kobe, Japan*; MICHIHIRO HIRATA, *Graduate School of Science, Kobe University, Kobe, Japan*; TAKASHI ISHIWATA, *Information Sciences, Hiroshima City University, Hiroshima, Japan*; EIZI HIROTA, *The Central Office, The Graduate University for Advanced Studies, Hayama, Kanagawa, Japan.*

Rotationally-resolved high-resolution fluorescence excitation spectra of the $\tilde{B}\ ^2E' \leftarrow \tilde{X}\ ^2A_2'$ electronic transition of $^{14}NO_3$ radical have been observed for 15860-15920 cm^{-1} region. Sub-Doppler excitation spectra were measured by crossing a single-mode laser beam perpendicular to a collimated radical beam, which was formed by the heat decomposition of $^{14}N_2O_5$; $^{14}N_2O_5 \rightarrow ^{14}NO_3 + ^{14}NO_2$. We have also measured the high-resolution fluorescence excitation spectra of the $^{14}NO_2\ \tilde{A}\ ^2B_2 \leftarrow \tilde{X}\ ^2A_1$ transition to distinguish the $^{14}NO_3$ signals from the $^{14}NO_2$ signals in the observed region. The typical linewidth was 30 MHz and the absolute wavenumber was calibrated with accuracy 0.0001 cm^{-1} by measurement of the Doppler-free saturation spectrum of iodine molecule and fringe pattern of the stabilized etalon. The observed rotational lines were too complicated to find any rotational series. In the observed spectra, only the rotational line pairs from the $\tilde{X}$ $^2A_2'(v'' = 0, K'' = 0, N'' = 1, F_1$ and $F_2)$ levels are assigned unambiguously by using the combination differences of the $\tilde{X}\ ^2A_2'$ state and measurement of the Zeeman splittings similar to the analysis of the 0-0 band at around 15100 cm^{-1} region. [a] [b] The observed results suggest the observed vibrationally excited states of the $\tilde{B}\ ^2E'$ state are also interacts with the other vibronic levels similar to the $\tilde{B}\ ^2E'(v' = 0)$ level.

---

[a]K. Tada, W. Kashihara, M. Baba, T. Ishiwata, E. Hirota, and S. Kasahara, *J. Chem. Physc.* **141**, 184307 (2014).

[b]K. Tada, T. Ishiwata, E. Hirota, and S. Kasahara, *J. Mol. Spectrosc.*, **321**, 23 (2016).

RF03 2:04 – 2:19

QUANTIFYING THE EFFECTS OF HIGHER ORDER JAHN-TELLER COUPLING TERMS ON A QUADRATIC JAHN-TELLER HAMILTONIAN IN THE CASE OF NO_3 AND Li_3.

HENRY TRAN, *Department of Chemistry and Biochemistry, The Ohio State University, Columbus, OH, USA*; JOHN F. STANTON, *Department of Chemistry, The University of Texas, Austin, TX, USA*; TERRY A. MILLER, *Department of Chemistry and Biochemistry, The Ohio State University, Columbus, OH, USA*.

The Jahn-Teller (JT) effect represents an enormous complication in the understanding of many molecules. We have been able to assign ~20 vibronic bands in the $\tilde{A}^2 E'' \leftarrow \tilde{X}^2 A_2'$ transition of NO_3 and determine the linear and quadratic JT coupling terms for ν_3 and ν_4, indicating strong and weak JT coupling along these modes respectively. It was found that the experimental results quantitatively disagree with ones determined from a vibronic Hamiltonian based on high-level *ab-initio* theory.[a] Typical analyses of experimental data use the quadratic JT Hamiltonian because limited measured levels tend to allow fitting only to coupling terms up to quadratic JT coupling. Hence, these analyses may neglect key contributions from cubic and quartic terms. To quantify this limitation, we have fit artificial spectra calculated with up to fourth order terms in the potential using a quadratic JT Hamiltonian and analyzed the results. The parameters chosen for this analysis are determined from *ab-initio* potentials for the $\tilde{A}$ state of NO_3 and $\tilde{X}$ state of Li_3 to gain further insight on these molecules. Our initial results concerning the limitations of the quadratic JT Hamiltonian will be presented.

[a]T. Codd, M.-W. Chen, M. Roudjane, J. F. Stanton, and T. A. Miller. Jet cooled cavity ringdown spectroscopy of the $\tilde{A}^2 E'' \leftarrow \tilde{X}^2 A_2'$ Transition of the NO_3 Radical. *J. Chem. Phys.*, 142:184305, 2015

RF04 2:21 – 2:36

ANALYSIS OF THE ROTATIONALLY RESOLVED, NON-DEGENERATE (a_1'') AND DEGENERATE (e') VIBRONIC BANDS IN THE $\tilde{A}^2 E'' \leftarrow \tilde{X}^2 A_2'$ TRANSITION OF NO_3.

HENRY TRAN, TERRY A. MILLER, *Department of Chemistry and Biochemistry, The Ohio State University, Columbus, OH, USA*.

The magnitude of the Jahn-Teller (JT) effect in NO_3 has been the subject of considerable research in our group and other groups around the world. The rotational contour of the 4_0^1 vibronic band was first described by Hirota and coworkers using an oblate symmetric top.[a] Deev *et al.* argued that an asymmetric top was required to describe the 2_0^1 band, although their spectrum was not completely rotationally resolved.[b] These discrepancies suggest that a rotational analysis will provide considerable experimental information on the geometry of NO_3. Our group has collected high-resolution, rotationally resolved spectra of the vibronic $\tilde{A}^2 E'' \leftarrow \tilde{X}^2 A_2'$ transitions. We have completed analysis of the 3_0^1 and $3_0^1 4_0^1$ parallel bands with a_1'' symmetry by using an oblate symmetric top with spin-rotation and centrifugal distortions. Several other parallel bands are now also reasonably understood. This analysis is consistent with a D_{3h} geometry for NO_3. In order to analyze the perpendicular bands with e' symmetry, we have adapted the oblate symmetric top Hamiltonian from the previous analysis to include spin-orbit coupling, coriolis coupling, and Watson Terms (JT distortions) that allow the oblate symmetric top Hamiltonian to transition continuously to the distorted limit of C_{2v} symmetry. Preliminary analysis of the 2_0^1 and $2_0^1 4_0^2$ bands has shown generally good agreement between model and experimental spectra. Our results indicate only modest JT distortions, although we do find evidence of multiple perturbations between these bands and high vibrational levels of the $\tilde{X}$ state. We will present our adapted Hamiltonian and the analysis of the 3_0^1, $3_0^1 4_0^1$, 2_0^1, and $2_0^1 4_0^2$ bands.

[a]E. Hirota, T. Ishiwata, K. Kawaguchi, M. Fujitake, N. Ohashi, and I. Tanaka. Near-infrared band of the nitrate radical NO_3 observed by diode laser spectroscopy. *J. Chem. Phys.*, 107:2829, 1997.

[b]A. Deev, J. Sommar, and M. Okumura. Cavity Ringdown Spectrum of the Forbidden $\tilde{A}^2 E'' \leftarrow \tilde{X}^2 A_2'$ Transition of NO_3: Evidence for static Jahn-Teller Distortion in the $\tilde{A}$ State. *J. Chem. Phys.*, 122:224305, 2005.

214

NEAR-INFRARED SPECTROSCOPY OF ETHYNYL RADICAL, C_2H

ANH T. LE, GREGORY HALL, TREVOR SEARS[a], *Division of Chemistry, Department of Energy and Photon Sciences, Brookhaven National Laboratory, Upton, NY, USA.*

The ethynyl radical, C_2H, is a reactive intermediate important in various combustion processes and also widely observed in the interstellar medium. In spite of extensive previous spectroscopic studies, the characterization of the near infrared transitions from the $\tilde{X}^2\Sigma^+$ state to the mixed vibrational overtone and $\tilde{A}^2\Pi$ states is incomplete. A strong band of C_2H at 7064 cm^{-1} was first observed in a neon matrix and assigned as the $\tilde{A}^2\Pi(002)^1 - \tilde{X}^2\Sigma^+$ transition by Forney et al.[b] Subsequent theoretical work of Tarroni and Carter[c] attributed the strong absorptions in this region to transitions terminating in two upper states, each a mixture of vibrationally excited $\tilde{X}$ states and different zero-order $\tilde{A}$-state bending levels: a $^2\Sigma^+$ symmetry combination of $\tilde{X}(0,2^0,3)$ and $\tilde{A}(0,3,0)^0\kappa$ and a $^2\Pi$ symmetry combination of $\tilde{X}(0,3^1,3)$ and $\tilde{A}(0,0,2)^1$. Transitions to them from the zero point level of the $\tilde{X}$ state are calculated to differ in energy by less than 10 cm^{-1} and to be within a factor of two in intensity. Diode laser transient absorption was used to record Doppler-limited spectra between 7020 and 7130 cm^{-1}, using 193 nm photolysis of CF_3C_2H as a source of C_2H. Two interleaved, rotationally resolved bands were observed, consistent with a $^2\Sigma$ - $^2\Sigma$ transition at 7088 cm^{-1} and a $^2\Pi$ - $^2\Sigma$ transition at 7108 cm^{-1}, in good accord with the Tarroni and Carter calculation. Progress on the assignment and fitting of the spectra will be reported.

Acknowledgements: Work at Brookhaven National Laboratory was carried out under Contract No. DE-SC0012704 with the U.S. Department of Energy, Office of Science, and supported by its Division of Chemical Sciences, Geosciences, and Biosciences.

[a]Also, Department of Chemistry, Stony Brook University, Stony Brook, New York 11794.
[b]D. Forney, M.E. Jacox, and W.E. Thompson, J. Mol. Spectrosc. 170, 178 (1995).
[c]R. Tarroni and S. Carter, Mol. Phys. 102, 2167 (2004).

STUDY OF INFRARED EMISSION SPECTROSCOPY FOR THE $B^1\Delta_g$-$A^1\Pi_u$ AND $B'^1\Sigma_g^+$-$A^1\Pi_u$ SYSTEMS OF C_2

JIAN TANG, WANG CHEN, KENTAROU KAWAGUCHI, *Graduate School of Natural Science and Technology , Okayama University, Okayama, Japan*; PETER F. BERNATH, *Department of Chemistry and Biochemistry, Old Dominion University, Norfolk, VA, USA.*

Recently, we carried out the perturbation analysis of C_2 spectra and identified forbidden singlet-triplet intersystem transitions,[a] which aroused further interest in other C_2 spectra for the many low-lying electronic states of this fundamental molecule. In 1988, the $B^1\Delta_g$-$A^1\Pi_u$ and $B'^1\Sigma_g^+$-$A^1\Pi_u$ band systems were discovered by Douay et al.,[b] who observed eight bands of the $B^1\Delta_g$-$A^1\Pi_u$ system with v up to 5 for the $B^1\Delta_g$ state and six bands of the $B'^1\Sigma_g^+$-$A^1\Pi_u$ system with v up to 3 for the $B'^1\Sigma_g^+$ state in the Fourier transform infrared emission spectra of hydrocarbon discharges. In the work presented here, we identified twenty-four bands of the two systems, among which the $B'^1\Sigma_g^+$ v = 4 and the $B^1\Delta_g$ v = 6, 7 and 8 vibrational levels involved in nine bands were studied for the first time. A direct global analysis with Dunham parameters was carried out satisfactorily for the $B^1\Delta_g$-$A^1\Pi_u$ system except for a small perturbation in the $B^1\Delta_g$ v = 6 level. The calculated rovibrational term energies up to $B^1\Delta_g$ v = 12 showed that the level crossing between the $B^1\Delta_g$ and $d^3\Pi_g$ states is responsible for many of the prominent perturbations in the Swan system observed previously.[c] Nineteen lines of the $B^1\Delta_g$-$a^3\Pi_u$ forbidden transitions were identified and the off-diagonal spin-orbit interaction constant A_{dB} between $d^3\Pi_g$ and $B^1\Delta_g$ was derived as 8.3(1) cm^{-1}. For the $B'^1\Sigma_g^+$-$A^1\Pi_u$ system, only individual band analyses for each vibrational level in the $B'^1\Sigma_g^+$ state could be done satisfactorily and Dunham parameters obtained from these effective parameters showed that the anharmonic vibrational constant $\omega_e x_e$ is anomalously small (nearly zero). Inspection of the RKR potential curves for the $B'^1\Sigma_g^+$ and $X^1\Sigma_g^+$ states revealed that an avoided crossing may occur around 30000 cm^{-1}, which is responsible for the anomalous molecular constants in these two states.[d]

[a]W. Chen, K. Kawaguchi, P. F. Bernath, and J. Tang, *J. Chem. Phys.,* **142,** 064317 (2015).
[b]M. Douay, R. Nietmann and P. F. Bernath, *J. Mol. Spectrosc.,* **131,** 261 (1988).
[c]A. Tanabashi, T. Hirao, T. Amano and P. F. Bernath, *Astrophys. J. Suppl. Ser.,* **169,** 472 (2007).
[d]W. Chen, K. Kawaguchi, P. F. Bernath, and J. Tang, *J. Chem. Phys.,* **144,** 064301 (2016).

RF07 3:12 – 3:27

A ZERO-ORDER PICTURE OF THE INFRARED SPECTRUM FOR THE METHOXY RADICAL: ASSIGNMENT OF STATES

BRITTA JOHNSON, EDWIN SIBERT, *Department of Chemistry, University of Wisconsin–Madison, Madison, WI, USA.*

The ground $\tilde{X}^2E$ vibrations of the methoxy radical have intrigued both experimentalists and theorists alike due to the presence of a conical intersection at the C_{3v} molecular geometry. This conical intersection causes methoxy's vibrational spectrum to be strongly influenced by Jahn-Teller vibronic coupling which leads to large amplitude vibrations and extensive mixing of the two lowest electronic states. This coupling combined with spin-orbit and Fermi couplings greatly complicates the assignments of states. Using the potential force field and calculated spectra of Nagesh and Sibert[1,2], we assign quantum numbers to the infrared spectrum. When the zero-order states are the diabatic normal mode states, there is sufficient mode mixing that the normal mode quantum numbers are poor labels for the final states. We define a series of zero-order Hamiltonians which include additional coupling elements beyond the normal mode picture but still allow for the assignment of Jahn-Teller quantum numbers. In methoxy, the two lowest frequency e modes, the bend (q_5) and the rock (q_6), are the modes with the strongest Jahn-Teller coupling. In general, a zero-order Hamiltonian which includes first-order Jahn-Teller coupling in q_6 is sufficient for most states of interest. Working in a representation which includes first-order Jahn-Teller coupling in q_6, we identify states in which additional coupling elements must be included; these couplings include first-order Jahn-Teller coupling in q_5, higher order Jahn-Teller coupling in q_5 and q_6 and, in the dueterated case, Jahn-Teller coupling which is modulated by the corresponding a modes.

[1] Nagesh, J.; Sibert, E. L. *J. Phys. Chem. A* **2012**, *116*, 3846–3855.

[2] Lee, Y.F.; Chou, W.T.; Johnson, B.A.; Tabor, D.P. ; Sibert, E.L.; Lee, Y.P. *J. Mol. Spectrosc.* **2015**, *310*, 57-67.

[2] Barckholtz, T. A.; Miller, T. A. *Int. Revs. in Phys. Chem.* **1998**, *17*, 435–524.

Intermission

RF08 3:46 – 4:01

INFRARED IDENTIFICATION OF THE CRIEGEE INTERMEDIATE $(CH_3)_2COO$

YI-YING WANG, *Applied Chemistry, National Chiao Tung University, Hsinchu, Taiwan*; YUAN-PERN LEE, *Applied Chemistry, National Chiao Tung University, Hsinchu, Taiwan, Institute of Atomic and Molecular Sciences, Academia Sinica, Taipei, Taiwan.*

The Criegee intermediates are carbonyl oxides that play critical roles in ozonolysis of alkenes in the atmosphere. We reported previously the mid-infrared spectra of the simplest Criegee intermediate CH_2OO.[a, b] and the methyl-substituted intermediate CH_3CHOO.[c] Here we report the transient infrared spectrum of $(CH_3)_2COO$, produced on UV photolysis of a mixture of $(CH_3)_2CI_2$, N_2, and O_2 in a flow reactor, using a step-scan Fourier-transform spectrometer. Guided by results of quantum-chemical calculations, rotational contours of the four observed bands are simulated successfully and provide definitive identification of $(CH_3)_2COO$. Although all observed bands of $(CH_3)_2COO$ contain hot bands from four vibrational modes of low energy, we were able to simulate the spectra satisfactorily. Observed bands with origins near 887, 1040, 1368, and 1422 cm^{-1} agree satisfactorily with corresponding anharmonic vibrational wavenumbers at 903, 1061, 1364, and 1422 cm^{-1} predicted with the B3LYP/aug-cc-pVTZ method. Furthermore, we could also estimate the rate coefficient of the self-reaction of $(CH_3)_2COO$. The direct infrared detection of $(CH_3)_2COO$ should prove useful for future field measurements and laboratory investigations of this Criegee intermediate.

[a]Y.-T. Su, Y.-H. Huang, H. A. Witek, Y.-P. Lee, *Science* **340**, 174 (2013).
[b]Y.-H. Huang, J. Li, H. Guo, Y.-P. Lee, *J. Chem. Phys.* **142**, 214301 (2015).
[c]H.-Y. Lin, Y.-H. Huang, X. Wang, J. M. Bowman, Y. Nishimura, H. A. Witek, Y.-P. Lee, *Nat. Comm.* **6**, 7012 (2015).

RF09

ANALYSES OF THE $\tilde{A}$-$\tilde{X}$ ELECTRONIC TRANSITIONS OF THE CH$_2$XOO·(X = I, Br, Cl) RADICALS

NEAL KLINE, MENG HUANG, TERRY A. MILLER, *Department of Chemistry and Biochemistry, The Ohio State University, Columbus, OH, USA.*

Cavity ringdown, near-infrared spectra have been previously observed following the photolysis of the dihalomethanes(CH$_2$XI, X = I, Br, Cl) in the presence of O$_2$ and N$_2$. In last year's Symposium[a], we presented evidence that all the spectra could be attributed to the $\tilde{A}$-$\tilde{X}$ electronic transition of the appropriate CH$_2$XOO· radical. We now present detailed analyses of these spectra. Similar spectral features have been observed for all radicals. The first strong transitions are located around 6800 cm^{-1}, and are assigned as associated with the origin. Other strong transitions are observed about 800 cm^{-1} blue of the origin, and have a multiple-peak structure similar to the corresponding origin bands. These bands are assigned to be the OO stretch of the $\tilde{A}$-$\tilde{X}$ electronic transitions, which are typically strong in the spectra of peroxy radicals, based on electronic structure calculations that provide vibrational frequencies and Franck-Condon factors. One-dimensional calculations of the internal torsion mode are applied to specifically explain the multiple-peak features in both the origin and OO stretch region as series of transitions including sequence bands and other hot bands from the vibrationally excited states of the low-frequency torsion mode in the $\tilde{X}$ state, which are significantly populated at room temperature. Additional bands can be assigned to fundamentals or combination bands of various other $\tilde{A}$ state modes.

[a]N. D. Kline, M. Huang, T. A. Miller, P. Lolur, R. Dawes, FD05, *70th International Symposium of Molecular Spectroscopy*(2015)

RF10

NOO PEROXY ISOMER EXPOSED WITH VELOCITY-MAP IMAGING[a]

BENJAMIN A LAWS, STEVEN J CAVANAGH, BRENTON R LEWIS, STEPHEN T GIBSON, *Research School of Physics and Engineering, Australian National University, Canberra, ACT, Australia.*

NO$_2$, a toxic gas formed in most combustion processes, plays an important role in the Earth's atmosphere due to its role in the production of both photochemical smog and tropospheric ozone. The existence of the peroxy radial, NOO, has been proposed, both as a collision reaction intermediate, and as a negative-ion in some discharge sources, in order to account for extended tails seen in some photoelectron spectra.[b]

In this work a velocity-mapped image of NO$_2^-$ photodetachment measured at 519 nm, shown, reveals high-energy electron structure, that persists at detachment energies lower than the electron affinity of ONO, 2.273 eV.[b] The central ring has the spectral signature of O$^-$, while the outer-ripples, that appear in character to be similar to NO− detachment, are, we propose due to the NOO$^-$ peroxy radical, which is also responsible for the presence of O$^-$. The photoelectron spectrum resolves the vibrational structure to characterize the neutral peroxy radical. The identification is further supported by *ab initio* calculations. The photoelectron angular distributions associated with the peroxy radical have a negative anisotropy parameter, opposite in sign to detachment from ONO$^-$.

[a]Research supported by the ARC DP160102585.
[b]K. M. Ervin and J. Ho and W. C. Lineberger, *J. Phys. Chem.* **92**, 5405 (1988). doi:10.1021/j100330a017

RF11 4:37 – 4:52

INFRARED LASER SPECTROSCOPY OF THE n-PROPYL AND i-PROPYL RADICALS IN HELIUM DROPLETS: SIGNIFICANT BEND-STRETCH COUPLING REVEALED IN THE CH STRETCH REGION

CHRISTOPHER P. MORADI, <u>GARY E. DOUBERLY</u>, *Department of Chemistry, University of Georgia, Athens, GA, USA*; DANIEL P. TABOR, EDWIN SIBERT, *Department of Chemistry, The Univeristy of Wisconsin, Madison, WI, USA.*

The n-propyl and i-propyl radicals were generated in the gas phase via pyrolysis of n-butyl nitrite ($CH_3(CH_2)_3ONO$) and i-butyl nitrite ($CH_3CH(CH_3)CH_2ONO$) precursors, respectively. Nascent radicals were promptly solvated by a beam of He nanodroplets, and the infrared spectra of the radicals were recorded in the C-H stretching region. In addition to three vibrations of n-propyl previously measured in an Ar matrix, we observe many unreported bands between 2800 and 3150 cm^{-1}, which we attribute to propyl radicals. The C-H stretching modes observed above 2960 cm^{-1} for both radicals are in excellent agreement with anharmonic frequencies computed using VPT2. Between 2800 and 2960 cm^{-1}, however, the spectra of n-propyl and i-propyl radicals become quite congested and difficult to assign due to the presence of multiple anharmonic resonances. Computations employing a local mode Hamiltonian reveal the origin of the spectral congestion to be strong coupling between the high frequency C-H stretching modes and the lower frequency bending/scissoring motions. The only significant local coupling is between stretches and bends on the same CH_2/CH_3 group.

RF12 4:54 – 5:09

INFRARED SPECTRUM OF FULVENALLENE AND FULVENALLENYL

<u>ALAINA R. BROWN</u>, JOSEPH T. BRICE, GARY E. DOUBERLY, *Department of Chemistry, University of Georgia, Athens, GA, USA.*

Fulvenallene (C_7H_6) and the fulvenallenyl (C_7H_5) radical are produced via thermal dissociation of phthalide in a continuous-wave SiC pyrolysis furnace. Prompt pick-up and solvation by helium droplets allows for well-resolved vibrational spectra of these species in the CH stretching region. The acetylenic CH stretch of the fulvenallenyl radical is a sensitive marker of the extent by which the unpaired electron is delocalized throughout the conjugated propargyl and cyclopentadienyl subunits. The nature of this electron delocalization is explored with spin density calculations at the CCSD(T)/ANO1 level of theory.

RF13 5:11 – 5:26

TWO-CENTER THREE-ELECTRON BONDING IN $ClNH_3$ REVEALED VIA HELIUM DROPLET INFRARED SPECTROSCOPY: ENTRANCE CHANNEL COMPLEX ALONG THE $Cl + NH_3 \rightarrow ClNH_2 + H$ REACTION

<u>PETER R. FRANKE</u>, CHRISTOPHER P. MORADI, *Department of Chemistry, University of Georgia, Athens, GA, USA*; MATIN KAUFMANN, *Physikalische Chemie II, Ruhr University Bochum, Bochum, Germany*; CHANGJIAN XIE, HUA GUO, *Department of Chemistry and Chemical Biology, University of New Mexico, Albuquerque, NM, USA*; GARY E. DOUBERLY, *Department of Chemistry, University of Georgia, Athens, GA, USA.*

Pyrolytic dissociation of Cl_2 is employed to dope helium droplets with single Cl atoms. Sequential addition of NH_3 to Cl-doped droplets leads to the formation of a complex residing in the entry valley to the substitution reaction, $Cl + NH_3 \rightarrow ClNH_2 + H$. Infrared Stark spectroscopy in the NH stretching region reveals symmetric and antisymmetric vibrations of a C_{3v} symmetric top. Frequency shifts from NH_3 and dipole moment measurements are consistent with a $ClNH_3$ complex containing a relatively strong two-center three-electron (2c-3e) bond. The nature of the 2c-3e bonding in $ClNH_3$ is explored computationally and found to be consistent with the complexation-induced blue shifts observed experimentally. Computations of interconversion pathways reveal nearly barrierless routes to the formation of this complex, consistent with the absence of two other complexes, NH_3Cl and $Cl-HNH_2$, which are predicted in the entry valley to the hydrogen abstraction reaction, $Cl + NH_3 \rightarrow HCl + NH_2$

RF14   5:28 – 5:43

JET-COOLED CHLOROFLUOROBENZYL RADICALS: SPECTROSCOPY AND MECHANISM

YOUNG YOON, <u>SANG LEE</u>, *Department of Chemistry, Pusan National University, Pusan, Korea.*

Whereas the benzyl radical, a prototypic aromatic free radical, has been the subject of numerous spectroscopic studies, halo-substituted benzyl radicals have received less attention, due to the difficulties associated with production of radicals from precursors. In particular, chloro-substituted benzyl radicals have been much less studied because of the weak visible emission intensity and weak C-Cl bond dissociation energy. The jet-cooled chlorofluorobenzyl radicals were generated in a technique of corona excited supersonic jet expansion using a pinhole-type glass nozzle for the vibronic assignments and measurements of electronic energies of the $D_1 \rightarrow D_0$ transition. The 2,4-,[a] 2,5-,[b] and 2,6-[c]chlorofluorobenzyl radicals were generated by corona discharge of corresponding precursors, chlorofluorotoluenes seeded in a large amount of helium carrier gas. The vibronic emission spectra were recorded with a long-path monochromator in the visible region. The emission spectra show the vibronic bands originating from two types of benzyl-type radicals, chlorofluorobenzyl and fluorobenzyl benzyl radicals, in which fluorobenzyl radicals were obtained by displacement of Cl by H produced by dissociation of methyl C-H bond. From the analysis of the spectra observed, we could determine the electronic energies in $D_1 \rightarrow D_0$ transition and vibrational mode frequencies at the D_0 state of chlorofluorobenzyl radicals, which show the origin band of the electronic transition to be shifted to red region, comparing with the parental benzyl radical. From the quantitative analysis of the red-shift, it has been found that the additivity rule can be applied to dihalo-substituted benzyl radicals. In this presentation, the dissociation process of precursors in corona discharge is discussed in terms of bond dissociation energy as well as the spectroscopic analysis of the radicals.

[a]C. S. Huh, Y. W. Yoon, and S. K. Lee, *J. Chem. Phys.* **136**, 174306 (2012).

[b]Y. W. Huh, S. Y. Chae, and S. K. Lee, *Chem. Phys. Lett.* **608**, 6 (2014).

[c]Y. W. Yoon, S. Y. Chae, M. Lim, and S. K. Lee, *Chem. Phys. Lett.* **637**, 148 (2015).

RF15   5:45 – 6:00

MECHANISM OF THE THERMAL DECOMPOSITION OF ETHANETHIOL AND DIMETHYLSULFIDE

WILLIAM FRANCIS MELHADO, <u>JARED CONNOR WHITMAN</u> , *Chemistry , Middlebury College , Middlebury , VT, USA* ; JESSICA KONG, *Chemistry, University of Washington, Seattle, WA, USA*; DANIEL EASTON ANDERSON, ANGAYLE (AJ) VASILIOU, *Chemistry , Middlebury College , Middlebury , VT, USA.*

Combustion of organosulfur contaminants in petroleum-based fuels and biofuels produces sulfur oxides (SO_x). These pollutants are highly regulated by the EPA because they have been linked to poor respiratory health and negative environmental impacts. Therefore much effort has been made to remove sulfur compounds in petroleum-based fuels and biofuels. Currently desulfurization methods used in the fuel industry are costly and inefficient. Research of the thermal decomposition mechanisms of organosulfur species can be implemented via engineering simulations to modify existing refining technologies to design more efficient sulfur removal processes. We have used a resistively-heated SiC tubular reactor to study the thermal decomposition of ethanethiol (CH_3CH_2SH) and dimethylsulfide (CH_3SCH_3). The decomposition products are identified by two independent techniques: 118.2 nm VUV photoionization mass spectroscopy and infrared spectroscopy. The thermal cracking products for CH_3CH_2SH are CH_2CH_2, SH, and H_2S and the thermal cracking products from CH_3SCH_3 are CH_3S, CH_2S, and CH_3.

RG. Clusters/Complexes
Thursday, June 23, 2016 – 1:30 PM
Room: 116 Roger Adams Lab

Chair: Brooks Pate, The University of Virginia, Charlottesville, VA, USA

RG01   1:30 – 1:45

CHARACTERIZATION OF AMMONIA-WATER CLUSTERS BY BROADBAND ROTATIONAL SPECTROSCOPY

LUCA EVANGELISTI, *Dipartimento di Chimica G. Ciamician, Università di Bologna, Bologna, Italy*; CRISTOBAL PEREZ, *CoCoMol, Max-Planck-Institut für Struktur und Dynamik der Materie, Hamburg, Germany*; BERHANE TEMELSO, *Department of Chemistry, Bucknell University, Lewisburg, PA, USA*; GEORGE C. SHIELDS, *Dean's Office, College of Arts and Sciences, and Department of Chemistry, Bucknell University, Lewisburg, PA, USA*; BROOKS PATE, *Department of Chemistry, The University of Virginia, Charlottesville, VA, USA*.

Neon carrier gas at 0.3 MPa of backing pressure is flowed over a room-temperature ammonia hydroxide solution before being expanded into a chirped-pulse Fourier transform microwave (CP-FTMW) spectrometer operating between 2 and 8 GHz. A dense spectrum was observed and the investigation allowed unambiguous assignment of the $(NH_3)_2(H_2O)_n$ with n=1,2 and $NH_3(H_2O)_n$ with n=2,3,4,5,6,8 with a signal to noise of at least 3:1. The structures show a cyclic arrangement for clusters with up four monomer and then move to a 3D arrangement. These clusters are of interest because of the different possibilities for hydrogen bond network related to the isolated water clusters. Calculations indicate that there are several possible low-energy isomers, with different levels of theory identifying different isomers as the global minimum. The evidence for the assignment and a discussion of the derived properties for the species are presented.

RG02   1:47 – 2:02

2OH OVERTONE SPECTROSCOPY OF WATER-CONTAINING VAN DER WAALS SPECIES

THOMAS VANFLETEREN, TOMAS FÖLDES, MICHEL HERMAN, JACQUES LIÉVIN, JÉROME LOREAU, *Service de Chimie Quantique et Photophysique, Université Libre de Bruxelles, Brussels, Belgium*; L. H. COUDERT, *LISA, CNRS, Universites Paris-Est Creteil et Paris Diderot, Creteil, France.*

We have used continuous-wave cavity ring-down spectroscopy to record part of the 2OH excitation range in an Ar/Kr supersonic expansion seeded with H_2O. Various bands were observed, and are being rotationally analyzed, of $Ar-H_2O$ and $Kr-H_2O$. The analysis of experimental linewidths allowed us to determine the mean upper state predissociation lifetime to be 3 ns for $Ar-H_2O$ and 4 ns for $Kr-H_2O$.

In this talk, the latest results concerning $Ar-H_2O$ will be presented. Several bands were identified and analyzed, highlighting some strong perturbations. The assignment of the many bands, as well as the perturbation processes, is complicated and is still in progress. The results of the analysis will be presented along with the perturbing effects of the dark states. Identification of these will be attempted using a multidimensional approach, based on the intramolecular potential energy surface of water monomer[a] and on the intermolecular potential energy surface of the complex, allowing us to evaluate the rovibrational energy levels of H_2O perturbed by the argon atom. Although several such intermolecular potentials are already available, like those reported by Makarewicz[b] and by Hou et al.,[c] none of them can be used in the present investigation as they were designed for vibrational states of the water monomer below the (101) state, involved in the present spectra.

A 6D intermolecular potential energy surface is currently being computed through *ab initio* calculations to deal with high lying states of the water monomer like the (101) state. With the help of this new surface and of the multidimensional approach, we are hoping to assign the bright and the dark states of the complex, near $7\,500$ cm^{-1}, involved in the present spectra.

[a]H. Partridge and D. W. Schwenke, *J. Chem. Phys.* **106** (1997) 4618.
[b]J. Makarewicz, *J. Chem. Phys.* **129** (2008) 184310.
[c]D. Hou, Y.-T. Ma, X.-L. Zhang, and H. Li, *J. Chem. Phys.* **144** (2016) 014301.

220

RG03   2:04 – 2:19

AN INFRARED SPECTROSCOPIC STUDY ON THE FORMATION OF THE HYDROGEN BONDED INCLUSION-STRUCTURES IN THE PROTONATED METNANOL WATER CLUSTERS

MARUSU KATADA, *Department of Chemistry, Tohoku University, Sendai, Japan*; PO-JEN HSU, *Institute of Atomic and Molecular Sciences, Academia Sinica, Taipei, Taiwan*; ASUKA FUJII, *Department of Chemistry, Tohoku University, Sendai, Japan*; JER-LAI KUO, *Institute of Atomic and Molecular Sciences, Academia Sinica, Taipei, Taiwan.*

We measured IR spectra of the protonated methanol–water mixed clusters ($H^+(CH_3OH)_n(H_2O)_1$, n =6 - 10) in the OH stretching vibrational region. Spectra of their Ar tagged clusters were also measured to explore hydrogen-bonded structure changes by the vibrational cooling. The temperature dependence of the isomer distribution was also examined by the harmonic superposition approximation (HSA) simulation. No essential change of the structures with the Ar tagging (lowering of temperature) was concluded in the size range of n = 8 - 10, indicating the remarkable stability of the inclusion structures in this size range. On the other hand, at n = 7, the large isomer distribution change with the Ar tagging is suggested. Moreover, at n = 6, the IR spectrum showed dramatic changes upon the Ar tagging. The protonated site switching from water to methanol well explained these observed changes.

RG04   2:21 – 2:36

STEPWISE INTERNAL ENERGY CONTROL FOR PROTONATED METHANOL CLUSTERS BY USING THE INERT GAS TAGGING

TAKUTO SHIMAMORI, *Department of Chemistry, Tohoku University, Sendai, Japan*; JER-LAI KUO, *Institute of Atomic and Molecular Sciences, Academia Sinica, Taipei, Taiwan*; ASUKA FUJII, *Department of Chemistry, Tohoku University, Sendai, Japan.*

Preferred isomer structures of hydrogen-bonded clusters should depend on their temperature because of the entropy term in the free energy. To observe such temperature dependence, we propose a new approach to control the internal energy (vibrational temperature) of protonated clusters in the gas phase. We performed IR spectroscopy of protonated methanol clusters, $H^+ (CH_3OH)_n$, n= 5 and 7, with the tagging by various inert gas species (Ar, CO_2, CO, CS_2, C_2H_2, and C_6H_6). We found that vibrational temperature of the tagged clusters raises with increase of the interaction energy with the tag species, and the observed cluster structures follow the theoretical prediction of the temperature dependence of the isomer population.

RG05   2:38 – 2:53

INFRARED ABSORPTION OF METHANOL-WATER CLUSTERS $M_n(H_2O)$, n = 1-4, RECORDED WITH THE VUV-IONIZATION/IR-DEPLETION TECHNIQUES

YU-FANG LEE, *Applied Chemistry, National Chiao Tung University, Hsinchu, Taiwan*; YUAN-PERN LEE, *Applied Chemistry, National Chiao Tung University, Hsinchu, Taiwan, Institute of Atomic and Molecular Sciences, Academia Sinica, Taipei, Taiwan.*

We investigated IR spectra in the CH- and OH-stretching regions of size-selected methanol-water clusters, $M_n(H_2O)$ with M representing CH_3OH and n = 1-4, in a pulsed supersonic jet by using the VUV (vacuum-ultraviolet)-ionization/IR-depletion technique. The VUV light at 118 nm served as the source of ionization in a time-of-flight mass spectrometer. The tunable IR laser served as a source of dissociation for clusters before ionization. Spectra of methanol-water clusters in the OH region show significant variations as the number of methanol molecules increase, whereas spectra in the CH region are similar. For $M(H_2O)$, absorption of a structure with H_2O as a proton donor was observed at 3570, 3682, and 3722 cm^{-1}, whereas that of methanol as a proton donor was observed at 3611 and 3753 cm^{-1}. For $M_2(H_2O)$, the OH-stretching band of the dangling OH of H_2O was observed at 3721 cm^{-1}, whereas overlapped bands near 3425, 3472, and 3536 cm^{-1} correspond to the OH-stretching modes of three hydrogen-bonded OH in a cyclic structure. For $M_3(H_2O)$, the dangling OH shifts to 3715 cm^{-1}, and the hydrogen-bonded OH-stretching bands become much broader, with a band near 3179 cm^{-1} having the smallest wavenumber. Scaled harmonic vibrational wavenumbers and relative IR intensities predicted for the methanol-water clusters with the M06-2X/aug-cc-pVTZ method are consistent with our experimental results. For $M_4(H_2O)$, observed spectrum agree less with theoretical predictions, indicating the presence of isomers other than the most stable cyclic one. Spectra of $M_n(H_2O)$ and M_{n+1} are compared and the cooperative hydrogen-bonding is discussed.

RG06 2:55 – 3:10

PREDICTING CERIUM + H₂O CLUSTER FORMATION WITH SIMULATED AND EXPERIMENTAL SPECTROSCOPY

<u>JOSEY E TOPOLSKI</u>, JARED O. KAFADER, MANISHA RAY, CAROLINE CHICK JARROLD, *Department of Chemistry, Indiana University, Bloomington, IN, USA.*

Ceria (CeO_2) has been established as a good support in heterogeneous catalysts for the water gas shift reaction. This study looks into cerium's reactivity with water, a water gas shift reagent, and aims to build an understanding of the three reactions which can occur: direct oxidation, -OH abstraction, and H_2O addition. Through the use of anion photoelectron spectroscopy and density functional theory calculations we have been able to determine that the reactivity is dependent on (1) the oxidation states of the metal centers, (2) the availability of 5d orbitals to form metal oxide bonds, and (3) the presence of electrons in the 6s* orbital. The results of this study can be used to inform design of catalytic materials for the water gas shift reaction.

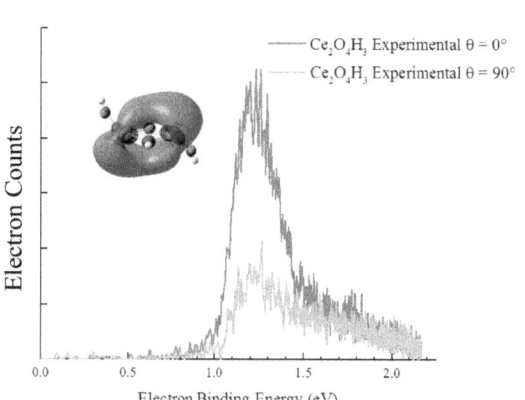

Intermission

RG07 3:29 – 3:44

STRONG QUANTUM COUPLING BETWEEN O-H⁺-O STRETCH AND FLANKING GROUP MOTIONS IN $(CH_3OH)_2H^+$ PART I: UNMASKING THE 800-1200 cm^{-1} PEAKS

<u>JAKE ACEDERA TAN</u>, JER-LAI KUO, *Institute of Atomic and Molecular Sciences, Academia Sinica, Taipei, Taiwan.*

Assigning the vibrational signatures between 800-1200 cm^{-1} is not a trivial task for $(CH_3OH)_2H^+$.[a] − [b] Such complication in the assignment arises due to the intermode coupling between O-H⁺-O stretch and flanking group motions. In this talk, we will examine the interactions between O-H⁺-O stretch and four flanking group motions: 1) out-of-phase C-O stretch, 2) in-plane CH_3 rock, 3) out-of-plane CH_3, and 4) O-O stretch. Vibrational interactions were investigated by solving a reduced-dimensional Schrödinger equation by means of Discrete Variable Representation (DVR) with harmonic oscillators as basis. Potential and dipole moment surfaces were constructed by scanning along normal mode coordinates using MP2 and CCSD(T) level. It was found out that the O-H⁺-O stretch strongly couples with the four above mentioned flanking motions, leading to intensity redistribution on peaks between 800-1200 cm^{-1}.[c]

[a] J.R. Roscioli, L.R. McCunn and M.A. Johnson. Science 2007, 316, 249

[b] T.D. Fridgen, L. Macaleese, T.B McMahon, J. Lemaire and P. Maitre. Phys. Chem. Chem. Phys. 2006, 8, 955-966

[c] J.A. Tan and J.-L. Kuo. J. Phys. Chem. A. 2015, 119, 11320-11328

RG08   **3:46 – 4:01**

STRONG QUANTUM COUPLING BETWEEN O-H$^+$-O STRETCH AND FLANKING GROUP MOTIONS IN $(CH_3OH)_2H^+$ Part II: TUNING THE COUPLING VIA ISOTOPOLOGUES

The vibrational coupling between O-H$^+$-O/O-D$^+$-O stretch and flanking group motions were explored in the following isotopologues: $(CH_3OH)_2H^+$, $(CD_3OH)_2H^+$, $(CH_3OD)_2D^+$, and $(CD_3OD)_2D^+$. At present only measurements for $(CH_3OH)_2H^+$ are available in the literature.[a][b] Reduced-dimensional calculations were performed by solving several vibrational Schrödinger equations using the method of Discrete Variable Representation (DVR) with harmonic oscillator as basis. Both potential and dipole moment surfaces were constructed at MP2/aug-cc-pVDZ by scanning along normal modes corresponding to: 1) O-H$^+$-O/O-D$^+$-O stretch, 2) out-of-phase C-O stretch, 3) in-plane CH_3/CD_3 rock, 4) out-of-plane CH_3/CD_3 rock, and 5) O-O stretch. It was found that vibrational states for isotopologues corresponding to O-H$^+$-O are more mixed than that of the O-D$^+$-O. Lastly, we proposed tentative assignments for the simulated spectrum and hope that experimental measurements will be available in the near future.

[a]J.R. Roscioli, L.R. McCunn and M.A. Johnson. Science 2007, 316, 249
[b]T.D. Fridgen, L. Macaleese, T.B McMahon, J. Lemaire and P. Maitre. Phys. Chem. Chem. Phys. 2006, 8, 955-966

RG09   **4:03 – 4:18**

INFRARED SPECTROSCOPY AND ANHARMONIC VIBRATIONAL ANALYSIS OF THE AR·H$^+$·AR PROTON-BOUND DIMER CATION

DAVID C McDONALD, *Chemistry, University of Georgia, Athens, GA, USA*; JAKE ACEDERA TAN, *Institute of Atomic and Molecular Sciences, Academia Sinica, Taipei, Taiwan*; JOSHUA H MARKS, *Chemistry, University of Georgia, Athens, GA, USA*; JER-LAI KUO, *Institute of Atomic and Molecular Sciences, Academia Sinica, Taipei, Taiwan*; MICHAEL A DUNCAN, *Department of Chemistry, University of Georgia, Athens, GA, USA.*

Ar·H$^+$·Ar is generated in a supersonic expansion via pulsed electrical discharge of hydrogen in argon. H$^+$Ar$_n$ clusters (n = 1-5) are extracted into a reflectron time-of-flight mass spectrometer and probed with infrared photodissociation spectroscopy (IRPS) in the 760-2550 cm^{-1} region. The limited number of atoms and vibrational coordinates make Ar·H$^+$·Ar a simple case for the fundamental study of anharmonic coupling in proton-bound dimer systems. Anharmonic theory suggests strong coupling between the shared proton stretch and the symmetric Ar stretching modes are the dominant features of the spectra.

RG10   **4:20 – 4:35**

THE FORMAMIDE$_2$-H$_2$O COMPLEX: STRUCTURE AND HYDROGEN BOND COOPERATIVE EFFECTS

SUSANA BLANCO, *Departamento de Química Física y Química Inorgánica / Grupo de Espectroscopía Molecular, Universidad de Valladolid, Valladolid, Spain*; PABLO PINACHO, *Departamento de Química Física y Química Inorgánica, Universidad de Valladolid, Valladolid, Spain*; JUAN CARLOS LOPEZ, *Departamento de Química Física y Química Inorgánica / Grupo de Espectroscopía Molecular, Universidad de Valladolid, Valladolid, Spain.*

The adduct formamide$_2$-H$_2$0 has been detected in a supersonic expansion and its rotational spectra in the 5-13 GHz frequency region characterized by narrow-band molecular beam Fourier transform microwave spectroscopy (MB-FTMW). The spectrum shows the hyperfine structure due to the presence of two ^{14}N-nuclei. This hyperfine structure has been analyzed and the determined quadrupole coupling constants together with the rotational constants have been a key for the identification of the adduct structure on the light of ab initio computations. The rotational parameters are consistent with the formation of a three body cycle thanks to the double proton acceptor/proton donor character of both formamide and water. The low value of the planar moment of inertia P_{cc} indicates that the heavy atom skeleton of the cluster is essentially planar. A detailed analysis of the results reveals the subtle effects of hydrogen bond cooperative effects in this system.

RG11 4:37 – 4:52

A STUDY OF THE FORMAMIDE-$(H_2O)_3$ COMPLEX BY MICROWAVE SPECTROSCOPY

PABLO PINACHO, *Departamento de Química Física y Química Inorgánica, Universidad de Valladolid, Valladolid, Spain*; JUAN CARLOS LOPEZ, SUSANA BLANCO, *Departamento de Química Física y Química Inorgánica / Grupo de Espectroscopía Molecular, Universidad de Valladolid, Valladolid, Spain.*

The adduct formamide-$(H_2O)_3$ has been detected in a supersonic expansion and its rotational spectrum characterized in the 5-13 GHz frequency region by narrow-band molecular beam Fourier transform microwave spectroscopy (MB-FTMW). The spectrum shows the hyperfine structure due to the presence of one ^{14}N-nuclei and small splittings due to a tunnelling motion of the complex. The spectra has been analyzed using a two-state Hamiltonian including Coriolis coupling terms to determine the vibrational spacing. The determined quadrupole coupling constants together with the rotational constants have been a key for the identification of the adduct structure on the light of ab initio computations. The rotational parameters are consistent with the formation of a four body cycle thanks to the double proton acceptor/proton donor character of both formamide and water. The rotational data are consistent with a non-planar heavy atom skeleton.

RG12 *Post-Deadline Abstract* 4:54 – 5:09

PURE ROTATIONAL SPECTRUM AND MOLECULAR GEOMETRY OF AN ISOLATED COMPLEX OF IMIDAZOLE AND UREA

SUSANA BLANCO, *Departamento de Química Física y Química Inorgánica / Grupo de Espectroscopía Molecular, Universidad de Valladolid, Valladolid, Spain*; JOHN C MULLANEY, CHRIS MEDCRAFT, NICK WALKER, *School of Chemistry, Newcastle University, Newcastle-upon-Tyne, United Kingdom*; ANTHONY LEGON, *School of Chemistry, University of Bristol, Bristol, United Kingdom.*

The investigation of the dynamics of biomolecules is crucial to understand biological processes. For this purpose, the initial research investigations on the conformational behavior of isolated biomolecules should go one further step by investigating the structure and conformation of complexes formed in supersonic jets by different biomolecules to model the interactions which take place in biological media. In this work, the imidazole-urea complex formed in a supersonic expansion has been investigated by using microwave spectroscopy. In parallel, the conformational space of the complex has been explored with ab initio calculations. The broadband microwave spectrum (8-18GHz frequency interval) has been recorded using a Chirped Pulse Fourier Transform Microwave spectrometer (CP-FTMW). The solid sample was formed by mixing pure samples of imidazole and urea within a solid copper matrix, and was vaporized using the second harmonic of a pulsed Nd:YAG laser. The analysis of the experimental data in the light of the theoretical predictions has allowed the unambiguous identification of the observed conformers in the microwave spectrum.

RG13 5:11 – 5:26

AN AB INITIO APPROACH TO ANALYZE FERMI RESONANCE IN AMMONIA CLUSTERS

KUN-LIN HO, *Institute of Atomic and Molecular Sciences, Academia Sinica, Taipei, Taiwan*; MARUSU KATADA, *Department of Chemistry, Tohoku University, Sendai, Japan*; JER-LAI KUO, *Institute of Atomic and Molecular Sciences, Academia Sinica, Taipei, Taiwan*; ASUKA FUJII, *Department of Chemistry, Tohoku University, Sendai, Japan.*

Anharmonic vibrational coupling among N-H stretching fundamental (ν_1 and ν_3) and N-H bending overtone ($2\nu_4$) vibrations in $(NH_3)_n$ (n = 1 to 5) are analyzed based a full dimensional Hamiltonian including third and quartic terms. In particular, we examine Fermi resonance between the symmetric N-H stretching (ν_1) and N-H bending overtone ($2\nu_4$) vibrations. As the cluster size increases, enhancement of the hydrogen bond strength makes ν_1 red-shifted while $2\nu_4$ blue-shifted. These shifts result in the crossing of the frequencies of ν_1 and $2\nu_4$ levels, and their energy order reverses between n = 3 to n = 4. Because the nature of Fermi resonance, although the zero-order ν_1 and $2\nu_4$ levels are shifted, the resultant mixed levels do not show remarkable changes in frequency. Instead, the major component of each mixed level largely changes and this causes significant redistribution of the intensity. Our results offer a solution to resolve puzzles on the intensity distribution and assignments of the Fermi mixing bands in the previously reported infrared spectra of $(NH_3)_n$.

RG14

MILLIMETER/SUBMILLIMETER SPECTRA OF WEAKLY-BOUND CLUSTERS

LUYAO ZOU, SUSANNA L. WIDICUS WEAVER, *Department of Chemistry, Emory University, Atlanta, GA, USA.*

Weakly-bound clusters are important for studying intermolecular interactions such as van der Waals forces and hydrogen bonding. The geometry and the effect of intermolecular force can be retrieved from their rovibrational spectra. Using a millimeter/submillimeter direct absorption spectrometer coupled with a fast-sweep detection technique, we are able to probe the ground vibrational state of weakly-bound molecules with high precision. With this spectrometer, we have fully characterized the pure rotational spectrum of *trans*-HO_3 up to 450 GHz. We have also identified lines in the *trans*-HO_3 spectrum as arising from the Ar-H_2O complex, and followed up with a full study of this cluster in the 200–850 GHz range. We are additionally studying proton-bound complexes that might have an impact in astrochemical environment, such as H_2–HCO^+. The experimental setup and the preliminary results for these complexes will be presented.

RH. Astronomy

Thursday, June 23, 2016 – 1:30 PM

Room: 274 Medical Sciences Building

Chair: Susanna L. Widicus Weaver, Emory University, Atlanta, GA, USA

RH01   1:30 – 1:40

DUO: A GENERAL MULTI-STATE PROGRAM FOR SOLVING THE NUCLEAR MOTION SCHRÖDINGER EQUATION FOR OPEN SHELL DIATOMIC MOLECULES

SERGEI N. YURCHENKO, *Department of Physics and Astronomy, University College London, Gower Street, London WC1E 6BT, United Kingdom*; LORENZO LODI, JONATHAN TENNYSON, *Department of Physics and Astronomy, University College London, Gower Street, London WC1E 6BT, United Kingdom*; ANDREY STOLYAROV, *Department of Chemistry, Moscow State University, Moscow, Russia.*

Although solution of the diatomic nuclear motion problem for a single, uncoupled potential energy curve is standard [1], there appears to be no general rovibronic program available for diatomics characterized by complex interactions between electronic states. We have therefore developed a new computational tool, DUO [2], to fill this gap. Duo is a flexible, user-friendly program capable of solving the Schrödinger equation for the nuclear motion of a general diatomic molecule with an arbitrary number and type of coupling between electronic states. The program allows one to build diatomic 'projects' from a set of pre-defined objects such as potential energy and dipole moment functions, spin-orbit, electronic angular momentum, spin-rotational, Λ-doubling terms etc. From these objects DUO computes energy levels, line positions and line intensities. Several analytic forms plus interpolation and extrapolation options are available for representation of the curves. DUO can refine potential energy and coupling curves to best reproduce reference data such as experimental energy levels or line positions. Duo is currently being used as a diatomic computational tool as part of the ExoMol project [3]; examples include AlO, ScH, CaO, VO, PO, NO, PS and C_2.

References

[1] R. J. Le Roy, LEVEL 8.0, University of Waterloo Chemical Physics Research Report CP-663, 2007.

[2] S. N. Yurchenko, L. Lodi, J. Tennyson, A. V. Stolyarov, Computer Phys. Comm., 2016, doi:10.1016/j.cpc.2015.12.021

[3] J. Tennyson, S. N. Yurchenko, Mon. Not. R. astr. Soc., 425, 21, 2012.

RH02 1:42 – 1:57

ExoMol: MOLECULAR LINE LIST FOR EXOPLANETS AND OTHER ATMOSPHERES

JONATHAN TENNYSON, *Department of Physics and Astronomy, University College London, London, IX, United Kingdom*; SERGEI N. YURCHENKO, *Department of Physics and Astronomy, University College London, Gower Street, London WC1E 6BT, United Kingdom.*

The discovery of extrasolar planets is one of the major scientific advances of the last two decades. Thousands of planets have now been detected and astronomers are beginning to characterize their composition and physical characteristics. To do this requires a huge quantity of spectroscopic data most of which are not available from laboratory studies. The ExoMol project [1] is generating a comprehensive solution to this problem by providing spectroscopic data on all the molecular transitions of importance in the atmospheres of exoplanets. These data are widely applicable to other problems such studies on cool stars, brown dwarfs and circumstellar environments as well as industrial and technological problems on earth. ExoMol employs a mixture of first principles and empirically tuned quantum mechanical methods to compute comprehensive and very large rotation–vibration and rovibronic line lists. Results span a variety of closed (NaH, SiO, PN, NaCl, KCl, CS) and open (BeH, MgH, CaH, AlO, VO) shell diatomics to triatomics (HCN/HNC, SO_2, H_2S, H_3^+), tetratomics (H_2CO, PH_3, SO_3, H_2O_2), plus methane [2] and nitric acid [3]. This has led directly to the detection of new species in the atmospheres of exoplanets [4]. A new comprehensive data release has just been completed [5]. Progress on and future prospects of the project will be summarised.

References

[1] J. Tennyson, S. N. Yurchenko, Mon. Not. R. astr. Soc., 425, 21, 2012.

[2] S. N. Yurchenko, J. Tennyson, J. Bailey, M. D. J. Hollis, G Tinetti, Proc. Nat. Acad. Sci., 111, 9379, 2014.

[3] A. I. Pavlyuchko, S. N. Yurchenko, J. Tennyson, Mon. Not. R. astr. Soc., 452, 1702, 2015.

[4] A. Tsiaras *et al*, Astrophys. J., in press. [5] J. Tennyson *et al*, J. Mol. Spectrosc., in press.

RH03 1:59 – 2:14

INFRARED SPECTROSCOPY OF HOT METHANE: EMPIRICAL LINE LISTS WITHIN THE 1 - 2 μm REGION

ANDY WONG, ROBERT J. HARGREAVES, PETER F. BERNATH, *Department of Chemistry and Biochemistry, Old Dominion University, Norfolk, VA, USA.*

Methane is one of the many hydrocarbons that is found in cool planetary atmospheres in our solar system. Its prominence also extents to hot sub-stellar environments such as brown dwarfs and hot Jupiter exoplanets. High resolution transmission spectra (0.02 cm^{-1}) have been recorded at eight different temperatures (between 294 - 1000 K) within the 1 - 2 μm region using a Fourier transform infrared spectrometer and tube furnace. From these observations, temperature dependent empirical line lists have been produced that include line position, intensity, lower state energy and possible quantum number assignments. Our line lists and spectra can be used to directly simulate the atmospheric spectra of brown dwarfs and exoplanets. These experimental line lists are also compared to predictions from *ab initio* variational calculations that are known to have diminished accuracy in the 1 - 2 μm region.

RH04 2:16 – 2:31

QUANTUM CHEMISTRY MEETS ROTATIONAL SPECTROSCOPY FOR ASTROCHEMISTRY: INCREASING MOLECULAR COMPLEXITY

CRISTINA PUZZARINI, *Dep. Chemistry 'Giacomo Ciamician', University of Bologna, Bologna, Italy.*

For many years, scientists suspected that the interstellar medium was too hostile for organic species and that only a few simple molecules could be formed under such extreme conditions. However, the detection of approximately 180 molecules in interstellar or circumstellar environments in recent decades has changed this view dramatically. A rich chemistry has emerged, and relatively complex molecules such as C_{60} and C_{70} are formed. Recently, researchers have also detected complex organic and potentially prebiotic molecules, such as amino acids, in meteorites and in other space environments. Those discoveries have further stimulated the debate on the origin of the building blocks of life in the universe.

Rotational spectroscopy plays a crucial role in the investigation of planetary atmosphere and the interstellar medium. Increasingly these astrochemical investigations are assisted by quantum-mechanical calculations of structures as well as spectroscopic and thermodynamic properties to guide and support observations, line assignments, and data analysis in these new and chemically complicated situations.[a] However, it has proved challenging to extend accurate quantum-chemical computational approaches to larger systems because of the unfavorable scaling with the number of degrees of freedom (both electronic and nuclear).

In this contribution, it is demonstrated that it is now possible to compute physicochemical properties of building blocks of biomolecules with an accuracy rivaling that of the most sophisticated experimental techniques. We analyze the spectroscopic properties of representative building blocks of DNA bases (uracil and thiouracil), of proteins (glycine and glycine dipeptide analogue), and also of PAH (phenalenyl radical and cation).

[a] V. Barone, M. Biczysko, C. Puzzarini 2015, Acc. Chem. Res., 48, 1413

RH05 2:33 – 2:48

MOLECULES IN LABORATORY AND IN INTERSTELLAR SPACE? [a]

VENKATESAN S. THIMMAKONDU, *Department of Chemistry, Birla Institute of Technology and Science, Pilani, K K Birla Goa Campus, Goa, Goa, India.*

In this talk, the quantum chemistry of astronomically relevant molecules will be outlined with an emphasis on the structures and energetics of C_7H_2 isomers, which are yet to be identified in space. Although more than 100's of isomers are possible for C_7H_2, to date only 6 isomers had been identified in the laboratory.[b,c,d] The equilibrium geometries of heptatriynylidene (**1**), cyclohepta-1,2,3,4-tetraen-6-yne (**2**), and heptahexaenylidene (**3**), which we had investigated theoretically will be discussed briefly.[e] While **1** and **3** are observed in the laboratory, **2** is a hypothetical molecule. The theoretical data may be useful for the laboratory detection of **2** and astronomical detection of **2** and **3**.

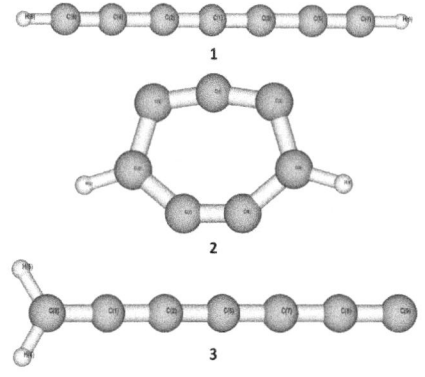

[a] THIS WORK IS SUPPORTED BY A RESEARCH GRANT (YSS/2015/00099) FROM SERB, DST, GOVERNMENT OF INDIA.

[b] Apponi, A. P.; McCarthy, M. C.; Gottlieb, C. A.; Thaddeus, P. Laboratory Detection of Four New Cumulene Carbenes: H_2C_7, H_2C_8, H_2C_9, and D_2C_{10}, *Astrophys. J.* 2000, 530, 357-361.

[c] Ball, C. D; McCarthy, M. C.; Thaddeus, P. Cavity Ringdown Spectroscopy of the Linear Carbon Chains HC_7H, HC_9H, $HC_{11}H$, and $HC_{13}H$. *J. Chem. Phys.* 2000, 112, 10149-10155.

[d] Dua, S.; Blanksby, S. J.; Bowie, J. H. Formation of Neutral C_7H_2 Isomers from Four Isomeric C_7H_2 Radical Anion Precursors in the Gas Phase. *J. Phys. Chem. A*, 2000, 104, 77-85.

[e] Thimmakondu, V. S. The equilibrium geometries of heptatriynylidene, cyclohepta-1,2,3,4-tetraen-6-yne, and heptahexaenylidene, *Comput. Theoret. Chem.* 2016, 1079, 1-10.

RH06   2:50 – 3:05

THE CENTER FOR ASTROCHEMICAL STUDIES AT THE MAX PLANCK INSTITUTE FOR EXTRATERRESTRIAL PHYSICS.

VALERIO LATTANZI, LUCA BIZZOCCHI, JACOB LAAS, BARBARA MICHELA GIULIANO, SILVIA SPEZZANO, CHRISTIAN ENDRES, PAOLA CASELLI, *The Center for Astrochemical Studies, Max-Planck-Institut für extraterrestrische Physik, Garching, Germany.*

The Center for Astrochemical Studies (CAS), at the Max Planck Institute for Extraterrestrial Physics (MPE) in Garching, has been founded to incorporate scientists with different background to elucidate the physical-chemical processes that lead to the formation of stars and planets. The CAS group includes experts in observations (including millimetre and sub-millimetre interferometry, radio and infrared telescopes), theory (physical processes and dynamics, gas-grain chemical processes and dust evolution, molecular astrophysics and collisional/rate coefficients), and laboratory. The latter is mainly focused on spectroscopic characterisation of molecular species relevant in space, including ions, radicals and astronomically complex organic molecules. In this talk the laboratory group of the CAS will be briefly presented, including current projects and planned experiments.

RH07   3:07 – 3:22

MILLIMETER/SUBMILLIMETER SPECTROSCOPY TO MEASURE THE BRANCHING RATIOS FOR METHANOL PHOTOLYSIS

MORGAN N McCABE, CARSON REED POWERS, SAMUEL ZINGA, SUSANNA L. WIDICUS WEAVER, *Department of Chemistry, Emory University, Atlanta, GA, USA.*

Methanol is one of the most abundant and important molecules in the interstellar medium, playing a key role in driving more complex organic chemistry both on grain surfaces and through gas-phase ion-molecule reactions. Methanol photolysis produces many radicals such as hydroxyl, methoxy, hydroxymethyl, and methyl that may serve as the building blocks for more complex organic chemistry in star-forming regions. The branching ratios for methanol photolysis may govern the relative abundances of many of the more complex species already detected in these environments. However, no direct, comprehensive, quantitative measurement of methanol photolysis branching ratios is available. Using a 193 nm excimer laser, the gas phase photolysis of methanol was studied in the (sub)millimeter range, where the rotational spectroscopic signatures of the photolysis products were probed. Here we present preliminary results from this experiment.

Intermission

RH08   3:41 – 3:56

HIGH-RESOLUTION SPECTROSCOPY OF THE $A^1\Pi(v'=0\text{-}10)-X^1\Sigma^+(v''=0)$ BANDS IN $^{13}C^{18}O$

JEAN LOUIS LEMAIRE, *CNRS, Institut des Sciences Moléculaires d'Orsay, Orsay, France*; MICHELE EIDELSBERG, *Meudon, Observatoire de Paris, Paris, France*; ALAN HEAYS, *Leiden Observatory, University of Leiden, Leiden, Netherlands*; LISSETH GAVILAN, *CNRS/INSU, UPMC Univ Paris 06, Paris, France*; STEVEN FEDERMAN, *Physics and Astronomy, University of Toledo, Toledo, OH, USA*; GLENN STARK, *Department of Physics, Wellesley College, Wellesley, MA, USA*; JAMES R LYONS, *School of Earth and Space Exploration, Arizona State University, Tempe, AZ, USA*; NELSON DE OLIVEIRA, DENIS JOYEUX, *DESIRS Beamline, Synchrotron SOLEIL, Saint Aubin, France.*

Ultraviolet spectrographs on space-borne astronomical facilities are used to study CO photochemistry primarily through observations of the $A - X$ Fourth Positive Band System. Absorption from $^{12}C^{16}O$, $^{13}C^{16}O$, $^{12}C^{18}O$, and $^{12}C^{17}O$ in diffuse interstellar clouds has been detected to date. While the necessary spectroscopic data are available to identify the isotopologues, measurements of oscillator strengths only exist for the most abundant variant, $^{12}C^{16}O$. In our ongoing experiments on the DESIRS beam-line at the SOLEIL Synchrotron, we are acquiring the necessary data on oscillator strengths and term values for other isotopologues. Here we present our latest results involving $A - X$ bands in $^{13}C^{18}O$.

RH09 3:58 – 4:13

HIGH-RESOLUTION INFRARED SPECTROSCOPY OF CARBON-SULFUR CHAINS: I. C_3S AND SC_7S

JOHN B DUDEK, *Department of Chemistry, Hartwick College, Oneonta, NY, USA*; THOMAS SALOMON, SVEN THORWIRTH, *I. Physikalisches Institut, Universität zu Köln, Köln, Germany.*

In the course of a recent 5 μm high-resolution infrared study of laser ablation products from carbon-sulfur targets, we have reinvestigated the ν_1 vibrational mode of the linear C_3S molecule complementing significantly the pioneering data originally reported by Takano and coworkers[a]. In addition, located within the R-branch of the C_3S vibrational mode, a weak new band is observed which exhibits very tight line spacing. On the basis of high-level quantum-chemical calculations, this feature is attributed to the linear SC_7S species, which stands for the first gas-phase spectroscopic detection of this heavy carbon-sulfur chain.

[a] S. Takano, J. Tang, and S. Saito 1996, J. Mol. Spectrosc. 178, 194

RH10 4:15 – 4:30

HIGH-RESOLUTION INFRARED SPECTROSCOPY OF CARBON-SULFUR CHAINS: II. C_5S AND SC_5S

SVEN THORWIRTH, THOMAS SALOMON, *I. Physikalisches Institut, Universität zu Köln, Köln, Germany*; JOHN B DUDEK, *Department of Chemistry, Hartwick College, Oneonta, NY, USA.*

Unbiased high-resolution infrared survey scans of the ablation products from carbon-sulfur targets in the 2100 to 2150 cm^{-1} regime reveal two bands previously not observed in the gas phase. On the basis of comparison against laboratory matrix-isolation work[a] and new high-level quantum-chemical calculations these bands are attributed to the linear C_5S and SC_5S clusters. While polar C_5S was studied earlier using Fourier-transform microwave techniques,[b,c] the present work marks the first gas-phase spectroscopic detection of SC_5S.

[a] H. Wang, J. Szczepanski, P. Brucat, and M. Vala 2005, Int. J. Quant. Chem. 102, 795
[b] Y. Kasai, K. Obi, Y. Ohshima, Y. Hirahara, Y. Endo, K. Kawaguchi, and A. Murakami 1993, ApJ 410, L45
[c] V. D. Gordon, M. C. McCarthy, A. J. Apponi, and P. Thaddeus 2001, ApJS 134, 311

RH11 4:32 – 4:47

ASTROCHEMISTRY LECTURE AND LABORATORY COURSES AT THE UNIVERSITY OF ILLINOIS: APPLIED SPECTROSCOPY

DAVID E. WOON, *Department of Chemistry, University of Illinois at Urbana-Champaign, Urbana, IL, USA*; BENJAMIN J. McCALL, *Departments of Chemistry and Astronomy, University of Illinois at Urbana-Champaign, Urbana, IL, USA.*

The Department of Chemistry at the University of Illinois at Urbana-Champaign offers two courses in astrochemistry, one lecture (Chem 450) and one laboratory (Chem 451). Both courses present the opportunity for advanced undergraduate and graduate students to learn about various spectroscopic concepts as they are applied toward an exotic subject, astrochemistry. In the lecture course, each student devotes a substantial fraction of the course work to one of the known astromolecules, building a wiki page for it during the semester, presenting a brief oral description about it in class, and then finally writing a paper about it. The course covers electronic, vibrational, and rotational spectroscopy, along with Einstein coefficients, line widths, and the interpretation of actual astronomical spectra. It also covers relevant reactions and reaction networks. Students learn to use pgopher for modeling rotational spectra. The lab course focuses on the methylidyne radical (CH). It begins with its chemistry and spectroscopy and then moves on to laboratory study of its electronic spectrum as observed in a butane flame and then collected with the university's 12" f/15 Brashear refracting telescope in the campus observatory built in 1896. Students learn to use IGOR to reduce CCD data.

RH12 4:49 – 5:04

ALMA DATA MINING TOOLKIT

DOUGLAS FRIEDEL, LESLIE LOONEY, *Department of Astronomy, University of Illinois at Urbana-Champaign, Urbana, IL, USA*; PETER J. TEUBEN, MARC W. POUND, KEVIN P. RAUCH, LEE MUNDY, *Department of Astronomy, University of Maryland, College Park, MD, USA*; ROBERT J HARRIS, *Department of Astronomy, University of Illinois at Urbana-Champaign, Urbana, IL, USA*; LISA XU, *NCSA, University of Illinois at Urbana-Champaign, Urbana, IL, USA.*

ADMIT (ALMA Data Mining Toolkit) is a Python based pipeline toolkit for the creation and analysis of new science products from ALMA data. ADMIT quickly provides users with a detailed overview of their science products, for example: line identifications, line 'cutout' cubes, moment maps, and emission type analysis (e.g., feature detection). Users can download the small ADMIT pipeline product (< 20MB), analyze the results, then fine-tune and re-run the ADMIT pipeline (or any part thereof) on their own machines and interactively inspect the results. ADMIT has both a web browser and command line interface available for this purpose. By analyzing multiple data cubes simultaneously, data mining between many astronomical sources and line transitions are possible. Users are also able to enhance the capabilities of ADMIT by creating customized ADMIT tasks satisfying any special processing needs. We will present some of the salient features of ADMIT and example use cases.

RH13 5:06 – 5:21

THE INFRARED DETECTION OF DEUTERATED PAHS IN HII REGIONS

KIRSTIN D DONEY, *Leiden Observatory, Sackler Laboratory for Astrophysics, Universiteit Leiden, Leiden, Netherlands*; ALESSANDRA CANDIAN, *Leiden Observatory, University of Leiden, Leiden, Netherlands*; TAMAMI MORI, TAKASHI ONAKA, *Department of Astronomy, The University of Tokyo, Tokyo, Japan*; XANDER TIELENS, *Leiden Observatory, University of Leiden, Leiden, Netherlands.*

The amount of deuterium locked up in polycyclic aromatic hydrocarbons (PAHs) has to date been an uncertain value. A near-infrared (NIR) spectroscopic survey of HII regions in the Milky Way, Large Magellanic Cloud (LMC), and Small Magellanic Cloud (SMC), obtained with the AKARI satellite, was performed to search for features indicative of deuterated PAHs in order to better constrain the D/H ratio of PAHs. Through comparison of the observed spectra with a calculated spectrum of deuterated PAHs the aromatic and (a)symmetric aliphatic C-D stretch vibrational modes were identified. In only six out of fifty-three of the observed sources, all of which are located in the Milky Way, emission features are seen between 4.4-4.8 μm that can be unambiguously attributed to deuterated PAHs. In all cases the aromatic C-D stretching feature is weaker than the aliphatic C-D stretching feature, which indicates that deuterium addition is favored over substitution. In addition, based on the weak or absent PAD features in most of the observed spectra, it is suggested that the mechanism for PAH deuteration in the ISM is uncommon.

RH14

LIGHT ON THE 3 μm EMISSION BAND FROM SPACE WITH MOLECULAR BEAM SPECTROSCOPY

ELENA MALTSEVA, *Van' t Hoff Institute for Molecular Sciences, University of Amsterdam, Amsterdam, Netherlands*; CAMERON J. MACKIE, *Leiden Observatory, Leiden University, Leiden, The Netherlands*; ALESSANDRA CANDIAN, ANNEMIEKE PETRIGNANI, XANDER TIELENS, *Leiden Observatory, University of Leiden, Leiden, Netherlands*; JOS OOMENS, *Institute for Molecules and Materials (IMM), Radboud University Nijmegen, Nijmegen, Netherlands*; XINCHUAN HUANG, *Carl Sagan Center, SETI Institute, Moutain View, CA, USA*; TIMOTHY LEE, *Space Science and Astrobiology Division, NASA Ames Research Center, Moffett Field, CA, USA*; WYBREN JAN BUMA, *Van' t Hoff Institute for Molecular Sciences, University of Amsterdam, Amsterdam, Netherlands*.

The majority of interstellar objects shows IR emission features also known as unidentified infrared (UIR) emission bands. These UIR bands are attributed to IR emission of highly-excited gaseous polycyclic aromatic hydrocarbons (PAHs). To understand the physical conditions and chemical evolution of the interstellar environment a precise identification of the emission carriers is desired. The 3 μm UIR feature is represented by a strong band at 3040 cm^{-1}, a plateau from 3150 to 2700 cm^{-1} and a number of weak features within this plateau. The 3040 cm^{-1} component is assigned to fundamental CH-stretch vibrations of PAHs, but there still remain many questions on the origin of the other features. In this work we have studied experimentally the 3 μm region of regular, hydrogenated and methylated PAHs (up to 5 rings), combining molecular beam techniques with IR-UV ion dip spectroscopy, and theoretically by density functional theory (DFT) calculations within the harmonic and anharmonic approximation. We find that (a) the 3 μm region of PAHs is dominated by Fermi resonances and thereby cannot be treated within the harmonic approximation; (b) the periphery structure of the molecules strongly affects the shape of the 3 μm band. In particular, the two-component emission interpretation can be explained by the presence of molecules with and without bay-hydrogens; (c) due to strong Fermi resonances of fundamental modes with combination bands regular PAHs can significantly contribute to the 3 μm plateau in the 3150-2950 cm^{-1}, while hydrogenated and methylated species are primarily responsible for features in the 2950-2750 cm^{-1} region.

RH15

TOP DOWN CHEMISTRY VERSUS BOTTOM UP CHEMISTRY

TAKESHI OKA, *Department of Astronomy and Astrophysics, Chemistry, The University of Chicago, Chicago, IL, USA*; ADOLF N. WITT, *Department of Physics and Astronomy, University of Toledo, Toledo, OH, USA*.

The idea of interstellar top down chemistry (TDC), in which molecules are produced from decomposition of larger molecules and dust in contrast to ordinary bottom up chemistry (BUC) in which molecules are produced synthetically from smaller molecules and atoms in the ISM, has been proposed in the chemistry of PAH [a,b] and carbon chain molecules [c,d] both for diffuse [a,c] and dense clouds [b,d]. A simple and natural idea, it must have occurred to many people and has been in the air for sometime [e]

The validity of this hypothesis is apparent for diffuse clouds in view of the observed low abundance of small molecules and its rapid decrease with molecular size on the one hand and the high column densities of large carbon molecules demonstrated by the many intense diffuse interstellar bands (DIBs) on the other. Recent identification of C_{60}^+ as the carrier of 5 near infrared DIBs with a high column density of 2×10^{13} cm^{-2} by Maier and others [f] confirms the TDC.

This means that the large molecules and dust produced in the high density high temperature environment of circumstellar envelopes are sufficiently stable to survive decompositions due to stellar UV radiaiton, cosmic rays, C-shocks etc. for a long time ($\geq 10^7$ year) of their migration to diffuse clouds and seems to disagree with the consensus in the field of interstellar grains [g]. The stability of molecules and aggregates in the diffuse interstellar medium will be discussed.

[a] Duley, W. W. 2006, Faraday Discuss. 133, 415
[b] Zhen, J., Castellanos, P., Paardekooper, D. M., Linnartz, H., Tielens, A. G. G. M. 2014, ApJL, 797, L30
[c] Huang, J., Oka, T. 2015, Mol. Phys. 113, 2159
[d] Guzmán, V. V., Pety, J., Goicoechea, J. R., Gerin, M., Roueff, E., Gratier, P., Öberg, K. I. 2015, ApJL, 800, L33
[e] L. Ziurys has sent us many papers beginning Ziurys, L. M. 2006, PNAS 103, 12274 indicating she had long been a proponent of the idea.
[f] Campbell, E. K., Holz, M., Maier, J. P., Gerlich, D., Walker, G. A. H., Bohlender, D, 2016, ApJ, in press
[g] Draine, B. T. 2003, ARA&A, 41, 241

RI. Metal containing
Thursday, June 23, 2016 – 1:30 PM
Room: B102 Chemical and Life Sciences

Chair: Anthony Merer, University of British Columbia, Vancouver, BC, Canada

RI01   1:30 – 1:45

HIGHLY UNSATURATED PLATINUM AND PALLADIUM CARBENES PtC_3 AND PdC_3 ISOLATED AND CHARAC-TERIZED IN THE GAS PHASE

DROR M. BITTNER, *School of Chemistry, Newcastle University, Newcastle-upon-Tyne, United Kingdom*; DANIEL P. ZALESKI, *Chemical Sciences and Engineering Division, Argonne National Laboratory, Argonne, IL, USA*; DAVID PETER TEW, *School of Chemistry, University of Bristol, Bristol, United Kingdom*; NICK WALKER, *School of Chemistry, Newcastle University, Newcastle-upon-Tyne, United Kingdom*; ANTHONY LEGON, *School of Chemistry, University of Bristol, Bristol, United Kingdom*.

Carbenes of platinum and palladium, PtC_3 and PdC_3, were generated in the gas phase through laser vaporization of a metal target in the presence of a low concentration of a hydrocarbon precursor undergoing supersonic expansion. Rotational spectroscopy and *ab initio* calculations confirm that both molecules are linear. The geometry of PtC_3 was accurately determined by fitting to the experimental moments of inertia of twenty-six isotopologues.

RI02   1:47 – 2:02

THE PURE ROTATIONAL SPECTRA OF FCPtF AND FPtI

DROR M. BITTNER, NICK WALKER, *School of Chemistry, Newcastle University, Newcastle-upon-Tyne, United Kingdom*; ANTHONY LEGON, *School of Chemistry, University of Bristol, Bristol, United Kingdom*.

Transitions measured by a chirped pulse Fourier transform microwave spectrometer in the frequency range 6.5-18.5 GHz have been fitted and tentatively assigned to the linear molecules FCPtF and FPtI, each in a $^1\Sigma$ electronic state. Laser ablation was used to introduce Pt into the gas phase from a metal rod with natural isotopic abundance. CF_3I was used as a source of C, F and I atoms. The products of reactions between the chemical precursors were cooled to a rotational temperature approaching 2K through supersonic expansion of the gaseous sample. Different isotopologues of Pt were observed. The spectra of other palladium and platinum containing complexes obtained in a similar way will be presented.

RI03   2:04 – 2:19

SPECTROSCOPIC STUDY OF LOCAL INTERACTIONS OF PLATINUM IN SMALL $[Ce_xO_y]Pt_{x'}^{-}$ CLUSTERS

MANISHA RAY, JARED O. KAFADER, CAROLINE CHICK JARROLD, *Department of Chemistry, Indiana University, Bloomington, IN, USA*.

Cerium oxide is a good ionic conductor, and the conductivity can be enhanced with oxygen vacancies and doping. This conductivity may play an important role in the enhancement of noble or coinage metal toward the water-gas shift reaction when supported by cerium oxide. The ceria-supported platinum catalyst in particular has received much attention because of higher activity at lower temperatures (LT) compared to the most common commercial LT-WGS catalyst. We have used a combination of anion photoelectron spectroscopy and density functional theory calculations to study the interesting molecular and electronic structures and properties of cluster models of ceria-supported platinum. $[Ce_xO_y]Pt_{x'}^{-}$ (x,x'=1,2 ; y≤2x') clusters exhibit evidence of ionic bonding possible because of the high electron affinity of Pt and the low ionization potential of cerium oxide clusters. In addition, Pt^- is a common daughter ion resulting from photodissociation of $[Ce_xO_y]Pt_{x'}^{-}$ clusters. Finally, several of the anion and neutral clusters have profoundly different structures. These features may play a role in the enhancement of catalytic activity toward the water-gas shift reaction.

RI04   2:21 – 2:36

ROTATIONAL SPECTROSCOPY OF ClZnCH$_3$ ($\tilde{X}^1A_1$): CHARACTERIZATION OF A MONOMERIC GRIGNARD-TYPE REAGENT

K. M. KILCHENSTEIN, *Department of Chemistry and Biochemistry; Department of Astronomy, Arizona Radio Observatory, University of Arizona, Tuscon, AZ, USA*; JIE MIN, *Department of Chemistry and Biochemistry, University of Arizona, Tucson, AZ, USA*; MATTHEW BUCCHINO, *Department of Chemistry and Astronomy, University of Arizona, Tucson, AZ, USA*; LUCY M. ZIURYS, *Department of Chemistry and Biochemistry; Department of Astronomy, Arizona Radio Observatory, University of Arizona, Tuscon, AZ, USA.*

The pure rotational spectrum of the organozinc halide, ClZnCH$_3$ ($\tilde{X}^1A_1$), has been measured using Fourier-transform microwave (FTMW) and millimeter-wave direct-absorption methods in the frequency range 10–296 GHz. This work is the first study of ClZnCH$_3$ by gas-phase spectroscopy. The molecule was created in a DC discharge from the reaction of zinc vapor, produced either by a Broida-type oven or by laser ablation, with chloromethane in what appears to be a metal insertion process. Rotational and chlorine quadrupole constants were determined for three zinc isotopologues. The Zn – Cl bond was found to be partly ionic and significantly shorter than in EtZnCl.

RI05   2:38 – 2:53

THRESHOLD IONIZATION OF La(C$_5$H$_8$) FORMED BY La-MEDIATED DEHYDROGENATION OF 1-PENTENE

WENJIN CAO, YUCHEN ZHANG, DONG-SHENG YANG, *Department of Chemistry, University of Kentucky, Lexington, KY, USA.*

La(C$_5$H$_8$) was formed by La reaction with 1-pentene (CH$_2$=CH-CH$_2$-CH$_3$) in a laser-vaporization supersonic molecular beam source and characterized with mass-analyzed threshold ionization (MATI) spectroscopy. The MATI spectrum displays an origin band at 38988 (5) cm^{-1} and three vibrational intervals of 130, 294, and 415 cm^{-1}. The La(C$_5$H$_8$) complex is identified as a five-membered metallacycle in C$_1$ point group, with the doublet and singlet being the lowest energy states of the neutral and cation, respectively. The energy at 38998 cm^{-1} corresponds to the adiabatic ionization energy of the complex, and the three vibration intervals in the order of the frequency increase are assigned to the terminal CH$_3$ torsion, asymmetric La-ligand stretch, and symmetric La-ligand stretch excitation of the ion. The La + 1-penetene reaction will also be compared with La reactions with other five-carbon hydrocarbon molecules, such as isoprene, 1-pentyne, and 1,4-pentadiene.

RI06   2:55 – 3:10

YTTRIUM-ASSISTED C-H AND C-C BOND ACTIVATION OF ETHYLENE PROBED BY MASS-ANALYZED THRESHOLD IONIZATION SPECTROSCOPY

JONG HYUN KIM, DONG-SHENG YANG, *Department of Chemistry, University of Kentucky, Lexington, KY, USA.*

The reaction between Y atom and ethylene (CH$_2$=CH$_2$) was performed in a laser-ablation supersonic molecular beam source. Y(C$_2$H$_2$), Y(C$_2$H$_4$), and Y(C$_4$H$_6$) were observed by time-of-flight mass spectrometry and investigated with mass-analyzed threshold ionization (MATI) spectroscopy and theoretical calculations. Y(C$_2$H$_2$) is formed by hydrogen elimination, Y(C$_2$H$_4$) by simple association, and La(C$_4$H$_6$) by C-C bond coupling and dehydrogenation. Both Y(C$_2$H$_2$) and Y(C$_2$H$_4$) have a C$_{2v}$ triangular structure with a C=C double bond in Y(C$_2$H$_2$) and a C-C single bond in Y(C$_2$H$_4$). Y(C$_4$H$_6$) has a five-membered metallacyclic structure (C$_s$) with Y binding to the two terminal carbon atoms of butene, which is the exactly same as that of Y(C$_4$H$_6$) formed in the Y + 1-butene reaction. For all three complexes, ionization has a small effect on the metal-carbon bond lengths because the rejected electron has basically a Y 5s character. The adiabatic ionization energies are measured to be 45679(5) cm^{-1} for Y(C$_2$H$_2$), 45603(5) cm^{-1} for Y(C$_2$H$_4$) and 43475(5) cm^{-1} for Y(C$_4$H$_6$). The metal-ligand stretching frequencies of the three complexes are also measured from the MATI spectra.

234

RI07   3:12 – 3:27

THE EFFECT OF ARGON TAGGING ON THE Ti$^+$(H$_2$O) SYSTEM OBSERVED THROUGH INFRARED SPEC-TROSCOPY

TIMOTHY B WARD, PROSSER CARNEGIE, MICHAEL A DUNCAN, *Department of Chemistry, University of Georgia, Athens, GA, USA.*

Ti$^+$(H$_2$O)Ar$_n$ clusters were produced in a laser vaporization/pulsed nozzle source. The clusters were then mass selected in a time-of-flight mass spectrometer and studied with infrared photodissociation spectroscopy in the OH stretching region. The spectra exhibits two bands, with the asymmetric band showing K-type rotational structure. Previous work has shown that most metal-water rare gas tagged systems adopt the C$_{2v}$ geometry. The asymmetric stretch also exhibits the well-knowm 3:1 ortho:para band intensity ratio in the K-type rotational structure. However the Ti$^+$(H$_2$O)Ar structure contains additional bands indicating a large spin-orbit splitting in the asymmetric stretch. Theory and PGOPHER simulations suggest that the binding of argon also leads to a change in the spin state from a quartet to a doublet in the titanium atom.

Intermission

RI08   3:46 – 4:01

THE MYSTERY OF THE ELECTRONIC SPECTRUM OF RUTHENIUM MONOPHOSPHIDE

ALLAN G. ADAM, RYAN M CHRISTENSEN, JACOB M DORE, RICARDA M. KONDER, *Department of Chemistry, University of New Brunswick, Fredericton, NB, Canada*; DENNIS W. TOKARYK, *Department of Physics, University of New Brunswick, Fredericton, NB, Canada.*

Using PH$_3$ as a reactant gas and ruthenium as the target metal in the UNB laser ablation spectrometer, the ruthenium monophosphide molecule (RuP) has been detected. Dispersed fluorescence experiments have been performed to determine ground state vibrational frequencies and the presence of any low-lying electronic states. Rotationally resolved spectra of two vibrational bands at 577nm and 592nm have been taken; the bands have been identified as 1-0 and 0-0 bands based on isotopic shifts. Ruthenium has seven stable isotopes and rotational transitions have been observed for six of the RuP isotopologues. RuP is isoelectronic to RuN so it is expected that RuP will have a $^2\Sigma^+$ ground state and low resolution spectra indicated a likely $^2\Sigma^+$ - $^2\Sigma^+$ electronic transition. Further investigation has led us to believe we are observing a $^2\Pi$ - $^2\Sigma^+$ transition but mysteriously some important rotational branches are missing. It is hoped that new data to be recorded on a second electronic system we have observed at 535nm will help shed light on this mystery.

RI09   4:03 – 4:18

LASER SPECTROSCOPY OF IRIDIUM MONOCHLORIDE

COLAN LINTON, *Department of Physics, University of New Brunswick, Fredericton, NB, Canada*; ALLAN G. ADAM, SAMANTHA FORAN, *Department of Chemistry, University of New Brunswick, Fredericton, NB, Canada*; TONGMEI MA, TIMOTHY STEIMLE, *Department of Chemistry and Biochemistry, Arizona State University, Tempe, AZ, USA.*

Iridium monochloride (IrCl) molecules have been produced in the gas phase using laser ablation sources at the University of New Brunswick (UNB) and Arizona State University (ASU). Low resolution laser induced fluorescence (LIF) spectra, obtained at UNB using a pulsed dye laser, showed three bands at 557, 545 and 534 nm which appeared to form an upper state vibrational progression. Dispersed fluorescence (DF) spectra, obtained by exciting each band at its band head frequency, showed a ground state vibrational progression extending from v=0 to 6. High resolution spectra (FWHM=0.006 cm^{-1}), taken using a cw ring dye laser, showed resolved rotational lines, broadened by unresolved Ir (I=3/2) hyperfine structure, in both the ^{193}Ir^{35}Cl and ^{191}Ir^{35}Cl isotopologues. Vibrational assignments of 0-0, 1-0 and 2-0 for the three bands were determined from the isotope structure and the rotational analysis showed the transition to be $^3\Phi_4$ - $^3\Phi_4$, similar to that previously observed in IrF. Higher resolution spectra (FWHM=0.001 cm^{-1}) of the 1-0 band, obtained at ASU, showed resolved hyperfine structure from which the magnetic and quadrupole hyperfine parameters in the ground and excited states were determined. The interpretation of the hyperfine parameters in terms of the electron configurations will be presented along with a comparison of the properties of IrCl and IrF.

RI10

LASER INDUCED FLUORESCENCE SPECTROSCOPY OF JET-COOLED CaOCa

MICHAEL N. SULLIVAN, DANIEL J. FROHMAN, MICHAEL HEAVEN, *Department of Chemistry, Emory University, Atlanta, GA, USA*; WAFAA M FAWZY, *Department of Chemistry, Murray State University, Murray, KY, USA.*

The group IIA metals have stable hypermetallic oxides of the general form MOM. Theoretical interest in these species is associated with the multi-reference character of the ground states. It is now established that the ground states can be formally assigned to the $M^+O^{2-}M^+$ configuration, which leaves two electrons in orbitals that are primarily metal-centered ns orbitals. Hence the MOM species are diradicals with very small energy spacings between the lowest energy singlet and triplet states. Previously, we have characterized the lowest energy singlet transition ($^1\Sigma_u^+ \leftarrow X^1\Sigma_g^+$) of BeOBe. In this study we obtained the first electronic spectrum of CaOCa. Jet-cooled laser induced fluorescence spectra were recorded for multiple bands that occured within the 14,800 - 15,900 cm^{-1} region. Most of the bands exhibited simple P/R branch rotational line patterns that were blue-shaded. Only even rotational levels were observed, consistent with the expected X $^1\Sigma_g^+$ symmetry of the ground state (^{40}Ca has zero nuclear spin). A progression of excited bending modes was evident in the spectrum, indicating that the transition is to an upper state that has a bent equilibrium geometry. Molecular constants were extracted from the rovibronic bands using PGOPHER. The experimental results and interpretation of the spectrum, which was guided by the predictions of electronic structure calculation, will be presented.

RI11

ELECTRONIC BANDS OF ScC IN THE REGION 620 - 720 NM

CHIAO-WEI CHEN, ANTHONY MERER, YEN-CHU HSU, *Institute of Atomic and Molecular Sciences, Academia Sinica, Taipei, Taiwan.*

ScC molecules have been observed by laser-induced fluorescence, following the reaction of laser-ablated scandium metal with acetylene under supersonic jet-cooled conditions. Rotational analyses have been carried out for about 40 bands of Sc^{12}C and Sc^{13}C in the region 14000 - 16000 cm^{-1}. Two lower states are found, with Ω = 3/2 and 5/2, indicating that the ground state is $^4\Pi_i$ or $^2\Delta$. As yet we cannot distinguish between these alternatives, but note that the ground state of the isoelectronic YC molecule[a] is $^4\Pi_i$. The ground state bond length in ScC is 1.95$_5$ Å, and the vibrational frequency is 712 cm^{-1}. At least eight electronic transitions occur in the region studied, the majority obeying the selection rule $\Delta\Omega$ = +1. Rotational perturbations are widespread, consistent with a high density of excited electronic states.

[a]B. Simard, P.A. Hackett and W.J. Balfour, Chem. Phys. Lett., **230**, 103 (1994).

RI12   4:54 – 5:09

LASER SPECTROSCOPY AND AB INITIO CALCULATIONS ON THE TaF MOLECULE

KIU FUNG NG, *Department of Chemistry, The University of Hong Kong, Hong Kong, Hong Kong*; WENLI ZOU, *Institute of Modern Physics, Northwest University, Xi'an, China*; WENJIAN LIU, *Department of Chemistry, Peking University, Beijing, China*; ALLAN S.C. CHEUNG, *Department of Chemistry, The University of Hong Kong, Hong Kong, Hong Kong.*

Electronic transition spectrum of the tantalum monoflouride (TaF) molecule in the spectral region between 448 and 520 nm has been studied using the technique of laser-ablation/reaction free jet expansion and laser induced fluorescence spectroscopy. TaF molecule was produced by reacting laser-ablated tantalum atoms with sulfur hexafluoride gas seeded in argon. Sixteen vibrational bands with resolved rotational structure have been recorded and analyzed, which were organized into six electronic transition systems and the ground state has been identified to be the $X^3\Sigma^-(0^+)$ state with bond length, r_o, and equilibrium vibrational frequency, ω_e, determined to be 1.8209 Å and 700.1 cm^{-1} respectively. In addition, four vibrational bands belong to another transition system involving lower state with $\Omega = 2$ component has also been analyzed. All observed transitions are with $\Delta\Omega = 0$. Least-squares fit of the measured line positions yielded molecular constants for the electronic states involved.

The Λ-S and Ω states of TaF were calculated at the state-averaged complete active space self-consistent field (SA-CASSCF) and the subsequent internally contracted multi-reference configuration interaction with singles and doubles and Davidson's cluster correction (MRCISD+Q) levels of theory with the active space of 4 electrons in 6 orbitals, that is, the molecular orbitals corresponding to Ta 5d6s are active. The spin-orbit coupling (SOC) is calculated by the state-interaction approach at the SA-CASSCF level via the relativistic effective core potentials (RECPs) spin-orbit operator, where the diagonal elements of the spin-orbit matrix are replaced by the above MRCISD+Q energies. The spectroscopic properties of the ground and many low-lying electronic states of the TaF molecule will be reported. With respect to the observed electronic states in this work, the calculated results are in good agreement with our experimental determinations. This work represents the first experimental investigation of the molecular structure of the TaF molecule.

RI13   5:11 – 5:26

SPECTROSCOPIC STUDY OF ThCl$^+$ BY TWO-PHOTON IONIZATION

JOSHUA BARTLETT, ROBERT A. VANGUNDY, MICHAEL HEAVEN, *Department of Chemistry, Emory University, Atlanta, GA, USA*; KIRK PETERSON, *Department of Chemistry, Washington State University, Pullman, WA, USA.*

Despite the irreplaceable role experimental data plays for evaluating the performance of computational predictions, diatomic actinide species have not received much spectroscopic attention. As an early actinide element, thorium-containing species are ideal candidates for these types of studies. The electronic structure is expected to be relatively simple compared to later actinides, and therefore allows straightforward assessment of calculations. Here, we have studied ThCl$^+$ for the first time via resonant two-photon ionization of jet-cooled ThCl produced by laser ablation of the metal reacted with dilute Cl$_2$. Laser-induced Fluorescence (LIF) spectra have been recorded for the neutral molecule from 16000 - 23500 cm^{-1} in search of a suitable intermediate state for subsequent two-photon ionization experiments. Monochromator dispersion of the fluorescence has recovered the ground state vibration and anharmonic constants of ThCl. Resonant Two-Photon Ionization (R2PI) within a time-of-flight mass spectrometer was used to confirm ThCl production, and Pulsed Field Ionization Zero Kinetic Energy photoelectron spectroscopy (PFI-ZEKE) has been performed to identify the ionization energy as well as several of the low-lying states of the ThCl$^+$ molecule. These constants have been predicted at the CASPT2 and CCSD(T) levels of theory, and a discussion of the calculations' performance will be presented alongside the recorded spectra.

RJ. Comparing theory and experiment
Thursday, June 23, 2016 – 1:30 PM
Room: 217 Noyes Laboratory

Chair: Marie-Aline Martin-Drumel, Harvard-Smithsonian CfA, Cambridge, MA, USA

RJ01   1:30 – 1:45

TO KINK OR NOT: THE SEARCH FOR LONG CHAIN CUMULENONES USING MICROWAVE SPECTRAL TAXONOMY

MICHAEL C McCARTHY, MARIE-ALINE MARTIN-DRUMEL, *Atomic and Molecular Physics, Harvard-Smithsonian Center for Astrophysics, Cambridge, MA, USA.*

Although cumulene carbenes terminated with sulfur up to H_2C_7S are known to possess C_{2v} geometries, the analogous oxygen species have only been characterized in the gas-phase up to H_2C_4O, and propadienone (H_2C_3O) and butatrienone (H_2C_4O) exhibit kinked heavy atom backbones. Using microwave spectral taxonomy, searches have been undertaken for pentatetrenone (H_2C_5O) and its isomers. Surprisingly, no evidence has been found for the cumulenone, but rotational lines of a bent-chain isomer, $HC(O)C_4H$, analogous in structure to propynal, $HC(O)CCH$, have been detected instead. In closely-related work, the sulfur analog $HC(S)C_4H$ has also been identified for the first time. This talk will provide a summary of our search procedure and experimental findings, quantum chemical calculations of isomeric stability and dipole moments, and prospects for detecting these longer chains in astronomical sources where c-C_3H_2O and $HC(O)CCH$ are known.

RJ02   1:47 – 2:02

CONFORMATION-SPECIFIC SPECTROSCOPY OF ALKYL BENZYL RADICALS: EFFECTS OF A RADICAL CENTER ON THE CH STRETCH INFRARED SPECTRA OF ALKYL CHAINS

JOSEPH A. KORN, TIMOTHY S. ZWIER, *Department of Chemistry, Purdue University, West Lafayette, IN, USA*; DANIEL P. TABOR, EDWIN SIBERT, *Department of Chemistry, University of Wisconsin–Madison, Madison, WI, USA.*

An important step in combustion processes is the abstraction of hydrogen to form alkyl benzyl radicals. In this talk we present the results of double resonance spectroscopy methods to explore the conformation-specific infrared spectroscopy of α-methylbenzyl, α-ethylbenzyl, and α-propylbenzyl radicals. A local mode-Hamiltonian model that includes Fermi resonance interactions will be described. This model enables the assignment of the alkyl CH stretch IR spectra of these molecules. This talk will contrast the alkyl chain results to their closed shell analogues, focusing on the the role of the radical site which leads to two important effects. First the CH-stretch frequencies of the β-carbons are shifted by approximately 50 cm^{-1}. Second, internal torsion about the C-C bond between the α-C and β-C atoms modulates these frequency shifts, this producing torsion-vibration mixing. The spectral consequences of this mixing are described.

RJ03   2:04 – 2:19

ANALYSIS OF THE SPECTRUM OF CH3OOH USING SECOND-ORDER PERTURBATION THEORY

LAURA C. DZUGAN, *Department of Chemistry and Biochemistry, The Ohio State University, Columbus, OH, USA*; ANNE B McCOY, *Department of Chemistry, University of Washington, Seattle, WA, USA*; AMITABHA SINHA, JAMIE MATTHEWS, *Department of Chemistry and Biochemistry, UC San Diego, San Diego, USA.*

In this study, we calculate the spectrum of the OH stretch overtone region in CH_3OOH using vibrational second-order perturbation theory. Comparison of the calculated to the experimental spectra shows very good agreement. With this in place, the goal of this study is to use second-order perturbation theory to investigate the following questions. To begin with, we explored the origins of the transition strength and found that both mechanical and electrical anharmonicities contribute to the overall intensity. Then we studied the amount of OO stretch and OH stretch character in the wavefunctions that correspond to the states that are accessed by the experiment. This is of interest because the VMP (vibrationally mediated photodissociation) action spectrum of CH_3OOH is obtained by detection of the OH radical following vibrational excitation of the overtones/combination bands and the subsequent photodissociation along the OO bond. Interestingly, OH is detected in its vibrationless state following excitation of the OH and CH stretch overtones and combination bands involving the OH stretch. In contrast, vibrationally excited OH is only detected following excitation of OH stretch overtones in methyl peroxide. To further understand the origins of the intensity in CH_3OOH, we also explored the effects of deuteration in the OH and CH overtone regions.

RJ04 2:21 – 2:36

DOING THE LIMBO WITH A LOW BARRIER: HYDROGEN BONDING AND PROTON TRANSFER IN HYDROXYFORMYLFULVENE

ZACHARY VEALEY, DEACON NEMCHICK, PATRICK VACCARO, *Department of Chemistry, Yale University, New Haven, CT, USA.*

Model compounds continue to play crucial roles for elucidating the ubiquitous phenomena of hydrogen bonding and proton transfer, often yielding invaluable insights into kindred processes taking place in substantially larger species. The symmetric double-minimum topography that characterizes the potential-energy landscape for an important subset of these systems allows unambiguous signatures of molecular dynamics (in the form of tunneling-induced bifurcations) to be extracted directly from spectral measurements. As a relatively unexplored member of this class, 6-hydroxy-2-formylfulvene (HFF) contains an intramolecular O–H$\cdots$O interaction that has participating atoms from the hydroxylic (donor) and ketonic (acceptor) moieties closely spaced in a quasi-linear configuration. This unusual arrangement suggests proton transduction to occur with minimal encumbrance, possibly leading to a pronounced dislocation of the shuttling hydron commensurate with the concepts of low-barrier hydrogen bonding (which are distinguished by great strength, short distance, and vanishingly small potential barriers). A variety of spectroscopic probes built primarily upon the techniques of laser-induced fluorescence and dispersed fluorescence have been enlisted to acquire the first vibronically resolved information reported for the ground $[\tilde{X}^1 A_1]$ and lowest-lying singlet excited $[\tilde{A}^1 B_2\ (\pi^*\pi)]$ electronic manifolds of HFF entrained in a cold supersonic free-jet expansion. These experimental findings will be discussed and compared to those obtained for related proton-transfer systems, with complimentary quantum-chemical calculations serving to unravel the unique bonding motifs and reactive pathways inherent to HFF.

RJ05 2:38 – 2:53

ROTATIONAL AND FINE STRUCTURE OF PSEUDO-JAHN-TELLER MOLECULES WITH C_1 SYMMETRY

JINJUN LIU, *Department of Chemistry, University of Louisville, Louisville, KY, USA.*

It has been found in our previous works that rotational and fine-structure analysis of spectra involving nearly degenerate electronic states may aid in interpretation and analysis of the vibronic structure, specifically in the case of pseudo-Jahn-Teller (pJT) molecules with C_s symmetry. The spectral analysis of pJT derivatives (isopropoxy[a] and cyclohexoxy[b]) of a prototypical JT molecule (the methoxy radical) allowed for quantitative determination of various contributions to the energy separation between the nearly degenerate electronic states, including the relativistic spin-orbit (SO) effect, the electrostatic interaction, and their zero-point energy difference. These states are coupled by SO and Coriolis interactions, which can also be determined accurately in rotational and fine structure analysis. Most recently, the spectroscopic model for rotational analysis of pJT molecules has been extended for analysis of molecules with C_1 symmetry, i.e., no symmetry. This model includes the six independently determinable components of the spin-rotation (SR) tensor and the three components of the SO and Coriolis interactions. It has been employed to simulate and fit high-resolution laser-induced fluorescence (LIF) spectra of jet-cooled alkoxy radicals with C_1 symmetry, including the 2-hexoxy and the 2-pentoxy radicals, as well as previously recorded LIF spectrum of the trans-conformer (defined by its OCCC dihedral angle) of the 2-butoxy radical.[c] Although the LIF spectra can be reproduced by using either the SR constants or SO and Coriolis constants, the latter simulation offers results that are physically more meaningful whereas the SR constants have to be regarded as effective constants. Furthermore, we will review the SO and Coriolis constants of alkoxy radicals that have been investigated, starting from the well-studied methoxy radical (CH_3O).

[a] J. Liu, D. Melnik, and T. A. Miller, J. Chem. Phys. 139, 094308 (2013)

[b] J. Liu and T. A. Miller, J. Phys. Chem. A 118, 11871-11890 (2014)

[c] L. Stakhursky, L. Zu, J. Liu, and T. A. Miller, J. Chem. Phys. 125, 094316 (2006)

RJ06

DISPERSED-FLUORESCENCE SPECTROSCOPY OF JET-COOLED CALCIUM ETHOXIDE RADICAL ($CaOC_2H_5$)

ANAM C. PAUL, MD ASMAUL REZA, JINJUN LIU, *Department of Chemistry, University of Louisville, Louisville, KY, USA.*

Metal-containing free radicals are important intermediates in metal-surface reactions and in the interaction between metals and organic molecules. In the present work, dispersed fluorescence (DF) spectra of the calcium ethoxide radical ($CaOC_2H_5$) have been obtained by pumping the $\tilde{A}^2A' \leftarrow \tilde{X}^2A'$ and the $\tilde{B}^2A'' \leftarrow \tilde{X}^2A'$ origin bands in its laser-induced fluorescence (LIF) spectrum. $CaOC_2H_5$ radicals were produced by 1064 nm laser ablation of calcium grains in the presence of ethanol under jet-cooled conditions. Dominant transitions in the vibrationally resolved DF spectra are well reproduced using Franck-Condon factors predicted by complete active space self-consistent (CASSCF) calculations. Differences in transition intensities between the $\tilde{A}^2A' \rightarrow \tilde{X}^2A'$ and the $\tilde{B}^2A'' \rightarrow \tilde{X}^2A'$ DF spectra are attributed to the pseudo-Jahn-Teller interaction between the $\tilde{A}^2A'$ and the $\tilde{B}^2A''$ states. Collision-induced population transfer between these two excited electronic states results in additional peaks in the DF spectra.

Intermission

RJ07

PHOTODETACHMENT OF O^- YIELDING $O(^1D_2, {}^3P)$ ATOMS, VIEWED WITH VELOCITY MAP IMAGING[a]

STEPHEN T GIBSON, BENJAMIN A LAWS, BRENTON R LEWIS, *Research School of Physics and Engineering, Australian National University, Canberra, ACT, Australia*; LY DUONG, *Research School of Astronomy and Astrophysics, Australian National University, Canberra, Australia.*

Electron photodetachment of $O^-(^2P_{3/2,1/2})$ is measured using velocity-map imaging at wavelengths near 350 nm, where detachment yields both $O(^1D_2)$ and $O(^3P_{2,1,0})$ atoms, simultaneously, producing slow (~ 0.1 eV) and fast electrons (~ 2 eV). The photoelectron spectrum resolves the fine-structure transitions, which together with the well known atomic fine-structure splittings,[b] and intensity ratios,[c] provide an excellent test of the spectral quality of the velocity-map imaging technique.

Although the photoelectron angular distribution for the two atomic limits have the same negative anisotropy sign, the energy dependence differs. The variation is qualitatively in accordance with R-matrix cross section calculations, that indicate a more gradual d-wave onset for the ^1D limit.[d] However, more exact evaluation is only possible with information about the matrix element phases.

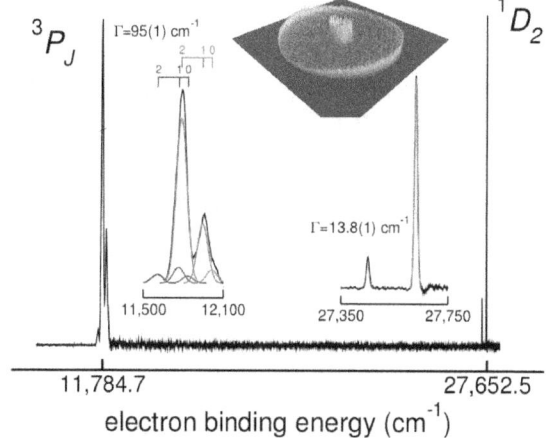

[a]Research supported by the Australian Research Council Discovery Project Grant DP160102585.

[b]physics.nist.gov/cgi-bin/ASD/energy1.pl

[c]O. Scharf and M. R. Godefried, arXiv:0808.3529v1

[d]O. Zatsarinny and K. Bartschat, *Phys. Rev. A*, **73**, 022714 (2006). doi:10.1103/PhysRevA.73.022714

RJ08

HIGH RESOLUTION VELOCITY MAP IMAGING PHOTOELECTRON SPECTROSCOPY OF THE BERYLLIUM OXIDE ANION, BeO^-

AMANDA REED DERMER, KYLE MASCARITOLO, MICHAEL HEAVEN, *Department of Chemistry, Emory University, Atlanta, GA, USA.*

The photodetachment spectrum of BeO^- has been studied using high resolution velocity map imaging photoelectron spectroscopy. The vibrational contours were imaged and compared with Franck-Condon simulations for the ground and excited states of the neutral. The electron affinity of BeO was measured for the first time, and anisotropies of several transitions were determined. Experimental findings are compared to high level *ab initio* calculations.

RJ09 4:03 – 4:18

CHARACTERIZING MOLECULAR STRUCTURE BY COMBINING EXPERIMENTAL MEASUREMENTS WITH DENSITY FUNCTIONAL THEORY COMPUTATIONS

JUAN M LOPEZ-ENCARNACION[a], *Department of Mathematics- Physics, University of Puerto Rico at Cayey, Cayey, Puerto Rico, USA.*

In this talk, the power and synergy of combining experimental measurements with density functional theory computations as a single tool to unambiguously characterize the molecular structure of complex atomic systems is shown. Here, we bring three beautiful cases where the interaction between the experiment and theory is in very good agreement for both finite and extended systems: 1) Characterizing Metal Coordination Environments in Porous Organic Polymers: A Joint Density Functional Theory and Experimental Infrared Spectroscopy Study[b]; 2) Characterization of Rhenium Compounds Obtained by Electrochemical Synthesis After Aging Process[c]; and 3) Infrared Study of $H(D)_2 + Co_4^+$ Chemical Reaction: Characterizing Molecular Structures.

[a] JMLE acknowledges the Puerto Rico NASA EPSCoR IDEAS-ER program for providing the travel funds
[b] J.M. López-Encarnación, K.K. Tanabe, M.J.A. Johnson, J. Jellinek, Chemistry–A European Journal 19 (41), 13646-13651

[c] A. Vargas-Uscategui, E. Mosquera, J.M. López-Encarnación, B. Chornik, R. S. Katiyar, L. Cifuentes, Journal of Solid State Chemistry 220, 17-21

RJ10 4:20 – 4:35

SPECTROSCOPIC STUDY OF TORSIONAL POTENTIALS, MOLECULAR STRUCTURE, NBO ANALYSIS AND OTHER MOLECULAR PARAMETERS OF SOME BIPYRIDINE-DICARBOXYLIC ACIDS USING FTIR AND FT-RAMAN SPECTRA AND THEORETICAL METHODS (DFT and IVP)

BYRU VENKATRAM REDDY, JYOTHI PRASHANTH, G. RAMANA RAO, *Department of Physics, KAKATIYA UNIVERSITY, WARANGAL, India.*

The Fourier Transform Infrared (FTIR) and Fourier Transform Raman (FT-Raman) spectra of 2,2'-bipyridine-3,3'-dicarboxylic acid (B3DA); 2,2'-bipyridine-4,4'-dicarboxylic acid (B4DA); and 2,2'-bipyridine-5,5'-dicarboxylic acid (B5DA) were measured in the range 4000-450 cm-1 and 4000-50 cm-1, respectively. Torsional potentials were evaluated at various angles of rotation around the C-C inter-ring bond for the three molecules. In order to arrive at the molecular conformation of lowest energy, this conformation was further optimized to get ground state geometry. Vibrational frequencies along with infrared and Raman intensities were computed. In the above calculations, DFT employing B3LYP functional with 6-311++G(d,p) basis set was used. The rms error between observed and calculated frequencies was 10.0, 10.9 and 10.2 cm-1 for B3DA, B4DA and B5DA, respectively. A 89-parameter modified valence force field was derived by solving inverse vibrational problem using Wilson's GF matrix method. The force constants were refined using 129 experimental frequencies of the three molecules in overlay least-squares technique. The average error between observed and computed frequencies was 11.32 cm-1. PED and eigen vectors calculated in the process were used to make unambiguous vibrational assignments of all the fundamental vibrations. The values of dipole moment, polarizability and hyperpolarizability were computed to determine the NLO behavior of these molecules. Stability of the molecules arising from hyper conjugative interactions, charge delocalization has been analyzed using natural bond orbital (NBO) analysis. The HOMO and LUMO energies and thermodynamic parameters were also evaluated. Charge density distribution and site of chemical reactivity of the molecule have been studied by mapping electron density isosurface with molecular electrostatic potential (MESP).

RJ11   *Post-Deadline Abstract*   **4:37 – 4:52**

ELECTRONIC STRUCTURE AND SPECTROSCOPY OF HBr and HBr$^+$

GABRIEL J. VAZQUEZ, *Instituto de Ciencias Fisicas, Universidad Nacional Autonoma de Mexico (UNAM), Cuernavaca, Morelos, Mexico*; H. P. LIEBERMANN, *Fachbereich C-Mathematik und Naturwissenschaften, Universität Wuppertal, Wuppertal, Germany*; H. LEFEBVRE-BRION, *Institut des Sciences Moléculaires d'Orsay, Université Paris-Sud, Orsay, France.*

We report preliminary ab initio electronic structure calculations of HBr and HBr$^+$. The computations were carried out employing the MRD-CI package, with a basis set of cc-pVQZ quality augmented with s–, p– and d–type diffuse functions. In a first series of calculations, without inclusion of spin–orbit splitting, potential energy curves of about 20 doublet and quartet electronic states of HBr$^+$, and about 30 singlet and triplet (valence and Rydberg) states of HBr were computed. This exploratory step provides a perspective of the character, shape, leading configurations, energetics, and asymptotic behaviour of the electronic states. The calculations taking into account spin-orbit are currently being performed. Our study focuses mainly on the Rydberg states and their interactions with the repulsive valence states and with the bound valence ion-pair state. In particular, the current calculations seek to provide information that might be relevant to the interpretation of recent REMPI measurements[a] which involve the interaction between the diabatic E$^1\Sigma^+$ Rydberg state and the diabatic V$^1\Sigma^+$ ion–pair state (which together constitute the adiabatic, double-well, B$^1\Sigma^+$ state). Several new states of both HBr and HBr$^+$ are reported.

---

[a]D. Zaouris, A. Kartakoullis, P. Glodic, P. C. Samartzis, H. R. Hródmarsson, Á. Kvaran, *Phys. Chem. Chem. Phys.*, **17**, 10468 (2015).

RJ12   **4:54 – 5:09**

THEORETICAL STUDY ON SERS OF WAGGING VIBRATIONS OF BENZYL RADICAL ADSORBED ON SILVER ELECTRODES

DE-YIN WU[a], YAN-LI CHEN[b], ZHONG-QUN TIAN[c], *College of Chemistry and Chemical Engineering, Xiamen University, Xiamen, China.*

Electrochemical surface-enhanced Raman spectroscopy (EC-SERS) has been used to characterize adsorbed species widely but reaction intermediates rarely on electrodes. In previous studies, the observed SERS signals were proposed from surface benzyl species due to the electrochemical reduction of benzyl chloride on silver electrode surfaces. In this work, we reinvestigated the vibrational assignments of benzyl chloride and benzyl radical as the reaction intermediate. On the basis of density functional theoretical (DFT) calculations and normal mode analysis, our systematical results provide more reasonable new assignments for both surface species. Further, we investigated adsorption configurations, binding energies, and vibrational frequency shifts of benzyl radical interacting with silver. Our calculated results show that the wagging vibration displays significant vibrational frequency shift, strong coupling with some intramolecular modes in the phenyl ring, and significant changes in intensity of Raman signals. The study also provides absolute Raman intensity in benzyl halides and discuss the enhancement effect mainly due to the binding interaction with respect to free benzyl radical.

---

[a]State Key Laboratory of Physical Chemistry of Solid Surfaces and Department of Chemistry, College of Chemistry and Chemical Engineering, Xiamen University, Xiamen, 361005, China

[b]State Key Laboratory of Physical Chemistry of Solid Surfaces and Department of Chemistry, College of Chemistry and Chemical Engineering, Xiamen University, Xiamen, 361005, China

[c]State Key Laboratory of Physical Chemistry of Solid Surfaces and Department of Chemistry, College of Chemistry and Chemical Engineering, Xiamen University, Xiamen, 361005, China

THEORETICAL STUDY ON SURFACE-ENHANCED RAMAN SPECTRA OF WATER ADSORBED ON NOBLE METAL CATHODES OF NANOSTRUCTURES

DE-YIN WU[a], RAN PANG[b], ZHONG-QUN TIAN, *College of Chemistry and Chemical Engineering, Xiamen University, Xiamen, China.*

The observed surface-enhanced Raman scattering (SERS) spectra of water adsorbed on metal film electrodes of silver, gold, and platinum nanoparticles were used to infer interfacial water structures. The basis is the change of the electrochemical vibrational Stark tuning rates and the relative Raman intensity of the stretching and bending modes. How it is not completely understood the reason why the relative Raman intensity ratio of the bending and stretching vibrations of interfacial water increases at the very negative potential region. Density functional theory calculations provide the conceptual model. The specific enhancement effect for the bending mode was closely associated with the water adsorption structure in a hydrogen bonded configuration through its H-end binding to surface sites with large polarizability due to strong cathodic polarization. The present theoretical results allow us to propose that interfacial water molecules exist on these metal cathodes with different hydrogen bonding interactions, the HO-H...Ag(Au) for silver and gold. In acidic solution, a surface electron–hydronium ion-pair was proposed as an adsorption configuration of interfacial water structures on silver and gold cathodes based on density functional theory (DFT) calculations. The EHIP is in the configuration of $H_3O^+(H_2O)_n e^-$, where the hydronium H_3O^+ and the surface electron is separated by water layers. The electron bound in the EHIP can first be excited under light irradiation, subsequently inducing a structural relaxation into a hydrated hydrogen atom. Thus, Raman intensities of the interfacial water in the EHIP species are signifcantly enhanced due to the cathodic polarization on silver and gold electrodes.

---

[a] State Key Laboratory of Physical Chemistry of Solid Surfaces and Department of Chemistry, College of Chemistry and Chemical Engineering, Xiamen University, Xiamen, 361005, China

[b] State Key Laboratory of Physical Chemistry of Solid Surfaces and Department of Chemistry, College of Chemistry and Chemical Engineering, Xiamen University, Xiamen, 361005, China

FA. Fundamental physics
Friday, June 24, 2016 – 8:30 AM
Room: 100 Noyes Laboratory

Chair: Shui-Ming Hu, University of Science and Technology of China, Hefei, China

FA01   8:30 – 8:45

TOWARD PRECISION MID-INFRARED SPECTROSCOPY ON THE OH RADICAL

ARTHUR FAST, *Dynamics at Surfaces, Max Planck Institute for Biophysical Chemistry, Göttingen, Germany*; JOHN FURNEAUX, *Homer L Dodge Department of Physics and Astronomy, University of Oklahoma, Norman, OK, USA*; SAMUEL MEEK, *Dynamics at Surfaces, Max Planck Institute for Biophysical Chemistry, Göttingen, Germany*.

Measurements of vibrational transitions in small molecules can be used to test for a possible time variation of the electron-proton mass ratio.[a] In our experiments, our goal is to measure two-photon $v = 2 \leftarrow v = 0$ vibrational transitions in the hydroxyl (OH) radical near 2×3500 cm^{-1} with a relative accuracy of 10^{-14}. Reaching this level of accuracy requires a mid-infrared laser with a linewidth of much less than 1 kHz, as well as the ability to compare the frequency of this laser with an absolute frequency standard.

To achieve the high short-term stability necessary for such a narrow linewidth, we are implementing a 532-nm CW reference laser by locking a frequency-doubled Nd:YAG laser to a molecular iodine transition using saturated absorption spectroscopy. Similar setups have demonstrated relative stabilities of around 10^{-14} at the one-second timescale.[b] The stability of this reference laser will then be transfered onto the idler of a 1064-nm-pumped optical parametric oscillator (OPO) using an optical frequency comb as a transfer oscillator. The frequency comb will also be used to measure the absolute optical frequencies of the various lasers and compare them to a GPS-linked radio frequency reference, providing long-term stability and absolute accuracy for the spectroscopic measurements.

---

[a]J.-P. Uzan. *Rev. Mod. Phys.* **75**, 403–455 (2003).

[b]Döringshoff, K., Mohle, K., Nagel, M., Kovalchuk, E. V., Peters, A: High performance iodine frequency reference for tests of the LISA laser system. *EFTF-2010 24th European Frequency and Time Forum (2010).*

FA02   8:47 – 9:02

STUDY OF LASER PRODUCED PLASMA OF LIMITER OF THE ADITYA TOKOMAK FOR DETECTION OF MOLECULAR BANDS

AWADHESH KUMAR RAI, *Department of Physics, Allahabad University, Allahabad, India.*

The tokamak wall protection is one of the prime concerns, and for this purpose, limiters are used. Graphite is commonly used as a limiter material and first wall material for complete coverage of the internal vacuum vessel surfaces of the tokamak. From the past few years, we are working to identify and quantify the impurities deposited on the different part of Aditya Tokamak in collaboration with the Scientists at Institute of Plasma Research, Ahmedabad, India using Laser Induced Breakdown Spectroscopy (LIBS) [1-3]. Laser induced breakdown spectroscopy (LIBS) spectra of limiter of Aditya Tokamak have been recorded in the spectral range of 200-900 nm in open atmosphere. Along with atomic and ionic spectral lines of the constituent elements of the limiter (1-3), LIBS spectra also give the molecular bands. When a high power laser beam is focused on the sample, laser induced plasma is produced on its surface. In early stage of the plasma Back ground continuum is dominated due to free-free or free-bound emission. Just after few nanoseconds the light from the plasma is dominated by ionic emission. Atomic emission spectra is dominated from the laser induced plasma during the first few microsecond after an ablation pulse where as molecular spectra is generated later when the plasma further cools down. For this purpose the LIBS spectra has been recorded with varying gate delay and gate width. The spectra of the limiter show the presence of molecular bands of CN and C2. To get better signal to background ratios of the molecular bands, different experimental parameters like gate delay, gate width, collection angle and collection point (spatial analysis off the plasama) of the plasma have been optimized. Thus the present paper deals with the variation of spectral intensity of the molecular bands with different experimental parameters. Keywords: Limiter, Molecular bands, C2, CN.

References: 1. Proof-of-concept experiment for On-line LIBS Analysis of Impurity Layer Deposited on ptical Window and Other Plasma Facing Components of Aditya Tokamak G. S. Maurya, R. Kumar, A. Kumar and A. K. Rai, Review of Scientific Instruments (In Press) 2. Analysis of deposited impurity material on the surface of optical window of the Tokamak using LIBS, (2014) G. S. Maurya, A. Jyotsana, R. Kumar, A. Kumar and A. K. Rai, Physica Scripta 89, 075601 3. Spatial analysis of impurities on the surface of flange and optical window of the Tokamak using Laser Induced Breakdown Spectroscopy, G. S. Maurya, A. Jyotsana, A. Kumar and A. K. Rai,(2014), Optics and Lasers in Engineering, 56, 13–18

244

FA03   9:04 – 9:19

CAVITY RING-DOWN SPECTROSCOPY OF HYDROGEN IN THE 784-852 NM REGION AND CORRESPONDING
LINE SHAPE IMPLEMENTATION INTO HITRAN

YAN TAN[a], JIN WANG, CUNFENG CHENG, AN-WEN LIU, SHUI-MING HU, *Hefei National Laboratory for Physical Science at Microscale, University of Science and Technology of China, Hefei, China*; PIOTR WCISLO, *Institute of Physics, Faculty of Physics, Astronomy and Informatics, Nicolaus Copernicus University, Torun, Poland*; ROMAN V KOCHANOV[b], IOULI E GORDON, LAURENCE S. ROTHMAN, *Atomic and Molecular Physics, Harvard-Smithsonian Center for Astrophysics, Cambridge, MA, USA.*

The hydrogen molecule as the most abundant neutral molecule in the universe is an important object of studies in different areas of science, especially astrophysics. The precision spectroscopy of the hydrogen molecule is particularly useful to verify the quantum electrodynamics theory (QED) in a molecular system. The electric quadrupole transitions of the second overtone of H_2 have been recorded with a high precision cavity ring-down spectrometer.[c] A total of eight lines including the extremely weak $S_3(5)$ line in the $784 - 852$ nm range have been observed. The line positions have been determined to an accuracy of 3×10^{-4} cm^{-1} and the line intensities were determined with a relative accuracy of about 1%. The deviations between the experimental and theoretical frequencies are less than 5×10^{-4} cm^{-1}, which is much smaller than the claimed theoretical uncertainty of 0.0025cm^{-1}. The data from this experiment along with other high-quality H_2 spectra have also been analyzed by the Hartmann-Tran[d] profile as a test case for incorporating[e] parametrization of this profile in the HITRAN[f] database. It was incorporated in the new relational structure of the HITRAN database (www.hitran.org) and into the HITRAN Application Programming Interface (HAPI)[g] for the case of H_2 spectra.

[a]Atomic and Molecular Physics Division , Harvard-Smithsonian Center for Astrophysics, USA

[b]Laboratory of Quantum Mechanics of Molecules and Radiative Processes, Tomsk State University, Russia

[c]Tan Y, Wang J, Cheng C-F, Zhao X-Q, Liu A-W, Hu S-M, J Mol Spectrosc 2014;300:60–4;

[d]Tran H, Ngo NH, Hartmann J-M, J Quant Spectrosc Radiat Transf 2013;129:199–203;

[e]Wcislo P, Gordon IE, Tran H, Tan Y, Hu S-M, Campargue A, et al., Accepted J Quant Spectrosc Radiat Transf HighRus Special Issue, 2015

[f]Rothman LS, Gordon IE, Babikov Y, Barbe A, Chris Benner D, Bernath PF, et al., J Quant Spectrosc Radiat Transf 2013;130:4–50;

[g]Kochanov RV, Gordon IE, Rothman LS, Wcislo P, Hill C, Wilzewski JS, Submitted to J Quant Spectrosc Radiat Transf HighRus Special Issue, 2015.

FA04   9:21 – 9:36

CARRIER DYNAMICS IN $CsPbBr_3$ NANOCRYSTALS IN PRESENCE OF ELECTRON AND HOLE ACCEPTORS: A
TIME RESOLVED TERAHERTZ SPECTROSCOPY STUDY.

SOHINI SARKAR, SNEHA BANERJEE, YETTAPU GURIVI REDDY, VIKASH KUMAR RAVI, ANGSHU-MAN NAG, PANKAJ MANDAL, *Department of Chemistry, Indian Institute of Science Education and Research, Pune, Maharshtra, India.*

Study of lead halide perovskites is a burgeoning field of research owing to their applications in solar cells and myriads of other light harvesting and emitting devices. In this work we have employed Terahertz time domain spectroscopy (THz-TDS) and time-resolved THz spectroscopy (TRTS) to study dielectric properties and carrier dynamics occurring within $CsPbBr_3$ perovskite nanocrystals (NCs) in presence of electron and hole acceptor molecules. The THz-TDS spectrum of $CsPbBr_3$ NCs features a strong and broad band with a peak around 3.4 THz which originates from multiple IR-active optical phonon modes of the nature of Pb-Br stretching and Br-Pb-Br bending vibrations. We observed very efficient electron and/or hole transfer in presence of either an electron or a hole acceptor, or both. Also, in presence of either an electron or hole acceptor the diffusion length reduces to half (4.1 μm) in comparison to parent NCs (9.2 μm). In presence of both, electron and hole acceptor molecules the diffusion length reduces to 0.6 μm. Considerable decrease in mobility values is also observed for the NCs in presence of electron and hole acceptor molecules. Details of the study will be discussed in the talk.

FA05 9:38 – 9:53

TOWARD ROTATIONAL STATE-SELECTIVE PHOTOIONIZATION OF ThF$^+$ IONS

YAN ZHOU, KIA BOON NG, DAN GRESH, WILLIAM CAIRNCROSS, MATT GRAU, YIQI NI, ERIC COR-NELL, JUN YE, *JILA, National Institute of Standards and Technology and Univ. of Colorado Department of Physics, University of Colorado, Boulder, Boulder, CO, USA.*

ThF$^+$ has been chosen to replace HfF$^+$ for a second-generation measurement of the electric dipole moment of the electron (eEDM). Compared to the currently running HfF$^+$ eEDM experiment, ThF$^+$ has several advantages: (i) the eEDM-sensitive state ($^3\Delta_1$) is the ground state, which facilitates a long coherence time [1]; (ii) its effective electric field (35 GV/cm) is 50% larger than that of HfF$^+$, which promises a direct increase of the eEDM sensitivity [2]; and (iii) the ionization energy of neutral ThF is lower than its dissociation energy, which introduces greater flexibility in rotational state-selective photoionization via core-nonpenetrating Rydberg states [3]. In this talk, we first present our strategy of preparing and utilizing core-nonpenetrating Rydberg states for rotational state-selective ionization. Then, we report spectroscopic data of laser-induced fluorescence of neutral ThF, which provides critical information for multi-photon ionization spectroscopy.

[1] D. N. Gresh, K. C. Cossel, Y. Zhou, J. Ye, E. A. Cornell, Journal of Molecular Spectroscopy, 319 (2016), 1-9

[2] M. Denis, M. S. Nørby, H. J. A. Jensen, A. S. P. Gomes, M. K. Nayak, S. Knecht, T. Fleig, New Journal of Physics, 17 (2015) 043005.

[3] Z. J. Jakubek, R. W. Field, Journal of Molecular Spectroscopy 205 (2001) 197–220.

Intermission

FA06 10:12 – 10:27

HIGH RESOLUTION GHZ AND THZ (FTIR) SPECTROSCOPY AND THEORY OF PARITY VIOLATION AND TUN-NELING FOR 1,2-DITHIINE (C$_4$H$_4$S$_2$) AS A CANDIDATE FOR MEASURING THE PARITY VIOLATING ENERGY DIFFERENCE BETWEEN ENANTIOMERS OF CHIRAL MOLECULES

SIEGHARD ALBERT, IRINA BOLOTOVA, ZIQIU CHEN, CSABA FÁBRI, LUBOS HORNY, MARTIN QUACK, GEORG SEYFANG, DANIEL ZINDEL, *Laboratory of Physical Chemistry, ETH Zurich, Zürich, Switzerland.*

We report high resolution spectroscopic results for 1,2-dithiine-(1,2-dithia-3,5-cyclohexadiene,C$_4$H$_4$S$_2$) in the Gigahertz and Terahertz spectroscopic ranges and exploratory theoretical calculations of parity violation and tunneling processes in view of a possible experimental determination of the parity violating energy difference $\Delta_{pv}E$ in this chiral molecule.[a,b] Theory predicts that the parity violating energy difference in the ground state ($\Delta_{pv}E\simeq11\times10^{-11}(hc)$ cm^{-1}) is in principle measurable as it is much larger than the calculated tunneling splitting for the symmetrical potential $\Delta_{\pm}E\simeq10^{-24}(hc)$ cm^{-1}. With a planar transition state for stereomutation at about 2500 cm^{-1} tunneling splittings become appreciable above 2300 cm^{-1}. This makes levels of well defined parity accessible to parity selection by available powerful infrared lasers and thus useful for one of the existing experimental approaches towards molecular parity violation.[c] The new GHz spectra lead to greatly improved ground state rotational parameters for 1,2-dithiine. These are used as starting point for the first successful analyses of high resolution interferometric Fourier Transform Infrared (FTIR, THz) spectra for the fundamentals ν_{17} (1308.873 cm^{-1} or 39.23903 THz), ν_{22} (623.094 cm^{-1} or 18.67989 THz) and ν_3 (1544.900 cm^{-1} or 46.314937 THz) for which highly accurate spectroscopic parameters are reported. The results are discussed in relation to current efforts to measure $\Delta_{pv}E$.[a–d]

[a]M. Quack , *Fundamental Symmetries and Symmetry Violations from High-resolution Spectroscopy*, Handbook of High Resolution Spectroscopy, M. Quack and F. Merkt eds.,John Wiley & Sons Ltd, Chichester, New York, 2001, vol. 1, ch. 18, pp. 659-722.

[b]S. Albert, I. Bolotova, Z. Chen, C. Fábri, L. Horný, M. Quack, G. Seyfang and D. Zindel,Proceedings of the 20th Symposium on Atomic, Cluster and Surface Physics (SASP 2016), Innsbruck University Press, 2016, pp. 127-130, ISBN:978-3-903122-04-8. and to be published

[c]P. Dietiker, E. Miloglyadov, M. Quack, A. Schneider and G. Seyfang, *J. Chem. Phys.* **143**, 244305 (2015).

[d]R. Prentner, M. Quack, J. Stohner and M. Willeke, *J. Phys. Chem. A* **119**, 12805-12822 (2015).

246

FA07 10:29–10:44

A GLOBAL MODEL FOR LONG-RANGE INTERACTION 'DAMPING FUNCTIONS'

PHILIP THOMAS MYATT, FREDERICK R. W. McCOURT, ROBERT J. LE ROY, *Department of Chemistry, University of Waterloo, Waterloo, ON, Canada.*

In recent years, 'damping functions', which characterize the weakening of inverse-power-sum long-range interatomic interaction energies with increasing electron overlap, have become an increasing important component of models for diatomic molecule interaction potentials.[a] However, a key feature of models for damping functions, their portability, has received little scrutiny. The present work set out to examine all available *ab initio* induction and dispersion damping function data[b] and to attempt to devise a 'global' scheme for diatomic molecule damping functions. It appears that while neutral (H, He, Li, and Ne, homonuclear and mixed) and anion (H$^-$ with H, He and Li) species obey (approximately) one common rule, proton plus neutral (H$^+$ with H, He and Li) and non-proton-cation plus neutral systems (He$^+$ and Li$^+$ with H, He and Li), must each be treated separately. However, for all three cases, a version of the Douketis-Scoles-Thakkar[c] (ionization potential)^{power} factor is a key scaling parameter.

[a]R.J. Le Roy, C. C. Haugen, J. Tao and Hui Li, *Mol. Phys.* **109**,435 (2011).

[b]P.J. Knowles and W.J. Meath,*J. Mol. Phys.* **60**, 1143 (1987); R.J. Wheatley and W.J. Meath,*J. Mol. Phys.* **80**, 25 (1993); R.J. Wheatley and W.J. Meath *J. Chem. Phys.* **179**, 341 (1994); R.J. Wheatley and W.J. Meath,*J. Chem. Phys.* **203**, 209 (1996).

[c]C. Douketis,G. Scoles, S. Marchetti, M. Zen and A. J. Thakkar, *J. Chem. Phys.* **76**, 3057 (1982).

FA08 10:46–11:01

DIRECT-POTENTIAL-FIT (DPF) ANALYSIS FOR THE $A\,^3\Pi_1 - X\,^1\Sigma^+$ SYSTEM OF I$^{35/37}$Cl.

SHINJI KOBAYASHI, NOBUO NISHIMIYA, TOKIO YUKIYA, MASAO SUZUKI, *Faculty of Engineering, Tokyo Polytechnic University, Atsugi, Japan*; ROBERT J. LE ROY, *Department of Chemistry, University of Waterloo, Waterloo, ON, Canada.*

The goal of this research is to obtain an optimal, portable, global description of, and summary of the dynamical properties of, the $A\,^3\Pi_1$ and $X\,^1\Sigma^+$ states of I$^{35/37}$Cl, by using 'direct potential fits' (DPFs) to all of the available spectroscopic data for this system to determine optimal analytic potential energy functions for these two states that represent all of those data (on average), within the experimental uncertainties. The DPF method compares observed spectroscopic data with synthetic data generated by solving the radial Schrödinger equation for the upper and lower level of every observed transition for some parameterized analytic potential function(s), and using least-squares fits to the data to optimize those parameters. The present work uses the Morse/Long-Range (MLR) potential function form because it is very flexible, can incorporate the correct theoretically known inverse-power-sum long-range behaviour, is everywhere continuous and differentiable to all orders, and has robust extrapolation properties at both large and small distances. The DPF approach also tends to require fewer fitting parameters than do traditional Dunham analyses, as well as having much more robust extrapolation properties in both the v and J domains. The present work combines the data for the $A\,^3\Pi_1$ and $X\,^1\Sigma^+$ states obtained in 1980 by Coxon *et al.*[a] using UV and near-infrared grating spectrometers, with our measurements in the 0.7-0.8μm region, obtained using a CW Ti:Sapphire Ring Laser.[b] The results of this study and our new fully analytic potential energy functions for the $A\,^3\Pi_1$ and $X\,^1\Sigma^+$ states of ICl will be presented.

[a] J.A. Coxon, R.M. Gordon and M.A. Wickramaaratchi, *J. Mol. Spectrosc.* **79** (1980) 363 and 380.

[b] T.Yukiya, N. Nishimiya and M. Suzuki, *J. Mol. Spectrosc.* **269** (2011) 193.

FB. Instrument/Technique Demonstration

Friday, June 24, 2016 – 8:30 AM

Room: 116 Roger Adams Lab

Chair: Christopher F. Neese, The Ohio State University, Columbus, OH, USA

FB01   8:30 – 8:45

DUAL EXCITATION-EMISSION PROPAGATION (DEEP) IMPACT- FTMW SPECTROMETER

DENNIS WACHSMUTH, DOMENICO PRUDENZANO, <u>JENS-UWE GRABOW</u>, *Institut für Physikalische Chemie und Elektrochemie, Gottfried-Wilhelm-Leibniz-Universität, Hannover, Germany.*

The in-phase/quadrature phase modulation passage-acquired coherence technique(IMPACT) Fourier-transform microwave (FT-MW) spectrometer utilizing two off-axis parabolic reflectors delivers broadband capabilities at a spectral resolution similar to the resolving power of the narrowband but more sensitive coaxial beam-resonator arrangement (COBRA) FT-MW spectroscopy.

Nevertheless, due to the signal pathway in the dual-path reflector arrangement, the high-frequency setup imposes a maximum applicable excitation power, thus limiting the polarization efficiency. Hence, less polar molecules were difficult to study.

In a novel approach this disadvantage could be circumvented by rotating of the field vector direction of the linearly polarized microwave radiation. The setup prevails the high spectral resolution but increases the sensitivity dramatically while allowing the utilisation of very high power tube amplifiers.

In this contribution we present the novel apparatus in detail as well as experimental results obtained with the modified spectrometer.

FB02   8:47 – 9:02

A CMOS MILLIMETER-WAVE TRANSCEIVER EMBEDDED IN A SEMI-CONFOCAL FABRY-PEROT CAVITY

<u>BRIAN DROUIN</u>, ADRIAN TANG, ERICH T SCHLECHT, EMILY BRAGEOT, *Jet Propulsion Laboratory, California Institute of Technology, Pasadena, CA, USA*; ADAM M DALY, *Chemistry and Biochemistry, University of Arizona, Tucson, AZ, USA*; QUN JANE GU, YU YE, RAN SHU, *Department of Electrical and Computer Engineering, University of California - Davis, Davis, CA, USA*; M.-C. FRANK CHANG, ROD M. KIM, *Electrical Engineering, University of California - Los Angeles, Los Angeles, CA, USA.*

The extension of radio-frequency CMOS circuitry into millimeter wavelengths promises the extension of spectroscopic techniques in compact, power efficient systems. We are now exploring the use of CMOS millimeter devices for low-mass, low-power instrumentation capable of remote or in-situ detection of gas composition during space missions. We have chosen to develop a Flygare-Balle type spectrometer, with a semi-confocal Fabry-Perot cavity to amplify the pump power of a mm-wavelength CMOS transmitter that is directly coupled to the planar mirror of the cavity. Since the initial report last year describing the designs, we have built a pulsed transceiver system at 89-104 GHz inside a 5 cm base length cavity and demonstrated cavity finesse up to 3000, allowing for modes with 30 MHz bandwidth and a sufficient cavity amplification factor for mW class transmitters. System and component testing revealed that the power-amplifier design (embedded in the chip) was faulty and the transceiver peak power is only 10 microwatts, which is insufficient for molecular excitation on the timescale of the gas residence time within the beam. An improved power amplifier circuit has been designed and is currently under fabrication, meanwhile, we have also developed a tunable synthesizer (embedded in the same chip) that allows for tuning over the full bandwidth at increments of 10 MHz. The presentation will cover these capabilities, describing the system and component tests, as well as any new developments.

FB03

FINITE-DIFFERENCE TIME-DOMAIN MODELING OF FREE INDUCTION DECAY SIGNAL IN CHIRPED PULSE MILLIMETER WAVE SPECTROSCOPY

ALEXANDER HEIFETZ, SASAN BAKHTIARI, HUAL-TEH CHIEN, *Nuclear Engineering Division, Argonne National Laboratory, Argonne, IL, USA*; KIRILL PROZUMENT, *Chemical Sciences and Engineering Division, Argonne National Laboratory, Argonne, IL, USA*; STEPHEN K GRAY, *Nanoscience and Technology Division, Argonne National Laboratory, Argonne, IL, USA*; RICHARD M WILLIAMS, *Detection Systems, Pacific Northwest National Laboratory, Richand, WA, USA.*

We have developed computational electrodynamics model of free induction decay (FID) signal in chirped pulse millimeter wave (CPMMW) spectroscopy. The computational model is based on finite-difference time-domain (FDTD) solution of Maxwell's equations in 1-D. Molecular medium is represented by two-level system derived using density matrix (DM) formulation. Each cell in the grid is assigned an independent set of DM equations, and thus acts as an independent source of induced polarization. Computer simulations with our 1-D model have shown that FID signal is propagating entirely in the forward direction. Intensity of FID radiation increases linearly along the cell length. These results can be explained analytically by considering phases of electromagnetic field radiated by each independent region of induced polarization. We show that there is constructive interference in the forward in forward direction, and destructive interference in backscattering direction. Results in this study are consistent with experimental observations that FID has been measured in the forward scattering direction, but not in backscattering direction.

FB04

HETERODYNE RECEIVER FOR LABORATORY SPECTROSOCPY OF MOLECULES OF ASTROPHYSICAL IMPORTANCE

NADINE WEHRES, FRANK LEWEN, CHRISTIAN ENDRES[a], MARIUS HERMANNS, STEPHAN SCHLEMMER, *I. Physikalisches Institut, Universität zu Köln, Köln, Germany.*

We present first results of a heterodyne receiver built for high-resolution emission laboratory spectroscopy of molecules of astrophysical interest. The room-temperature receiver operates at frequencies between 80 and 110 GHz, consistent with ALMA band 3. Many molecules have been identified in the interstellar and circumstellar medium at exactly these frequencies by comparing emission spectra obtained from telescopes to high-resolution laboratory absorption spectra. Taking advantage of the recent progresses in the field of mm/submm technology in the astronomy community, we have built a room-temperature emission spectrometer making use of heterodyne receiver technology at an instantaneous bandwidth of currently 2.5 GHz. The system performance, in particular the noise temperature and systematic errors, is presented. The proof-of-concept is demonstrated by comparing the emission spectrum of methyl cyanide to respective absorption spectra and to the literature. Future prospects as well as limitations of the new laboratory receiver for the spectroscopy of complex organic molecules or transient species in discharges will be discussed.

[a] current address: Max-Planck-Institute for Extraterrestrial Physics, Garching, Germany

FB05

^{13}C-TRIPLY LABELED ETHYL CYANIDE SUBMILLIMETERWAVE STUDY WITH LILLE'S FAST SCAN DDS-BASED SPECTROMETER

A. PIENKINA, R. A. MOTIYENKO, L. MARGULÈS, *Laboratoire PhLAM, UMR 8523 CNRS - Université Lille 1, Villeneuve d'Ascq, France*; HOLGER S. P. MÜLLER, *I. Physikalisches Institut, Universität zu Köln, Köln, Germany*; J.-C. GUILLEMIN, *Institut des Sciences Chimiques de Rennes, UMR 6226 CNRS - ENSCR, Rennes, France.*

This study of the ^{13}C-triply labeled species of ethyl cyanide (CH$_3$CH$_2$CN) follows our recent work on the three ^{13}C-doubly-labeled that allowed their detection[a] in the line survey recently obtained with ALMA (EMoCA)[b]. The detection of isotopologues could improve the knowledge of the astrochemistry. The other goal is to clean the surveys from the lines of known molecules in order to detect new ones, this is especially important for the abundant complex organic molecules like ethyl cyanide. As in the case of the doubly substitued species, no spectroscopic studies exist up to now for ^{13}CH$_3^{13}$CH$_2^{13}$CN, the first predictions were thus obtained from scaled ab initio calculations. The spectra were recorded and analyzed up to 1 THz. More than 5500 lines were fitted with quantum numbers J and K_a up to 95 and 25 respectively.

The spectra were obtained with the new version of the Lille's solid state spectrometers. This new version used Direct Digital Synthesizer in order to speed up acquisition time. We constructed a spectrometer covering a decade, from 150 to 1500 GHz, it scans the full range in 24 hours with high sensitivity and accuracy.

This work was supported by the CNES and the Action sur Projets de l'INSU, PCMI. This work was also done under ANR-13-BS05-0008-02 IMOLABS

[a]Margulès, L.; *et al.* 2015, 69th *International Symposium on Molecular Spectroscopy*, **RI06**
[b]Belloche, A.; *et al.* 2014, *Science*, **345**, 1584

Intermission

FB06

THE KASSEL LABORATORY ASTROPHYSICS THZ SPECTROMETRS

JOHANNA CHANTZOS, DORIS HERBERTH, PIA KUTZER, CHRISTOPH MUSTER, GUIDO W FUCHS, *Physics Department, University of Kassel, Kassel, Germany*; THOMAS GIESEN, *Institute of Physics, University Kassel, Kassel, Germany.*

We present a brief overview of the recently established laboratory astrophysics group in Kassel/Germany with a focus on our THz technology. After an outline of our laboratory equipment and recent projects the talk will focus on our new fast spectral scan technique for molecular jet experiments.

Here, a new test setup for broadband fast sweep spectrometry in the MW to submm wavelength region has been realized and can be applied to identify transient molecules in a supersonic jet. An arbitrary waveform generator (AWG) is used to generate chirped pulses with a linear frequency sweep in the MHz regime. Pulse durations are of a few microseconds. These pulses are up-converted in frequency, e.g. into the 50 GHz microwave frequency range utilizing a synthesizer, or using a synthesizer plus standard amplifier multiplier chain (AMC) to reach the 100-300 GHz region. As test, NH_3 has been measured between 18-26 GHz in a supersonic jet of 500 μs duration. Acetonitrile (CH_3CN) was tested in the (90-110) GHz range. The spectrometer is capable of providing fast, broadband and low-noise measurements. Experiments with non-stabel molecular production conditions can greatly benefit from these advantages. The setup enables the study of Van-der-Waals-clusters, as well as carbon chain molecules and small metal-containing refractory molecules when combined with appropriate molecule sources.

FB07   10:24 – 10:39

DETERMINING CONCENTRATIONS AND TEMPERATURES IN SEMICONDUCTOR MANUFACTURING PLAS-
MAS VIA SUBMILLIMETER ABSORPTION SPECTROSCOPY

YASER H. HELAL, CHRISTOPHER F. NEESE, FRANK C. DE LUCIA, *Department of Physics, The Ohio State University, Columbus, OH, USA*; PAUL R. EWING, *Applied Materials, Austin, TX, USA*; ANKUR AGARWAL, BARRY CRAVER, PHILLIP J. STOUT, MICHAEL D. ARMACOST, *Applied Materials, Sunnyvale, CA, USA*.

Plasmas used in the manufacturing processes of semiconductors are similar in pressure and temperature to plasmas used in studying the spectroscopy of astrophysical species. Likewise, the developed technology in submillimeter absorption spectroscopy can be used for the study of industrial plasmas and for monitoring manufacturing processes. An advantage of submillimeter absorption spectroscopy is that it can be used to determine absolute concentrations and temperatures of plasma species without the need for intrusive probes. A continuous wave, 500 – 750 GHz absorption spectrometer was developed for the purpose of being used as a remote sensor of gas and plasma species. An important part of this work was the optical design to match the geometry of existing plasma reactors in the manufacturing industry. A software fitting routine was developed to simultaneously fit for the background and absorption signal, solving for concentration, rotational temperature, and translational temperature. Examples of measurements made on inductively coupled plasmas will be demonstrated. We would like to thank the Texas Analog Center of Excellence/Semiconductor Research Corporation (TxACE/SRC) and Applied Materials for their support of this work.

FB08   10:41 – 10:56

AUTOFIT AND THE SPECTRUM OF EUGENOL

ERIKA RIFFE, SAWYER WELDEN, EMMA COCKRAM, KATHERINE ERVIN, STEVEN SHIPMAN, *Department of Chemistry, New College of Florida, Sarasota, FL, USA*; CAMERON M FUNDERBURK, GORDON G BROWN, *Department of Science and Mathematics, Coker College, Hartsville, SC, USA*; SUSANNA L. WIDI-CUS WEAVER, *Department of Chemistry, Emory University, Atlanta, GA, USA*.

The rotational spectrum of eugenol, the primary constituent in clove oil, was obtained via chirped-pulse Fourier transform microwave spectroscopy from 3-8 GHz in a supersonic expansion on a sample that was extracted from cloves via steam distillation. Ab initio calculations indicate that this molecule possesses several conformers with energies that are only a few hundred wavenumbers above that of the global minimum conformation, due to different relative orientations of the molecule's methoxy and allyl groups. Eugenol's spectrum was analyzed with a new version of the Autofit software that has been designed to run in cluster computing environments. Here we will present the results of this study, including benchmarking results for the new version of Autofit.

FB09   10:58 – 11:13

COHERENT NONLINEAR TERAHERTZ SPECTROSCOPY OF HALOMETHANE LIQUIDS

IAN A FINNERAN, RALPH WELSCH, *Division of Chemistry and Chemical Engineering, California Institute of Technology, Pasadena, CA, USA*; MARCO A. ALLODI, *Department of Chemistry, The Institute for Biophysical Dynamics, and The James Franck Institute, The University of Chicago, Chicago, IL, USA*; THOMAS F. MILLER III, GEOFFREY BLAKE, *Division of Chemistry and Chemical Engineering, California Institute of Technology, Pasadena, CA, USA*.

The low-energy terahertz motions of liquids greatly influence their behavior, but are not fully understood. Here, we present results from a recently developed heterodyne-detected Terahertz Kerr Effect (TKE) spectrometer, using an intense picosecond terahertz pump pulse, followed by a weak near-infrared femtosecond probe pulse. In the responses of several halomethane liquids, we find evidence for terahertz intramolecular vibrational coupling and the excitation of intermolecular motions. The experimental results are further supported by reduced density matrix and molecular dynamics simulations. With modest improvements in sensitivity, we expect this technique to be applicable to hydrogen-bonded liquids and amorphous solids.

FB10   11:15 – 11:30

SUB-THZ VIBRATIONAL SPECTROSCOPY FOR ANALYSIS OF OVARIAN CANCER CELLS

JEROME P. FERRANCE, *R&D, Vibratess LLC, Charlottesville, VA, USA*; IGOR SIZOV, *computational modeling, Vibratess LLC, Charlottesville, VA, USA*; AMIR JAZAERI, *M D Anderson Cancer Center, University of Texas, Austin, USA*; AARON MOYER, *instrument development, Vibratess LLC, Charlottesville, VA, USA*; BORIS GELMONT, *Department of Electrical and Computer Engineering, University of Virginia, Charlottesville, VA, USA*; TATIANA GLOBUS, *R&D, Vibratess LLC, Charlottesville, VA, USA*.

Sub-THz vibrational spectroscopy utilizes wavelengths in the submillimeter-wave range (1.5-30 cm^{-1}), beyond those traditionally used for chemical and biomolecular analysis. This low energy radiation excites low-frequency internal molecular motions (vibrations) involving hydrogen bonds and other weak connections within these molecules. The ability of sub-THz spectroscopy to identify and quantify biological molecules is based on detection of signature resonance absorbance at specific frequencies between 0.05 and 1 THz, for each molecule. The long wavelengths of this radiation, mean that it can even pass through entire cells, detecting the combinations of proteins and nucleic acids that exist within the cell.

This research introduces a novel sub-THz resonance spectroscopy instrument with spectral resolution sufficient to identify individual resonance absorption peaks, for the analysis of ovarian cancer cells. In vitro cell cultures of SK-OV-3 and ES-2 cells, two human ovarian cancer subtypes, were characterized and compared with a normal non-transformed human fallopian tube epithelial cell line (FT131). A dramatic difference was observed between the THz absorption spectra of the cancer and normal cell sample materials with much higher absorption intensity and a very strong absorption peak at a frequency of 13 cm^{-1} dominating the cancer sample spectra. Comparison of experimental spectra with molecular dynamic simulated spectroscopic signatures suggests that the high intensity spectral peak could originate from overexpressed mi-RNA molecules specific for ovarian cancer. Ovarian cancer cells are utilized as a proof of concept, but the sub-THz spectroscopy method is very general and could also be applied to other types of cancer.

FB11   *Post-Deadline Abstract*   11:32 – 11:47

THE DATABASE FOR ASTRONOMICAL SPECTROSCOPY - UPDATES, ADDITIONS AND PLANS FOR SPLATA-LOGUE FOR ALMA FULL SCIENCE OPERATIONS

ANTHONY REMIJAN, *ALMA, National Radio Astronomy Observatory, Charlottesville, VA, USA*; NATHAN A SEIFERT, *Department of Chemistry, University of Alberta, Edmonton, AB, Canada*; BRETT A. McGUIRE, *ALMA, National Radio Astronomy Observatory, Charlottesville, VA, USA*.

For the past 10 years, Splatalogue has been constantly updated, modified and enhanced in order to make molecular spectroscopy data readily available to the astronomical community. Splatalogue is fully integrated into the ALMA Observing Tool, the ALMA data reduction and analysis package (CASA) and several enhanced tools being developed through the ALMA development program including the next generation CASA viewer (CARTA) and the ALMA Data Mining Toolkit (ADMIT). In anticipation for ALMA full science operations, a number of improvements have taken place over the past year to the Splatalogue database including, but not limited too, additions to Splatalogue from the JPL and CDMS line lists, improvements and reconciliation of the Lovas/NIST Catalog assigning NRAO recommended rest frequencies to every astronomically detected transition, including recent astronomical surveys to the list of transitions detected in space and finally, improved search and display features as requested by the astronomical community. Splatalogue is planning for the next 10 years of development and welcomes any and all contributions to improving the data integrity and availability to the scientific community.

FC. Chirped pulse
Friday, June 24, 2016 – 8:30 AM
Room: 274 Medical Sciences Building

Chair: Justin L. Neill, BrightSpec, Inc., Charlottesville, VA, USA

FC01   8:30 – 8:45

IT IS ALL ABOUT PHASE AND IT IS NOT STAR TREK

ROBERT W FIELD, DAVID GRIMES, TIMOTHY J BARNUM, STEPHEN COY, *Department of Chemistry, MIT, Cambridge, MA, USA*; YAN ZHOU, *JILA, National Institute of Standards and Technology and Univ. of Colorado Department of Physics, University of Colorado, Boulder, Boulder, CO, USA.*

The marriage of chirped pulse millimeter-wave spectroscopy with a buffer gas cooled molecular beam source has yielded an increase in spectral velocity (number of resolution elements per unit time) of a factor of one million! But it gets even better. Essential information is encoded not just in the frequencies of the transitions, but also in the relative intensities and especially phases of the transitions. Transitions between Rydberg states of atoms and molecules are an ideal test ground for techniques that fully exploit these newly accessible observables.

FC02   8:47 – 9:02

OBSERVATION OF SUPERRADIANCE IN MMWAVE SPECTROSCOPY OF RYDBERG STATES: BAD IS THE NEW GOOD

DAVID GRIMES, TIMOTHY J BARNUM, *Department of Chemistry, MIT, Cambridge, MA, USA*; YAN ZHOU, *JILA, National Institute of Standards and Technology and Univ. of Colorado Department of Physics, University of Colorado, Boulder, Boulder, CO, USA*; STEPHEN COY, ROBERT W FIELD, *Department of Chemistry, MIT, Cambridge, MA, USA.*

The 10^6 increase in spectral velocity allowed by the combination of chirped pulse millimeter-wave spectroscopy with a buffer gas cooled molecular beam source qualitatively expands the classes of possible experiments. As an example, it allows for investigation of cooperative radiation effects (such as superradiance and subradiance) in large samples of atoms in Rydberg states in a single shot. However, these same effects can present obstacles to the application of the full increase in spectral velocity to high-resolution spectroscopic experiments through both frequency (chirps, broadenings, and shifts) and intensity effects.

FC03   9:04 – 9:19

INTENSITIES OF WEAKLY-ALLOWED RYDBERG-RYDBERG TRANSITIONS MEASURE CORE MULTIPOLES: WHY AND HOW

STEPHEN COY, TIMOTHY J BARNUM, DAVID GRIMES, ROBERT W FIELD, *Department of Chemistry, MIT, Cambridge, MA, USA.*

A Rydberg electron in a molecular Rydberg state interacts with the ion core on several different distance and time scales. At closest approach, the electron feels the atomic structure of the core, and the lumpy structure due to the shorter-range multipoles. In the Quantum Defect Model, this is described by the short-range K matrix of Altunata et al, or the core defects of Chernov et al. At longer range the core dipole modifies the electron orbital angular momentum because it has the same radial dependence as the centrifugal potential. At greater distance, stroboscopic effects between core rotation and electron motion appear, and finally the inverse Born-Oppenheimer rection of Zon's group.

Core multipole moments contribute to quantum defects over a wide intermediate range. The point Coulomb-dipole problem is separable and solvable, and the radial part is found to have the same form as the hydrogenic solutions, but with continuous effective values of the n,L quantum numbers primarily dependent on the state energy. In contrast, the angular part shows L-mixing that is due primarily to the dipole moment, and then higher multipoles. We describe how the angular part of the wave function leads to strong dependence of the intensities of weak transition on the core dipole and near independence of short range core properties.

Chirped-pulse microwave techniques record positions and intensities of many transitions in a single chirp, allowing the instrument function to be calibrated, and intensities to be measured.

FC04 9:21 – 9:36

CPMMW SPECTROSCOPY OF RYDBERG STATES OF NITRIC OXIDE

TIMOTHY J BARNUM, CATHERINE A. SALADRIGAS, DAVID GRIMES, STEPHEN COY, *Department of Chemistry, MIT, Cambridge, MA, USA*; EDWARD E. EYLER, *Department of Physics, University of Connecticut, Storrs, CT, USA*; ROBERT W FIELD, *Department of Chemistry, MIT, Cambridge, MA, USA*.

The spectroscopy of Rydberg states of NO has a long history [1], stimulating both experimental and theoretical advances in our understanding of Rydberg structure and dynamics. The closed-shell ion-core ($^1\Sigma^+$) and small NO^+ dipole moment result in regular patterns of Rydberg series in the Hund's case (d) limit, which are well-described by long-range electrostatic models (e.g., [2]). We will present preliminary data on the core-nonpenetrating Rydberg states of NO (orbital angular momentum, $\ell \geq 3$) collected by chirped-pulse millimeter-wave (CPmmW) spectroscopy. Our technique directly detects electronic free induction decay (FID) between Rydberg states with $\Delta n^* \approx 1$ in the region of $n^* \sim$ 40-50, providing a large quantity (12 GHz bandwidth in a single shot) of high quality (resolution $\sim$ 350 kHz) spectra. Transitions between high-ℓ, core-nonpenetrating Rydberg states act as reporters on the subtle details of the ion-core electric structure.

[1] Huber KP. Die Rydberg-Serien im Absorptions-spektrum des NO-Moleküls. *Helv. Phys. Acta* **3**, 929 (1961).

[2] Biernacki DT, Colson SD, Eyler EE. Rotationally resolved double resonance spectra of NO Rydberg states near the first ionization limit. *J. Chem. Phys.* **88**, 2099 (1988).

FC05 9:38 – 9:48

A 75–110 GHz CP-FTmmW SPECTROMETER FOR REACTION DYNAMICS AND KINETICS STUDIES

DANIEL P. ZALESKI, KIRILL PROZUMENT, *Chemical Sciences and Engineering Division, Argonne National Laboratory, Argonne, IL, USA*.

A BrightSpec chirped-pulsed Fourier transform millimeter-wave spectrometer operating in the 75–110 GHz spectral region has been installed at Argonne National Laboratory. The instrument has been tailored for chemical reaction dynamics and kinetics studies, and the arrangement allows for easy alternation between a room temperature flow cell and a supersonic expansion. The molecular beam is equipped with a pyrolysis nozzle for monitoring reaction products. Benchmark measurements in the flow cell will be presented along with early pyrolysis observations.

FC06 9:50 – 10:05

MICROWAVE SPECTRAL TAXONOMY AND ASTRONOMICAL SEARCHES FOR VIBRATIONALLY-EXCITED C_2S AND C_3S

BRETT A. McGUIRE, *NAASC, National Radio Astronomy Observatory, Charlottesville, VA, USA*; MARIE-ALINE MARTIN-DRUMEL, *Atomic and Molecular Physics, Harvard-Smithsonian Center for Astrophysics, Cambridge, MA, USA*; JOHN F. STANTON, *Department of Chemistry, The University of Texas, Austin, TX, USA*; MICHAEL C McCARTHY, *Atomic and Molecular Physics, Harvard-Smithsonian Center for Astrophysics, Cambridge, MA, USA*.

C_2S and C_3S are common interstellar species, and have relatively simple reaction chemistries. For these reasons, they frequently serve as probes of chemical evolution and physical conditions in rich astronomical sources. Because their rotational lines are often conspicuous there, detection of C_2S and C_3S in vibrationally-excited states might provide additional insight into formation pathways and excitation conditions. However, knowledge of the vibrational satellite transitions of both species is incomplete. Here, we report laboratory measurements of rotational spectra of vibrationally-excited C_2S and C_3S obtained from two microwave spectral taxonomy studies, in which CS_2 alone or in combination with a hydrocarbon precursor (acetylene or diacetylene), were produced using an electrical discharge. For C_3S, these studies, in combination with high-level quantum chemical calculations, greatly extend previous microwave measurements, while for C_2S, satellite transitions from several vibrational states have been observed for the first time. On the basis of precise laboratory rest frequencies, renewed searches for these transitions can be undertaken with confidence in publicly-available astronomical line surveys.

Intermission

FC07   10:24 – 10:34

PHOTODISSOCIATION OF ISOXAZOLE AND PYRIDINE STUDIED USING CHIRPED PULSE MICROWAVE SPECTROSCOPY IN PULSED UNIFORM SUPERSONIC FLOWS

NUWANDI M ARIYASINGHA, *Department of Chemistry, university of Missouri, Columbia, MO, USA*; BAPTISTE JOALLAND, *Departmnt de Physique Moleculaire, Institut de Physique de rennes, Bat 11C, Campus de Beaulieu, France*; ALEXANDER M MEBEL, *Department of Chemistry and Biochemistry, Florida International University, Miami, Florida, USA*; ARTHUR SUITS, *Department of Chemistry, university of Missouri, Columbia, MO, USA.*

Chirped - Pulse Fourier-transform microwave spectroscopy in uniform supersonic flows (Chirped- Pulse/Uniform Flow: CPUF) has been applied to study the photodissociation of two atmospherically relevant N containing heterocyclic compounds; pyridine and isoxazole. Products were detected using rotational spectroscopy. HC_3N, HCN were observed for pyridine and CH_3CN, HCO and HCN were observed for isoxazole and we report the first detection of HNC for both of the systems. Key points in potential energy surface were explored and compared with the experimental observations. Branching ratios were calculated for all the possible channels and will be presented.

FC08   10:36 – 10:46

CHIRPED PULSE MICROWAVE SPECTROSCOPY IN PULSED UNIFORM SUPERSONIC FLOWS: OBSERVATION OF K-DEPENDENT RATES IN THE CL + PROPYNE REACTION

NUWANDI M ARIYASINGHA, BERNADETTE M. BRODERICK, JAMES O. F. THOMPSON, ARTHUR SUITS, *Department of Chemistry, university of Missouri, Columbia, MO, USA.*

Chirped-Pulse Fourier-transform microwave spectroscopy in uniform supersonic flows (CPUF) has been applied to study the reaction of Cl atoms with propyne. The approach utilizes broad-band microwave spectroscopy to extract structural information with MHz resolution and near universal detection, in conjunction with a Laval flow system, which offers thermalized conditions at low temperatures and high number densities. Our previous studies have exploited this approach to obtain multichannel product branching fractions in a number of polyatomic systems, with isomer and often vibrational level specificity. This report highlights an additional capability of the CPUF technique: here, the state-specific reactant depletion is directly monitored on a microsecond timescale. In doing so, a clear dependence on the rotational quantum number K in the rate of the reaction between Cl atoms and propyne is revealed. Future prospects for the technique will be discussed.

FC09   10:48 – 11:03

SOME SIGNAL PROCESSING TECHNIQUES FOR USE IN BROADBAND TIME DOMAIN MICROWAVE SPECTROSCOPY

S. A. COOKE, *Natural and Social Science, Purchase College SUNY, Purchase, NY, USA.*

At the present time, in the typical broadband, time domain microwave spectroscopy experiment each free induction decay (FID) collected is on the order of 10^6 data points in length with a sampling rate on the order of 10^{-12} seconds per point. Traditionally, the FID is processed using a fast Fourier transform algorithm (FFT) with the resulting power spectrum used in ensuing spectral analyses. For use with the FFT algorithm we have implemented some pre- and post-processing techniques to improve the signal quality. These techniques include the use of Lissajous plots to ensure phase stability in signal addition, novel windowing functions, and also automated broadband phase corrections which allow the absorption spectrum to be used as a more highly resolved version of the traditional power spectrum (see figure). We have also implemented alternatives to the FFT algorithm for time domain signal processing including Hankel singular valued decomposition, a maximum entropy method, and wavelet transformations. Although these techniques are unlikely to be used in place of a fast Fourier transform we will demonstrate how each of these techniques may be used to augment the traditional FFT algorithm in regards to spectral analysis.

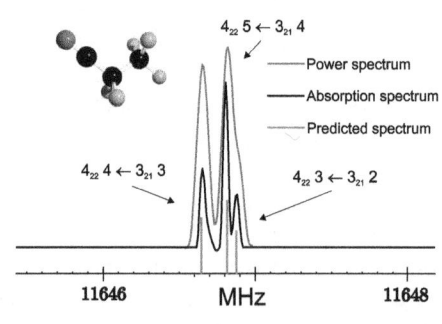

FC10   **11:05 – 11:20**

FREQUENCY BAND PERFORMANCE COMPARISONS FOR ROOM-TEMPERATURE CHIRPED PULSE MILLIME-TER WAVE SPECTROSCOPY

JUSTIN L. NEILL, BRENT HARRIS, ROBIN PULLIAM, MATT MUCKLE, *BrightSpec Labs, BrightSpec, Inc., Charlottesville, VA, USA*; BROOKS PATE, *Department of Chemistry, The University of Virginia, Charlottesville, VA, USA.*

We present a performance comparison between chirped pulse millimeter wave spectrometers operating over 75-110, 260-290, and 520-580 GHz. For molecules at room temperature, the line strength has an approximately ν^3 dependence until the peak of the Boltzmann distribution (typically in the submillimeter) is reached. However, we find competitive performance for 75-110 GHz spectrometers–with an average sensitivity drop of approximately 3-5 in equal measurement time, compared to a 260-290 GHz instrument with the same excitation power and measurement cell length. The narrower linewidth and lower line density at lower frequency, moreover, increase the usable dynamic range at 75-110 GHz by a factor of approximately 3-10 before reaching the confusion limit, giving better performance for extracting weak lines in a strong forest. This talk will discuss the reasons for and implications of these differences in performance for applications of chirped pulse millimeter wave spectroscopy.

FC11   **11:22 – 11:32**

CHIRPED-PULSE FOURIER TRANSFORM MICROWAVE SPECTROSCOPY OF DIFLUOROBENZALDEHYDES

GORDON G BROWN, SYDNEY A GASTER, DEONDRE L PARKS, BRANDON J YARBROUGH, *Department of Science and Mathematics, Coker College, Hartsville, SC, USA.*

The pure rotational spectra of 2,3-difluorobenzaldehyde and 2,6-difluorobenzaldehyde were measured on a chirped-pulsed Fourier transform microwave (CP-FTMW) spectrometer in the 3 – 8 GHz frequency range. The spectra were analyzed to find the spectroscopic constants of the molecule. For 2,3-difluorobenzaldehyde, the more stable *trans* conformer was observed, while the less stable *cis* conformer was not observed. ^{13}C isotopologues were observed in natural abundance for 2,6-difluorobenzaldehyde but not for 2,3-difluorobenzaldehyde. The rotational constants and centrifugal distortion constants have been compared to *ab initio* calculations.

FC12   **11:34 – 11:49**

MICROWAVE SPECTROSCOPY AND STRUCTURE DETERMINATION OF H_2S – MI (M=Cu,Ag,Au)

CHRIS MEDCRAFT, *School of Chemistry, Newcastle University, Newcastle-upon-Tyne, United Kingdom*; ANTHONY LEGON, *School of Chemistry, University of Bristol, Bristol, United Kingdom*; NICK WALKER, *School of Chemistry, Newcastle University, Newcastle-upon-Tyne, United Kingdom.*

A series of hydrogen sulphide-metal iodide complexes (H_2S-MI, M=Cu, Ag and Au) have been measured via chirped pulse Fourier transform microwave spectroscopy between 7.5-18 GHz. The complexes were generated in a supersonic expansion via laser ablation of the metal and decomposition of CF_3I. Experimental structures were obtained by least squares fitting of structural parameters to the rotational constants of deuterium and metal (^{63}Cu / ^{65}Cu and ^{107}Ag / ^{109}Ag) isotopologues. Interestingly K_{-1}=1 transitions were observed in the spectra containing D_2S, these were not observed in previous studies of similar molecules (H_2S-MCl). This allowed for the determination of an extra rotational constant and, consequently, extra structural information could be obtained. The structures are compared to high level coupled cluster theory calculations.

FD. Dynamics and kinetics

Friday, June 24, 2016 – 8:30 AM

Room: B102 Chemical and Life Sciences

Chair: Anh T. Le, Brookhaven National Laboratory, Upton, Ny, USA

FD01 8:30 – 8:45

COOLING OF ELECTRONICALLY-EXCITED He$_2$ MOLECULES IN A MICROCAVITY PLASMA JET

RUI SU, THOMAS J. HOULAHAN, JR., J. GARY EDEN, *Department of Electrical and Computer Engineering, University of Illinois at Urbana-Champaign, Urbana, IL, USA.*

Helium dimers in the $d^3\Sigma_u^+$ excited electronic state with potential energy >24 eV and radiative lifetime of 25 ns have been generated in a microcavity plasma jet and rotationally cooled by supersonic expansion in vacuum. The dynamic process of cooling is recorded by imaging the axis of expansion onto the slit of Czerny-Turner spectrometer, yielding spatial-temporal spectrograms of $d^3\Sigma_u^+ \rightarrow b^3\Pi_g$ $(v', v'') = (0, 0)$ emission. Analysis of the data shows the spatial-temporal evolution of the rotational temperature to be a damped sinusoid that reaches a minimum value of 100K. This reproducible behavior is attributed to the reflection of electrons from a virtual cathode located downstream of the nozzle and indicates that the spatially-averaged electron density is 10^8 cm^{-3}. We present this observed rotational temperature oscillation during the supersonic cooling process as an example of the potential of our supersonic microplasma expansion as a tool to explore physical dynamics in diatomic molecules having high excitation energies and small lifetimes.

FD02 8:47 – 9:02

NON-ADIABATIC DYNAMICS OF ICN$^-$(Ar)$_n$ and BrCN$^-$(Ar)$_n$

BERNICE OPOKU-AGYEMAN, *Department of Chemistry and Biochemistry, The Ohio State University, Columbus, OH, USA*; ANNE B McCOY, *Department of Chemistry, University of Washington, Seattle, WA, USA.*

We investigate the dynamics of the photodissociation of ICN$^-$(Ar)$_n$ and BrCN$^-$(Ar)$_n$ following electronic excitation to states that dissociate into X$^-$ + CN and X* + CN$^-$ (X = I or Br) using classical dynamics approaches. Observations made from previous experiments and calculations of these anions demonstrated that non-adiabatic effects are important in the photodissociation process and are reflected in the branching ratios of the photoproducts.[a,b] The addition of an argon atom is expected to shift the relative energies of these excited states, thereby altering the product branching. Interestingly, experimental studies show that electronically exciting ICN$^-$ solvated with even a single argon atom leads to a small fraction of the products recombine to form ICN$^-$.[a]

In this study, the dynamics are carried out using classical mechanics, treating the non-adiabatic effect with a surface hopping algorithm.[c] We assess the accuracy of this approach by first calculating the branching ratios for the bare anions and comparing the results to those from quantum dynamics calculations.[a,b] Once the results from both the quantum and classical dynamics are shown to be consistent, the classical dynamics simulations are extended to the argon solvated anions.

[a]S. Case, E. M. Miller, J. P. Martin, Y. J. Lu, L. Sheps, A. B. McCoy, and W. C. Lineberger, Angew. Chem., Int. Ed. 51, 2651 (2012).

[b]B. Opoku-Agyeman, A. S. Case, J. H. Lehman, W. Carl Lineberger and A. B. McCoy, J. Chem Phys. 141, 084305 (2014).

[c]J. C. Tully, J. Chem Phys. 93, 1061 (1990).

FD03 9:04 – 9:19

ROVIBRATIONAL LEVELS AND INELASTIC SCATTERING OF THE H_2O-Ar CLUSTER IN FULL AND REDUCED DIMENSIONALITY

STEVE ALEXANDRE NDENGUE, MOUMITA MAJUMDER, RICHARD DAWES, *Department of Chemistry, Missouri University of Science and Technology, Rolla, MO, USA*; FABIEN GATTI, *Charles Gerhardt Institute, University Montpellier 2, Montpellier, France*; HANS-DIETER MEYER, *Theoretical Chemistry Institute, University of Heidelberg, Heidelberg, Germany.*

The Water-Argon cluster is an important system of fundamental and practical interest. It is for example known to be one of the simplest systems capable of manifesting "hydrophobic interactions" and as such is an ideal candidate for the study of those interactions. On the fundamental level, it is a model system for the description of the intermolecular potential, rovibrational states and inelastic scattering of an atom and an asymmetric top van der Waal complex and thus may serve as a test to perform similar work on other systems. Additionally, the description of the H_2O-Ar intermolecular interaction is an important initial step to a deeper understanding of the static and dynamical properties of condensed phases such H_2O doped in large $(Ar)_N$ clusters. We investigate in this work the H_2O-Ar cluster on a global potential energy surface recently generated. We thus compute the rovibrational energy levels of the cluster in the rigid rotor approximation and in full dimensionality using the MCTDH improved relaxation method and compare our results with available experimental measurements and previous calculations. Then, we present inelastic scattering cross-sections of H_2O+Ar collisions obtained in the rigid rotor approximation using time-independent method and time-dependent method, and compare where available results with previous calculations. Finally, we will discuss the extension of the scattering calculations to the full dimensional case and the prospect of studying rovibrational relaxation within accurate time-dependent quantum calculations on similar systems or clusters.

FD04 9:21 – 9:36

TWO-TONE FREQUENCY MODULATION ABSORPTION SPECTROSCOPY STUDYING HO_2 FORMATION FROM THE OXIDATION OF TETRAHYDROFURAN

MING-WEI CHEN, IVAN ANTONOV, BRANDON ROTAVERA, LEONID SHEPS, CRAIG A. TAATJES, *Combustion Research Facility, Sandia National Laboratories, Livermore, CA, USA.*

HO_2 is a key radical produced by the chain-termination step in the low-temperature combustion of fuel molecules. Time-resolved direct IR absorption can be used for quantitative measurements of HO_2 concentration as a function of time in chemical reactions. In this work, one HO_2 rotational transition line of O-H stretch overtone band (2,0,0)-(0,0,0) has been used to study the kinetics of HO_2 formation from the tetrahydrofuran oxidation. Tetrahydrofuran is an interesting example for studying the oxidation of cyclic either, because of the two oxidation reaction pathways started from hydrogen abstraction on different C-H bonds. The Cl initiated tetrahydrofuran oxidation was performed in a Herriott-type multipass cell to reduce the complexity of the reaction mechanism. The measurements were taken under the temperature ranging from 500K to 750K, to study the chain-termination kinetics at various temperatures. Additionally, the kinetics of HO_2 formation from the oxidation of tetrahydrofuran was also found to be influenced by different O_2 concentration.

FD05

PYROLYSIS OF TROPYL RADICAL (C_7H_7) AND BENZYL RADICAL ($C_6H_5CH_2$) IN A HEATED MICRO-REACTOR

GRANT BUCKINGHAM, <u>BARNEY ELLISON</u>, JESSICA P PORTERFIELD, *Department of Chemistry and Biochemistry, University of Colorado, Boulder, CO, USA*; JOHN W DAILY, *Department of Mechanical Engineering, University of Colorado Boulder, Boulder, CO, USA*; MUSAHID AHMED, *UXSL, Chemical Sciences Division, Lawrence Berkeley National Laboratory, Berkeley, CA, USA*; DAVID ROBICHAUD, MARK R NIMLOS, *Biomass Molecular Science , National Renewable Energy Laboratory , Golden, CO, USA.*

Benzyl radical ($C_6H_5CH_2$) is a crucial intermediate in the combustion and pyrolysis of substituted aromatic species that are common both in modern gasoline and potential future biofuels. The decomposition of benzyl radical is complicated and has been shown by isotopic labeling to require interesting isomerizations pathways. To better understand these pathways, a set of C_7H_7 radicals has been studied in a heated micro-reactor. Through multiple experiments, it has be shown that benzyl radical and cycloheptatrienyl (tropyl) radical (c-C_7H_7) do not interconvert, even at temperatures where both have completely thermally decomposed. To confirm this, tropyl radical has been studied directly and its pyrolysis is quite simple, only cyclopentadienyl radical (c-C_5H_5) and acetylene (HCCH) are formed. Cyclopentadienyl radical then decomposes to acetylene and propargyl radical ($HCCCH_2$). These products have all been identified through use of tunable synchrotron radiation by confirming their respective photoionization spectra. Matrix isolation infrared (IR) spectroscopy has also been used to identity these products. A previously unanswered question in benzyl radical decomposition has been addressed by studying the pyrolysis of 2,5-norbornadiene, which indicates benzyl radical may decompose through a norbornadiene-like bicyclic radical intermediate. This pathways successfully predicts the correct isotopically labeled products observed previously during ^{13}C labeled benzyl pyrolysis.

Intermission

FD06

SINGLE PHOTON INITIATED DECOMPOSITION REARRANGEMENT REACTIONS (SPIDRR) OF ORGANIC MOLECULES MEDIATED BY THE Ni^+ CATION

<u>DARRIN BELLERT</u>, ADAM MANSELL, ZACHARY THEIS, MICHAEL GUTIERREZ, *Chemistry Department, Baylor University, Waco, TX, USA.*

The Bellert group at Baylor University has developed a novel method for performing single photon initiated decomposition rearrangement reactions (SPIDRR) of organic molecules mediated by a transition metal cation. The advantage that SPIDRR affords is the direct measurement of first order microcanonical rate constants, k(E), determined at resolved internal energies. Furthermore, the SPIDRR technique measures kinetic details of exothermic reactions where product production is limited only by submerged activation barriers (kinetic barriers that are at energies below the separated reactant limit). Thus, such reactions approach unit efficiency, are thermodynamically driven, and are of greater relevance to catalytic research. Direct measurements of k(E) values extend to isotopically labelled species that provide direct measurement of the kinetic isotope effect (KIE), furnishing unique insight into the mechanistic details of a reaction.

This talk presents results from the visible photon initiated, Ni^+ induced decarbonylation reaction of propionaldehyde. Here a rather unique energy dependent behavior of the measured rate constants was observed and attributed to a dynamic competition between parallel reaction coordinates available to the photo-excited precursor. RRKM calculations in concert with high level DFT is used to support and further experimental results.

FD07 **10:29 – 10:44**

FLUORESCENCE MICROSPECTROSCOPY FOR TESTING THE DIMERIZATION HYPOTHESIS OF BACE1 PROTEIN IN CULTURED HEK293 CELLS

SPENCER GARDEEN, JOSEPH L. JOHNSON, <u>AHMED A HEIKAL</u>, *Chemistry and Biochemistry, University of Minnesota Duluth, Duluth, MN, USA.*

Alzheimer's Disease (AD) is a neurodegenerative disorder that results from the formation of beta-amyloid plaques in the brain that trigger the known symptoms of memory loss in AD patients. The beta-amyloid plaques are formed by the proteolytic cleavage of the amyloid precursor protein (APP) by the proteases BACE1 and gamma-secretase. These enzyme-facilitated cleavages lead to the production of beta-amyloid fragments that aggregate to form plaques, which ultimately lead to neuronal cell death. Recent detergent protein extraction studies suggest that BACE1 protein forms a dimer that has significantly higher catalytic activity than its monomeric counterpart. In this contribution, we examine the dimerization hypothesis of BACE1 in cultured HEK293 cells using complementary fluorescence spectroscopy and microscopy methods. Cells were transfected with a BACE1-EGFP fusion protein construct and imaged using confocal, and differential interference contrast to monitor the localization and distribution of intracellular BACE1. Complementary fluorescence lifetime and anisotropy measurements enabled us to examine the conformational and environmental changes of BACE1 as a function of substrate binding. Using fluorescence correlation spectroscopy, we also quantified the diffusion coefficient of BACE1-EGFP on the plasma membrane as a means to test the dimerization hypothesis as a fucntion of substrate-analog inhibition. Our results represent an important first towards examining the substrate-mediated dimerization hypothesis of BACE1 in live cells.

FD08 **10:46 – 11:01**

ULTRAFAST TRANSIENT ABSORPTION SPECTROSCOPY INVESTIGATION OF PHOTOINDUCED DYNAMICS IN NOVEL DONOR-ACCEPTOR CORE-SHELL NANOSTRUCTURES FOR ORGANIC PHOTOVOLTAICS

<u>JACOB STRAIN</u>, ABDELQADER JAMHAWI, *Department of Chemistry, University of Louisville, Louisville, KY, USA*; THULITHA M ABEYWICKRAMA, WENDY LOOMIS, HEMALI RATHNAYAKE, *Chemistry, Western Kentucky University, Bowling Green, KY, USA*; JINJUN LIU, *Department of Chemistry, University of Louisville, Louisville, KY, USA.*

Novel donor-acceptor nanostructures were synthesized via covalent synthesis and/or UV cross-linking method. Their photoinduced dynamics were investigated with ultrafast transient absorption (TA) spectroscopy. These new nanostructures are made with the strategy in mind to reduce manufacturing steps in the process of fabricating an organic photovoltaic cell. By imitating the heterojunction interface within a fixed particle domain, several fabrication steps can be bypassed reducing cost and giving more applicability to other film deposition methods. Such applications include aerosol deposition and ink-jet printing. The systems that were studied by TA spectroscopy include PDIB core, PDIB-P3HT core-shell, and PDIB-PANT core-shell which range in size from 60 to 130 nm. Within the experimentally accessible spectra range there resides a region of ground state bleaching, stimulated emission, and excited-state absorption of both neutrals and anions. Control experiments have been carried out to assign these features. At high pump fluences the TA spectra of PDIB core alone also indicate an intramolecular charge separation. The TA spectroscopy results thus far suggest that the core-shells resemble the photoinduced dynamics of a standard film although the particles are dispersed in solution, which indicates the desired outcome of the work.

FD09 11:03–11:08

DIFFUSION ASSISTED ELECTRON INJECTION TO CDS QUANTUM DOTS: A CONCLUSIVE FITTING WITH STATIC QUENCHING COLLINS KIMBALL MODEL

SUBHADIP GHOSH, *School of Chemical Sciences , National Institute of Science Education and Research, Bhubaneswar, India.*

Static quenching Collins Kimball (SQCK) diffusion model convincingly explained bimolecular photo induced electron transfer (PET) kinetics from electron donor molecule N-methyl aniline (NMA) to a series of CdS quantum dots (QDs) of different sizes [1-2]. A small change in chemical driving force (0.01 eV) causes almost three times acceleration of PET rate (k = 8.30×10^9 M^{-1} S^{-1}) within the larger size QDs (5.4 nm) as compared to that (k = 2.74×10^9 M^{-1} S^{-1}) observed among the smaller (3.8 nm) QD particles. On further analysis, we studied temperature dependence of PET rate. From an Arrhenius type fit of PET rate as a function of temperature, we found accelerated PET rate in larger size QDs is due to lower activation barrier and higher value of electronic coupling matrix between the product state and reactant state. Time evolution of the position dependent sink term of SQCK model clearly identified different regimes (static, non-stationary and stationary) associated with the bimolecular PET kinetics in solution phase [2]

[1] Litniewski, M.; Gorecki, J. Molecular Dynamics Tests of the Smoluchowski–Collins–Kimball. Phys. Chem. Chem. Phys. 2004, 6, 72-83.

[2] Bhowmik, A.; Kaur, H.; Koley, S.; Jana, S.; Ghosh, S. Diffusion Assisted Bimolecular Electron Injection to CdS Quantum Dots: Existence of Different Regimes in Time Dependent Sink Term of Collins Kimball Model. J. Phys. Chem. C Accepted (DOI: 10.1021/acs.jpcc.5b11169)

FD10 11:10–11:20

ULTRAFAST EXTREME ULTRAVIOLET ABSORPTION SPECTROSCOPY OF METHYLAMMONIUM LEAD IODIDE PEROVSKITE

MAX A VERKAMP, MING-FU LIN, ELIZABETH S RYLAND, JOSH VURA-WEIS, *Department of Chemistry, University of Illinois at Urbana-Champaign, Urbana, IL, USA.*

Methylammonium lead iodide (perovskite) is a leading candidate for use in next-generation solar cell devices. However, the photophysics responsible for its strong photovoltaic qualities are not fully understood. Ultrafast extreme ultraviolet (XUV) absorption was used to investigate electron and hole dynamics in perovskite by observing transitions from a common inner-shell level (I 4d) to the valence and conduction bands. Ultrashort (30 fs) pulses of XUV radiation with a broad spectrum (40-70 eV) were generated via high-harmonic generation using a tabletop instrument. Transient absorption measurements with visible pump and XUV probe directly observed the relaxation of charge carriers in perovskite after above-band excitation in the femtosecond and picosecond time ranges.

FE. Synchrotron

Friday, June 24, 2016 – 8:30 AM

Room: 217 Noyes Laboratory

Chair: Sven Thorwirth, University of Cologne, Cologne, Germany

FE01   8:30 – 8:45

THE SOLEIL VIEW ON PROTOTYPICAL ORGANIC NITRILES: SELECTED VIBRATIONAL MODES OF ETHYL CYANIDE, C_2H_5CN, AND SPECTROSCOPIC ANALYSIS USING AN AUTOMATED SPECTRAL ASSIGNMENT PROCEDURE (ASAP)

CHRISTIAN ENDRES, PAOLA CASELLI, *The Center for Astrochemical Studies, Max-Planck-Institut für extraterrestrische Physik, Garching, Germany*; MARIE-ALINE MARTIN-DRUMEL, MICHAEL C Mc-CARTHY, *Atomic and Molecular Physics, Harvard-Smithsonian Center for Astrophysics, Cambridge, MA, USA*; OLIVIER PIRALI, *AILES beamline, Synchrotron SOLEIL, Saint Aubin, France*; NADINE WEHRES, STEPHAN SCHLEMMER, SVEN THORWIRTH, *I. Physikalisches Institut, Universität zu Köln, Köln, Germany*.

Vibrational spectra of small organic nitriles, propionitrile and n-butyronitrile, have been investigated at high spectral resolution at the French national synchroton facility SOLEIL using Fourier-transform far-infrared spectroscopy (< 700 cm^{-1}). The *Automated Spectral Assignment Procedure* (ASAP)[a] has been used for line assignement and accurate determination of rotational level energies, in particular, of the $\nu_{20}=1$ and the $\nu_{12}=1$ states of propionitrile. The analysis does not only confirm the applicability of the ASAP in the treatment of (dense) high-resolution infrared spectra but also reveals some of its limitations which will be discussed in some detail.

[a]M. A. Martin-Drumel, C. P. Endres, O. Zingsheim, T. Salomon, J. van Wijngaarden, O. Pirali, S. Gruet, F. Lewen, S. Schlemmer, M. C. McCarthy, and S. Thorwirth 2015, J. Mol. Spectrosc. 315, 72

FE02   8:47 – 9:02

FAR-INFRARED SPECTROSCOPY OF *SYN*-VINYL ALCOHOL

PAUL RASTON, *Department of Chemistry and Biochemistry, James Madison University, Harrisonburg, Virginia, USA*; HAYLEY BUNN, *School of Chemistry and Physics, The University of Adelaide, Adelaide, South Australia, Australia*.

Vinyl alcohol has been extensively studied in both the microwave[a,b] and mid-IR[c,d] spectral regions, where 9 out of 15 vibrational modes have been identified. Here we present the first far-IR spectrum of vinyl alcohol, collected below 700 cm^{-1} at the Australian Synchrotron. The high resolution (0.001 cm^{-1}) spectrum reveals the ν_{11} and ν_{15} fundamentals of syn-vinyl alcohol at 489 cm^{-1} and 407 cm^{-1}, in addition to two hot bands of the ν_{15} mode at 369 cm^{-1} and 323 cm^{-1}. High J transitions in the R-branch of the ν_{15} band were found to be perturbed by an a-axis Coriolis interaction with the nearby ν_{11} state. The ν_{15} torsional mode of *syn*-vinyl alcohol was fit using a Watson's A-reduced Hamiltonian to yield rotational, centrifugal distortion, and Coriolis coupling parameters.

[a]S. Saito, Chem. Phys. Lett. 42, 3 (1976)
[b]M. Rodler et al., J. Am. Chem. Soc. 106, 4029 (1948)
[c]Y. Koga et al., J. Mol. Spec. 145, 315 (1991)
[d]D-L. Joo et al., J. Mol. Spec. 197, 68 (1999)

FE03   9:04–9:19

FAR-INFRARED SPECTROSCOPY OF *ANTI*-VINYL ALCOHOL

PAUL RASTON, *Department of Chemistry and Biochemistry, James Madison University, Harrisonburg, Virginia, USA*; HAYLEY BUNN, *School of Chemistry and Physics, The University of Adelaide, Adelaide, South Australia, Australia.*

Vinyl alcohol can exist in two rotameric forms, known as *syn*- and *anti*- vinyl alcohol, where *syn* is the most stable. Both rotamers have been observed in the interstellar medium towards Sagittarius B2(N) making them of particular astrophysical importance[a]. Vinyl alcohol has been subject to various spectroscopic investigations, however, the *anti* rotamer has only been obsvered in the microwave region[b]. We report the high resolution (0.001 cm^{-1}) FTIR spectrum of *anti*-vinyl alcohol collected at the infrared beamline facility of the Australian Synchrotron. Vinyl alcohol was produced via the pyrolysis of 2-chloroethanol at $900°C$[c], and its far infrared spectrum reveals the presence of the strong ν_{15} fundamental and hot band of *anti*-vinyl alcohol. Rotational and centrifugal distortion constants of this higher energy rotamer have since been determined for the ν_{15} and $2\nu_{15}$ states, and the ground state constants have been refined.

[a]B. E. Turner, A. J. Apponi, ApJ 561, 207 (2001)
[b]M. Rodler, J. Mol. Spec. 114, 23 (1985)
[c]D-L Joo, et al., J. Mol. Spec. 197, 68 (1999)

FE04   9:21–9:36

A COMBINED GIGAHERTZ AND TERAHERTZ SYNCHROTRON-BASED FOURIER TRANSFORM INFRARED (TERAHERTZ) SPECTROSCOPIC INVESTIGATION OF META- AND ORTHO-D-PHENOL: OBSERVATION OF TUNNELING SWITCHING

ZIQIU CHEN, SIEGHARD ALBERT, CSABA FÁBRI, ROBERT PRENTNER, MARTIN QUACK, *Laboratory of Physical Chemistry, ETH Zurich, Zürich, Switzerland.*

Tunneling switching is of fundamental interest for certain experiments aiming at detecting parity violation in chiral molecules.[a,b] A particularly intriguing recent development is the theoretical prediction of prototypical tunneling switching in meta- and ortho-D-phenol (C_6H_4DOH) as opposed to phenol (C_6H_5OH)[c] where only tunneling dominates the dynamics: For meta and ortho-D-phenol at low energy, tunneling is completely suppressed due to isotopic substitution, which introduces an asymmetry in the effective potential including zero point vibrational energy in the lowest quasiadiabatic channel. This effectively localizes the molecular wavefunction at either the *syn* or *anti* structure of meta- and ortho-D-phenol. At higher torsional states of meta- and ortho-D-phenol, tunneling becomes dominant, thus switching the dynamics to a delocalized quantum wavefunction.The pure rotational spectra of the meta- and ortho-D-phenol were recorded between 60 and 110 GHz using an experimental setup[d] which we have improved somewhat whereas the rotationally resolved vibrational spectra in the THz and infrared region were collected in the range of 200 to 1000 cm^{-1}using synchrotron-based FTIR spectroscopy.[e]The detailed assignment of the new GHz spectra including excited vibrational states, whereas previously only microwave spectra of the ground state were known,[f] shall be discussed, in terms of the experimental evidence demonstrating tunneling switching in the first overtone of the torsional vibration of meta-D-phenol.

[a]M. Quack and M. Willeke, *J. Phys .Chem. A* **110**, 3338-3348 (2006).
[b]M. Quack, *Adv. Chem. Phys* **157**, 247-291 (2014).
[c]S. Albert, Ph. Lerch, R. Prentner and M. Quack, *Angew. Chem. Int. Ed.* **52**, 346-349 (2013).
[d]M. Suter and M. Quack, *Appl. Opt* **54**, 4417-4431 (2015).
[e]S. Albert, Ph. Lerch, R. Prentner and M. Quack, *68th International Symposium on Molecular Spectroscopy*, Columbus, Ohio, USA, June 17-21, paper TG09 (2013).
[f]T. Pedersen, N. W. Larsen and L. Nygaard, *J. Mol. Struc.* **4**, 59-77 (1969).

Intermission

FE05 9:55 – 10:10

FAR INFRARED SYNCHROTRON SPECTRUM OF TRIMETHLYENE OXIDE

OMAR MAHASSNEH, JENNIFER VAN WIJNGAARDEN, *Department of Chemistry, University of Manitoba, Winnipeg, MB, Canada.*

Rotationally-resolved vibrational spectra of trimethlyene oxide (c-C_3H6O) from 650 through 1200 cm^{-1}were recorded using far infrared synchrotron radiation at the Canadian Light Source with better than 0.001 cm^{-1}resolution. The observed bands correspond to at least eight different fundamental vibrations in this region. Due to the low frequency ring puckering motion,[a] the observed rovibrational pattern of each band is congested with hot-combination bands that originate in the first two excited ring puckering states (52.9 cm^{-1}, 142.6 cm^{-1}). The ongoing analysis of the strong *b-type* bands corresponding to asymmetric in-plane CO stretching (ν_{10}:1008 cm^{-1})[b] will be discussed along with the identification of allowed Coriolis interactions arising from nearby energy levels related to in-plane CC stretching (ν_9: 940 cm^{-1}, ν_3: 1033 cm^{-1}).

[a]G. Moruzzi *et al.*, J. Mol. Spectrosc. **219, 152 (2003).**

[b]Bánhegyi *et al.* Spectrochim. Acta. **39A, 761 (1983).**

FE06 10:12 – 10:27

WAKEFIELDS IN COHERENT SYNCHROTRON RADIATION

BRANT E. BILLINGHURST, J. C. BERGSTROM, C. BARIBEAU, *EFD, Canadian Light Source Inc., Saskatoon, Saskatchewan, Canada*; T. BATTEN, *CID, Canadian Light Source Inc., Saskatoon, Canada*; L. DALLIN, *Accelerator Operations, Canadian Light Source Inc., Saskatoon, Canada*; TIM E MAY, *EFD, Canadian Light Source Inc., Saskatoon, Saskatchewan, Canada*; J. M. VOGT, *CID, Canadian Light Source Inc., Saskatoon, Canada*; WARD A. WURTZ, *Accelerator Operations, Canadian Light Source Inc., Saskatoon, Canada*; ROBERT L. WARNOCK, *Stanford Synchrotron Radiation Lightsource, SLAC National Accelerator Laboratory, Menlo Park, CA, USA*; D. A. BIZZOZERO, *Department of Mathematics and Statistics, University of New Mexico, Albuquerque, New Mexico, USA*; S. KRAMER, *NSLS, Brookhaven National Laboratory, Upton, New York, USA*; K. H. MICHAELIAN, *CanmetENERGY, Natural Resources Canada, Edmonton, Alberta, Canada.*

When the electron bunches in a storage ring are sufficiently short the electrons act coherently producing radiation several orders of magnitude more intense than normal synchrotron radiation. This is referred to as Coherent Syncrotron Radiation (CSR). Due to the potential of CSR to provide a good source of Terahertz radiation for our users, the Canadian Light Source (CLS) has been researching the production and application of CSR. CSR has been produced at the CLS for many years, and has been used for a number of applications. However, resonances that permeate the spectrum at wavenumber intervals of 0.074 cm^{-1}, and are highly stable under changes in the machine setup, have hampered some experiments. Analogous resonances were predicted long ago in an idealized theory. Through experiments and further calculations we elucidate the resonance and wakefield mechanisms in the CLS vacuum chamber. The wakefield is observed directly in the 30–110 GHz range by rf diodes. These results are consistent with observations made by the interferometer in the THz range. Also discussed will be some practical examples of the application of CSR for the study of condensed phase samples using both transmission and Photoacoustic techniques.

FE07

ULTRAFAST MOLECULAR THREE-ELECTRON COLLECTIVE AUGER DECAY

RAIMUND FEIFEL[a], *Department of Physics, Faculty of Science, University of Gothenburg, Gothenburg, Sweden.*

A new class of many-electron Auger transitions in atoms was initially proposed over 40 years ago[b], but the first tentative evidence for its real existence was only adduced by Lee et al.[c] in 1993, on the basis of the resonant Auger spectrum of Kr. Using a multi-electron coincidence technique with synchrotron radiation, we unambiguously showed very recently that the transition suggested by Lee et al. in Kr really does take place, but with a rather small branching ratio[d]. Related inter-atomic three-electron transitions in rare gas clusters were recently predicted by Averbukh and Kolorenč[e] and demonstrated by Ouchi et al.[f]. From consideration of the energy levels involved it seems that the basic three-electron process could occur in molecules too, wherever a double inner-valence shell vacancy lies at a higher energy than the molecular triple ionisation onset. Experiments on CH_3F reveal for the first time the existence of this new decay pathway there[g], and calculations show that despite its three-electron nature, its effective oscillator strength is orders of magnitudes higher than in atoms, allowing an efficient competition with both molecular dissociation and two-electron decay channels on the ultrafast time scale. The dramatic enhancement of the molecular three-electron Auger transition can be explained in terms of a partial breakdown of the molecular orbital picture of ionisation. We predict that the collective decay pathway will be significant in a wide variety of heteroatomic molecules ionised by extreme UV and soft X-rays, particularly at Free-Electron-Lasers where double inner-shell vacancies can be created efficiently by two-photon transitions.

[a]raimund.feifel@physics.gu.se

[b]G.N. Ogurtsov et al., Sov. Phys. Tech. Phys. 15, 1656 (1971) and V.V. Afrosimov et al., JETP Lett. 21, 249 (1975).

[c]I. Lee, R. Wehlitz, U. Becker and M. Ya. Amusia, J. Phys. B: At. Mol. Opt. Phys. 26, L41 (1993).

[d]J.H.D. Eland, R.J. Squibb, M. Mucke, S. Zagorodskikh, P. Linusson, and R. Feifel, New J. Phys. 17, 122001 (2015).

[e]V. Averbukh and P. Kolorenč, Phys. Rev. Lett. 103, 183001 (2009).

[f]T. Ouchi et al., Phys. Rev. Lett. 107, 053401 (2011).

[g]R. Feifel et al., Phys. Rev. Lett. 116, 073001 (2016).

FE08

$^{14}NH_3$ LINE POSITIONS AND INTENSITIES IN THE FAR-INFRARED: COMPARISON OF FT-IR MEASUREMENTS TO EMPIRICAL HAMILTONIAN MODEL PREDICTIONS

KEEYOON SUNG, *Jet Propulsion Laboratory, California Institute of Technology, Pasadena, CA, USA*; SHAN-SHAN YU, *Molecular Spectroscopy, Jet Propulsion Laboratory, Pasadena, CA, USA*; JOHN PEARSON, *Jet Propulsion Laboratory, California Institute of Technology, Pasadena, CA, USA*; OLIVIER PIRALI, *AILES beamline, Synchrotron SOLEIL, Saint Aubin, France*; F. KWABIA TCHANA, *CNRS, Université Paris Est Créteil et Paris Diderot, LISA, Créteil, Val de Marne, France*; LAURENT MANCERON, *Beamline AILES, Synchrotron SOLEIL, Saint-Aubin, France.*

We have analyzed multiple spectra of high purity (99.5%) normal ammonia sample recorded at room temperatures using the FT-IR and AILES beamline at Synchrotron SOLEIL, France. More than 2830 line positions and intensities are measured for the inversion-rotation and rovibrational transitions in the $50 - 660$ cm^{-1} region. Quantum assignments were made for 2047 transitions from eight bands including four inversion-rotation bands (gs(a-s), ν_2(a-s), $2\nu_2$(a-s), and ν_4(a-s)) and four ro-vibrational bands ($\nu_2 - $ gs, $2\nu_2 - $ gs, $\nu_4 - \nu_2$, and $2\nu_2 - \nu_4$), as well as covering more than 300 lines of $\Delta K = 3$ forbidden transitions. Out of the eight bands, we note that $2\nu_2 - \nu_4$ has not been listed in the HITRAN 2012 database. The measured line positions for the assigned transitions are in an excellent agreement (typically better than 0.001 cm^{-1}) with the predictions from the empirical Hamiltonian model [S. Yu, J.C. Pearson, B.J. Drouin, et al.(2010)] in a wide range of J and K for all the eight bands. The comparison with the HITRAN 2012 database is also satisfactory, although systematic offsets are seen for transitions with high J and K and those from weak bands. However, differences of 20% or so are seen in line intensities for allowed transitions between the measurements and the model predictions, depending on the bands. We have also noticed that most of the intensity outliers in the Hamiltonian model predictions belong to transitions from gs(a-s) band. We present the final results of the FT-IR measurements of line positions and intensities, and their comparisons to the model predictions and the HITRAN 2012 database.[a]

[a]Research described in this paper was performed at the Jet Propulsion Laboratory and California Institute of Technology, under contracts and cooperative agreements with the National Aeronautics and Space Administration.

FE09

INFRARED SPECTROSCOPY OF THE H_2/HD/D_2-O_2 VAN DER WAALS COMPLEXES

PAUL RASTON, *Department of Chemistry and Biochemistry, James Madison University, Harrisonburg, Virginia, USA*; <u>HAYLEY BUNN</u>, *School of Chemistry and Physics, The University of Adelaide, Adelaide, South Australia, Australia.*

Hydrogen is the most abundant element in the universe and oxygen is the third, so understanding the interaction between the two in their different forms is important to understanding astrochemical processes. The interaction between H_2 and O_2 has been explored in low energy scattering experiments[a,b] and by far infrared synchrotron spectroscopy of the van der Waals complex[c]. The far infrared spectra suggest a parallel stacked average structure with seven bound rotationally excited states. Here, we present the far infrared spectrum of HD/D_2-O_2 and the mid infrared spectrum of H_2-O_2 at 80 K, recorded at the infrared beamline facility of the Australian Synchrotron. We observed 'sharp' peaks in the mid infrared region, corresponding to the end over end rotation of H_2-O_2, that are comparatively noisier than analogous peaks in the far infrared where the synchrotron light is brightest. The larger reduced mass of HD and D_2 compared to H_2 is expected to result in more rotational bound states and narrower bands. The latest results in our ongoing efforts to explore this system will be presented.

[a]Y. Kalugina, et al., Phys. Chem. Chem. Phys. 14, 16458 (2012)

[b]S. Chefdeville et al. Science 341, 1094 (2013)

[c]H. Bunn et al. ApJ 799, 65 (2015)

FE10   *Post-Deadline Abstract*

VACUUM ULTRAVIOLET SPECTROSCOPY OF THE LOWEST-LYING ELECTRONIC STATE IN SUB-CRITICAL AND SUPERCRITICAL WATER

<u>TIMOTHY W MARIN</u>, *Chemistry, Benedictine University, Lisle, IL, USA*; IRENEUSZ JANIK, DAVID M BARTELS, DAN CHIPMAN, *Radiation Laboratory, University of Notre Dame, Notre Dame, IN, USA.*

We report vacuum ultraviolet absorption spectra for the lowest-lying electronic state of high-temperature and supercritical water, where spectra were measured from room temperature up to the critical temperature, and as a function of density above the critical temperature. Spectra are seen to redshift with increasing temperature, demonstrating gradual breakdown of the hydrogen bond network. Above the critical temperature, tuning the density gives direct insight into the extent of hydrogen bonding in the supercritical regime. The known gas-phase monomer spectrum can be duplicated in the low-density limit, with negligible contribution from hydrogen bonding. With increasing density, the spectrum blue shifts as small water clusters form, increasing the number of hydrogen bonds lowering the ground-state energy. The presence of vibrational structure inherent to the lowest-density gas-phase limit spectrum gradually diminishes with increasing density, giving a reasonable measure of the extent of water monomers having unperturbed electronic structure as a function of density.

AUTHOR INDEX

Rice, Johnathan S – WH07
Richardson, Jeremy O – TI06
Rieker, Greg B – TB03, TB04
Riffe, Erika – FB08
Rijs, Anouk – TC02, TF08, TH12
Rivera-Rivera, Luis A. – TE03, TK15
Roadman, Danny – TC09
Robertson, Evan Gary – MG10
Robichaud, David – WB04, FD05
Rodgers, M T – MH14, MH15, WD05, WD06
Rodríguez, Miguel A. – WC06
Rodriguez Castillo, Sarah – WI10
Roenitz, Kevin – TC09
Rosario, Hoimonti – WB02
Roscioli, Joseph R – TB02
Ross, Stephen Cary – MF09, WF12
Rotavera, Brandon – FD04
Rotger, Maud – TJ10, TJ11
Rothman, Laurence S. – TG12, FA03
Roy, Harrison – MH15, WD06
Roy, P. – MG11
Rudd, Lydia – TC09
Rugango, Rene – TD09
Rupasinghe, Priyanka – WB05
Russkikh, Irina – WK12
Ryland, Elizabeth S – TC01, FD10

S

Sabbah, Hassan – WI10
Saha, Debasis – TE12
Sakai, Nami – WH09
Saladrigas, Catherine A. – FC04
Salomon, Thomas – RH09, RH10
Sander, Stanley P. – MG01
Sanov, Andrei – TG11
Sanz, M. Eugenia – WC10, WC11
Sarkar, Sohini – TE12, FA04
Sasada, Hiroyuki – TJ09, WK15
Schlecht, Erich T – FB02
Schlemmer, Stephan – MF01, MF02, WD02, WD09, WH01, WI03, WI04, WI05, FB04, FE01
Schmiedt, Hanno – MF01, MF02
Schmutz, Hansjürg – WI06
Schnell, Melanie – TH09, TH12, TI05, TI07, WC01, WC12
Schnierer, Rico – TH02
Schooss, Detlef – WI09
Schwaab, Gerhard – MK01, TC04
Schwan, Raffael – MK01, TC04
Schwartz, Aaron Z. A. – WJ05
Schwartz, George – MG07
Schwerdtfeger, Peter – WJ13
Sears, Trevor – WE10, RF05
Seifert, Nathan A – MI13, MJ04,

MJ05, WK04, FB11
Semeria, Luca – RA01
Senent, Maria Luisa – WA07
Serebrennikova, Olga – WK12
Sergachev, Ilia – WI07
Serrato III, Agapito – TF12
Seyfang, Georg – FA06
Sharma, Sandeep – TK07
Sheets, Donal – WJ14
Shen, Linhan – MG04, MG05
Shepherd, James J. – TK14
Sheps, Leonid – FD04
Sheybani-Deloui, S. – WG10, WG13
Shields, George C. – TI06, RG01
Shimamori, Takuto – RG04
Shingledecker, Christopher N – WH10
Shipman, Steven – FB08
Shorter, Joanne – TB02
Shu, Gang – TD09
Shu, Ran – FB02
Shy, Jow-Tsong – MH02
Sias, Eric – WK10
Sibert, Edwin – MJ08, TI02, RF07, RF11, RJ02
Silbaugh, Matthew J – MK08
Simao, Alcides – WC05
Simon, Aude – WI10
Singer, James – MI04
Singh, Vipin Bahadur – TJ04, TJ05
Sinha, Amitabha – RJ03
Sizov, Igor – FB10
Smith, CJ – WF03, WG12
Smith, Houston Hartwell – TH15
Smith, James E. T. – TD03, TD04, TD06
Smith, Mary Ann H. – TG06, WB02, WB03, WJ01, WJ02
Sonnenberger, Andrew – TC09
Sowards, John – WK10
Spada, Lorenzo – MJ04, MJ05, TI08, WE04, WK04
Spaun, Ben – TB05, TC08, TF04
Spezzano, Silvia – RH06
Springer, Sean D. – TK15
Srikantaiah, Sree – WK09
Srivastava, Santosh Kumar – TJ04
Stahl, Wolfgang – MI01, WE03, WF05, WF10, WF11
Stanton, John F. – MF07, MI11, MJ13, TD02, WK03, RF03, FC06
Stark, Glenn – RH08
Stearns, Jaime A. – MH07, WK11
Steber, Amanda – TH09, TH12, TI07
Steimle, Timothy – MA04, WE08, WE09, RI09
Steinmetz, Vincent – WD06
Stevens, Philip S. – WB01

Stewart, Jacob – TB07
Stoffels, Alexander – WD02, WI04
Stolyarov, Andrey – RH01
Storm, Shaye – WH12
Stout, Phillip J. – FB07
Strain, Jacob – FD08
Strelnikova, Evgeniya – WK12
Strom, Aaron I. – MK14
Stroscio, Gautam – WB05
Su, Rui – FD01
Suhm, Martin A. – MJ07, MJ09
Suits, Arthur – FC07, FC08
Sullivan, Michael N. – RI10
Sun, Ming – WK01
Sun, Yu Robert – TE08
Sung, Keeyoon – TG06, WB04, WJ01, WJ02, FE08
Suter, Jonathan D – TB06
Suzuki, Masao – FA08
Svarovskaya, Lidiya – WK12
Swann, William C – TB03

T

Taatjes, Craig A. – FD04
Tabor, Daniel P. – MJ08, TI02, RF11, RJ02
Tada, Kohei – RF02
Takano, Shuro – WH09
Tan, Jake Acedera – RG07, RG08, RG09
Tan, Yan – TE08, FA03
Tandon, Hannah K. – WG09
Tang, Adrian – FB02
Tang, Jian – RF06
Tao, Lei-Gang – TE08
Tari, Ozlem – TK04
Tashkun, Sergey – TK03
Taubman, Matthew S – TB06
Temelso, Berhane – TI06, RG01
Tennyson, Jonathan – TK03, TK11, RH01, RH02
Tercero, Belén – TH01, WF04, WF06, WF07
Teuben, Peter J. – WH12, RH12
Tew, David Peter – WC08, RI01
Thapaliya, Bishnu P. – MF05, TF05, WA03
Theis, Zachary – FD06
Thibault, Franck – WB02
Thimmakondu, Venkatesan S. – RH05
Thomas, Javix – MJ06
Thompson, Christopher – MG10
Thompson, James O. F. – FC08
Thompson, Michael C – WG01, WG02
Thorwirth, Sven – WD02, WK05,

BrightSpec

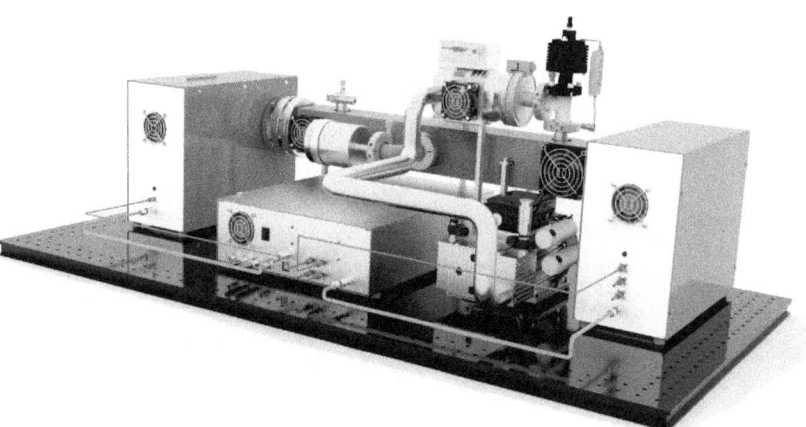
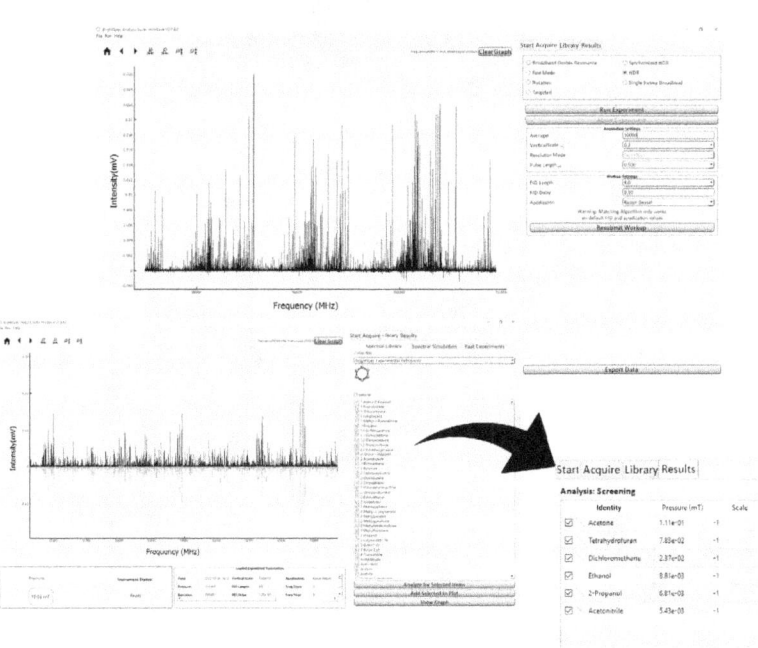

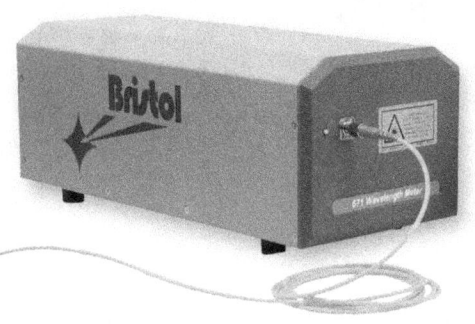
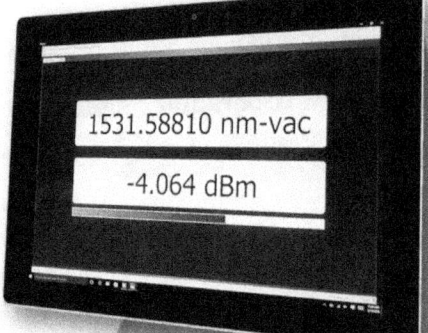
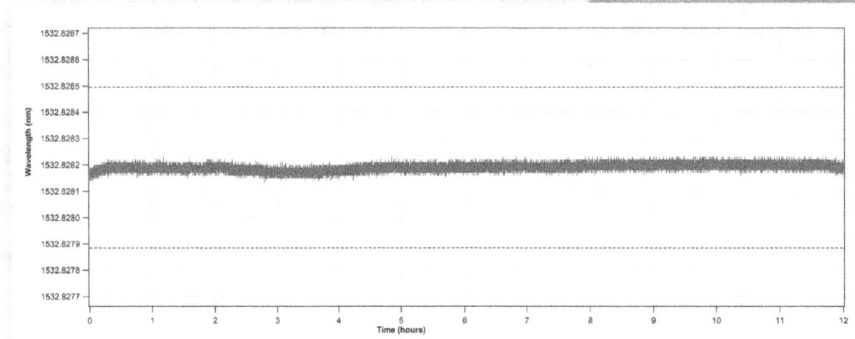

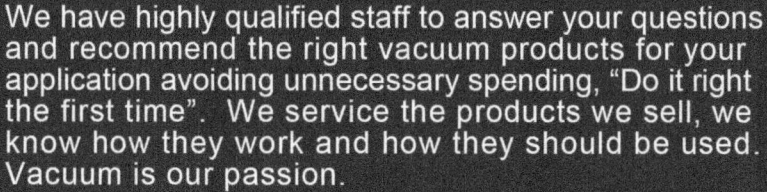

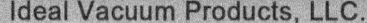

Journal of Molecular Spectroscopy

Give your research the advantage it deserves

The *Journal of Molecular Spectroscopy* presents experimental and theoretical articles on all subjects relevant to molecular spectroscopy and its modern applications. An international medium for the publication of some of the most significant research in the field, the Journal of Molecular Spectroscopy is an invaluable resource for astrophysicists, chemists, physicists, engineers, and others involved in molecular spectroscopy research and practice.

Six reasons to publish in the *Journal of Molecular Spectroscopy*

- ➢ Fast editorial times (total time from submission until final decision < 5 weeks)
- ➢ High refereeing standards
- ➢ Fast journal publication times (final corrected article online < 8 weeks)
- ➢ Wide dissemination: over 7,000 institutes worldwide have access to the journal
- ➢ High exposure of your article online; >130,000 article downloads per year on www.sciencedirect.com
- ➢ High archival value; cited half-life >10.0 years

Feature articles

The *Journal of Molecular Spectroscopy* publishes peer reviewed Feature articles - these articles will give an overview of areas of particular significance in molecular spectroscopy. They may review and consolidate an area of theoretical development or a collection of experimental data, in each case offering some new insights. The articles may also summarize the present status of a rapidly developing and/or evolving field. All the articles should serve as introductions to areas of spectroscopy other than one's specialty and should be particularly valuable to students entering the field.

Feature articles will be solicited by invitation of the Editor. However, the Editor invites you to suggest, with a reasonable level of detail, a topics that could be of interest. Self-nominations by potential authors are particularly encouraged.

For more information
www.elsevier.com/locate/jms

THE JOURNAL OF PHYSICAL CHEMISTRY

The Most Influential Journals in Physical Chemistry

EDITOR-IN-CHIEF: **George C. Schatz,** *Northwestern University*

The Journal of Physical Chemistry A

DEPUTY EDITOR:
Anne B. McCoy,
University of Washington

Journal Scope:
- Isolated Molecules, Clusters, Radicals, and Ions
- Environmental Chemistry, Geochemistry, and Astrochemistry
- Quantum Chemistry

The Journal of Physical Chemistry B

DEPUTY EDITOR:
Joan-Emma Shea,
University of California, Santa Barbara

Journal Scope:
Biophysical Chemistry, Biomaterials, Liquids, and Soft Matter

The Journal of Physical Chemistry C

DEPUTY EDITOR:
Catherine J. Murphy,
University of Illinois at Urbana-Champaign

Journal Scope:
Energy Conversion and Storage, Optical and Electronic Devices, Interfaces, Nanomaterials, and Hard Matter

The Journal of Physical Chemistry Letters

DEPUTY EDITOR:
Gregory Scholes,
Princeton University

Journal Scope:
Significant scientific advances in physical chemistry, chemical physics, and materials science

Follow us on Facebook and Twitter @JPhysChem

Quantel Laser Welcomes You to the
International Symposium on Molecular Spectroscopy
71st Meeting – June 20-24 2016

Sponsor of the Women's Networking Reception –
Wednesday June 22nd at 5:30pm

Quantel Laser
601 Haggerty Lane
Bozeman, MT 59715
1-877-QUANTEL

www.quantel-laser.com

Call for Award Nominations

Visit www.coblentz.org for more information

COBLENTZ
SOCIETY

COBLENTZ
SOCIETY

ABB Bomem-Michelson Award: ABB sponsors the Bomem-Michelson Award to honor scientists whom have advanced the technique(s) of vibrational, molecular, Raman, or electronic spectroscopy. Contributions may be theoretical, experimental, or both. The recipient must be actively working and at least 37 years of age. The nomination should include a resume of the candidate's career as well as a synopsis of the special research achievements that make the candidate an eligible nominee for the ABB sponsored Bomem-Michelson Award. Nominations for the award are open between February 1st and **May 1st** each year. Further information regarding the ABB Bomem-Michelson Award can be found at www.coblentz.org/awards/the-bomem-michelson-award.

Coblentz Award: The Coblentz Award is presented annually to an outstanding young molecular spectroscopist under the age of 40. The candidate must be under the age of 40 on January 1st of the year of the award. Nominations should include a detailed description of the nominee's accomplishments, a curriculum vitae and as many supporting letters as possible. Annual updates of files of nominated candidates are encouraged. Nominations for the Coblentz Award are open between January 3rd and **July 15th** each year. Further information regarding the Coblentz Award is available at www.coblentz.org/awards/the-coblentz-award.

Craver Award: The Craver Award is presented annually to an outstanding young molecular spectroscopist whose efforts are in the area of applied analytical vibrational spectroscopy. The candidate must be under the age of 45 on January 1st of the year of the award. The work may include any aspect of (near-, mid-, or far-infrared) IR, THz, or Raman spectroscopy in applied analytical vibrational spectroscopy. Nominees are welcome from academic, government, or industrial research. Nominations must include a detailed description of the nominee's accomplishments, curriculum vitae or resume, and a minimum of three supporting letters. Nominations for the Craver Award are open between March 30th and **August 30th** each year. Further information about the Craver Award is available at www.coblentz.org/awards/the-craver-award.

Ellis R. Lippincott Award: The Ellis R. Lippincott Award is presented annually in recognition of significant contributions and notable achievements in the field of vibrational spectroscopy. The medal is jointly sponsored by the Coblentz Society, the Optical Society of America and the Society for Applied Spectroscopy. Recipients must have made significant contributions to vibrational spectroscopy as judged by their influence on other scientists. Because innovation was a hallmark of the work of Ellis R. Lippincott, this quality in the contributions of candidates will be carefully appraised. Nominations for the award are open between January 1st and **October 1st** each year. Nominations should be submitted to: Lippincott Award Chairperson, awards@osa.org. Further information regarding the Ellis R. Lippincott Award is available at www.coblentz.org/awards/the-lippincott-award.

Honorary Membership: The Coblentz Society awards honorary memberships in the Society to people who have made outstanding contributions to the field of vibrational spectroscopy or any other field related to the purposes of the Society. Nominations close on **February 1st** each year, with awards announced at the Annual Members Meeting at Pittcon and presented at FACSS. Send your nomination for 2016 to Dr. Mark Druy, Coblentz Society President at madruy@gmail.com.

A Very Special Thanks to Our Sponsors!

ISMS MEETING VENUE INFORMATION

All contributed talks will be held in the Chemistry complex (and immediately adjoining buildings). The plenary talks will be held across the quad (about 600') in Foellinger Auditorium.

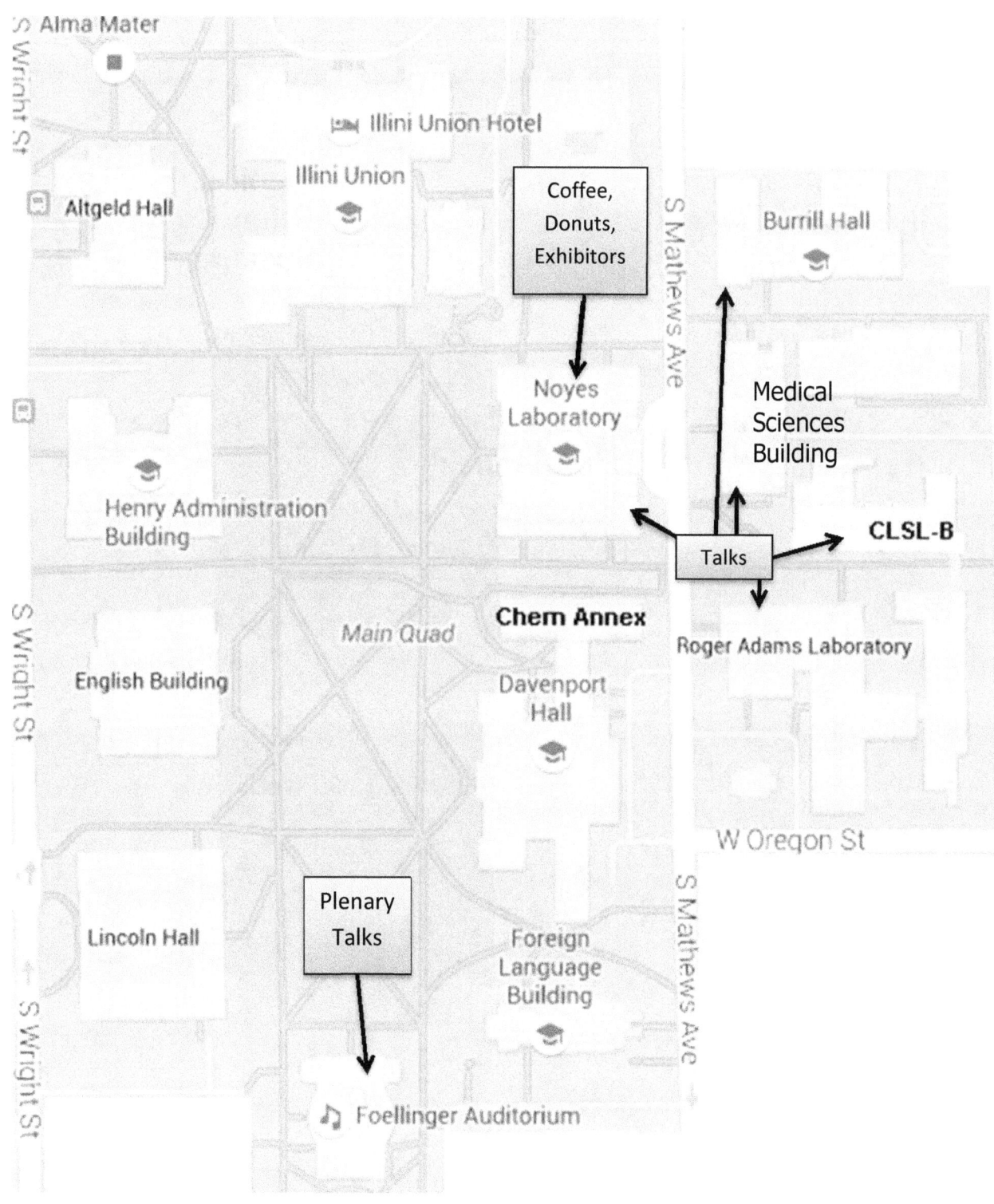

ACCESSIBLE ENTRANCES

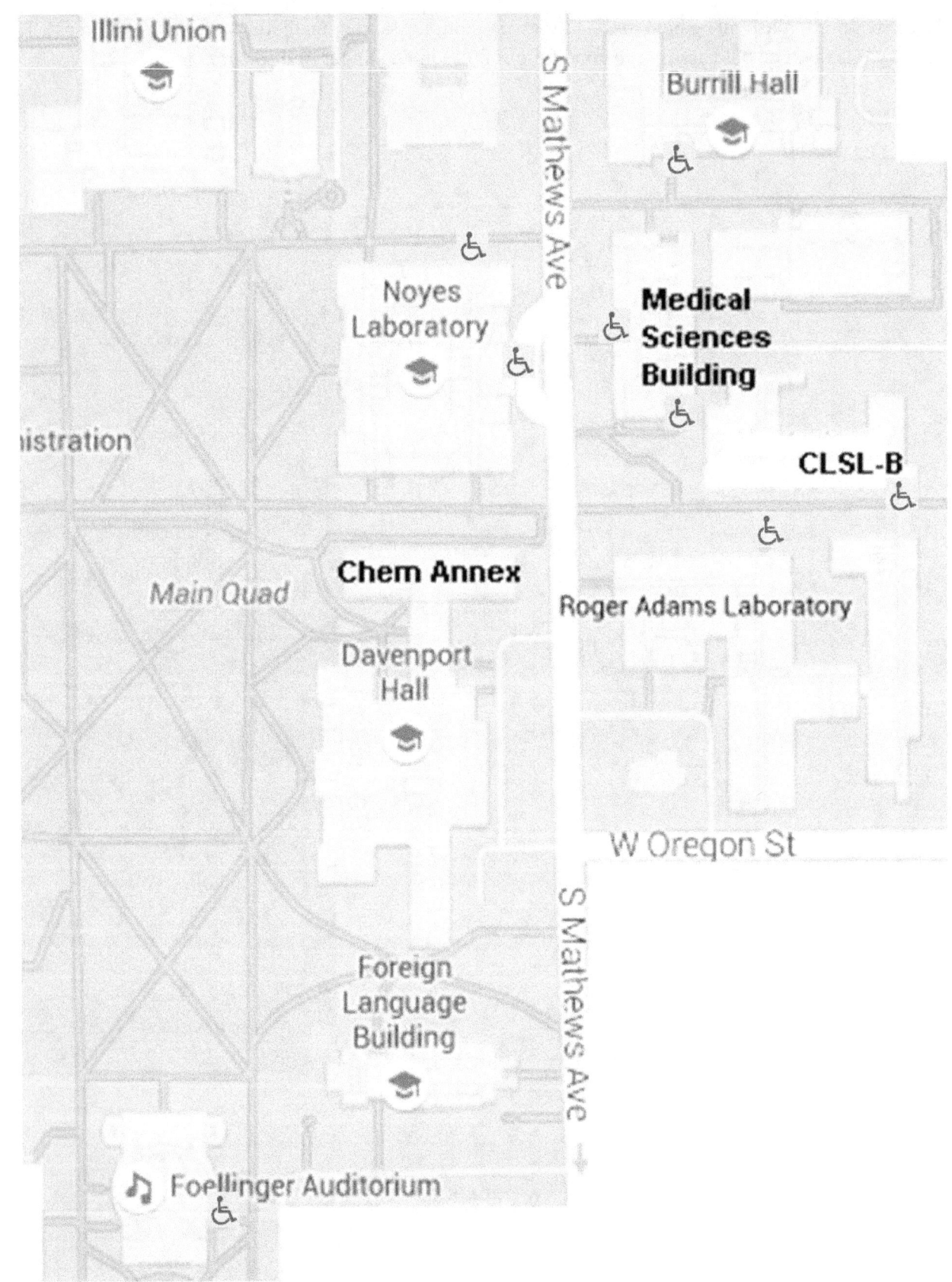

NOYES LABORATORY (NL)

Noyes Laboratory houses our Registration and Exhibitor/Refreshment Rooms (163/165), the Computer Lab (151), two lecture halls (NL 100 and NL 217), and the Chemistry Library.

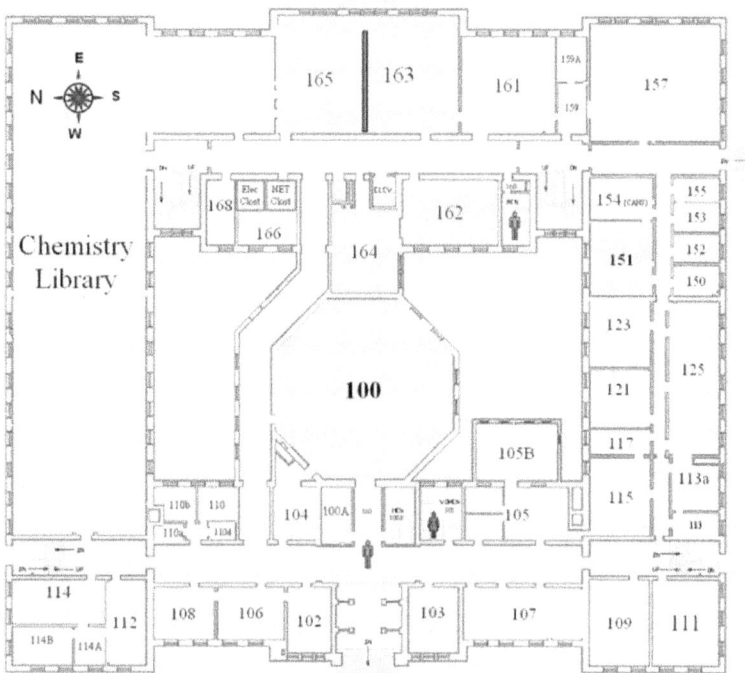

Noyes Laboratory - 1st Floor

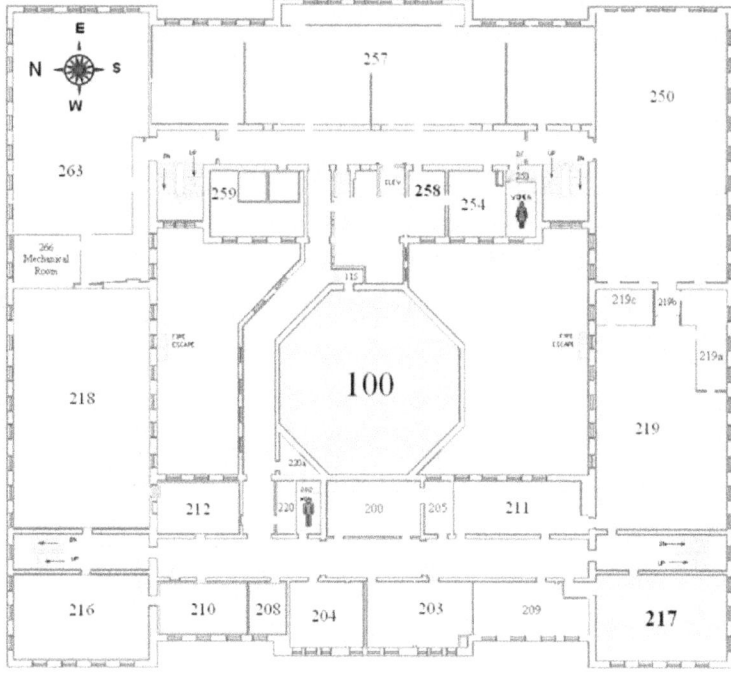

Noyes Laboratory - 2nd Floor

ROGER ADAMS LABORATORY (RAL)

Roger Adams Laboratory is across the street to the east of Chemistry Annex. It has one lecture hall (RAL 116). Please note that in Roger Adams Lab, the ground level is called "Ground" and the First Floor is equivalent to the Second Floor in the other buildings.

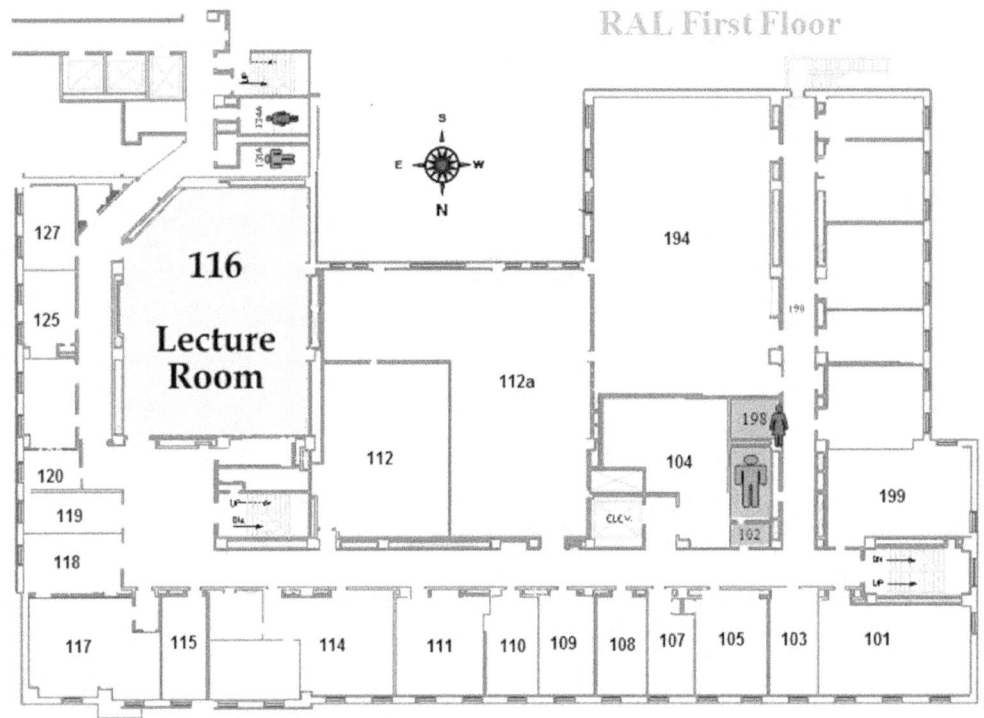

CHEMICAL AND LIFE SCIENCES (CLSL)

CLSL is a multi-wing building located across the street to the east of Noyes Laboratory. The lecture hall (CLSL B102) is in the B wing across the pedestrian walkway to the northeast of Roger Adams.

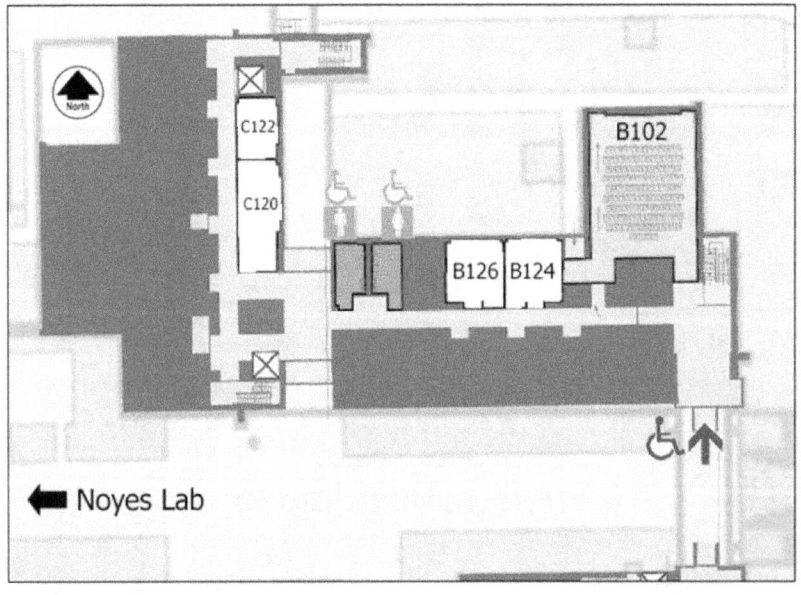

MEDICAL SCIENCES BUILDING (MSB)

Medical Sciences is across the pedestrian walkway to the north of RAL. It has one lecture hall (274).

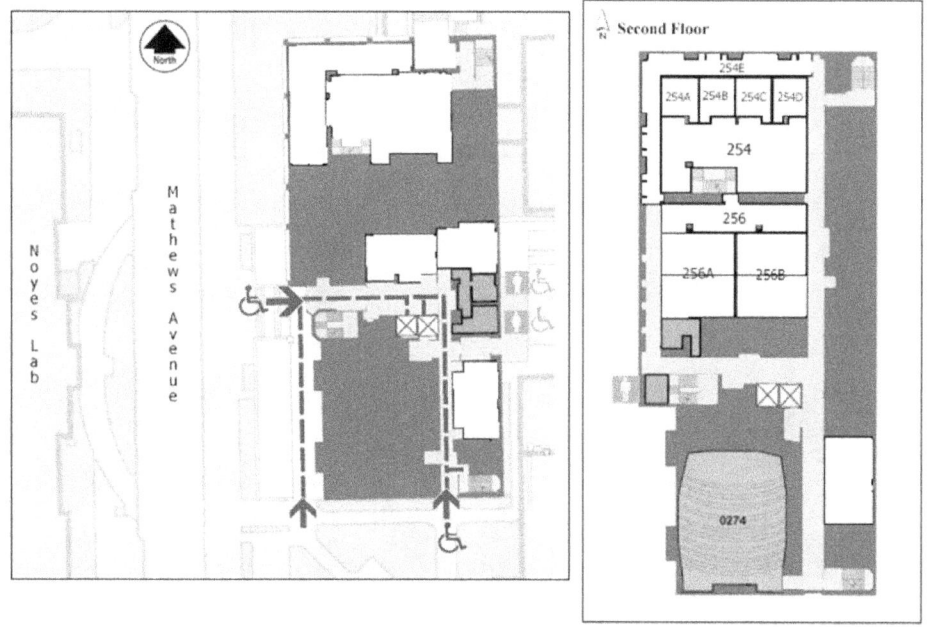

BURRILL HALL

Burrill Hall is due north of Medical Sciences. It has one lecture hall (140).

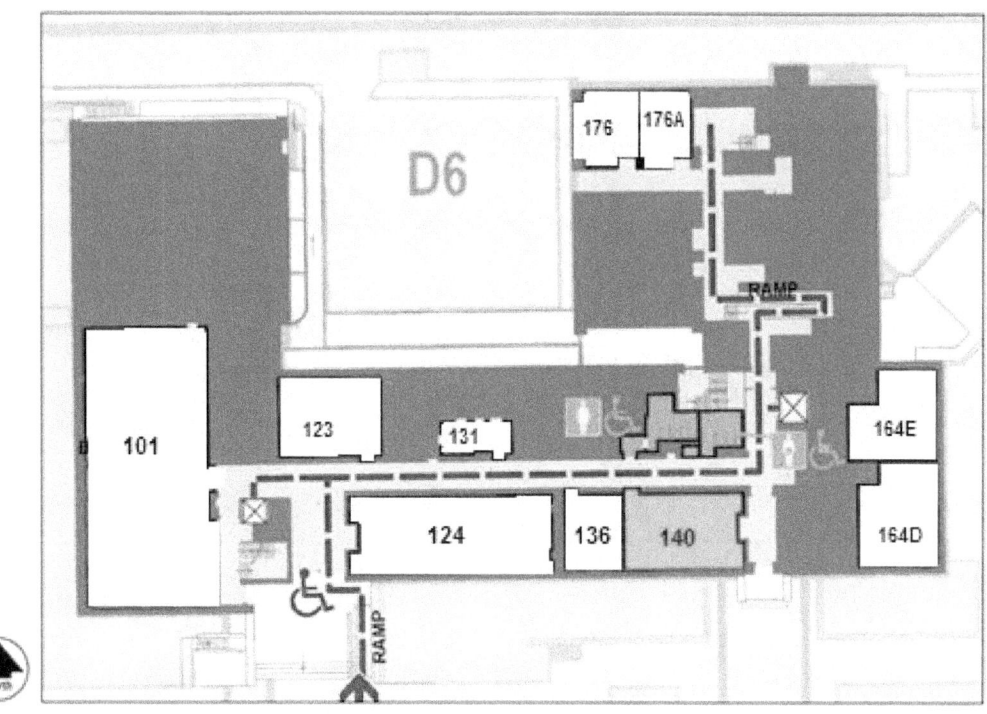

Foellinger Auditorium (Plenary and Intermission)

Foellinger Auditorium is located at the south end of the Quad. The main doors on the north (quad) side will open at 8:10 AM (the side ADA/wheelchair door will be open around 8:00 AM). There is seating on the main level and the upper balcony. There is no elevator in the building.

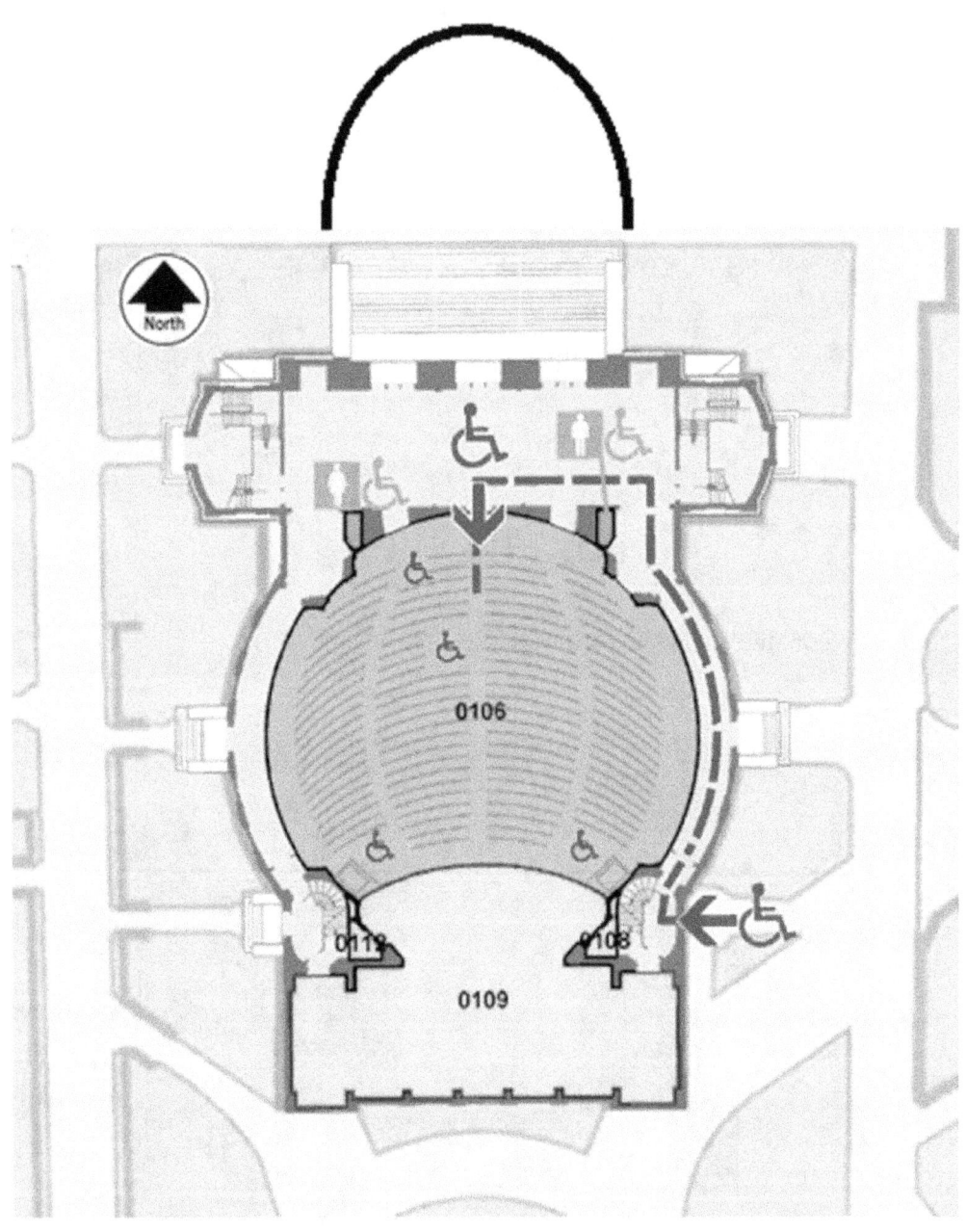

PARKING (E14) TO BOUSFIELD DORM

If you purchase a parking permit and are staying at the dorm, you will park in lot E14 (any spot). E14 is nearly due south of Bousfield Hall Dorm.

Parking enforcement begins at 6:00 AM on Monday, so you will need to have your car in lot E14 with your permit displayed before then. There are many parking meters on E. Peabody Drive (and in the lot across from Bousfield) if you wish to park closer for short periods (25 cents/15 minutes – generally between 6 AM and 6 PM, but check the meter because some go until 9 PM).

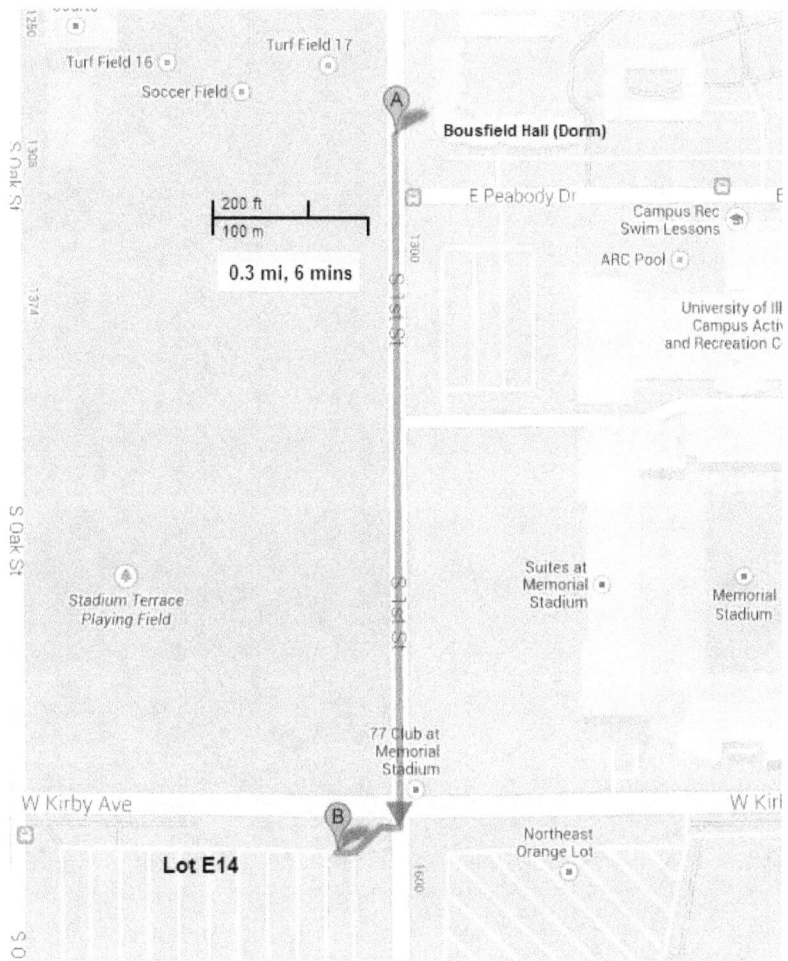

BOUSFIELD/NUGENT DORM to MEETING VENUE (walking)

Bousfield & Nugent Halls are just under a mile (15-20 minute walk) from the main symposium buildings

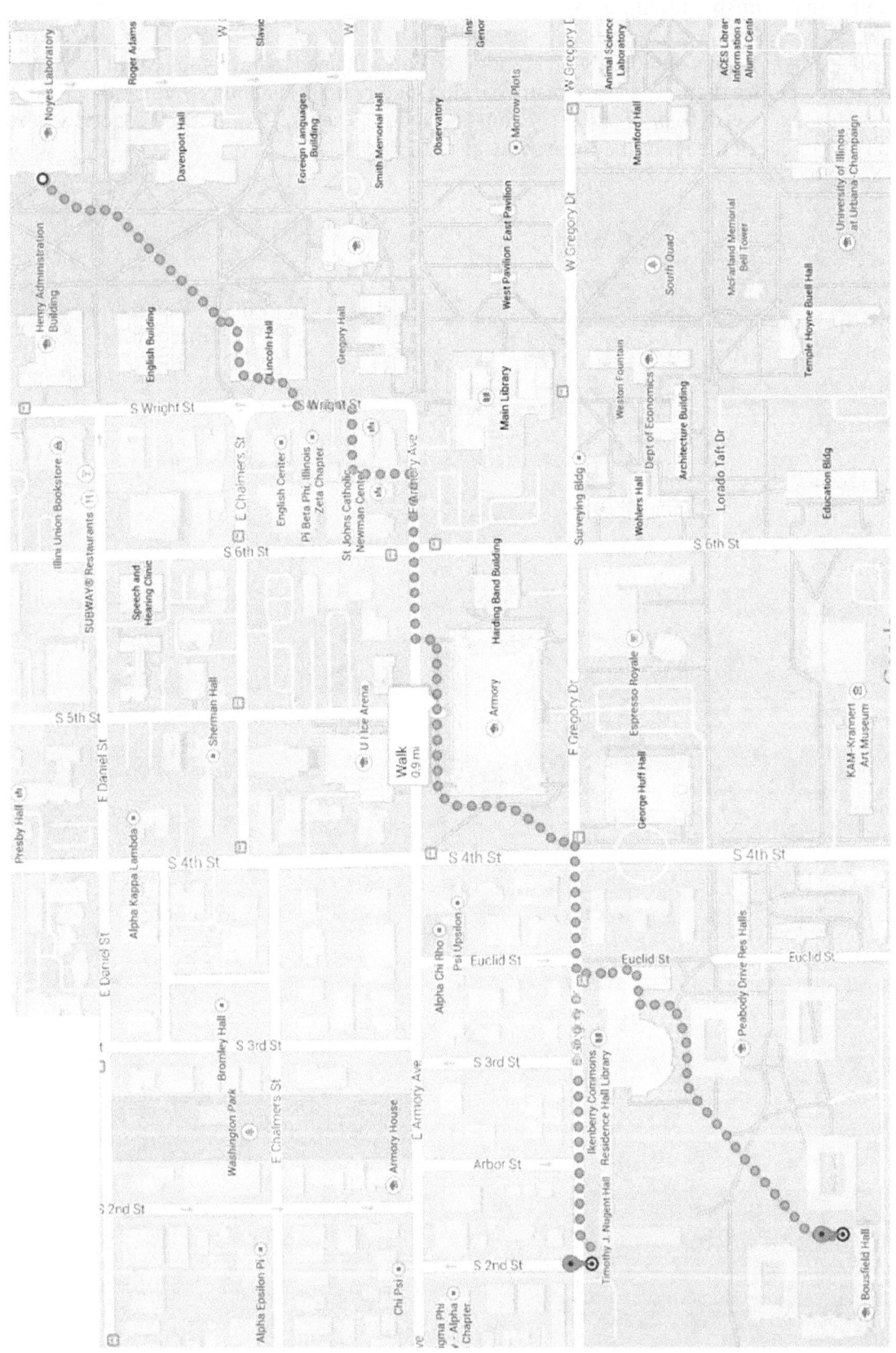

BOUSFIELD/NUGENT DORM to MEETING VENUE (bus)

There is convenient and free bus service between Bousfield/Nugent Dorms and 1 block from the meeting venue. The Yellow Line picks up on the corner of First and Peabody (Bousfield), and also on Gregory Drive (Nugent) in front of Ikenberry Commons, and drops off at the Wright Street Terminal (just outside of the Henry Administration Building). Return locations are the same but across the street. The Yellow Line will also take you to downtown Champaign, but you will need to pay for your return (only iStops are free). Approximately every 10 minutes during the day.

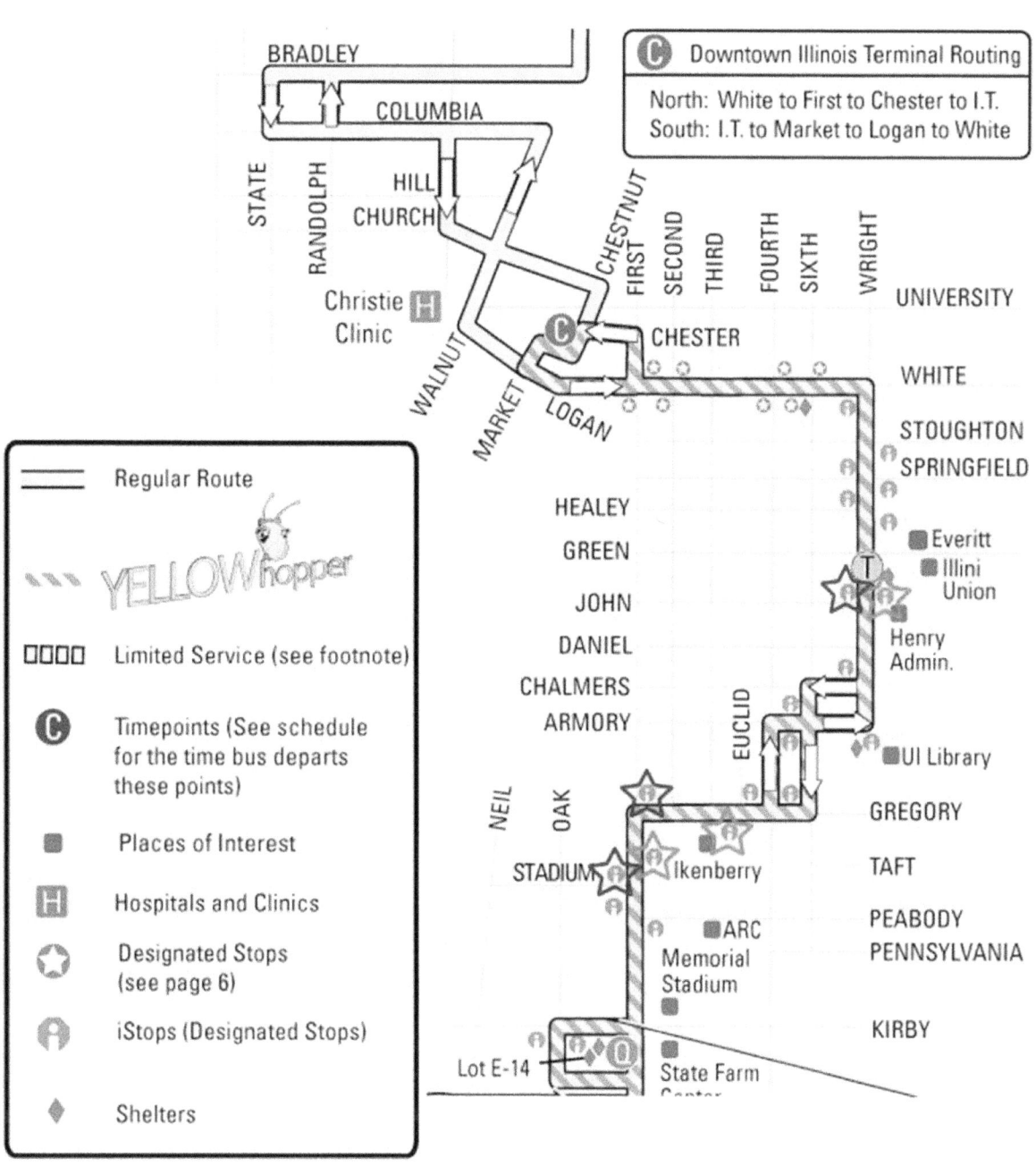

The Gold Line picks up on the corner of First and Peabody, and also on Gregory Drive in front of Ikenberry Commons and drops off at the Krannert Center (across the street from CLSL-B).   Return locations are across the street. Runs every ~10 minutes during the day (offset from the Yellow Line by 5 minutes).

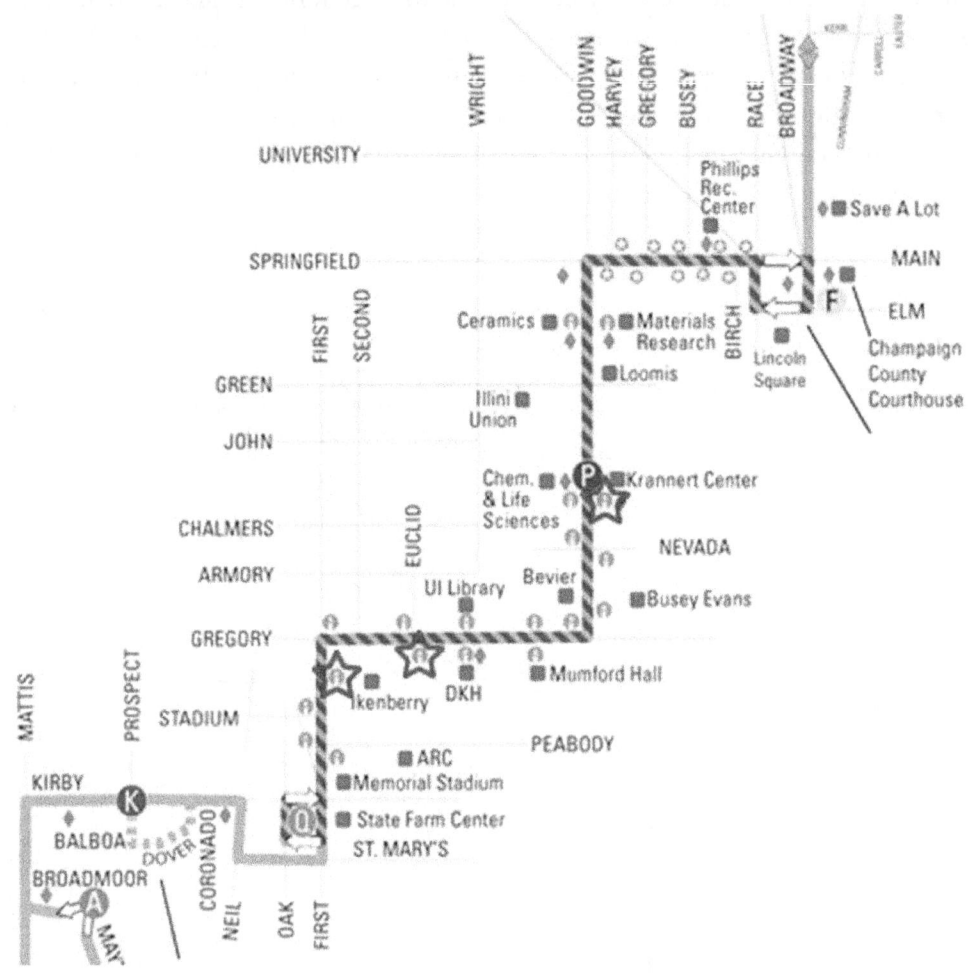

Bus Stops (Yellow Line = Left Arrow, Gold Line = Right Arrow,
Foellinger Auditorium (Plenary) and Noyes Lab = Stars)

A: Alice Campbell Alumni Center
B: Bousfield Hall (Dorm)
D: Nugent Hall (Dorm)
F: Foellinger Auditorium (Plenary)
G: Green Street (Restaurants)
H: Hampton Inn
I: Ikenberry Commons (Picnic)
K: Burrill Hall (Talks)
M: Medical Sciences Building (Talks)
N: Noyes Lab (Talks/Donuts/Coffee)
P: Parking Lot (C6/C16/E14/F23)
R: Roger Adams Lab (Talks)
S: Chem Life Sciences B (Talks)
U: Illini Union (Hotel, Restaurants)
Z: iHotel

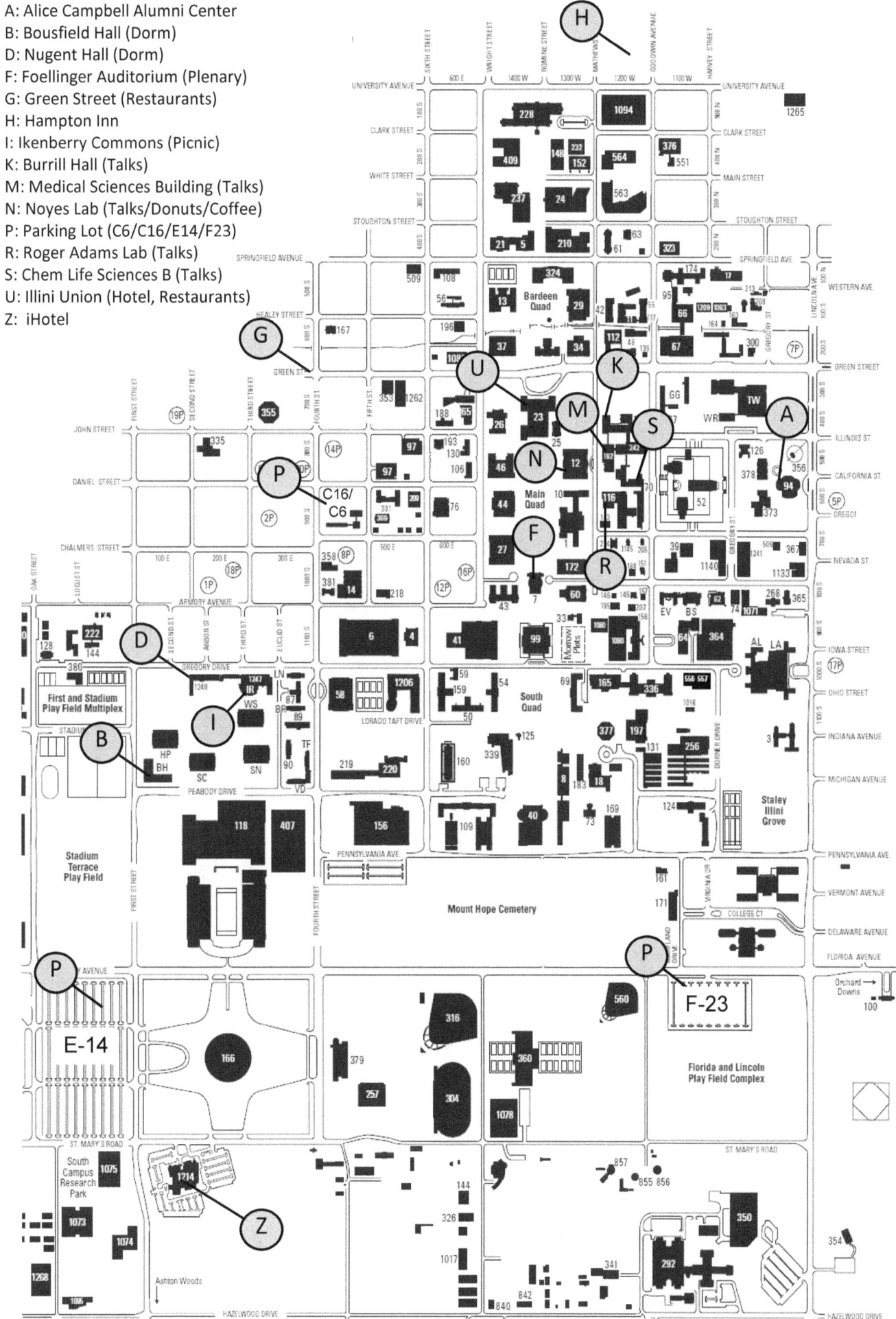

NOTES

NOTES

www.ingramcontent.com/pod-product-compliance
Lightning Source LLC
Chambersburg PA
CBHW080651190526
45169CB00006B/2063